BIOMATERIALS AND STEM CELLS IN REGENERATIVE MEDICINE

BIOMATERIALS AND STEM CELLS IN REGENERATIVE MEDICINE

Edited by
Murugan Ramalingam
Seeram Ramakrishna
Serena Best

CRC Press
Taylor & Francis Group
Boca Raton London New York

CRC Press is an imprint of the
Taylor & Francis Group, an **informa** business

CRC Press
Taylor & Francis Group
6000 Broken Sound Parkway NW, Suite 300
Boca Raton, FL 33487-2742

First issued in paperback 2017

ISBN-13: 978-1-4398-7925-2 (hbk)
ISBN-13: 978-1-138-07677-8 (pbk)

Library of Congress Cataloging-in-Publication Data

Biomaterials and stem cells in regenerative medicine / editors, Murugan Ramalingam,
    Seeram Ramakrishna, Serena Best.
        p. cm.
    Includes bibliographical references and index.
    ISBN 978-1-4398-7925-2 (hardback)
    1. Biomedical materials. 2. Stem cells--Therapeutic use. 3. Regenerative
medicine--Materials. I. Ramalingam, Murugan. II. Ramakrishna, Seeram. III. Best,
Serena.

    R857.M3B56854 2012
    610.28--dc23   2012014564

Visit the Taylor & Francis Web site at
http://www.taylorandfrancis.com

and the CRC Press Web site at
http://www.crcpress.com

Contents

Preface

Regenerative medicine has developed extremely rapidly during the last few years. Alongside this, the ideas, aspirations, and expectations of cell biologists, materials scientists and engineers, chemists, biochemists, and clinicians involved in the innovation of new techniques, treatments, and therapies have flourished. Work on the utilization of biomaterials and stem cells for the treatment of chronic disease has revealed the huge potential that they offer, and applications now range from the treatment of localized defects and disease to developments for the repair and replacement of whole organs. However, it is only relatively recently that understanding has begun to develop about the nature of the cellular environment required for optimal tissue repair and regeneration.

This book consists of 25 chapters and seeks to address the range of different types of applications for which biomaterials and stem cell therapy can be utilized and describes the recent work that has been performed to offer suitable cell scaffolds and substrates for tissue repair. Chapters 1 through 4 describe polymeric systems for stem cell delivery, and Chapters 5 through 12 discuss the potential of membranes and porous scaffolds in tissue repair, including myocardial, periodontal, ophthalmic, and bone tissues. Chapters 13 through 16 consider the optimization of the interaction between stem cells and biomaterial substrates, and Chapters 17 through 22 consider the source and nature of stem cells for tissue engineering applications. Finally, Chapters 23 through 25 discuss the clinical translation of stem cell–based tissue engineering for regenerative medicine.

This book is intended for a wide audience, including graduate students, researchers, professors, and industrial experts working in the fields of biomaterials, stem cells, and tissue engineering. It is a great asset for all those multidisciplinary communities and will act as a valuable benchmark for the current status of clinically relevant research and development in the field of stem cells and regenerative medicine, supplying information relevant to both biomaterials and scaffolds, and also to the biological and stem cells communities.

Editors

Murugan Ramalingam, PhD, is an associate professor of biomaterials and tissue engineering at the Institut National de la Santé et de la Recherche Médicale U977, Faculté de Chirurgie Dentaire, Université de Strasbourg (UdS), France. Concurrently, he holds an adjunct associate professorship at Tohoku University (Japan). Prior to joining the UdS, he was an assistant professor at the WPI-Advanced Institute for Materials Research, Japan. He has also worked at the National Institute of Standards and Technology (United States) and the National Institutes of Health (United States) under the U.S. National Academies Associateship program. He received his PhD (biomaterials) from the University of Madras. He has also undergone training in ethical and policy issues on stem cells at Harvard University and in operations management at the University of Illinois–Chicago. His current research interests are focused on the development of multiphase biomaterials through conventional to nanotechnology to biomimetic approaches, cell patterning, stem cell differentiation, and tissue engineering. He has authored over 125 publications, including peer-reviewed journal papers, conference proceedings, book chapters, authored books, edited books, and patents relevant to biomaterials and tissue engineering. He serves on the editorial boards of multiple biomaterials and tissue engineering–related journals, including as editor in chief of the *Journal of Bionanoscience* and the *Journal of Biomaterials and Tissue Engineering*. He is a recipient of CSIR fellowship (India), SMF fellowship (Singapore), NRC fellowship (United States), and National Professeur des Universités (France).

Seeram Ramakrishna, FREng, FNAE, FAIMBE, is a professor of materials engineering at the National University of Singapore. He pioneered translucent biomaterials and devices, which are now marketed globally. He specializes in the design, processing, and validation of biomimetic scaffolds for the regeneration of various tissues. He is acknowledged as number one in the field of materials by electrospinning. Ramakrishna has authored 5 books and over 400 international journal papers, which have attracted approximately 14,000 citations with an h-index of 54. He has delivered more than 100 plenary and keynote lectures worldwide. He is ranked 27th in the world among biocompatible materials experts by Elsevier Science. Thomson Reuters Web of Knowledge Essential Science Indicators (ESI) places him among the top 1% of materials scientists worldwide (ESI rank is 30). Ramakrishna is also an elected fellow of major engineering academies and professional societies in Singapore, ASEAN, India, the United Kingdom, and the United States.

Serena Best, PhD, is a professor at the Cambridge Centre for Medical Materials within the Department of Materials Science and Metallurgy, University of Cambridge. She graduated with a PhD from the University of London in 1990. She then took an academic position within the IRC in biomedical materials, based at Queen Mary, University of London, where she headed the bioceramics research activity prior to her current position. She has more than 130 publications in the field of biomaterials, and her work has contributed to the formation of two successful spin-off companies. Professor Best is an editor of the *Journal of Materials Science: Materials in Medicine* and also sits on several editorial boards, including the *Journal of the Royal Society Interface*. Her specific interests currently encompass the development of bioactive ceramics, coatings, and composites for various biomedical applications.

Contributors

Sheikh R. Ahmad
Department of Engineering and
    Applied Science
Centre for Applied Laser
    Spectroscopy
Cranfield University
Swindon, United Kingdom

Jordan E. Anderson
Department of Bioengineering
Swanson School of Engineering
and
McGowan Institute for Regenerative
    Medicine
University of Pittsburgh
Pittsburgh, Pennsylvania

Anthony Atala
Wake Forest Institute for
    Regenerative Medicine
Wake Forest School of Medicine
Winston-Salem, North Carolina

Harold S. Bernstein
Eli and Edythe Broad Center of
    Regeneration Medicine and Stem
    Cell Research
and
Cardiovascular Research Institute
and
Department of Pediatrics
University of California, San
    Francisco
San Francisco, California

Marco C. Bottino
Dental Biomaterials Division
Department of Restorative Dentistry
School of Dentistry
Indiana University
Indianapolis, Indiana

Mark Bradley
School of Chemistry
University of Edinburgh
Edinburgh, United Kingdom

Peter M. Brett
Division of Biomaterials and Tissue
    Engineering
Eastman Dental Institute
University College London
London, United Kingdom

Roger A. Brooks
Orthopaedic Research Unit
University of Cambridge
Cambridge, United Kingdom

Casey K. Chan
Division of Bioengineering
National University of Singapore
Singapore

Guo-Qiang Chen
Tsinghua University
Beijing, China

Chao-Min Cheng
Institute of Nanoengineering and
    Microsystems
National Tsing Hua University
Hsinchu, Taiwan

Katarzyna Cholewa-Kowalska
Faculty of Materials Science and
    Ceramics
Department of Glass Technology
    and Amorphous Coatings
AGH University of Science and
    Technology
Krakow, Poland

Kelley J. Colopietro
McGowan Institute for Regenerative
   Medicine
University of Pittsburgh
Pittsburgh, Pennsylvania

Luca Dainese
Department of Cardiovascular
   Surgery
University of Milan
Monzino Cardiology Centre
Milan, Italy

Bridget M. Deasy
Department of Bioengineering
Swanson School of Engineering
and
McGowan Institute for Regenerative
   Medicine
University of Pittsburgh
Pittsburgh, Pennsylvania

Ioana Demetrescu
Faculty of Applied Chemistry and
   Materials Science
Polytechnic University of
   Bucharest
Bucharest, Romania

Chunhua Deng
Department of Urology
First Affiliated Hospital
Sun Yat-Sen University
Guangzhou, China

Yanan Du
Department of Biomedical
   Engineering
School of Medicine
Tsinghua University
Beijing, China

Nicholas R. Forsyth
Keele University
Staffordshire, United Kingdom

Iain R. Gibson
School of Medical Sciences
Institute of Medical Sciences
University of Aberdeen
Aberdeen, United Kingdom

Linda G. Griffith
Department of Biological
   Engineering
Massachusetts Institute of
   Technology
Cambridge, Massachusetts

Anna Guarino
Cardiovascular Tissue Bank of
   Milan
Monzino Cardiology Centre
Milan, Italy

Summer Hanson
Division of Plastic and
   Reconstructive Surgery
Department of Surgery
School of Medicine and Public Health
University of Wisconsin-Madison
Madison, Wisconsin

David C. Hay
Medical Research Council
Centre for Regenerative Medicine
University of Edinburgh
Edinburgh, United Kingdom

Peiman Hematti
Department of Medicine
University of Wisconsin-Madison
School of Medicine and Public
   Health
Carbone Cancer Center
University of Wisconsin
Madison, Wisconsin

Daniela Ionita
Faculty of Applied Chemistry and
   Materials Science
Polytechnic University of Bucharest
Bucharest, Romania

Esmaiel Jabbari
Biomimetic Materials and Tissue
    Engineering Laboratories
Department of Chemical
    Engineering
University of South Carolina
Columbia, South Carolina

Sunyoung Joo
Wake Forest Institute for
    Regenerative Medicine
Wake Forest School of Medicine
Winston-Salem, North Carolina

Victoria Kearns
Department of Eye and Vision
    Science
University of Liverpool
Liverpool, United Kingdom

Ferdous Khan
School of Chemistry
University of Edinburgh
Edinburgh, United Kingdom

Jaehyun Kim
Wake Forest Institute for
    Regenerative Medicine
Wake Forest School of Medicine
Winston-Salem, North Carolina

In Kap Ko
Wake Forest Institute for
    Regenerative Medicine
Wake Forest School of Medicine
Winston-Salem, North Carolina

Randall J. Lee
Cardiovascular Research Institute
and
Department of Medicine
and
Institute for Regeneration Medicine
University of California, San
    Francisco
San Francisco, California

Sang Jin Lee
Wake Forest Institute for
    Regenerative Medicine
Wake Forest School of Medicine
Winston-Salem, North Carolina

Xiaokang Li
Department of Biomedical
    Engineering
School of Medicine
Tsinghua University
Beijing, China

Yong Li
Department of Bioengineering
Swanson School of Engineering
and
McGowan Institute for Regenerative
    Medicine
University of Pittsburgh
Pittsburgh, Pennsylvania

Susan Liao
Division of Materials Technology
School of Materials Science and
    Engineering
Nanyang Technological University
Singapore

Guihua Liu
Wake Forest Institute of
    Regenerative Medicine
Wake Forest University
Winston-Salem, North Carolina

Sharon Mason
Department of Eye and Vision
    Science
University of Liverpool
Liverpool, United Kingdom

Claire N. Medine
Medical Research Council
Centre for Regenerative Medicine
University of Edinburgh
Edinburgh, United Kingdom

Barbara Micheli
Cardiovascular Tissue Bank of
  Milan
Monzino Cardiology Centre
Milan, Italy

Prabha D. Nair
Division of Tissue Engineering and
  Regeneration Technologies
Sree Chithra Tirunal Institute for
  Medical Sciences and Technology
Trivandrum, India

Luong T.H. Nguyen
Center for Nanofibers and
  Nanotechnology
National University of Singapore
Singapore

Alejandro Nieponice
University of Favaloro
Buenos Aires, Argentina

and

McGowan Institute for Regenerative
  Medicine
University of Pittsburgh
Pittsburgh, Pennsylvania

Serge Ostrovidov
Advanced Institute for Materials
  Research
Tohoku University
Sendai, Japan

Anna M. Osyczka
Faculty of Biology and Earth
  Sciences
Department of Cell Biology and
  Imaging
Jagiellonian University
Krakow, Poland

Elzbieta Pamula
Faculty of Materials Science and
  Ceramics
Department of Biomaterials
AGH University of Science and
  Technology
Krakow, Poland

Salvatore Pernagallo
School of Chemistry
University of Edinburgh
Edinburgh, United Kingdom

Maurizio Pesce
Laboratory of Cardiovascular Tissue
  Engineering
Monzino Cardiology Centre
Milan, Italy

Cristian Pirvu
Faculty of Applied Chemistry and
  Materials Science
Polytechnic University of
  Bucharest
Bucharest, Romania

Gianluca Polvani
Department of Cardiovascular
  Surgery
University of Milan
Monzino Cardiology Centre
Milan, Italy

Francesca Prandi
Laboratory of Cardiovascular Tissue
  Engineering
Monzino Cardiology Centre
Milan, Italy

Seeram Ramakrishna
Center for Nanofibers and
  Nanotechnology
National University of Singapore
Singapore

and

King Saud University
Riyadh, Saudi Arabia

Murugan Ramalingam
Advanced Institute for Materials
  Research
Tohoku University
Sendai, Japan

and

National Institute of Health and
  Medical Research
Faculty of Dental Surgery
University of Strasbourg
Strasbourg, France

Nirmal S. Remya
Division of Tissue Engineering and
  Regeneration Technologies
Sree Chithra Tirunal Institute
  for Medical Sciences and
  Technology
Trivandrum, India

Carissa Ritner
Cardiovascular Research Institute
University of California, San
  Francisco
San Francisco, California

Melanie Rodrigues
Department of Pathology
University of Pittsburgh
Pittsburgh, Pennsylvania

Azadeh Seidi
Technology Center
Okinawa Institute of Science and
  Technology
Okinawa, Japan

Carl Sheridan
Department of Eye and Vision
  Science
University of Liverpool
Liverpool, United Kingdom

Rosalind Stewart
Department of Eye and Vision
  Science
University of Liverpool
Liverpool, United Kingdom

Mariusz Szuta
Department of Cranio-Maxillofacial,
  Oncological and Reconstructive
  Surgery
College of Medicine
Jagiellonian University
Krakow, Poland

James Zhenggui Tang
Department of Pharmacy School of
  Applied Sciences
University of Wolverhampton
West Midlands, United Kingdom

Giulio Tessitore
Department of Cardiovascular
  Surgery
University of Milan
Monzino Cardiology Centre
Milan, Italy

Vinoy Thomas
Department of Physics
Center for Nanoscale Materials and
  Biointegration
University of Alabama at
  Birmingham
Birmingham, Alabama

Marcus J. Tillotson
Division of Biomaterials and Tissue
  Engineering
Eastman Dental Institute
University College London
London, United Kingdom

Olga Tura
Medical Research Council
Centre for Regenerative Medicine
University of Edinburgh
Edinburgh, United Kingdom

Maria Cristina Vinci
Laboratory of Cardiovascular Tissue
  Engineering
Monzino Cardiology Centre
Milan, Italy

Yogesh K. Vohra
Department of Physics
Center for Nanoscale Materials and
  Biointegration
University of Alabama at
  Birmingham
Birmingham, Alabama

Alan Wells
Department of Pathology
and
McGowan Institute of Regenerative
  Medicine
University of Pittsburgh
Pittsburgh, Pennsylvania

Rachel Williams
Department of Eye and Vision
  Science
University of Liverpool
Liverpool, United Kingdom

Claudia Wittkowske
Department of Mechanical
  Engineering
Institute of Medical and Polymer
  Engineering
Technical University of Munich
Munich, Germany

Odessa Yabut
Cardiovascular Research Institute
University of California, San
  Francisco
San Francisco, California

Rui Yao
Department of Biomedical
  Engineering
School of Medicine
Tsinghua University
Beijing, China

James J. Yoo
Wake Forest Institute for
  Regenerative Medicine
Wake Forest School of Medicine
Winston-Salem, North Carolina

Jiashing Yu
Department of Chemical
  Engineering
National Taiwan University
Taipei, Taiwan

Rong Zhang
School of Chemistry
University of Edinburgh
Edinburgh, United Kingdom

Yuanyuan Zhang
Wake Forest Institute of
  Regenerative Medicine
Wake Forest University
Winston-Salem, North Carolina

1

Identification and Application of Polymers as Biomaterials for Tissue Engineering and Regenerative Medicine

Claire N. Medine, Ferdous Khan, Salvatore Pernagallo, Rong Zhang, Olga Tura, Mark Bradley, and David C. Hay

CONTENTS

1.1 Introduction

Biomaterials are defined as a material or a combination of materials, synthetic or natural in origin, which can be used to repair, replace or model tissues and organs in vitro and in vivo (Clare 2001; Helmus and Tweden 1995; Shi 2003). With advances in the fields of chemistry, biology, and physics, an entirely new, collaborative area of research has identified new biomaterials with improved biological performance (Binyamin et al. 2006). The biomaterials can be classified according to their performance and criteria, such as polymers, metals, ceramics, and composites. Polymers have now contributed to improved quality of life for millions of people worldwide (Binyamin et al. 2006; Dobrzanski 2006; Langer and Tirrell 2004). They represent the most significant class of biomaterials in medical application, as they can be inert, chemically diverse, processed with ease, and some polymers are resorbable. Polymers can encompass a wide range of physical and chemical properties; they can be used directly, combined with other materials or coated onto surfaces, and they are readily functionalized and can be degraded by the body after a desired period. Additionally, polymers come in many different forms including solids, fibers, films, and gels which provide not only support but also the correct structural architecture essential for tissue regeneration (Nair and Laurencin 2007). As a result, biopolymers are currently the materials of choice for thousands of medical applications.

In this chapter, we focus on synthetic and natural polymers for specific biological/biomedical applications. Some advances will be highlighted, with the use of polymers to support functional endothelial, hepatocyte, and osteocyte differentiation in vitro and in vivo.

1.2 Polymer Biomaterials

1.2.1 High-Throughput Approach to the Design, Fabrication, and Discovery of Polymers

The traditional approach to biomaterials discovery has been the iterative, synthesis, screening, and testing of materials. Traditional methods of screening, identification, and testing of new polymers were notoriously slow and low throughput. In recent years, automated and parallel screening of polymers has grown enormously. The use of high-throughput approaches, such as microarraying, has allowed the rapid screening of chemically diverse polymers and offers an important tool for both novel material discovery and the identification of correlations between performance and structure (Hook et al. 2010; Michael et al. 2004). A number of printing techniques have been developed that allow the fabrication of microarrays containing several thousand polymer materials. In this section, we focus on the contact printing of preformed polymers and inkjet fabrication techniques.

1.2.1.1 Polymer Microarray Fabrication

1.2.1.1.1 Contact Printing

Contact printing is the approach typically used for dispensing presynthesized polymers in the fabrication of polymer microarrays. The main factors affecting the shape and uniformity of the printed polymer are the solvent; the substrate; control of humidity; and a variety of parameters including inking time, stamping time, number of stamps per spot, and washing conditions. These printing conditions dramatically influence the quality and reproducibility of the final polymer microarray. A variety of solvents and solvent mixtures have been investigated (Tourniaire 2006). As a result of this study, *N*-methyl-pyrrolidinone (NMP) was selected due to its reproducible nature. Its boiling point (202°C/1 atm) also reduces solvent evaporation, which allows large numbers of polymers and microarrays to be printed in a single run.

Several materials have been investigated for this application, which include glass, gold-coated glass slides, and aluminum and perfluoroalkylthiol-modified slides as well as plastic surfaces. However, gold surfaces were unable to prevent nonspecific cellular adhesion, and for this reason, dip-coating aminoalkylsilane slides with a thin film of agarose were chosen. Importantly, agarose is readily sterilized by UV irradiation and does not dissolve in most organic solvents (Tourniaire 2006). Organic soluble materials must be avoided; otherwise blends will naturally result.

Polymer printing is carried out using a contact microarrayer with solid or split pins (Figure 1.1A). This arrayer allows the control of a variety of parameters including inking time, stamping time, number of stamps per spot, and washing conditions. These printing conditions will influence the quality and reproducibility of the final polymer microarray (Figure 1.1B) (Tourniaire 2006; Tourniaire et al. 2006).

1.2.1.1.2 Inkjet Printing

Polymer microarrays have been fabricated using inkjet printing which was first used to prepare a hydrogel polymer array using water soluble acrylate

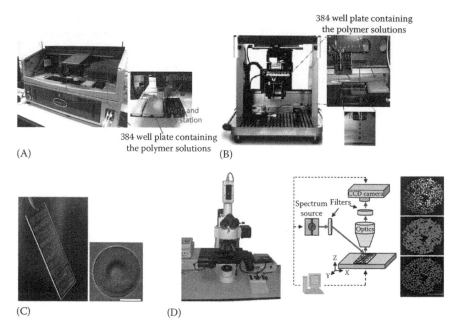

FIGURE 1.1
(A) Q-Array Mini contact microarrayer. (B) Arrayer: Microdrop printing system (Microdrop Gmbh, Germany). (C) Left: image of a high-density polymer microarray with over 2000 polymer spots. The right image is of a single polymer spot taken by a phase contrast microscope (100 mm scale bar). (D) From left to right: Nikon 50i fluorescence microscope with a motorized X–Y–Z stage, general scheme of an IMSTAR HCS device equipped with the Pathfinder™ software package, Pathfinder software automated cell quantification. Fluorescent images of K562 cells grown on a representative PA374 spot. From top to bottom: DAPI channel and Pathfinder software automatic cell quantification. Scale bar 100 μm.

monomers (Zhang et al. 2008). Inkjet printers (Figure 1.1C) are automated noncontact dispensing systems which use electrical actuators to eject picoliter volumes of liquid from micron-sized apertures onto a substrate (e.g., a glass slide) in a desired pattern (Delaney et al. 2009). To fabricate polymer microarrays, aqueous solutions of an initiator and monomers are deposited sequentially onto the same position, with a solution containing a catalyst to initiate the polymerization being printed subsequently. This drop by drop mixing approach requires control of humidity in the printing chamber to prevent the rapid evaporation of water from the printed monomer spots before hardening. This approach is used to create an array comprising over 2000 different polymers from 19 monomers (Zhang et al. 2009) (Figure 1.1B and C).

Monomers which are not water soluble can also be used to fabricate polymer microarrays to broaden the range of printed polymers by independently printing both a photoinitiator and monomers (or a mixture of initiator and monomers) in an organic solvent onto agarose-coated glass slides, pre-coated with a thin layer of mineral paraffin oil. The printed monomer droplets sink and settle down onto the agarose layer and with polymerization subsequently by exposing to UV light (Liberski et al. 2008). Using this approach, a polymer microarray of 1100 polymers can be fabricated in situ on a single microscope glass slide.

1.2.1.2 Polymer Library Screening and Selection

Most of the detection and imaging systems used in the field of polymer microarrays rely on fluorescence. With the drive towards increasing throughput, such systems have required automation for both detection and image analysis. High-resolution high-content screening (HCS) systems (down to 0.2 µm) based on a fluorescence microscope fitted with a motorized stage, allowing the automated scanning of arrays with multiple images of each spot captured from each channel (Figure 1.1D). HCS and automated image analysis (using, e.g., the Pathfinder™ software package) allows the rapid analysis of multiple parameters (including cell number, shape and size, and fluorescent intensities) from hundreds of images in order to provide accurate and meaningful interpretations of various assays (Figure 1.1D).

1.2.2 Synthetic Polymers for Application in Tissue Regeneration

1.2.2.1 Polyacrylates

Acrylate polymers are classified as nondegradable polymers. They are typically transparent and elastic and show resistance to breakage. There are a number of common acrylates whose basic chemical formula is shown in Figure 1.2A. These polymers can be prepared from a huge range of acrylic monomers, as illustrated in Figure 1.2B.

Polyacrylates have multiple applications in the modern field of biomedicine. Chalmers' discovery of the methacrylate polymers led to the development of PMMA by Crawford in the mid-1930s. PMMA was found to be biocompatible with high mechanical strength and low wearing rate suitable for bone replacement. It has been used to prepare hard contact lenses,

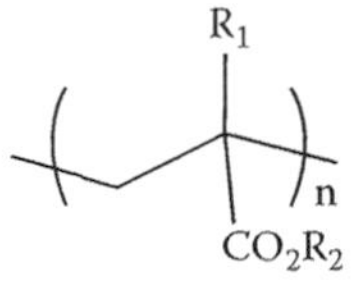

R_1	R_2	Name
H	H	Poly(acrylic acid)
CH_3	H	Poly(methacrylic acid)
H	CH_3	Poly(methyl acrylate)
CH_3	CH_3	Poly(methyl metacrylate)
CH_3	C_2H_5	Poly(methyl ethacrylate)
CH_3	C_2H_4OH	Poly(2-hydroxyethyl methacrylate)

(A)   (B)

FIGURE 1.2
(A) Basic chemical structure of poly(acrylates) and (B) Some examples.

artificial teeth, dental fillings, and bone cements (Cui et al. 1998; Heikkilä et al. 1996; Mizomoto 2004). While poly(hydroxyethyl methacrylate) (PHEMA) is a hydrophilic, soft sponge polymer which has been used in the preparation of soft contact lenses, keratoprostheses, artificial nerve conduits, and artificial cartilage (Jiang et al. 2010; Tabesh et al. 2009; Yañez et al. 2008). Wyre and Downes (2000) demonstrated that poly(ethyl methacrylate)/tetrahydrofurfuryl methacrylate (PEMA/THFMA) enhanced chondrocyte growth in vitro and was a suitable candidate material for cartilage repair. To avoid an operation during artificial cartilage implantation, loosely cross-linked co-polymer of N-isopropylacrylamide (NIPAAm) and acrylic acid (AAc) showed pliability in an aqueous solution at room temperature facilitating injection into a cartilage defect and gel formation in situ, which supported bovine articular chondrocyte viability for at least 28 days in vitro (Stile et al. 1999). To address the nondegradable problem, degradable components such as ethylene glycol, lactide, and ε-caprolactone can be chemically bonded to the acrylate polymer system. For liver regeneration, galactose-derivatized polyacrylamide hydrogels were found to support the growth and function of primary rat hepatocytes (Cho et al. 2006). Three polyacrylates have also been identified to support attachment, identity, and function of human-embryonic-stem-cell-derived hepatic endoderm (Hay et al. 2011).

The polyacrylates have a broad range of properties which make them suitable for supporting various cell cultures as an extracellular matrix (ECM) (Baker 2011). Poly(N-isopropylacrylamide-co-acrylic acid) [p(NIPAAm-co-AAc)] cross-linked with an acrylated peptide (Li et al. 2006) and poly(methyl vinyl ether-alt-maleic anhydride) (PMVE-alt-MA) (Brafman et al. 2010) were found to support short-term hESC culture, while poly[2-(methacryloyloxy)ethyl dimethyl-(3-sulfopropyl) ammonium hydroxide] (PMEDSAH) (Villa-Diaz et al. 2010) was reported to support long-term self-renewal of two hESC lines (over 20 passages), although the colonies had to be manually picked during cellular passaging. To identify polyacrylates for the culture of specific cell lines, we developed polymer hydrogel microarrays via in situ polymerization of acrylate monomers on glass slides using an inkjet printer (Zhang et al. 2008, 2009) which permitted the passaging of cells with just mild temperature reduction, thereby avoiding enzyme mediated passaging which has been blamed for causing genetic abnormalities in hESCs (Caisander et al. 2006; Stephenson et al. 2010).

1.2.2.2 Polyurethanes

Polyurethanes (PUs) are a class of polymeric materials with a wide range of properties which make them a good candidate as functional materials for applications in biomedical science (Tare et al. 2009). PUs are block co-polymers synthesized by reacting a polyol with a diisocyanate and optionally a chain extender (forming carbamate or urethane linkages (–NH–CO–O–) between monomers). PUs are often composed of so-called hard and soft segments and have been extensively used in the manufacturing of biocompatible

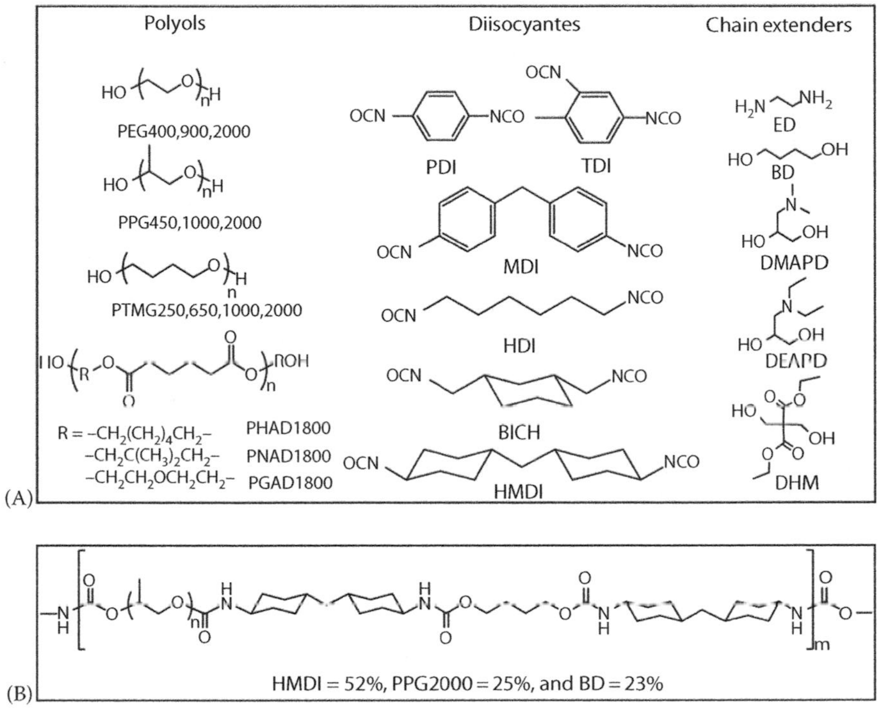

(A)

(B)   HMDI = 52%, PPG2000 = 25%, and BD = 23%

FIGURE 1.3
PUs consist of a polyol and a diisocyanate with or without a chain extender. (A) The most common ingredients used in PU synthesis and (B) an example of a PU structure which consists of PPG2000 (25%) as a polyol, HMDI (52%) as a diisocyanate and BD (23%) as a chain extender.

prosthesis and medical devices. A wide variety of PUs can be obtained using various polyols, diisocyanates, and chain extenders (a list of common reactants used for the synthesis of PUs is given in Figure 1.3A, and an example of a PU structure is shown in Figure 1.3B).

By using diverse monomers, PUs can give a wide range of mechanical, physical, and biological properties. Thus, PUs have been widely used in medical applications such as tubing, storage bags, dressings, pacemaker coatings, and precisely controllable drug release from a biodegradable stent coating (Gogolewski 1989; Santerre et al. 2005). In recent years, advances have been made in the synthesis of PUs, improving their mechanical resistance to abrasion, their tissue compatibility, and the ways in which they are processed. Researchers have sought to alleviate the difficulties caused by biodegradability by optimizing the ratio between diisocyanate components and polyols in the polymeric chain (Eberhart et al. 1999; Gogolewski 1989; Santerre and Labow 1997). Most of the PUs used in medicine have polyester, polyether, or polycarbonate (PC) as soft segments due to their mechanical properties (Ravi et al. 2009). For example, a polyester-based PU identified through high-throughput screening of a polymer microarray was found to support long-term culture of hepatocytes derived from hESCs (Hay et al. 2011), which could benefit the application of hepatocytes in regenerative medicine. PUs could also be used as artificial nerve guide conduits (NGCs) especially at joint regions due to the elasticity and flexible nature of PUs (Jiang et al. 2010).

In recent years, much attention has been focused on the application of biodegradable PUs as scaffolds for bone and cartilage regeneration (Puppi et al. 2010). To this end, PUs made from diols of poly(lactic acid) (PLA), poly(glycolic acid) (PGA) or poly(ε-caprolactone) (PCL) as well as their copolymers and linear aliphatic diisocyanates (Santerre et al. 2005) are of interest. It has been found that the degradation of these PUs starts from the hydrolysis of polyester linkages to generate α-hydroxy acids, urethane, and urea both in vitro and in vivo. The acidic degradation products are of concern due to their potential to cause local inflammation (Bruin et al. 1988; Guelcher 2008; Tatai et al. 2007). The architecture of scaffolds is another factor which can influence skeletal tissue regeneration. Porous scaffolds of PUs mimicking the native bone structures have been shown to significantly promote regeneration of new bone tissues (Gogolewski and Gorna 2007; Grad et al. 2003; Henry et al. 2009).

1.2.2.3 Other Synthetic Polymers

In addition to the polyacrylates and PUs, other various types of polymers have been applied in tissue engineering. A small (nonexhaustive) list of polymers used in tissue engineering applications is presented in Table 1.1.

1.3 Polymers for Skeletal Tissue Repair and Regeneration

Regeneration of skeletal tissue is a major clinical need, to replace or restore the function of diseased of traumatized bone. The field of skeletal tissue regeneration aims to combine progenitor cell biology with biocompatible scaffolds, to initiate repair and regeneration of bone tissue. To date, metals and ceramics have been used most extensively as bone implants. Although highly effective, these materials display a number of drawbacks, including stress-shielding problems due to their high elastic modulus (Fan et al. 2004), lack of degradability in a biological environment, and difficulty in processing and manufacturing (Chapekar 2000; Crane et al. 1995; Liu and Ma 2004; Yaszemski et al. 1996).

To overcome the disadvantages of metals or ceramic-based implants, polymers have been used clinically (Kenny and Buggy 2003; Middleton and Tipton 2000). The benefits of using polymer biomaterials include their ease

TABLE 1.1

A List of Polymers Used for Tissue Engineering Applications

Polymers	Structures	Application	Comments	References
PE		Orthopedic implants, such as cartilage scaffold	Nonbiodegradable. Need surface coating to improve the biocompatibility	Daculsi et al. (1999), Barone et al. (2011)
ePTFE		Vascular grafts, orthopedic implants, NGCs, and facial plastic surgery	Nonbiodegradable. There are time limits (~5 year) as implants. Protein coating on the surface may be needed	Ravi et al. (2009), Barone et al. (2011), Seal et al. (2001), Lin et al. (2010)
PVDF (piezoelectric polymer)		Vascular grafts, NGCs	Nonbiodegradable. Need modification to obtain right mechanical and chemical properties. Promote nerve regeneration	Seal et al. (2001), Jiang et al. (2010)
Poly(vinyl alcohol) (PVA)		Artificial cartilage	Nonbiodegradable. Need modification to increase its mechanical, biocompatible properties. Suitable for small-scale cartilage replacement or joint resurfacing	Seal et al. (2001)
PE oxide (PEO)		Artificial cartilage, ECM for hepatocytes, spinal cord repair	Biodegradable. It's biocompatible but modification (cross-linking, galactose grafting, etc.) needed to improve its mechanical, biological, and physical properties. Low risk of immune response after implantation	Seal et al. (2001), Jia et al. (2009), Cho et al. (2006), Tabesh et al. (2009)
Poly(dimethyl siloxane) (PDMS)		NGCs	Nonbiodegradable. Need further operation to remove the tube. Improve the nerve regeneration. Thin wall and good flexibility	Seal et al. (2001)

(continued)

TABLE 1.1 (continued)

A List of Polymers Used for Tissue Engineering Applications

Polymers	Structures	Application	Comments	References
Polysulfone		NGCs	Nonbiodegradable. Need another operation to remove the tube. Need surface modification to improve the nerve regeneration	Seal et al. (2001)
Polyphosphazene		NGCs, cartilage and bone regeneration, hepatocyte matrix	Biodegradable. Decreased immune response, less scar tissue formed on the tube wall, improved tissue regeneration; promote liver cells proliferation	Seal et al. (2001), Puppi et al. (2010)
Poly(glycolic acid) (PGA)		NGCs, vascular graft, cartilage and bone regeneration, liver, urologic tissue reconstruction, and intestine repair	Biodegradable. Coated with collagen to improve nerve regeneration; grafting with polydioxanone to improve mechanical properties	Ravi et al. (2009), Seal et al. (2001), Puppi et al. (2010), Marler et al. (1998)
Poly((L)-lactic acid) (PLA, PLLA)		Cartilage regeneration, NGCs	Biodegradable. Promote nerve regeneration	Seal et al. (2001), Puppi et al. (2010), Jiang et al. (2010)
PCL		Arterial substitute, nerve regeneration	Biodegradable. Promote the vascular regeneration. Coated with collagen for cell binding to promote nerve regeneration	Ravi et al. (2009), Tabesh et al. (2009), Puppi et al. (2010)
PLGA		NGCs; screws in bone replacement; liver and cartilage repair	Biodegradable, coated with inosine to improve nerve regeneration	Seal et al. (2001), Puppi et al. (2010), Cho et al. (2006), Marler et al. (1998)

Polymer	Structure	Application	Properties	References
Poly(L-lactide-co-e-caprolactone) (PCL)		NGCs, cartilage regeneration	Biodegradable. Filled with Matrigel or treated with O_2 plasma to improve the nerve regeneration	Seal et al. (2001), Jiang et al. (2010), Puppi et al. (2010)
PET (Dacron)		Bone regeneration, vascular grafts, liver regeneration	Nonbiodegradable. Need surface coating for cellular binding. Higher risk of causing inflammatory response	Seal et al. (2001), Cho et al. (2006)
Polypyrrole (PP)		Nerve regeneration	Nonbiodegradable. Composites mixed with hyaluronic acid or coating on PET to promote the nerve regeneration	Seal et al. (2001), Tabesh et al. (2009)
Poly(phosphoester) (PPE)		Nerve regeneration	Biodegradable, promote the axon regeneration	Jiang et al. (2010)
Poly(glycerol sebacic acid) (PGS)		NGCs	Biodegradable, tough elastomer. Support Schwann cell culture	Tabesh et al. (2009)
PPF		Bone regeneration	Biodegradable with reduced inflammatory response	Puppi et al. (2010)
Poly(1,4-butylene succinate) (PBSu)		Bone regeneration	Biodegradable with reduced inflammatory response	Puppi et al. (2010)

ePTFE, expanded polytetrafluoroethylene; PVDF, Poly(vinylidenedifluoride).

of manufacture, the ability to design complex shapes, the ability to mass produce cost effectively, and their availability with a wide range of physical and mechanical properties. The polymers used for orthopedic application can either be from synthetic or natural origins. In this section we will discuss the use of both synthetic and natural polymers for bone and cartilage tissue regeneration.

1.3.1 Bone Tissue Regeneration

The synthetic polymers utilized in bone tissue engineering have been classified into two main classes: biodegradable and nonbiodegradable synthetic polymers. Polymer synthesis allows for direct control over chemical composition, the ability to modify chemical structure, and therefore, they generally exhibit predictable and reproducible physical and mechanical properties and degradation rate. Polymer versatility has led to the development of biodegradable and biocompatible polymers, which are extremely useful in medical applications including bone tissue regeneration.

One of the most common polymers used as a scaffold biomaterial has been the saturated poly-α-hydroxy esters, which include PLA, PCL, and poly(glycolic acid) (PGA), as well as their copolymers (poly(lactic-co-glycolide) [PLGA]) (Khan et al. 2010; Kohn and Langer 1996; Mano et al. 2004; Seal et al. 2001). PLA can be obtained in three forms: L-PLA (PLLA), D-PLA (PDLA), and racemic mixture of D,L-PLA (PDLLA). These polymers have a high biocompatibility, and the ability to degrade into harmless monomer units, via hydrolytic degradation through de-esterification. Following degradation, the monomer units are removed by natural pathways. The body already contains highly regulated mechanisms for completely removing monomer components of lactic and glycolic acids. These polymers have a useful range of mechanical properties (depending on the copolymer ratio). PCL is also an important member of the aliphatic polyester family. Recently, the blending of PLA and PCL has received substantial attention for skeletal repair and regeneration (Khan et al. 2010). It has been used to effectively entrap antibiotics and thus a construct made with PCL being used to enhance bone in growth and regeneration in the treatment of bone defects (Pitt et al. 1981). Polypropylene fumarate (PPF) is an unsaturated linear polyester, and the degradation products, propylene glycol and fumaric acid, are biocompatible and readily removed from the body. PPF has been suggested for use as a scaffold for guided tissue regeneration, often as part of an injectable bone replacement composite (Yaszenski et al. 1995).

For bone repair and replacement, a number of specific requirements need to be met by the polymer scaffold materials including biocompatibility, osteoconduction or induction, temporary mechanical support, controlled degradation, and adequate interstitial fluid flow. Initial attempts at creating alternatives to conventional bone grafts (allografts and xenografts) were to develop synthetic bone replacements. Numerous investigations led to the development of a number of bone void filling materials and graft extending materials. Additionally, investigations from multiple laboratories has resulted in a long list of biomaterials with osteoconductive properties (Devin et al. 1996; Khan et al. 2010; Kim et al. 2011; Liu and Ma 2004).

Further investigation revealed that modifications of scaffold architecture and material properties could improve bone growth. This led to a research thrust aimed at creating scaffold materials with biomimetic properties that would mimic the role of the ECM in many cell functions, including adhesion, migration, and proliferation. For example, median pore size has been found to influence conductive properties of tissue engineering scaffolds (Robinson et al. 1995). The surface properties of polymeric scaffolds will influence cell function including texture, hydrophobicity, charge and chemical composition. Engineering of biomaterial surfaces and bulk properties will allow bio-specific interactions between appropriate cell types and scaffold materials.

Kim et al. (2011) investigated the fabrication of photo cross-linked PPF and diethyl fumarate (DEF) scaffolds, and showed that these polymers support rapid BMSC osteoblastic differentiation, and stimulated dramatic BMSC responses that promoted rapid bone tissue regeneration.

For certain orthopedic applications where high stiffness is necessary, the nonbiodegradable polymers can be used, having the advantage that the stiffness of these polymeric materials is much closer to the stiffness of bone in contrast to metals or ceramics (Burg et al. 2000; Eschbach 2000). Tensile strength and the elastic modulus of polymers make it possible to apply them as bone implants. The most important synthetic nondegradable polymers in bone tissue engineering are polyethylene (PE), polypropylene (PP), selected PUs, polytetrafluoroethylene (PTFE), poly(vinyl chloride) (PVC), polyamides (PAs), poly(methyl methacrylate) (PMMA), polyoxymethylene (POM, polyacetal resin), PC, poly(ethylene terephthalate) (PET), poly(ether ether ketone) (PEEK), and polysulfone (PSU). These polymers are also considered to be bio-stable in the human body and have been applied medically, ranging from PTFE vascular grafts to ultrahigh molecular weight polyethylene (UHMWPE) acetabular cups (Mano et al. 2004; Wang 2003).

1.3.2 Cartilage Tissue Regeneration

For cartilage tissue engineering applications, the requirements of the physical and biological properties of the polymer scaffolds are different compared to that of bone. Scaffolds fabricated with both synthetic and natural materials in a variety of physical forms such as fibers, meshes, and gels have been applied to cartilage tissue engineering (Temenoff and Mikos 2000). Solid scaffolds provide a substrate on which cells can attach and proliferate, while liquid and gel scaffolds function to physically entrap the cells. An example of a solid scaffold is poly(glycolic acid) (PGA). PGA meshes have been successfully used to engineer cartilage both in vitro and in vivo (Freed et al. 1995). Recently, two solid scaffold systems, PCL and PGA, were used for the chondrogenesis of MSCs with some success (Huang et al. 2002). Several studies (Solchaga et al. 1999, 2002) have demonstrated that hyaluronic-acid-based scaffolds also have positive tissue forming abilities using chondrocytes and MSCs.

Hydrogels are a class of scaffolds that have been investigated in tissue engineering applications and include alginate, pluronics, chitosan (CS), and fibrin glue as examples. Fibrin glue is a biological gel that has been used to encapsulate chondrocytes, but the resulting gel is often weak, and there is little control of network formation (Silverman et al. 1999). Alginate, a polysaccharide, forms an ionic network in the presence of divalent or multivalent ions. Many groups have investigated the activity and biological properties of cells entrapped in alginate in vitro (Bouhadir et al. 2001). Alginate has also been examined in vivo for use in craniofacial cartilage replacement and as cartilage plugs to prevent vesicoureteral reflux (Paige et al. 1996). Researchers have also modified alginate with adhesive peptides in order to encapsulate anchorage dependent and independent cells (Rowley et al. 1999; Shin et al. 2004). Alginate and agarose (or ionic and thermoresponsive polymers in general) provide little control over the gelation process; particularly in a clinical setting once cross-linking is induced, by the addition of an ionic solution or a temperature change, the process cannot be stopped or accelerated. Thus, the need for a new biomaterial or method for cell encapsulation that provides control over gel formation and shape maintenance led to the development of the photopolymerization system for tissue engineering applications (Elisseeff et al. 1999). These initial studies investigated photopolymerization and tissue regeneration in a nondegradable system.

In the past decade, the CS-based materials have drawn considerable attention in the application of cartilage tissue regeneration (Khan et al. 2009; Lu et al. 1999; Suh and Matthew 2000). Lu et al. (1999) have demonstrated that the injection of CS solution into the knee articular cavity of rats significantly increases the density of chondrocyte in the knee cartilage, indicating a potential choice of material for articular cartilage repair.

CS has been used to synthesize hydrogels by complexation or aggregation (Berger et al. 2004) and has also proven useful as a cell-carrier substance. Chenite et al. (2000) developed an injectable thermosensitive hydrogel based on the neutralization of CS by adding β-glycerophosphate. One of the drawbacks of using the CS-glycerophosphate system is the limited mechanical properties of hydrogels. An alternative approach is needed to develop a more rigid texture gel which would be advantageous (Couto et al. 2009; Hoemann et al. 2007). Recently, we developed a hydrogel by CS and polyethylenimine (PEI) blending which showed rheological stability, and supported cellular growth with chondrogenic characteristics (Khan et al. 2009).

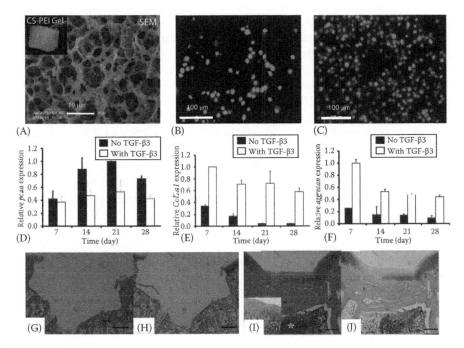

FIGURE 1.4

(A) Hydrogels prepared by blending solutions of CS and PEI (gel top left) and SEM image. Human skeletal cells were cultured up to 4 weeks in the CS–PEI hydrogel. Micrographs of cells cultured on hydrogels on (B) culture day 7 and (C) day 21. (D–F) Analysis of chondrogenic gene expression (*Pcna, Col2a1*, and *Aggrecan*) by fetal skeletal cells cultured within hydrogels over a course of 28 days with and without TGF β3. Relative gene expression levels were normalized to β-*Actin* expression (served as a house-keeping gene). The group with the highest expression was assigned a value of 1, and expression levels in the remaining groups were determined relative to that group (data presented with a mean ± SD for n=3). (G, H) Histological appearances of the osteochondral defect in the patellar groove immediately after preparing a defect. (I, J) Photomicrograph of histological sections for osteochondral defects in the patellar groove at 12 weeks after implantation of CS-hyaluronic acid hydrogel, which is shown filled with hyaline-like cartilage or a combination of hyaline-like cartilage and fibrocartilage growth. Note the evidence of CS–hyaluronic acid hybrid fibers remaining in the defect (denoted with an asterisk). (G, I) HE staining and (H, J) safranin-O staining. Scale bars 1 mm (G–J) and 200 µm (inset of I). (A through F: Reprinted with permission from Khan, F. et al., *Angew. Chem. Int.* Ed., 48, 978, 2009; G through J: From Kasahara, Y. et al., *J. Biomed. Mater. Res. A*, 86(1), 127, 2008.)

When the human skeletal cells, derived from predominantly cartilaginous fetal femora, were cultured within the CS/PEI hydrogels, the cells maintained chondrocyte-like morphology (Figure 1.4) (Kasahara et al. 2008; Khan et al. 2009), and the characteristic functional features were similar to those of normal cartilage.

The CS-based gels also act as an ideal cell-carrier substance which most closely mimics the naturally occurring environment in the articular cartilage matrix. Suh and Matthew (2000) demonstrated that cartilage-specific

ECM components such as type II collagen and glycosaminoglycan (GAG) play a critical role in regulating expression of the chondrocytic phenotype and in supporting chondrogenesis both in vitro and in vivo. The CS has a structural similarity to GAGs present in native cartilage ECM (Berger et al. 2004), and it is for this reason that CS-based materials remain to be the most suitable candidate for cartilage tissue regeneration.

However, the application of such gels or scaffolds in tissue engineering depends mainly on the characteristic features of the CS and other polymers. The key parameters influencing the characteristic properties of CS-based biomaterials are molecular weight (M_w), degree of deacetylation (DD) of CS, gel and scaffold fabrication methods, and the formulation and characteristic properties of functional additives to be incorporated with CS (Berger et al. 2004).

1.4 Polymers for Endothelial Cell Growth and Function

1.4.1 Selected Biopolymers to Support the Culture and Differentiation of EPCs

Cardiovascular disease is the leading cause of morbidity and mortality in the Western world despite the existence of various treatment options (http://www.who.int/mediacentre/factsheets/fs317/en/index.html). Traditionally, vascular repair in the adult was thought only to occur by the proliferation and migration of preexisting mature endothelial cells (ECs) from the adjacent vasculature (Risau 1995). More recently, the existence of bone-marrow-derived endothelial progenitor cells (EPCs) capable in de novo vascularization has been identified (Asahara et al. 1997, 1999; Shi et al. 1998). EPCs in response to injury are mobilized and recruited to the ischemic sites where they contribute to new-vessel formation.

Since the discovery of circulating EPCs in peripheral blood (Asahara et al. 1997, 1999; Shi et al. 1998), therapeutic strategies using autologous stem cells to facilitate the repair and regeneration of ischemic tissue have been under intensive investigation (Hakuno et al. 2002; Kalka et al. 2000; Orlic et al. 2001). However, these strategies involving local or systemic infusion, or direct implantation of progenitor cells, are limited by poor homing and engraftment into the target tissue which result in loss of cell phenotype and viability.

Therefore, with the realization that cell therapy alone is insufficient for successful tissue regeneration; the application of bioengineering research by the development of synthetic biopolymer matrices as defined environments for EPC growth has the potential to provide solutions to these limitations. Well-designed scaffolds, together with synthetic biopolymers as cell carriers, could promote greater cellular engraftment and provide a template to guide the formation of new tissue. Controlling cellular microenvironments by engineering 3D biomaterials introduces the advantage of generating specific matrices with instructive cues to promote cell attachment and with signals that control EPC propagation and synchronize their differentiation (Chan and Mooney 2008; Mooney and Vandenburgh 2008). Importantly, these polymers can be manufactured to GMP standards and provide a resource for the construction of extra-corporeal devices.

1.4.2 Use of Selected Polymers to Enhance the Biocompatibility of Intravascular Stents and Conduits by Supporting EPCs Adhesion and Differentiation

Endothelization of the vascular graft is an essential step during graft healing. Confluent EC monolayers on prosthetic grafts provide the optimal naturally occurring antithrombogenic and blood-compatible surface (Piterina et al. 2009). The development of an endothelial monolayer on the luminal surface of synthetic vascular grafts would prevent thrombus formation by inhibiting platelet accumulation on the graft surface and hence improve long-term patency rates (Piterina et al. 2009).

Spontaneous in vivo endothelization of vascular grafts occurs by the endothelization of both mature ECs originating from surrounding host vascular tissue and by the migration of circulating EPCs mobilized from the bone marrow. However, the material properties and surface structure of the graft may not always provide a favorable microenvironment for their recruitment and may inhibit the rate of attachment and proliferation of ECs (Hristov and Weber 2004; Yasu 2009). Indeed multiple in vivo clinical studies reveal that prosthetic vascular grafts often remain mostly without an endothelium, even after decades of implantation (Bordenave et al. 2005; Zilla et al. 2007).

The success of a vascular graft is shown to depend mostly upon the intrinsic properties to the graft material. Thus, the identification of a novel synthetic biopolymer which specifically promotes EC attachment would be of great benefit to the success of vascular grafts.

1.4.2.1 Stent-Based Therapy

Adverse cardiac events after percutaneous coronary intervention (PCI) continue to be problematic despite advances in stent design and adjunctive pharmacotherapy (Padfield et al. 2010). The treated coronary artery segment inevitably undergoes significant mechanical trauma after intervention. Endothelial denudation by rigid stent struts and high-pressure balloon inflations disturbs vascular function and initiates an intensive local inflammatory response. Vascular injury may result in neo-intimal hyperplasia, in-stent restenosis, and the potentially fatal complication of acute stent thrombosis (De la Torre-Hernandez et al. 2008). These complications arise, in part, as sequelae of the vascular trauma that leave the treated segment of vessel denuded of its endothelium with a consequent disruption of normal vascular function.

Complete re-endothelialization takes place approximately 3 months after stent implantation. By that time, the vascular injury has already triggered both local (Farb et al. 1999) and systemic (Mills et al. 2009) inflammation with a typical cellular response of a rapid influx of neutrophils followed by the migration of monocytes and macrophages into the vessel wall. Cytokine and growth factor release stimulates the migration and proliferation of intimal smooth muscle cells and fibroblasts, and the recruitment of adventitial myofibroblasts, encouraging neo-intimal hyperplasia and in-stent restenosis (Wilcox et al. 2001). Rapid re-endothelialization is therefore of critical importance to restore normal vascular function, reduce vascular inflammation, and prevent adverse remodeling after PCI (Kipshidze et al. 2004).

Current strategies to prevent complications following PCI include impregnating stents with either antibodies or integrin-binding peptides. These functionalized stents display molecules that are directed toward proteins on the surface of EPCs. They are designed to attract EPCs and therefore promote rapid re-endothelialization. While some encouraging results have been derived from these studies, the first randomized, controlled trial recently demonstrated a trend toward increased restenosis with the capture stent when compared with a standard chromium-cobalt stent (Cervinka 2009). Restenosis with capture stents may occur as a consequence of nonspecific binding with non-EPCs. Ongoing studies will directly compare the safety and efficacy of this new generation of capture stents. Other current strategies to reduce the incidence of complications following percutaneous intervention are based on suppressing cellular proliferation. Drug-eluting stents which release a drug to block cell proliferation have dramatically reduced the incidence of early in-stent restenosis, but local antiproliferative therapy seems to interfere with vascular healing and prevent formation of a functional endothelial layer (Muldowney et al. 2007). Stents pre-coated with an endothelial specific biopolymer could facilitate the migration of adjacent mature cells and/or engraftment of circulating EPCs to the site of the stent to facilitate rapid re-endothelization. Importantly, endothelial-specific biopolymers could also be used to coat 3D scaffolds, providing a resource for the construction of intravascular devices with potential novel applications in vascular intervention or surgery.

Thus, adopting polymers that support EPC adhesion has the potential to produce a new generation of stents which avoid the natural recognition system of the body. A stent, which is not toxic or inflammatory, improves endothelial coverage while reducing the formation of stent neo-intima is of major interest. Implantation of such devices in preclinical models of angioplasty and stenting may have enormous clinical potential.

1.5 Biopolymer Matrices as Defined Support Systems for Stem-Cell-Derived Hepatocyte-Like Cells

1.5.1 Polymers Supporting Functional Hepatocyte-Like Cells Derived from Pluripotent Stem Cells

Hepatocytes are responsible for the metabolism of most ingested chemical compounds. Therefore, these cells represent the most appropriate model for the evaluation of integrated drug metabolism, toxicity/metabolism correlations, mechanisms of hepatotoxicity, and the interactions (inhibition and induction) of xenobiotics and drug metabolizing enzymes.

Currently, adult human hepatocytes represent the gold standard for predictive drug toxicology, but their routine deployment is hindered by the scarcity of donor tissue. Moreover, primary hepatocytes are expensive due to their limited availability, are generally sourced from lower quality fragments of tissue and thus exhibit variability in lifespan and function. As a consequence, rodent liver models have been widely applied in preclinical drug testing; however, their lack of predictability and high phenotypic variation make them an unreliable predictive model for humans (Guillouzo 1998), and many drug trials have had to be terminated because such models fail to predict human drug toxicity.

Other alternatives include human hepatoma lines; however, these do not represent a metabolically competent and physiologically relevant cell model for predicting human liver toxicity (Allain et al. 2002; Cai et al. 2000; Wege et al. 2003). As such there is an urgent need for reliable and predict able in vitro models. The ability to efficiently generate functional hepatocyte-like cells (HLCs) from a renewable source of genotypically stable pluripotent stem cells has significant potential to meet the needs of predictive toxicology. Moreover, pluripotent stem cells can undergo unlimited self-renewal, retaining the potential to differentiate into all somatic cells (Reubinoff et al. 2000; Thomson et al. 1998). These abilities make them an attractive source of human cells for use in cell-based modeling and drug discovery.

Numerous groups have successfully generated efficient levels of HLCs from hESCs (Agarwal et al. 2008; Baharvand et al. 2006; Basma et al. 2009; Dalgetty et al. 2009; Duan et al. 2007; Fletcher et al. 2008; Greenhough et al. 2010; Hannoun et al. 2010; Hay et al. 2007, 2008a,b; Lavon et al. 2004; Payne et al. 2011; Touboul et al. 2010). We have developed a robust and highly efficient protocol for the generation of HLCs from pluripotent stem cells from hESC and iPSC lines (Hay et al. 2008a, 2011; Sullivan et al. 2010). The yield of hepatocytes using this procedure is typically in the region of 90%. These hESC-derived HLCs exhibit endocrine and exocrine function comparable to that of primary human adult hepatocytes, with iPSC models less functional, but improving.

Although we are able to generate efficient numbers of functional hepatocytes using this method, there is still room for improvement in hepatocyte function and viability. One such example is stem-cell-derived HLC short-term viability and diminishing function in vitro similar to that associated with primary human hepatocytes. Therefore, we explored novel strategies to extend the lifespan of functional stem-cell-derived HLCs. By fusing cutting edge pluripotent stem cell models (Hay et al. 2008a; Sullivan et al. 2010) and high-throughput polymer screening technologies (Pernagallo et al. 2009; Thaburet et al. 2004; Tourniaire et al. 2006), we have developed a novel extracellular support matrix capable of supporting cell-specific function and viability (Hay et al. 2011).

Our polymer library screen consisted of a library of 380 PUs and polyacrylates. We identified three polyacrylates and three PUs that facilitated HLC replating and maintenance of cellular identity (Hay et al. 2011). Our synthetic approach permitted the long-term culture of functional stem-cell-derived HLCs for 15 days in vitro and exhibited greater epithelial marker expression and liver-specific function compared to matrigel cultures. Moreover, stem-cell-derived HLCs exhibited superior drug inducibility to primary human hepatocytes (Hay et al. 2011). Importantly, the PU when coated on a clinically approved bio-artificial liver (BAL) matrix also supported long-term hepatocyte function and growth (Figure 1.5), demonstrating that we have made significant headway toward the goal of high fidelity and stable in vitro human models, and may have clinical application in a BAL device.

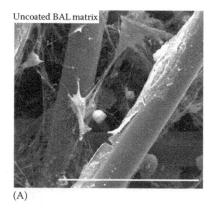
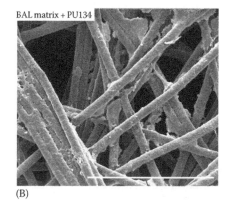

(A)   (B)

FIGURE 1.5
Replating of hESC-derived hepatic endoderm onto uncoated and PU 134 coated BAL matrix. At day 24 in culture, the cells were fixed and examined by electron microscopy. (A) Uncoated BAL matrix with hESC-derived hepatic endoderm attached. Scale bar represents 50 μm. (B) BAL matrix coated with polymer 134 with hESC-derived hepatic endoderm attached. Scale bars represents 200 μm.

1.5.2 Defined Extracellular Polymer Matrices for Stem Cell Product Manufacture

We have previously shown that pluripotent stem cells can be differentiated efficiently to HLCs (Hay et al. 2008a; Sullivan et al. 2010). While these experiments provide proof of concept that this is a feasible approach, the manner in which the cells were cultured is not amenable to automated large-scale manufacture. In order to scale up our technology, cost effectively, it is important that the culture conditions are defined. At present, hESCs and iPSCs are maintained on extracellular matrices and in medium containing animal products which lack definition. Therefore, the use of polymer support matrices offers a number of advantages over current biological extracellular matrices employed and will facilitate the defined culture of pluripotent stem cell populations. Synthetic polymer matrices provide a cost-effective and scalable model suitable for industry and can be synthesized to GMP standards which may be critical to the translation of cell models to tissue engineering and regenerative medicine.

1.5.3 Application of Stem-Cell-Derived HLC in Drug Screening and BAL Construction

The use of stem-cell-derived HLC models, maintained on synthetic polymer matrices which maintain long-term hepatic function, offers numerous potentials. Pharmaceutical companies must test drugs for toxicity in humans, and animal models are costly and time consuming. Additionally, animal models are very different to humans in numerous ways, and despite being considered the most representative model for humans, results obtained from animal models do not always extrapolate to humans. Therefore, stem-cell-derived HLCs with long term function could be used to assess drug toxicity and safety more reliably. As such stem-cell-derived liver models represent a new and uniform predictive tool for pharmaceutical companies to understand drug metabolism, and toxicity. It is expected that the pharmaceutical industry and regulators will embrace cell-based assays derived from stem cells once they become commercially available.

An inexhaustible supply of functional human hepatocytes derived from pluripotent stem cells would also benefit the development of human-based extracorporeal BAL devices for the treatment of liver failure. The aim of BAL devices is to provide both liver detoxification and synthetic functions, to ultimately bridge patients to liver transplantation or allow the native liver to recover from liver injury. At present, porcine hepatocytes are often used to fuel these BALs, due to high availability and low costs; however, these have been prohibited in many European countries. There is a high need for humanized BAL devices which are instead fueled with human hepatocytes. Primary human hepatocytes are not used for BAL construction due to their limited availability and stability ex vivo. Human ESC- and iPSC-derived hepatocytes represent a scalable source of cells for BAL construction.

1.6 Conclusion

The combination of polymer chemistry and cell biology has led to significant advances in the identification and understanding of synthetic matrices which provide cellular support, aiding cellular differentiation and tissue formation. Polymer microarray screening has been key in this process, playing a major role in the unbiased discovery of novel surfaces that are likely to have a significant impact on modern medicine.

References

Agarwal S., Holton K., and Lanza R. 2008. Efficient differentiation of functional hepatocytes from human embryonic stem cells. *Stem Cells* 26(5):1117–1127.

Allain J.E., Dagher I., Mahieu-Caputo D., Loux N., Andreoletti M., Westerman K., Briand P., Franco D., Leboulch P., and Weber A. 2002. Immortalization of a primate bipotent epithelial liver stem cell. *Proc Natl Acad Sci USA* 99:3639–3644.

Asahara T., Masuda H., Takahashi T., Kalka C., Pastore C., Silver M., Kearne M., Magner M., and Isner J.M. 1999. Bone marrow origin of endothelial progenitor cells responsible for postnatal vasculogenesis in physiological and pathological neovascularization. *Circ Res* 85:221–228.

Asahara T., Murohara T., Sullivan A., Silver M., Van Der Zee R., Li T., Witzenbichler B., Schatteman G., and Isner J.M. 1997. Isolation of putative progenitor endothelial cells for angiogenesis. *Science* 275:964–966.

Baharvand H., Hashemi S.M., Kazemi Ashtiani S., and Farrokhi A. 2006. Differentiation of human embryonic stem cells into hepatocytes in 2D and 3D culture systems in vitro. *Int J Dev Biol* 50(7):645–652.

Baker M. 2011. Stem cells in culture: Defining the substrate. *Nat Methods* 8(4):293–297.

Barone D.T.J., Raquez J.M., and Dubois P. 2011. Bone-guided regeneration: from inert biomaterials to bioactive polymer (nano)composites. *Polym Adv Technol* 22:463–475.

Basma H., Soto-Gutiérrez A., Yannam G., Liu L., Ito R., Yamamoto T., Ellis E. et al. 2009. Differentiation and transplantation of human embryonic stem cell-derived hepatocytes. *Gastroenterology* 136(3):990–999.

Berger J., Reist M., Mayer J.M., Felt O., and Gurny R. 2004. Structure and interactions in chitosan hydrogels formed by complexation or aggregation for biomedical applications. *Eur J Pharm Biopharm* 57:35–52.

Binyamin G., Shafi B.M., and Mery C.M. 2006. Biomaterials: A primer for surgeons. *Semin Pediatr Surg* 15(4):276–283.

Bordenave L., Fernandez P., Rémy-Zolghadri M., Villars S., Daculsi R., and Midy D. 2005. *In vitro* endothelialized ePTFE prostheses: Clinical update 20 years after the first realization. *Clin Hemorheol Microcirc* 33:227–234.

Bouhadir K.H., Lee K.Y., Alsberg E., Damn K.L., Anderson K.W., and Mooney D.J. 2001. Degradation of partially oxidized alginate and its potential application for tissue engineering. *Biotechnol Prog* 17:945–950.

Brafman D.A., Chang C.W., Fernandez A., Willert K., Varghese S., and Chien S. 2010. Long-term human pluripotent stem cell self-renewal on synthetic polymer surfaces. *Biomaterials* 31:9135–9144.

Bruin P., Veenstra G.J., Nijenhuis A.J., and Pennings A.J. 1988. Design and synthesis of biodegradable poly(ester-urethane) elastomer networks composed of non-toxic building blocks. *Makromol Chem Rapid Commun* 9:589–594.

Burg K.J.L., Porter S., and Kellam J.F. 2000. Biomaterial developments for bone tissue engineering. *Biomaterials* 21:2347–2359.

Cai J., Ito M., Westerman K.A., Kobayashi N., Leboulch P., and Fox I.J. 2000. Construction of a non-tumorigenic rat hepatocyte cell line for transplantation: Reversal of 410 hepatocyte immortalization by site-specific excision of the SV40 T antigen. *J Hepatol* 33:701–708.

Caisander G., Park H., Frej K., Lindqvist J., Bergh C., Lundin K., and Hanson C. 2006. Chromosomal integrity maintained in five human embryonic stem cell lines after prolonged *in vitro* culture. *Chromosome Res* 14:131–137.

Cervinka P. 2009. A randomized comparison of Genous stent vs chromium–cobalt stent for treatment of ST-elevation myocardial infarction. A 6-month clinical, angiographic and IVUS follow-up. GENIUS-STEMI trial. Paper presented at *American College of Cardiology 2009 Scientific Sessions and i2 Summit*, March 2009, Orlando, FL.

Chan G. and Mooney D.J. 2008. New materials for tissue engineering: Towards greater control over the biological response. *Trends Biotechnol* 26.382–392.

Chapekar M.S. 2000. Tissue engineering: Challenges and opportunities. *J Biomed Mater Res* 53:617–620.

Chenite A., Chaput C., Wang D., Combes C., Buschmann M.D., Hoemann C.D., Leroux J.C., Atkinson B.L., Binette F., and Selmani A. 2000. Novel injectable neutral solutions of chitosan form biodegradable gels in situ. *Biomaterials* 21:2155–2161.

Cho C.S., Seo S.J., Park I.K., Kim S.H., Kim T.H., Hoshiba T., Harada I., and Akaike T. 2006. Galactose-carrying polymers as extracellular matrices for liver tissue engineering. *Biomaterials* 27:576–585.

Clare A.G. 2001. Materials science vs. ceramic engineering—Parasitic or symbiotic. *Am Ceram Soc Bull* 80(9):61.

Couto D.S., Hong Z., and Mano J.F. 2009. Development of bioactive and biodegradable chitosan-based injectable systems containing bioactive glass nanoparticles. *Acta Biomater* 5:115–123.

Crane G.M., Ishaug S.L., and Mikos A.G. 1995. Bone tissue engineering. *Nat Med* 1:1322–1324.

Cui L.L., Jiang J., Xia Z.G., Chen G.J., and Wang Z.Z. 1998. Charge storage and transport in polymethylmethacrylate (PMMA) film. *J Electrostat* 44(1–2):61–65.

Daculsi G., Weiss P., Bouler J.M., Gauthier O., Millot F., and Aguado E. 1999. Biphasic calcium phosphate/hydrosoluble polymer composites: A new concept for bone and dental substitution biomaterials. *Bone* 25(2):59S–61S.

Dalgetty D.M., Medine C.N., Iredale J.P., and Hay D.C. 2009. Progress and future challenges in stem cell-derived liver technologies. *Am J Physiol Gastrointest Liver Physiol* 297(2):G241–G248.

De la Torre-Hernández J.M., Alfonso F., Hernández F., Elizaga J., Sanmartin M., Pinar E., Lozano I. et al. 2008. Drug eluting stent thrombosis: Results from the multicenter Spanish registry ESTROFA (Estudio ESpanol sobre TROmbosis de stents FArmacoactivos). *J Am Coll Cardiol* 51:986–990.

Delaney J.T., Smith P.J., and Schubert U.S. 2009. Inkjet printing of proteins. *Soft Matter* 5(24):4866–4877.

Devin J.E., Attawia M.A., and Laurencin C.T. 1996. Three-dimensional degradable porous polymer-ceramic matrices for use in bone repair. *J Biomater Sci Polymer Ed* 7:661–669.

Dobrzanski L.A. 2006. Significance of materials science for the future development of societies. *J Mater Process Technol* 175(1–3):133–148.

Duan Y., Catana A., Meng Y., Yamamoto N., He S., Gupta S., Gambhir S., and Zern M.A. 2007. Differentiation and enrichment of hepatocyte-like cells from human embryonic stem cells in vitro and in vivo. *Stem Cells* 25(12):3058–3068.

Eberhart A., Zhang Z., Guidoin R., Laroche G., Guay L., De la Faye D., Batt M., and King M.W. 1999. A new generation of polyurethane vascular prostheses: Rara Avis or Ignis Fatuus? *J Biomed Mater Res* 48(4):546–558.

Elisseeff J., Anseth K., Sims D., Randolph M., and Langer R. 1999. Transdermal photopolymerization for minimally invasive implantation. *Proc Natl Acad Sci USA* 96:3104–3107.

Eschbach L. 2000. Nonresorbable polymers in bone surgery. *Injury, Int J Care Injured* 31:S-D22–S-D27.

Fan J.P., Tsui C.P., Tang C.Y., and Chow C.L. 2004. Influence of interphase layer on the overall elastoplastic behaviors of HA/PEEK biocomposite. *Biomaterials* 25:5363–5373.

Farb A., Sangiorgi G., Carter A.J., Walley V.M., Edwards W.D., Schwartz R.S., and Virmani R. 1999. Pathology of acute and chronic coronary stenting in humans. *Circulation* 99:44–52.

Fletcher J., Cui W., Samuels K., Black J.R., Currie I.S., Terrace J.D., Payne C. et al. 2008. The inhibitory role of stromal cell mesenchyme on human embryonic stem cell hepatocyte differentiation is overcome by Wnt3a treatment. *Cloning Stem Cells* 10(3):331–340.

Freed L. and Vunjak-Novakovic G. 1995. Tissue engineering of cartilage. In: Bronzind J, editor. The Biomedical Engineering Handbook. Boca Raton: CRC, 1995, pp. 1778–96.

Gogolewski S. 1989. Selected topics in biomedical polyurethanes. A review. *Colloid Polym Sci* 267(9):757–785.

Gogolewski S. and Gorna K. 2007. Biodegradable polyurethane cancellous bone graft substitutes in the treatment of iliac crest defects. *J Biomed Mater Res* 80A:94–101.

Grad S., Kupcsik L., Gorna K., Gogolewski S., and Alini M. 2003. The use of biodegradable polyurethane scaffolds for cartilage tissue engineering: Potential and limitations. *Biomaterials* 24:5163–5171.

Greenhough S., Medine C.N., and Hay D.C. 2010. Pluripotent stem cell derived hepatocyte like cells and their potential in toxicity screening. *Toxicology* 278:250–255.

Guelcher S.A. 2008. Biodegradable polyurethanes: Synthesis and applications in regenerative medicine. *Tissue Eng B* 14:3–17.

Guillouzo A. 1998. Liver cell models in in vitro toxicology. *Environ Health Perspect* 106(Suppl. 2):511–532.

Hakuno D., Fukuda K., Makino S., Konishi F., Tomita Y., Manabe T., Suzuki Y., Umezawa A., and Ogawa S. 2002. Bone marrow-derived regenerated cardiomyocytes (CMG Cells) express functional adrenergic and muscarinic receptors. *Circulation* 105:380–386.

Hannoun Z., Fletcher J., Greenhough S., Medine C.N., Samuel K., Sharma R., Pryde A. et al. 2010. The comparison between conditioned media and serum free media in human embryonic stem cell culture and differentiation. *Cellular Reprogramming* 12(2):133–140.

Hay D.C., Fletcher J., Payne C., Terrace J.D., Gallagher R.C.J., Snoeys J., Black J.R. et al. 2008a. Highly efficient differentiation of hESCs to functional hepatic endoderm requires activinA and Wnt3a signalling. *Proc Natl Acad Sci USA* 105(34):12301–12306.

Hay D.C., Pernagallo S., Diaz-Mochon J.J., Medine C.N., Greenhough S., Hannoun Z., Schrader J. et al. 2011. Unbiased screening of polymer libraries to define novel substrates for functional hepatocytes with inducible drug metabolism. *Stem Cell Res* 6:92–102.

Hay D.C., Zhao D., Fletcher J., Hewitt Z., Black J., McLean D., Elcombe C.R., Ross J.A., Wolf R., and Cui W. 2008b. Efficient differentiation of hepatocytes from human embryonic stem cells exhibiting markers recapitulating liver development in vivo. *Stem Cells* 26(4):894–902.

Hay D.C., Zhao D., Ross A., Mandalam R., Lebkowski J., and Cui W. 2007. Direct differentiation of human embryonic stem cells to hepatocyte-like cells exhibiting functional activities. *Cloning Stem Cells* 9(1):51–62.

Heikkilä J.T., Aho A.J., Kangasniemi I., and Yli-Urpo A. 1996. Polymethylmethacrylate composites: Disturbed bone formation at the surface of bioactive glass and hydroxyapatite. *Biomaterials* 17(18):1755–1760.

Helmus M.N. and Tweden K. 1995. Materials selection. In *Encyclopedic Handbook of Biomaterials and Bioengineering*, ed. D.L. Wise, Part A Vol. 1, pp. 27–59. New York: CRC Press.

Henry J.A., Burugapalli K., Neuenschwander P., and Pandit A. 2009. Structural variants of biodegradable polyesterurethane *in vivo* evoke a cellular and angiogenic response that is dictated by architecture. *Acta Biomater* 5:29–42.

Hoemann C.D., Chenite A., Sun J., Hurtig M., Serreqi A., Lu Z., Rossomacha E., and Buschmann M.D. 2007. Cytocompatible gel formation of chitosan-glycerol phosphate solutions supplemented with hydroxyl ethyl cellulose is due to the presence of glyoxal. *J Biomed Mater Res, Part A* 83:521–529.

Hook A.L., Anderson D.G., Langer R., Williams P., Davies M.C., and Alexander M.R. 2010. High throughput methods applied in biomaterial development and discovery. *Biomaterials* 31(2):187–198.

Hristov M. and Weber C. 2004. Endothelial progenitor cells: Characterization, pathophysiology, and possible clinical relevance. *J Cell Mol Med* 8:498–508.

http://www.who.int/mediacentre/factsheets/fs317/en/index.html

Huang Q., Goh J.C., Hutmacher D.W., and Lee E.H. 2002. *In vivo* mesenchymal cell recruitment by a scaffold loaded with transforming growth factor beta1 and the potential for in situ chondrogenesis. *Tissue Eng* 8:469–482.

Jia X.Q. and Kiick K.L. 2009. Hybrid multicomponent hydrogels for tissue engineering. *Macromol Biosci* 9:140–156.

Jiang X., Lim S.H., Mao H.Q., and Chew S.Y. 2010. Current applications and future perspectives of artificial nerve conduits. *Exp Neurol* 223(1):86–101.

Kalka C., Masuda H., Takahashi T., Kalka-Moll W.M., Silver M., Kearney M., Li T., Isner J.M., and Asahara T. 2000. Transplantation of ex vivo expanded endothelial progenitor cells for therapeutic neovascularization. *Proc Natl Acad Sci USA* 97:3422–3427.

Kasahara Y., Iwasaki N., Yamane S., Igarashi T., Majima T., Nonaka S., Harada K., Nishimura S., and Minami A. 2008. Development of mature cartilage constructs using novel three-dimensional porous scaffolds for enhanced repair of osteochondral defects. *J Biomed Mater Res A* 86(1):127–136.

Kenny S.M. and Buggy M. 2003. Bone cements and fillers: A review. *J Mater Sci Mater Med* 14:923–938.

Khan F., Tare R.S., Kanczler J.M., Oreffo R.O.C., and Bradley M. 2010. Strategies for cell manipulation and skeletal tissue engineering using high-throughput polymer blend formulation and microarray techniques. *Biomaterials* 31(8):2216–2228.

Khan F., Tare R.S., Oreffo R.O.C., and Bradley M. 2009. Versatile biocompatible polymer hydrogels: Scaffolds for cell growth. *Angew Chem Int Ed* 48:978–982.

Kim K., Dean D., Wallace J., Breithaupt R., Antonios G., Mikos A.G., and Fisher J.P. 2011. The influence of stereolithographic scaffold architecture and composition on osteogenic signal expression with rat bone marrow stromal cells. *Biomaterials* 32(15):3750–3763.

Kipshidze N., Dangas G., Tsapenko M., Moses J., Leon M.B., Kutryk M., and Serruys P. 2004. Role of the endothelium in modulating neointimal formation: Vasculoprotective approaches to attenuate restenosis after percutaneous coronary interventions. *J Am Coll Cardiol* 44:733–739.

Kohn J. and Langer R. 1996. Bioresorbable and bioerodible materials. In *Biomaterials Science: An Introduction to Materials in Medicine*, eds. B.D. Ratner, A.S. Hoffman, F.J. Schoen, and J.E. Lemons, pp. 64–72. New York: Academic Press.

Langer R. and Tirrell D.A. 2004. Designing materials for biology and medicine. *Nature* 428(6982):487–492.

Lavon N., Yanuka O., and Benvenisty N. 2004. Differentiation and isolation of hepatic like cells from human embryonic stem cells. *Differentiation* 72(5):230–238.

Li Y.J., Chung E.H., Rodriguez R.T., Firpo M.T., and Healy K.E. 2006. Hydrogels as artificial matrices for human embryonic stem cell self-renewal. *J Biomed Mater Res* 79A:1–5.

Liberski A.R., Tizzard G.J., Diaz-Mochon J.J., Hursthouse M.B., Milnes P., and Bradley M. 2008. Screening for polymorphs on polymer microarrays. *J Comb Chem* 10:24–27.

Lin J.D., Chen X.P., Zhang W.Q., Xu L.G., and Zheng X.Y. 2010. Temporal Augmentation Using a Polytetrafluoroethylene Implant With the Assistance of an Endoscope. *Aesth Plast Surg* 34(6):701–704.

Liu X. and Ma P.X. 2004. Polymeric scaffolds for bone tissue engineering. *Ann Biomed Eng* 32:477–486.

Lu J.X., Prudhommeaux F., Meunier A., Sedel L., and Guillemin G. 1999. Effects of chitosan on rat knee cartilages. *Biomaterials* 20(20):1937–1944.

Mano J.F., Sousa R.A., Boesel L.F., Neves N.M., and Reis R.L. 2004. Bioinert, biode-
    gradable and injectable polymeric matrix composites for hard tissue replace-
    ment: State of the art and recent developments. *Compos Sci Technol* 64:789–817.
Marler J.J., Upton J., Langer R., and Vacanti J.P. 1998. Transplantation of cells in matri-
    ces for tissue regeneration. *Adv Drug Delivery Rev* 33(1–2):165–182.
Michael A.R.M., Richard H., and Ulrich S.S. 2004. Combinatorial methods, automated
    synthesis and high-throughput screening in polymer research: The evolution
    continues. *Macromol Rapid Commun* 25(1):21–33.
Middleton J.C. and Tipton A.J. 2000. Synthetic biodegradable polymers as orthopedic
    devices. *Biomaterials* 21:2335–2346.
Mills N.L., Tura O., Padfield G.J., Millar C., Lang N.N., Stirling D., Ludlam C., Turner
    M.L., Barclay G.R., and Newby D.E. 2009. Dissociation of phenotypic and func-
    tional endothelial progenitor cells in patients undergoing percutaneous coro-
    nary intervention. *Heart* 95:2003–2008.
Mizomoto H. 2004. The synthesis and screening of polymer libraries using a high
    throughput approach. PhD thesis, University of Southampton, Southampton, U.K.
Mooney D.J. and Vandenburgh H. 2008. Cell delivery mechanisms for tissue repair.
    Cell Stem Cell 2:205–213.
Muldowney J.A. III, Stringham J.R., Levy S.E., Gleaves L.A., Eren M., Piana R.N.,
    and Vaughan D.E. 2007. Antiproliferative agents alter vascular plasminogen
    activator inhibitor-1 expression: A potential prothrombotic mechanism of drug-
    eluting stents. *Arterioscler Thromb Vasc Biol* 27:400–406.
Nair L.S. and Laurencin C.T. 2007. Biodegradable polymers as biomaterials. *Prog
    Polym Sci* 32(8–9):762–798.
Orlic D., Kajstura J., Chimenti S., Jakoniuk I., Anderson S.M., Li B., Pickel J. et al. 2001.
    Bone marrow cells regenerate infarcted myocardium. *Nature* 410:701–705.
Padfield G., Newby D.E., and Mills N.L. 2010. Understanding the role of endothe-
    lial progenitor cells in percutaneous coronary intervention. *Am Coll Cardiol*
    55:1553–1565.
Paige K., Cima L., Yaremchuck M., Schloo B., Vacanti J., and Vacanti C. 1996. De
    novo cartilage generation using calcium alginate-chondrocyte constructs. *Plast
    Reconstr Surg* 97:168–178.
Payne C.M., Samuel K., Pryde A., King J., Brownstein D., Schrader J., Medine C.N.
    et al. 2011. Persistence of functional hepatocyte like cells in immune compro-
    mised mice. *Liver Int* 31(2):254–262.
Pernagallo S., Diaz-Mochon J.J., and Bradley M. 2009. A cooperative polymer-DNA
    microarray approach to biomaterial investigation. *Lab Chip* 9:397–403.
Piterina A.V., Cloonan A.J., Meaney C.L., Davis L.M., Callanan A., Walsh M.T., and
    McGloughlin T.M. 2009. ECM-based materials in cardiovascular applications:
    Inherent healing potential and augmentation of native regenerative processes.
    Int J Mol Sci 10:4375–4417.
Pitt C.G., Gratzel M.M., and Kimmel G.L. 1981. Aliphatic polyesters. 2. The degra-
    dation of poly(DL-lactide), poly(e-caprolactone) and their copolymers *in vivo*.
    Biomaterials 2:215–220.
Puppi D., Chiellini F., Piras A.M., and Chiellini E. 2010. Polymeric materials for bone
    and cartilage repair. *Prog Polym Sci* 35:403–440.
Ravi S., Qu Z., and Chaikof E.L. 2009. Polymer materials for tissue engineering of
    arterial substitutes. *Vascular* 17:S45–S54.

Reubinoff B.E., Pera M.F., Fong C.Y., Trounson A., and Bongso A. 2000. Embryonic stem cell lines from human blastocysts: Somatic differentiation in vitro. *Nat Biotechnol* 18:399–404.

Risau W. 1995. Differentiation of endothelium. *FASEB J* 9:926–933.

Robinson B.P., Hollinger J.O., Szachowicz E.H., and Brekke J. 1995. Calvarial bone repair with porous D,L-polylactide. *Otolaryngol Head Neck Surg* 112:707–713.

Rowley J.A., Madlambayan G., and Mooney D.J. 1999. Alginate hydrogels as synthetic extracellular matrix materials. *Biomaterials* 20:45–53.

Santerre J.P. and Labow R.S. 1997. The effect of hard segment size on the hydrolytic stability of polyether-urea-urethanes when exposed to cholesterol esterase. *J Biomed Mater Res* 36(2):223–232.

Santerre J.P., Woodhouse K., Laroche G., and Labow R.S. 2005. Understanding the biodegradation of polyurethanes: From classical implants to tissue engineering materials. *Biomaterials* 26(35):7457–7470.

Seal B.L., Otero T.C., and Panitch A. 2001. Polymeric biomaterials for tissue and organ regeneration. *Mater Sci Eng* 34:147–230.

Shi D. 2003. *Biomaterials and Tissue Engineering*. Berlin, Germany: Springer, Chapters 1, 2, and 5.

Shi Q., Rafii S., Wu M.H., Wijelath E.S., Yu C., Ishida A., Fujita Y. et al. 1998. Evidence for circulating bone marrow-derived endothelial cells. *Blood* 92:362–367.

Shin H., Zygourakis K., Farach-Carson M.C., Yaszemski M.J., and Mikos A.G. 2004. Modulation of differentiation and mineralization of marrow stromal cells cultured on biomimetic hydrogels modified with Arg-Gly-Asp containing peptides. *J Biomed Mater Res* 69A:535–543.

Silverman R., Passaretti D., Huang W., Randolph M., and Yaremchuk M. 1999. Injectable tissue-engineered cartilage using a fibrin glue polymer. *Plast Reconstr Surg* 103:1809–1818.

Solchaga L.A., Dennis J.E., Goldberg V.M., and Caplan A.I. 1999. Hyaluronic acid-based polymers as cell carriers for tissue-engineered repair of bone and cartilage. *J Orthop Res* 17:205–213.

Solchaga L.A., Gao J., Dennis J.E., Awadallah A., Lundberg M., Caplan A.I., and Goldberg V.M. 2002. Treatment of osteochondral defects with autologous bone marrow in a hyaluronan-based delivery vehicle. *Tissue Eng* 8:333–347.

Stephenson E., Ogilvie C.M., Patel H., Cornwell G., Jacquet L., Kadeva N., Braude P., and Ilic D. 2010. Safety paradigm: Genetic evaluation of therapeutic grade human embryonic stem cells. *J R Soc Interface* 7:S677–S688.

Stile R.A., Burghardt W.R., and Healy K.E. 1999. Synthesis and characterization of injectable poly(N-isopropylacrylamide)-based hydrogels that support tissue formation *in vitro*. *Macromolecules* 32:370–379.

Suh J.K.F. and Matthew H.W.T. 2000. Application of chitosan-based polysaccharide biomaterials in cartilage tissue engineering: A review. *Biomaterials* 21:2589–2598.

Sullivan G.J., Hay D.C., Park I.H., Fletcher J., Hannoun Z., Payne C.M., Dalgetty D. et al. 2010. Generation of functional human hepatic endoderm from human iPS cells. *Hepatology* 51(1):329–335.

Tabesh H., Amoabediny G.H., Nik N.S., Heydari M., Yosefifard M., Siadat S.O.R., and Mottaghy K. 2009. The role of biodegradable engineered scaffolds seeded with Schwann cells for spinal cord regeneration. *Neurochem Intern* 54:73–83.

Tare R.S., Khan F., Tourniaire G., Morgan S.M., Bradley M., and Oreffo R.O.C. 2009. A microarray approach to the identification of polyurethanes for the isolation of human skeletal progenitor cells and augmentation of skeletal cell growth. *Biomaterials* 30(6):1045–1055.

Tatai L., Moore T.G., Adhikari R., Malherbe F., Jayasekara R., Griffiths I., and Gunatillake P.A. 2007. Thermoplastic biodegradable polyurethanes: The effect of chain extender structure on properties and in-vitro degradation. *Biomaterials* 28:5407–5417.

Temenoff J.S. and Mikos A.G. 2000. Review: Tissue engineering for regeneration of articular cartilage. *Biomaterials* 21:431–440.

Thaburet J.-F.O., Mizomoto H., and Bradley M. 2004. High-throughput evaluation of the wettability of polymer libraries. *Macromol Rapid Commun* 25:366–370.

Thomson J.A., Itskovitz-Eldor J., Shapiro S.S., Waknitz M.A., Swiergiel J.J., Marshall V.S., and Jones J.M. 1998. Embryonic stem cell lines derived from human blasto-cysts. *Science* 282:1145–1147.

Touboul T., Hannan N.R., Corbineau S., Martinez A., Martinet C., Branchereau S., Mainot S. et al. 2010. Generation of functional hepatocytes from human embryonic stem cells under chemically defined conditions that recapitulate liver development. *Hepatology* 51(5):1754–1765.

Tourniaire G. 2006. Polymer microarray-development and applications. PhD thesis, University of Edinburgh, Edinburgh, U.K.

Tourniaire G., Collins J., Campbell S., Mizomoto H., Ogawa S., Thaburet J.-F., and Bradley M. 2006. Polymer microarrays for cellular adhesion. *Chem Commun* 20:2118–2120.

Villa-Diaz L.G., Nandivada H., Ding J., Nogueira-de-Souza N.C., Krebsbach P.H., O'Shea K.S., Lahann J., and Smith G.D. 2010. Synthetic polymer coatings for long-term growth of human embryonic stem cells. *Nat Biotechnol* 28(6):581–583.

Wang M. 2003. Developing bioactive composite materials for tissue replacement. *Biomaterials* 24:2133–2151.

Wege H., Chui M.S., Le H.T., Strom S.C., and Zern M.A. 2003. *In vitro* expansion of human hepatocytes is restricted by telomere-dependent replicative aging. *Cell Transplant* 12:897–906.

Wilcox J.N., Okamoto E.I., Nakahara K.I., and Vinten-Johansen J. 2001. Perivascular responses after angioplasty which may contribute to postangioplasty restenosis: A role for circulating myofibroblast precursors? *Ann N Y Acad Sci* 947:68–90.

Wyre R.M. and Downes S. 2000. An *in vitro* investigation of the PEMA/THFMA polymer system as a biomaterial for cartilage repair. *Biomaterials* 21(4):335–343.

Yañez F., Concheiro A., and Alvarez-Lorenzo C. 2008. Macromolecule release and smoothness of semi-interpenetrating PVP-pHEMA networks for comfortable soft contact lenses. *Eur J Pharm Biopharm* 69(3):1094–1103.

Yasu T. 2009. Differentiation of endothelial progenitor cells. *Circ J* 73:1199–1200.

Yaszenski M.J., Payne R.G., Hayes W.C., Langer R., Aufdemorte T.B., and Mikos A.G. 1995. The ingrowth of new bone tissue and initial mechanical properties of a degrading polymeric composite scaffold. *Tissue Eng* 1:41–52.

Yaszemski M.J., Payne R.G., Hayes W.C., Langer R., and Mikos A.G. 1996. The evolution of bone transplantation: Molecular, cellular, and tissue strategies to engineer human bone. *Biomaterials* 7:175–185.

Zhang R., Liberski A., Khan F., Diaz-Mochon J.J., and Bradley M. 2008. Inkjet fabrication of hydrogel microarrays using in situ nanolitre-scale polymerisation. *Chem Commun* 11:1317–1319.

Zhang R., Liberski A., Sanchez-Martin R., and Bradley M. 2009. Microarrays of over 2000 hydrogels—Identification of substrates for cellular trapping and thermally triggered release. *Biomaterials* 30:6193–6201.

Zilla P., Bezuidenhout D., and Human P. 2007. Prosthetic vascular grafts: Wrong models, wrong questions and no healing. *Biomaterials* 28:5009–5027.

2

Hydrogel as Stem Cell Niche for In Vivo Applications in Regenerative Medicine

Xiaokang Li, Claudia Wittkowske, Rui Yao, and Yanan Du

CONTENTS

2.1 Introduction

Diseases and injuries can lead to severe and irreversible damages to tissues and organs. Organ transplantation is currently the most effective way to restore tissue function and improve the survival rate of patients. However, over 98,000 people are currently on the waiting list for organ transplantation, and the number is rapidly growing due to the huge shortage of organ donors [1,2]. The field of regenerative medicine holds great promise to meet this challenge by replacing damaged tissues with functional engineered counterparts. The success in clinical application using engineered tissues requires reliable cell sources with the ability to regrow into specific cellular components of targeted tissues and biomaterials to support cellular functions.

With the unique properties of self-renewal and pluripotency, stem cells have emerged as the most promising cell source for tissue engineering and regenerative medicine. According to their origin, stem cells generally include embryonic stem cells (ESCs) and adult stem cells as well as the recently discovered induced pluripotent stem cells (iPSCs) [3]. Currently, adult stem cells are the most commonly used cell source for regenerative medicine which do not pose same ethical issue as ESCs and can be readily harvested from adult tissues, such as bone marrow and adipose tissue.

To enhance the efficacy of stem-cell-based applications, a well-defined microenvironment, often referred to as stem cell niche, is indispensable which includes cellular and acellular components (e.g., matrix and soluble factors) [4]. Due to the resemblance to the native extracellular matrix (ECM), hydrogels have been extensively applied as the matrix component of the stem cell niche. Hydrogels are polymer networks with high water content exhibiting biocompatibility, biodegradability, and tunable mechanical properties which are ideal for cellular support and tissue regeneration [5].

In this chapter, natural and synthetic hydrogels used in stem cell engineering were first reviewed followed by exemplary studies utilizing stem-cell-incorporated hydrogels for regenerating various tissues, including cartilage, adipose, bone, vasculature, etc. In general approaches for hydrogel-mediated stem cell engineering (Figure 2.1), stem cells are first harvested from embryos or adult tissues with human or nonhuman origins and expanded in vitro to obtain sufficient cell numbers. The cells are subsequently mixed with hydrogel precursor solution, which is either applied to the patients or animals by direct injection (in the aqueous form or as microspheres) or by implantation (as solidified constructs with cells embedded in 3D polymerized hydrogel scaffolds). The differentiation of stem cells into the desired cell lineage can be induced either in vitro or directly in vivo after implantation, which is largely regulated by biophysical and biochemical cues. The hydrogel-mediated delivery of stem cells provides both a physical barrier to stabilize the implanted cells at the defect site and microenvironment for new tissue formation with defined structures and functions. As an emerging field, numerous investigations have been conducted in recent years to examine the effects of hydrogel on stem cell fate in vitro as contrasted by limited cases that have been extended to in vivo studies. Here we focus on hydrogel-mediated stem cell engineering for in vivo studies, which are of vital importance to future clinical applications in regenerative medicine.

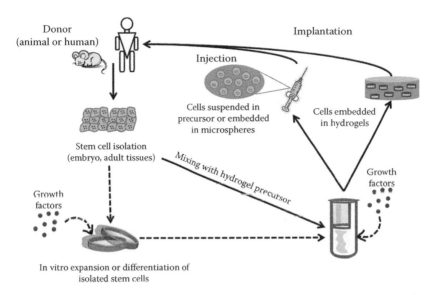

FIGURE 2.1
Schematic of basic strategies for hydrogel based stem cell engineering.

2.2 Hydrogels as Scaffolding Materials

Hydrogels have been widely used in tissue engineering, thanks to their structural and compositional resemblances to the ECM, their extensive framework for supporting cellular proliferation, and the convenience for delivery in a minimally invasive manner [2,6]. Hydrogels with various compositions and forms have been employed as space-filling agents, as 3D constructs for organizing cells, and as scaffolds for delivering biomolecules [1]. Integrated with microfabrication approaches, dimension-defined hydrogels can be made with precision in micro- or nanoscale to better mimic the exquisite architectures of stem cell niche [7]. Depending on their origins, hydrogels can be categorized into natural hydrogels, synthetic hydrogels, and the hybrid combining these two. Hydrogels utilized in tissue engineering and regenerative medicine are required to meet a number of design criteria including biocompatibility, controlled degradation, gelation conditions compatible with physiological applications, and tailored bioactivity (i.e., adhesiveness and specific ligand-receptor interactions) and mechanical stiffness [1,2,7,8]. Tables 2.1 and 2.2 summarize the origin and chemical structure of natural and synthetic hydrogels used in stem cell engineering.

TABLE 2.1

Origin and Chemical Structure of Natural Hydrogels Used in Stem
Cell Engineering

Hydrogel	Origin	Chemical Structure
Collagen	Naturally occurring proteins as major components of ECM in mammalian tissues	Three polypeptide chains twined around one another to form triple helix
HA (also hyaluronan)	GAG	Repeating β-1,4-D-glucuronic acid-β-1,3-N-acetyl-D-glucosamine disaccharide
Matrigel® (by BD Bioscience)	EHS mouse sarcoma cells	Protein mixture secreted by EHS mouse tumor
Fibrin gel	Formed by reaction of fibrinogen and thrombin	Fibrinogen and thrombin
Alginate	Marine brown algae	(1–4)-linked β-D-mannuronic acid (M) and α-L-guluronic acid (G) monomers
Cellulose	Cell wall of green plants, many algae, and oomycetes	(1–4)-linked β-D-glucose units

Sources: Drury, J.L. and Mooney, D.J., *Biomaterials*, 24(24), 4337, 2003; Slaughter, B.V. et al., *Adv. Mater.*, 21(32–33), 3307, 2009; Mather, M.L. and Tomlins, P.E., *Regen. Med.*, 5(5), 809, 2010; Hoffman, A.S., *Ann. N.Y. Acad. Sci.*, 944(1), 62, 2001; Lee, K.Y. and Mooney, D.J., *Chem. Rev.*, 101(7), 1869, 2001; Burdick, J.A. and Prestwich, G.D., *Adv. Mater.*, 23(12), H41, 2011.
See Glossary for definition of terms.

2.2.1 Natural Hydrogels

Hydrogels derived from natural polymers have been extensively used in stem cell engineering applications as they either are components of natural ECM or exhibit properties similar to the matrix components in native tissues. As shown in Table 2.1, collagen, for example, is one major ECM protein of mammalian tissue [9,10]. Similarly, hyaluronic acid (HA) is universally present in mammalian tissue as the simplest glycosaminoglycan (GAG) [9]. Alginate and cellulose can be extracted from plants and are widely used for in vivo applications [11,12]. Other examples for natural hydrogels include Matrigel®,

TABLE 2.2

Origin and Chemical Structure of Synthetic Hydrogels Used in Stem
Cell Engineering

Hydrogel	Unit	Molecular Weight (g/mol)
PEG/PEO (chemically synonymous, but different molecular weight)		<20,000 (PEG) >20,000 (PEO)
PLA		Wide range
PLGA		Wide range

Sources: Drury, J.L. and Mooney, D.J., *Biomaterials*, 24(24), 4337, 2003; Slaughter, B.V. et al., *Adv. Mater.*, 21(32–33), 3307, 2009; Mathcr, M.L. and Tomlins, P.E., *Regen. Med.*, 5(5), 809, 2010; Hoffman, A.S., *Ann. N.Y. Acad. Sci.*, 944(1), 62, 2001; Lee, K.Y. and Mooney, D.J., *Chem. Rev.*, 101(7), 1869, 2001; Burdick, J.A. and Prestwich, G.D., *Adv. Mater.*, 23(12), H41, 2011.
See Glossary for definition of terms

which is a mixture of proteins secreted by Engelbreth-Holm-Swarm (EHS) mouse sarcoma cells and fibrin gel, which is formed by the clotting reaction of fibrinogen and thrombin [13,14]. Potential problems of applying natural hydrogels for regenerative medicine may arise from undefined compositions, batch-to-batch variation, and risk of immune rejection for animal-derived materials.

2.2.2 Synthetic Hydrogels

Synthetic hydrogels are chemically defined polymers with well-tunable properties such as chemical composition, molecular weights, block structures, cross-linking density, mechanical strength, and degradability [2]. The ability to control and reproduce these properties offers great advantages to achieve optimal and consistent performance in tissue engineering. As shown in Table 2.2, synthetic polymers such as polyethylene glycol (PEG), polyethylene oxide (PEO), polylactic acid (PLA), poly(lactic-co-glycolic acid) (PLGA), or polyethylene(glycol)diacrylate (PEGDA) [15–18] have been frequently used in stem-cell-based tissue engineering applications. Synthetic polymers also prevent the potential contamination caused by pathogen originated from animal-derived matrices.

2.2.3 Cross-Linking Strategies

Hydrogels keep their structural integrity while remaining insoluble in aqueous solutions due to physical or chemical cross-linking between individual polymer chains. Physical treatments such as UV radiation and temperature alterations (warming or cooling) can polymerize gel precursor resulting in the formation of hydrogels with covalent linkages between adjacent polymeric chains. For example, PEGDA, a diacrylated derivative of PEG, can be photocross-linked by UV light to form biocompatible PEGDA hydrogels with tunable stiffness spanning large ranges. Widely used thermal cross-linkable materials include Matrigel, collagen, and HA, which are present as precursor solution in lower temperatures and form hydrogel when heated [19].

Hydrogel precursor can also be cross-linked by chemical reactions [20]. For example, the divalent metal cations (i.e., Ca^{2+}) can induce the cross-linking of alginate during chelation. Fibrin gel is formed by the enzymatic polymerization of fibrinogen in the presence of thrombin as seen in the clotting process during wound healing [8]. For certain hydrogels, the polymerization can be triggered by either physical or chemical strategies. For example, cellulose hydrogel can be formed by heating a precursor solution of cellulosic or exposing the precursor to high-energy radiation. Besides physical treatments, chemical cross-linking with aldehydes or epichlorohydrin can also generate stiff cellulose networks [20].

2.3 Hydrogel-Based Stem Cell Engineering for Regenerative Medicine

As one of the most well-studied and widely applied biomaterials in tissue engineering, hydrogels have been proved to provide an ECM-mimicking microenvironment as stem cell niche, to regulate stem cell fate both in vitro and in vivo. The integration of hydrogels and stem cells offers both tunable biomaterials and unlimited cell sources which opens new opportunities to regenerate viable tissues and organs. Hydrogel-mediated stem cell engineering is particularly powerful in regenerating soft tissues, such as cartilage and adipose tissue, in vivo which have demonstrated great potential for clinical success in the near future. Attempts have also been made to regenerate other tissues with more complex structures and functions (i.e., bone, nerves, vasculature, muscle, and retina) indicating its wide application for regenerative medicine.

2.3.1 Cartilage Regeneration

Articular cartilage is a nonvascularized, noninnervated connective tissue, which is highly hydrated and composed of chondrocytes embedded in type II collagen and GAGs [1]. Cell density in cartilage tissue is extremely low which limits its capacity for self-repair. Damage to the articular cartilage caused by osteoarthritis is becoming a significant clinical and social problem due to the increase in the elderly population. Tissue-engineered cartilages represent promising alternatives for cartilage repair. Since the availability of chondrocytes in human is limited and their proliferation capacity decreases with age [21], most studies in cartilage repair focus on using adult stem cells as cell source. For instance, mesenchymal stem cells derived from bone marrow (BMSCs) exhibited higher chondrogenic potential than stem cells isolated from other sources such as adipose tissue [12,22,23].

With similar hydration and softness as cartilage tissue, hydrogels have been the most commonly used scaffold in cartilage tissue engineering. Additionally, hydrogels can be injected in a minimally invasive manner directly in the cartilage defect site. As summarized in Table 2.3, a variety of hydrogels (e.g., fibrin, agarose, gelatin, collagen, and alginate) have already been fabricated to serve as scaffolds to achieve chondrogenesis [22,24,41].

Growth factors, such as transforming growth factor-β (TGF-β), fibroblast growth factor (FGF), and insulin-like growth factor-I (IGF-I), play important roles in the lineage-specific differentiation of BMSCs into chondrocytes which has already been well demonstrated in vitro [21,25,26,41]. Incorporation of growth factors into hydrogel was also shown to enhance the chondrogenesis of the implanted stem cells in vivo. In one example, TGF-β1 was coupled to HA which has been used to load BMSCs for cartilage repair in a mice model and a rabbit model, respectively. The growth-factor-loaded hydrogel enhanced the formation of cartilaginous matrix in both animal models [27] (Figure 2.2I). In another case, MSCs adhered to the surface of TGF-β3 functionalized PLGA microspheres, which were injected subcutaneously and intramuscularly in mice. Cartilage-like tissue was formed after 4 weeks as indicated by upregulation of cartilage-specific markers at the mRNA and protein levels [28].

One problem of using growth factors for in vivo studies is the short half-life which hinders their in vivo efficacy. To prolong the activity of growth factors, DNA plasmids encoding corresponding growth factors were incorporated into the hydrogels and were transfected to the stem cells to maintain sustained expression. Wang et al. [29] established a composite construct for BMSCs which comprised a PLGA sponge/ fibrin hydrogel scaffold loaded with plasmid DNA encoding TGF-β1. Twelve weeks postimplantation, successful repair of the full-thickness cartilage defect could be achieved in a rabbit model (Figure 2.2II).

Hydrogel-mediated stem cell engineering for cartilage repair has already been applied in a clinical trial. In this study, hBMSCs were embedded in a collagen gel and transplanted into the articular cartilage defect in patients suffering from knee osteoarthritis. Cartilage-like tissue could be observed 42 weeks after transplantation. However, the clinical improvement was not significantly different in the stem-cell-transplanted group compared to

TABLE 2.3

Hydrogel-Mediated Stem Cell Engineering for Cartilage Regeneration In Vivo

Cell Source	Hydrogel	In Vivo Model	Results	References
MSC	Collagen	Implantation Full-thickness cartilage defect Rabbits	Greater chondrogenic potential of MSCs from synovium and bone marrow than other sources Synovium-derived MSCs had greatest proliferation potential	[31]
hASC	Cross-linked HA with TGF-β1	Injection Subcutaneously Nude mice Implantation Full-thickness defect of knee cartilage Rabbits	TGF-β1-conjugated hydrogel produced more type II collagen Formation of cartilaginous matrix and maintenance of original volume in the TGF-β1-conjugated hydrogel	[27]
hBMSC	PLGA microspheres with TGF-β3	Injection Intramuscularly and subcutaneously SCID mice	Formation of histologically resembling cartilage, staining positive for chondrocyte markers	[28]
BMSC	Fibrin gel, plasmid DNA encoding TGF-β1 and PLGA sponge	Implantation Full-thickness cartilage defect Rabbits	Repair of cartilage defects Good integration with surrounding tissue	[29]
BMSC	Collagen and collagen-alginate	Implantation Subcutaneously Rabbits	Chondrogenic differentiation of BMSCs	[12]
BMSC	PLGA	Implantation Joint cavity Sheep	Expression of type II collagen but no differentiation into bone-like cells after implanting to the normal joint cavity	[32]
BMSC	Collagen	Implantation Articular cartilage defect in knee joints Humans	Width, thickness, stiffness, and cell morphology of newly formed tissue were better resembling native cartilage in cell-transplanted group than in cell-free control group No significant difference in clinical outcome among different experiment groups	[30]

See Glossary for definition of terms.

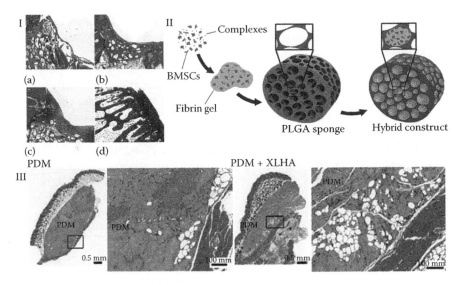

FIGURE 2.2
Hydrogel-mediated stem cell engineering for regenerating cartilage and adipose tissue.

the cell-free control group [30]. This result indicates that further scientific studies and technical improvements are required before the integration of stem cells and hydrogels can be truly beneficial for cartilage regeneration in clinical studies.

2.3.2 Adipose Tissue Regeneration

The shortage of adipose tissue in many clinical situations, such as traumatic injury, oncologic surgery, or congenital defects, poses a big challenge for plastic and reconstructive surgeons. Currently, transplantation of autologous adipose tissue is the most widely adopted method for soft tissue reconstruction [33]. However, the clinical outcome of adipose tissue transplantation is often unpredictable due to implant resorption caused by a lack of vascularization [34,35]. Tissue-engineered adipose grafts with proper vascularization and scaffolding support could reduce implant resorption and thus start gaining more attention [34]. Although readily available, mature adipocytes are not suitable for engineering adipose tissue because of their limited proliferative capabilities. Preadipocytes, the progenitors which show better proliferation and specialized differentiation into adipocytes, are an alternative [34], but still require exogenous endothelial cells to improve the vascularization of the engineered adipose tissue.

With the ability to differentiate into adipocytes and vascular cells, stem cells are an ideal cell source for engineering vascularized adipose tissue [36]. Since adipose tissue is largely available, harvesting ASCs from adipose tissue is more convenient than from bone marrow. However, depending on the location of the donor tissue as well as the age and gender of the patient, ASCs vary in metabolic activity and in their capacity for proliferation and differentiation [34].

Hydrogel-based scaffolds for adipose tissue engineering are expected to recapitulate the physical and architectural features of the native adipose tissue. These features include tunable softness and appropriate pore size facilitating proliferation and differentiation of stem cells into adipocytes [34]. As summarized in Table 2.4, a variety of hydrogels, including those for clinical use, such as type I collagen sponges, nonwoven PGA, and HA [37], have already been successfully applied for regenerating adipose tissues in vivo.

TABLE 2.4

Hydrogel-Mediated Stem Cell Engineering for Adipose Tissue Regeneration In Vivo

Cell Source	Hydrogel	In Vivo Model	Results	References
hMSC, hASC	Collagen, PLA, or silk fibroin	Implantation Muscle pouch Rats	Attachment of hMSCs and hASCs on all scaffolds Rapid collagen and PLA scaffold degeneration Silk fibroin scaffolds provided longer-term structural integrity	[40]
hBMSC	PEGDA	Implantation Dorsum Immunodeficient mice	Expression of adipogenic markers (Oil Red O staining) and adipocyte specific genes Maintenance of volume	[38]
ASC	Type I collagen sponge, nonwoven PGA, or HA gel	Implantation Subcutaneously Athymic mice	Type I collagen sponge provided best scaffold for generation of adipose tissue	[37]
hASC	PDM and cross-linked HA	Implantation Subcutaneously Athymic mice	Maintenance of volume Differentiation into mature adipocytes Implant integrated with host tissue and established vascularization	[35]
ASC	PLGA microspheres	Injection Subcutaneously Nude mice	Enhanced tissue regeneration and adipogenic differentiation as indicated by Oil Red O staining Complete differentiation of ASCs into adipocytes was confirmed using RT-PCR	[39]
MSC	PLGA microspheres	Injection Neck Nude mice	Microsphere diameter of 100–150 µm formed fat pads resembling the native tissue	[39]

See Glossary for definition of terms.

In one study, hydrogel constructs consisting of cross-linked PEGDA were shown to provide favorable microenvironment to promote the differentiation of the encapsulated BMSCs into adipocytes in mice. Furthermore, the implanted constructs maintained their volume and architecture for 4 weeks [38]. Hybrid scaffolds which combine functionalities of two or multiple material components have been applied as well. For example, combination of placental decellular matrix (PDM) as the backbone for structural maintenance and cross-linked HA for achieving a high degree of vascularization led to a positive effect in terms of angiogenesis and adipogenesis in vivo [35] (Figure 2.2III).

Minimizing postoperation scars is extremely important in plastic and reconstructive surgery. Therefore, injectable hydrogels are preferred compared to sheet- or sponge-type scaffolds, which tend to leave scars due to the incision during transplantation. However, most injectable hydrogels are jellylike and are insufficient for providing desired rigidity. A possible solution might be the use of injectable microspheres (i.e., PLGA microspheres), which can load stem cells and exhibit enhanced mechanical strength [39]. In addition, controlled degradation of hydrogels is vital for maintaining long-term integrity of adipose tissue in vivo. Rapidly degraded hydrogels (i.e., collagen) have been shown to lack the ability to support the growth of the adipose tissues during the 4 week implantation period, while slower-degraded biomaterials (i.e., silk) improved the long-term maintenance of the engineered adipose tissue [40].

2.3.3 Bone Regeneration

Birth defects, tumor resections, and serious injury can all lead to bone damage, which is normally repaired by transplantation of autologous bone grafts in clinical practice. However, the limited availability of autologous bone and the need for a secondary surgery to harvest the bone from the donor both cause critical problems [42]. Tissue-engineered bone grafts based on stem cells can potentially overcome these problems and offer better therapeutic strategies to repair bone defects. Adult BMSCs can easily differentiate into osteocytes; hence, they are the preferred cell source when engineering bone tissue [43].

To resemble the rigidity of bone, the scaffold utilized in bone tissue engineering is expected to present sufficient mechanical strength. Hydrogels, however, do not possess the mechanical properties to be used alone as scaffold in load bearing applications such as bone regeneration [1]. Therefore, most hydrogel-based applications in bone tissue engineering are in combination with stiffer materials such as bioceramics (e.g., coral and hydroxyapatite-tricalcium phosphate) [44]. In vitro studies combining hydroxyapatite with natural hydrogels such as collagen [45] or chitosan [46] have already showed promising results. The successful differentiation of MSCs to bone in vivo has been demonstrated utilizing different hydrogels (i.e., fibrin glue, alginate, collagen I) in combination with a stiff, porous scaffold made of tricalcium phosphate ceramic [42]. Extensive bone formation particularly in collagen I hydrogel has been achieved as indicated by mineralization, secretion of bone-specific matrix, and gene expression of bone-specific markers. In this study, the stem-cell-loaded constructs were implanted into the nonload bearing subcutaneous dorsal pockets of nude mice, which is often regarded as a standard technique for evaluating the in vivo ossification. Since mechanical stimulation plays a critical role in bone formation [44], implantation into a load-bearing site to replace the function of bone is necessary to provide an appropriate mechanical stimulation for stem cell development. In another in vivo study, BMSCs were embedded in a bone-like composite consisting of three different materials (nanohydroxyapatite, collagen, and PLA) and were implanted in a radial defect. The composite showed an enhanced and accelerated bone formation in vivo compared to the use of fresh-frozen allogeneic bone [47].

The encapsulation of growth factors in hydrogels can further improve the formation of bone tissue. Kim et al. performed an in situ polymerization on an HA-based hydrogel to encapsulate human mesenchymal stem cells (hMSCs) and bone morphogenetic protein-2 (BMP-2) [48]. The histological results demonstrated that the hydrogel construct loaded with hMSCs and BMP-2 showed the highest expression of osteocalcin and mature bone formation in vivo compared to hydrogels with only BMP-2 or hMSCs as well as the blank control (Figure 2.3I). Besides growth factors, the encapsulation of oxygen carriers in the hydrogel construct can additionally enhance bone formation by supplying newly grown tissues directly with oxygen [49] (Table 2.5).

2.3.4 Vascular Regeneration

Engineering blood vessels for providing sufficient vascularization in engineered tissue constructs is one of the main objectives in vascular tissue engineering. Endothelial progenitor cells (EPCs), which are committed to differentiate into endothelial cells, are one of the most widely studied cell source

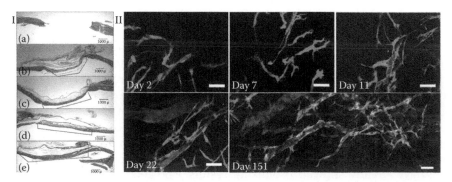

FIGURE 2.3
Hydrogel-mediated stem cell engineering for regenerating bone and vasculature.

TABLE 2.5

Hydrogel-Mediated Stem Cell Engineering for Bone Regeneration In Vivo

Cell Source	Hydrogel	In Vivo Model	Results	References
MSC	PEGDA	Implantation Dorsum Immunodeficient mice	Mandibular condyles formed de novo and showed cartilaginous and osseous phenotypes	[43]
BMSC	PEO	Injection Subcutaneously Experimental mice	Hard nodes formed in mice with MSC Hydrogel had been absorbed in mice without MSC Histological examination showed trabecular bone and neocartilage	[50]
BMSC	Nanohydroxyapatite/ collagen/PLA (nHAC/PLA)	Implantation Radial defect Rabbits	nHAC/PLA with MSC enhanced and accelerated bone formation	[47]
MSC	Fibrin glue, alginate, and collagen I hydrogel applied to β-TCP scaffolds	Implantation Subcutaneously Nude mice	Collagen I induced greater mineralization, secretion of bone-specific matrix, and gene expression of bone-specific markers than fibrin glue and alginate	[42]
hMSC	HA with BMP-2	Implantation Calvarial defect Rats	BMP-2 and MSCs induced highest expression of osteocalcin and mature bone formation with vascular markers	[48]
MSC	Fibrin gel with oxygen carriers	Implantation Subcutaneously, radial bone defect, lumbar paravertebral muscle C3H/HeN mice	Increase in bone formation, cell survival, and osteocalcin activity	[49]

See Glossary for definition of terms.

for vascular tissue engineering [51,52]. Meanwhile, the rapid advancement in using pluripotent ESCs and adult stem cells opens new possibilities for vascular regeneration. Both human ESCs and MSCs, for example, have the ability to differentiate into smooth muscle cells or endothelial cells, which are present in the media and intima, respectively, in the native blood vessels [36].

To regenerate vasculature from stem cells, it is often required to first differentiate the stem cells into endothelial cells in vitro followed by implantation in an animal model. The vasculogenesis is tightly regulated by both biochemical and biophysical cues, including growth factors, matrices, oxygen concentration (during hypoxia), and hydrodynamic shear [11,53]. Incorporation of growth factors into the hydrogel offers a facile method to induce the differentiation of stem cells into vascular lineage (i.e., endothelial cells). However, direct delivery of growth-factor-loaded hydrogels in vivo can potentially cause unwanted side effects, such as vascular leakage, due to uncontrolled release of the growth factors at the target site. Therefore, hydrogel-based stem cell delivery systems which can provide sustained stimulation of vasculogenesis are preferable. For example, Kasper et al. demonstrated in vitro that mechanically stimulated MSCs can secrete more angiogenic growth factors which lead to enhanced vasculogenesis compared to unstimulated MSCs [54].

Vascular networks have been established in animal models originated from hydrogel constructs incorporated with hESCs-derived endothelial cells. Wang et al. coimplanted hESCs-derived endothelial cells and mouse mesenchymal precursor cells embedded in fibronectin-collagen hydrogel into mice. Functional vasculatures could be observed 11 days after implantation, which were successfully integrated into the host vascular system and maintained stable for more than 150 days (Figure 2.3II) [55]. In another study, Matrigel was used as scaffold for transplanting hESCs-derived endothelial-like cells into nude mice. The formation of nascent microvessels could be observed which contain mouse blood cells and support blood flow indicating successful integration with the host circulation (Table 2.6) [56].

2.3.5 Hydrogel-Mediated Stem Cell Engineering for Regeneration of Other Tissues

Stem-cell-based tissue engineering can ideally regenerate all the tissue types. In this section, we provide examples on hydrogel-based stem cell engineering for nerve, retina, and muscle regeneration. Although, only preliminary results have been obtained, the great diversity of applications in this field demonstrates its great potential for regenerative medicine in the future (Table 2.7).

2.3.5.1 Nerve Regeneration

The nervous system, consisting of the central nervous system (CNS) and peripheral nervous system (PNS), possesses different abilities for self-repair. Neurons of the PNS exhibit a greater capacity to regrow than those of the CNS after damage.

For PNS repair, formation of new axons across the injury site is the key requirement [57]. A graft is needed to bridge the transected nerve when the

TABLE 2.6

Hydrogel-Mediated Stem Cell Engineering for Vascular Regeneration In Vivo

Cell Source	Hydrogel	In Vivo Model	Results	References
hBMSC	Alginate with VEGF	Implantation Subcutaneously or femoral artery ligation SCID mice or NOD mice	Enhanced blood vessel formation in the presence of VEGF and mechanical stimulation	[11]
hESC	Fibronectin-collagen	Implantation Cranial window SCID mice	Differentiation of hESCs into endothelial cells Formation of branching blood vessels Integration with host vasculature	[55]
hESC	Matrigel®	Injection Subcutaneously Nude mice	Differentiation of hESCs into endothelial and smooth muscle cells Formation of human microvasculature Functional integration with host vasculature	[56]

See Glossary for definition of terms.

injury is extensive. Autologous nerve grafts (autografts) have been widely used and considered as the gold standard in clinical surgery for PNS regeneration. However, several limitations still exist for autografts, such as extra incision, sacrifice of the donor nerve, and the risk of neuroma formation [58]. Therefore, synthetic grafts, namely, nerve conduits, provide a promising alternative to autografts [59]. Both natural and synthetic hydrogels (e.g., HA, chitosan, laminin, or PEG) were fabricated into nerve conduits and were proved effective for guiding and supporting the ingrowth of novel neurites from transected nerves [59–61]. Furthermore, hydrogels incorporating multipotent neural stem cells (NSCs), which have the potential to differentiate into neurons, astrocytes, and oligodendrocyte [62], showed enhanced PNS regeneration compared to hydrogels alone. For example, Zhang et al. encapsulated NSCs in an HA-collagen hydrogel conduit and achieved better reinnervation of damaged facial nerve compared to blank hydrogel in a rabbit model.

In contrast, for CNS repair, it is nearly impossible to restore the lost functions which usually lead to CNS degenerative disorders such as Parkinson's and Alzheimer's diseases. Regeneration in the CNS based on hydrogels alone usually shows limited efficacy [63,64], and it requires multiple cell types to function synergistically. Therefore, the use of multipotent NSCs [62], which can be harvested from multiple regions in mammalian brain at different

TABLE 2.7

Hydrogel-Mediated Stem Cell Engineering for Neural, Retina, and Muscle Regeneration

Cell Source	Hydrogel	In Vivo Model	Results	References
Nerve regeneration				
NSC	HA and collagen composite with neurotrophin-3	Implantation Facial nerve damage Rabbits	Repair of PNS Reinnervations of damaged facial nerve on the defect site	[62]
Neural progenitor cell (NPC)	HA-heparin-collagen	Implantation Cortical photothrombotic stroke C57BL/6 mice	Repair of CNS Cells survived twofold in the hydrogel matrix compared to cells without hydrogel Hydrogel reduced infiltration of inflammatory cells by forming zone around transplanted stem cells	[70]
Retina regeneration				
Retinal stem-progenitor cell (RSPC)	HAMC	Injection Mice	Cells survived better and distributed evenly in subretinal space in mice	[71]
Mouse ESC	Matrigel®	No in vivo	3D neural retinal tissues self-formed in vitro	[75]
Muscle regeneration				
MuSC	PEG	Implantation Immunodeficient mice	Substrate elasticity is a critical factor of MuSCs fate in culture MuSCs cultured on hydrogel substrates mimicking the rigidity of native muscle self-renewed in vitro and regenerated novel muscle tissues in vivo	[76]

See Glossary for definition of terms.

ages [65,66], shows great promise in CNS regeneration. Transplantation of NSCs alone was proved to promote functional restoration in animal models [67–69]. However, the transplanted cells integrated poorly into the adult CNS due to the lack of various bioactive cues which induce cell migration and formation of new axons or synapses [61]. Hydrogels incorporated with bioactive cues can serve as stem cell niche for CNS repair. For example, Zhong et al. [70] fabricated HA hydrogel containing bioactive heparin sulfate and

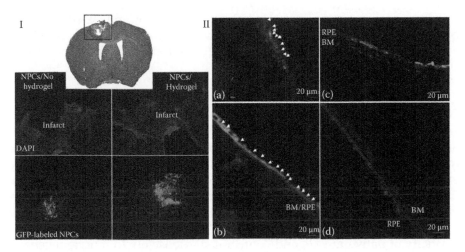

FIGURE 2.4
Hydrogel-mediated stem cell engineering for regenerating nerve and retina. (From Zhong, J. et al., *Neurorehabil. Neural Repair*, 24(7), 636, 2010; Ballios, B.G. et al., *Biomaterials*, 31(9), 2555, 2010.)

encapsulated NSCs to form a stem cell-hydrogel complex. High survival of embedded cells in vivo was demonstrated, resulting in regrowth of nerve tissues in a mouse stroke model (Figure 2.4I).

2.3.5.2 Retina Regeneration

Vision loss, often caused by the damage of the retina (i.e., retinitis pigmentosa, diabetic retinopathy, age-related macular degeneration), has a big impact on the quality of life and can affect the entire age spectrum [71]. The present therapy toward retinal degeneration focuses on pharmacological treatments, but this approach is unable to restore the impaired retina. An alternative strategy based on cell therapy relies on transplantation of retinal cells (fetal retinal pigmented epithelium cells and neural retinal cells) directly to the damaged regions of the retina [72–74], which usually suffers from poor cellular survival, distribution, and tissue integration. Ballios et al. overcame these limitations by developing a hyaluronan and methylcellulose hybrid hydrogel (HAMC) as stem cell delivery system, which was biodegradable and easily injectable in a minimally invasive manner [71]. This novel hydrogel-based delivery system supported retinal stem cell survival and proliferation in vitro and enabled contiguous distribution of stem cells in vivo with reduced cell aggregation (Figure 2.4I). Another study conducted by Eiraku et al. heralded a new therapeutic method for regenerating damaged retina [75]. An optic cup structure was reconstructed in 3D Matrigel system using mouse ESC aggregates, and fully stratified 3D neural retinal tissue sheets were formed spontaneously in vitro. The formation of retinal sheets in vitro recapitulated the complex morphogenesis of retinal anlage in vivo (Figure 2.4II). Therefore, this approach opens novel avenue to create retinal tissue using stem-cell-based tissue engineering.

2.3.5.3 Muscle Tissue Regeneration

Treatment of muscle degenerations, such as muscular dystrophy, requires new muscle tissue formation. Muscle stem cells (MuSCs) are an ideal source to regrow muscle tissue. However, this ability is lost when MuSCs are expanded in vitro for clinical use. Gilbert et al. [76] found that substrate elasticity was a critical factor to regulate MuSC self-renewal in vitro culture by varying the rigidity of the PEG hydrogel substrate. MuSCs expanded with high efficiency in vitro when cultured on hydrogel substrates mimicking the elasticity of native muscle and contributed extensively to muscle regeneration when subsequently transplanted into mice.

2.4 Summary and Outlook

Hydrogels provide a versatile and robust platform to maintain stem cell self-renewal and induce lineage-specific differentiation. As an ideal matrix component for stem cell niche, hydrogels can be easily customized by incorporating biochemically functional moieties or adjusting mechanical properties and physical states. Hydrogel-mediated stem cell engineering has been proved to be effective in regenerating various tissue types (e.g., cartilage, adipose, bone, vascular, and nerves) in vivo mainly in animal models. Clinical trials involving both hydrogel and stem cell components have been also conducted for soft tissue regeneration. Despite the great potential in this field, a number of challenges are faced related to common problems involved with stem cell therapy, such as how to achieve efficient and reliable differentiation of stem cells with high specificity, how to prevent the risk of teratocarcinoma formation and unwanted immune responses [77], and how to integrate the regenerated tissue with the host tissues. The encapsulation of stem cells in hydrogels might be beneficial to overcome these challenges by providing microenvironment to locally improve the differentiation efficiency and as a protecting zone to isolate from the immunocytes [70] as well as facilitating the remodeling of the regenerated tissue. With the future advancement in this field, more in vivo studies are expected to conduct using hydrogel-mediated stem cell engineering for regenerating various tissues. And their therapeutic efficacy and long-term safety will be ultimately assessed in clinical trials.

Glossary of Terms

ASC	adipose-derived stem cell
BMP	bone morphogenetic protein
BMSC	bone-marrow-derived mesenchymal stem cell
CNS	central nervous system
ECM	extracellular matrix
EPC	endothelial progenitor cell
FGF	fibroblast growth factor
GAG	glycosaminoglycan
HA	hyaluronic acid
hESC	human embryonic stem cell
hMSC	human mesenchymal stem cell
IGF	insulin-like growth factor
iPSC	induced pluripotent stem cell
MSC	mesenchymal stem cell
MuSC	muscle stem cell
nHAC	nanohydroxyapatite/collagen
NPC	neural progenitor cell
NSC	neural stem cell
PCL	polycaprolactone
PDM	placental decellular matrix
PEG	polyethylene glycol
PEGDA	polyethylene glycol diacrylate
PEO	polyethylene oxide
PGA	polyglycolic acid
PLA	poly lactic acid
PLGA	poly(lactic-co-glycolic acid)
PNS	peripheral nervous system
TCP	tricalcium phosphate ceramic
TGF	transforming growth factor
VEGF	vascular endothelial growth factor

References

1. Drury, J.L. and D.J. Mooney, Hydrogels for tissue engineering: Scaffold design variables and applications. *Biomaterials*, 2003. **24**(24): 4337–4351.
2. Slaughter, B.V. et al., Hydrogels in regenerative medicine. *Advanced Materials*, 2009. **21**(32–33): 3307–3329.
3. Wu, S.M. and K. Hothedlinger, Harnessing the potential of induced pluripotent stem cells for regenerative medicine. *Nature Cell Biology*, 2011. **13**(5): 497–505.

4. Mohyeldin, A., T. Garzon-Muvdi, and A. Quinones-Hinojosa, Oxygen in stem cell biology: A critical component of the stem cell niche. *Cell Stem Cell*, 2010. **7**(2): 150–161.

5. Liu, S.Q. et al., Synthetic hydrogels for controlled stem cell differentiation. *Soft Matter*, 2010. **6**(1): 67–81.

6. Mather, M.L. and P.E. Tomlins, Hydrogels in regenerative medicine: Towards understanding structure-function relationships. *Regenerative Medicine*, 2010. **5**(5): 809–821.

7. Hoffman, A.S., Hydrogels for biomedical applications. *Ann. N.Y. Acad. Sci.*, 2001. **944**(1): 62–73.

8. Lee, K.Y. and D.J. Mooney, Hydrogels for tissue engineering. *Chemical Reviews*, 2001. **101**(7): 1869–1879.

9. Alberts, B., *Molecular Biology of the Cell*, 3rd edn., 1994, New York: Garland Pub. Vol. xliii, 1294 [67] p.

10. Lee, C.H., A. Singla, and Y. Lee, Biomedical applications of collagen. *International Journal of Pharmaceutics*, 2001. **221**(1–2): 1–22.

11. Lee, K.Y. et al., Controlled growth factor release from synthetic extracellular matrices. *Nature*, 2000. **408**(6815): 998–1000.

12. Zheng, L. et al., Chondrogenic differentiation of mesenchymal stem cells induced by collagen-based hydrogel: An in vivo study. *Journal of Biomedical Materials Research Part A*, 2009. **9999A**: NA.

13. Zhao, H.G. et al., Fabrication and physical and biological properties of fibrin gel derived from human plasma. *Biomedical Materials*, 2008. **3**(1): 015001.

14. Ye, Q. et al., Fibrin gel as a three dimensional matrix in cardiovascular tissue engineering. *European Journal of Cardio-Thoracic Surgery*, 2000. **17**(5): 587–591.

15. Bryant, S.J. and K.S. Anseth, Controlling the spatial distribution of ECM components in degradable PEG hydrogels for tissue engineering cartilage. *Journal of Biomedical Materials Research Part A*, 2003. **64A**(1): 70–79.

16. Rice, M.A. and K.S. Anseth, Encapsulating chondrocytes in copolymer gels: Bimodal degradation kinetics influence cell phenotype and extracellular matrix development. *Journal of Biomedical Materials Research Part A*, 2004. **70A**(4): 560–568.

17. Salinas, C.N. and K.S. Anseth, The enhancement of chondrogenic differentiation of human mesenchymal stem cells by enzymatically regulated RGD functionalities. *Biomaterials*, 2008. **29**(15): 2370–2377.

18. Nuttelman, C.R., S.M. Henry, and K.S. Anseth, Synthesis and characterization of photocrosslinkable, degradable poly(vinyl alcohol)-based tissue engineering scaffolds. *Biomaterials*, 2002. **23**(17): 3617–3626.

19. Burdick, J.A. and G.D. Prestwich, Hyaluronic acid hydrogels for biomedical applications. *Advanced Materials*, 2011. **23**(12): H41–H56.

20. Sannino A., D. Christian, and M. Madaghiele, Biodegradable cellulose-based hydrogels: Design and applications. *Materials Letters*, 2009. **2**(2): 353–373.

21. Chen, G. et al., Chondrogenic differentiation of mesenchymal stem cells in a leakproof collagen sponge. *Materials Science and Engineering C: Biomimetic and Supramolecular Systems*, 2008. **28**(1): 195–201.

22. Mauck, R.L. and J.A. Burdick, Engineering Cartilage Tissue. In *Tissue Engineering*, Pallua, N. and Suschek, C.V. (eds.), Springer-Verlag Berlin Heidelberg, 2011. 493–520.

23. Chung, C. and J.A. Burdick, Engineering cartilage tissue. *Advanced Drug Delivery Reviews*, 2008. **60**(2): 243–262.

24. Hunziker, E., Articular cartilage repair: Basic science and clinical progress. A review of the current status and prospects. *Osteoarthritis and Cartilage*, 2002. **10**(6): 432–463.

25. Li, J. and M. Pei, Optimization of an in vitro three-dimensional microenvironment to reprogram synovium-derived stem cells for cartilage tissue engineering. *Tissue Engineering Part A*, 2011. **17**(5–6): 703–712.

26. Indrawattana, N. et al., Growth factor combination for chondrogenic induction from human mesenchymal stem cell. *Biochemical and Biophysical Research Communications*, 2004. **320**(3): 914–919.

27. Jung, H.H., K. Park, and D.K. Han, Preparation of TGF-beta 1-conjugated biodegradable pluronic F127 hydrogel and its application with adipose-derived stem cells. *Journal of Controlled Release*, 2010. **147**(1): 84–91.

28. Bouffi, C. et al., The role of pharmacologically active microcarriers releasing TGF-β3 in cartilage formation in vivo by mesenchymal stem cells. *Biomaterials*, 2010. **31**(25): 6485–6493.

29. Wang, W. et al., In vivo restoration of full-thickness cartilage defects by poly(lactide-co-glycolide) sponges filled with fibrin gel, bone marrow mesenchymal stem cells and DNA complexes. *Biomaterials*, 2010. **31**(23): 5953–5965.

30. Wakitani, S., Human autologous culture expanded bone marrow mesenchymal cell transplantation for repair of cartilage defects in osteoarthritic knees. *Osteoarthritis and Cartilage*, 2002. **10**(3): 199–206.

31. Koga, H. et al., Comparison of mesenchymal tissues-derived stem cells for in vivo chondrogenesis: Suitable conditions for cell therapy of cartilage defects in rabbit. *Cell and Tissue Research*, 2008. **333**(2): 207–215.

32. Chen, J. et al., In vivo chondrogenesis of adult bone-marrow-derived autologous mesenchymal stem cells. *Cell and Tissue Research*, 2005. **319**(3): 429–438.

33. Cherubino, M. and K.G. Marra, Adipose-derived stem cells for soft tissue reconstruction. *Regenerative Medicine*, 2009. **4**(1): 109–117.

34. Gomillion, C.T. and K.J.L. Burg, Stem cells and adipose tissue engineering. *Biomaterials*, 2006. **27**(36): 6052–6063.

35. Flynn, L. et al., Adipose tissue engineering in vivo with adipose-derived stem cells on naturally derived scaffolds. *Journal of Biomedical Materials Research Part A*, 2009. **89A**(4): 929–941.

36. Pittenger, M.F. et al., Multilineage potential of adult human mesenchymal stem cells. *Science*, 1999. **284**(5411): 143–147.

37. Itoi, Y. et al., Comparison of readily available scaffolds for adipose tissue engineering using adipose-derived stem cells. *Journal of Plastic Reconstructive and Aesthetic Surgery*, 2010. **63**(5): 858–864.

38. Alhadlaq, A., M. Tang, and J.J. Mao, Engineered adipose tissue from human mesenchymal stem cells maintains predefined shape and dimension: Implications in soft tissue augmentation and reconstruction. *Tissue Engineering*, 2005. **11**(3–4): 556–566.

39. Choi, Y. et al., Adipogenic differentiation of adipose tissue derived adult stem cells in nude mouse. *Biochemical and Biophysical Research Communications*, 2006. **345**(2): 631–637.

40. Mauney, J.R. et al., Engineering adipose-like tissue in vitro and in vivo utilizing human bone marrow and adipose-derived mesenchymal stem cells with silk fibroin 3D scaffolds. *Biomaterials*, 2007. **28**(35): 5280–5290.

41. Park, K.S. et al., Chondrogenic differentiation of bone marrow stromal cells in transforming growth factor-beta(1) loaded alginate bead. *Macromolecular Research*, 2005. **13**(4): 285–292.

42. Weinand, C. et al., Comparison of hydrogels in the in vivo formation of tissue-engineered bone using mesenchymal stem cells and beta-tricalcium phosphate. *Tissue Engineering*, 2007. **13**(4): 757–765.

43. Alhadlaq, A. and J.J. Mao, Tissue-engineered neogenesis of human-shaped mandibular condyle from rat mesenchymal stem cells. *Journal of Dental Research*, 2003. **82**(12): 951–956.

44. Reichert, J.C. and D.W. Hutmacher, Bone tissue engineering. *Tissue Engineering*, 2011. 431–456.

45. Liao, S. et al., A three-layered nano-carbonated hydroxyapatite/collagen/PLGA composite membrane for guided tissue regeneration. *Biomaterials*, 2005. **26**(36): 7564–7571.

46. Zhang, L. et al., Preparation and in vitro investigation of chitosan/nano-hydroxyapatite composite used as bone substitute materials. *Journal of Materials Science: Materials in Medicine*, 2005. **16**(3): 213–219.

47. Zhou, D.S., Repair of segmental defects with nano-hydroxyapatite/collagen/PLA composite combined with mesenchymal stem cells. *Journal of Bioactive and Compatible Polymers*, 2006. **21**(5): 373–384.

48. Kim, J. et al., Bone regeneration using hyaluronic acid-based hydrogel with bone morphogenic protein-2 and human mesenchymal stem cells. *Biomaterials*, 2007. **28**(10): 1830–1837.

49. Kimelman-Bleich, N. et al., The use of a synthetic oxygen carrier-enriched hydrogel to enhance mesenchymal stem cell-based bone formation in vivo. *Biomaterials*, 2009. **30**(27): 4639–4648.

50. Chen, F., Injectable bone. *British Journal of Oral and Maxillofacial Surgery*, 2003. **41**(4): 240–243.

51. Kaushal, S. et al., Functional small-diameter neovessels created using endothelial progenitor cells expanded ex vivo. *Nature Medicine*, 2001. **7**(9): 1035–1040.

52. Xu, Q., Circulating progenitor cells regenerate endothelium of vein graft atherosclerosis, which is diminished in ApoE-deficient mice. *Circulation Research*, 2003. **93**(8): 76e–86e.

53. Hanjaya-Putra, D. and S. Gerecht, Vascular engineering using human embryonic stem cells. *Biotechnology Progress*, 2009. **25**(1): 2–9.

54. Kasper, G. et al., Mesenchymal stem cells regulate angiogenesis according to their mechanical environment. *Stem Cells*, 2007. **25**(4): 903–910.

55. Wang, Z.Z. et al., Endothelial cells derived from human embryonic stem cells form durable blood vessels in vivo. *Nature Biotechnology*, 2007. **25**(3): 317–318.

56. Ferreira, L.S. et al., Vascular progenitor cells isolated from human embryonic stem cells give rise to endothelial and smooth muscle-like cells and form vascular networks in vivo. *Circulation Research*, 2007. **101**(3): 286–294.

57. Suri, S. and C.E. Schmidt, Cell-laden hydrogel constructs of hyaluronic acid, collagen, and laminin for neural tissue engineering. *Tissue Engineering Part A*, 2010. **16**(5): 1703–1716.

58. Cao, J.I. et al., The use of laminin modified linear ordered collagen scaffolds loaded with laminin-binding ciliary neurotrophic factor for sciatic nerve regeneration in rats. *Biomaterials*, 2011. **32**(16): 3939–3948.

59. Bellamkonda, R.V., Peripheral nerve regeneration: An opinion on channels, scaffolds and anisotropy. *Biomaterials*, 2006. **27**(19): 3515–3518.

60. Gumera, C., B. Rauck, and Y.D. Wang, Materials for central nervous system regeneration: Bioactive cues. *Journal of Materials Chemistry*, 2011. **21**(20): 7033–7051.

61. Nisbet, D.R. et al., Neural tissue engineering of the CNS using hydrogels. *Journal of Biomedical Materials Research Part B: Applied Biomaterials*, 2008. **87B**(1): 251–263.

62. Zhang, H. et al., Implantation of neural stem cells embedded in hyaluronic acid and collagen composite conduit promotes regeneration in a rabbit facial nerve injury model. *Journal of Translational Medicine*, 2008. **6**: 67.

63. Wang, A. et al., Induced pluripotent stem cells for neural tissue engineering. *Biomaterials*, 2011. **32**(22): 5023–5032.

64. Willerth, S.M., Neural tissue engineering using embryonic and induced pluripotent stem cells. *Stem Cell Research and Therapy*, 2011. **2**(2): 17.

65. Park, K.I., Y.D. Teng, and E.Y. Snyder, The injured brain interacts reciprocally with neural stem cells supported by scaffolds to reconstitute lost tissue. *Nature Biotechnology*, 2002. **20**(11): 1111–1117.

66. Perale, G. et al., Engineering injured spinal cord with bone marrow-derived stem cells and hydrogel-based matrices: A glance at the state of the art. *Journal of Applied Biomaterials and Biomechanics*, 2008. **6**(1): 1–8.

67. Jeong, S.W. et al., Human neural stem cell transplantation promotes functional recovery in rats with experimental intracerebral hemorrhage. *Stroke*, 2003. **34**(9): 2258–2263.

68. Chu, K. et al., Human neural stem cells can migrate, differentiate, and integrate after intravenous transplantation in adult rats with transient forebrain ischemia. *Neuroscience Letters*, 2003. **343**(2): 129–133.

69. Kelly, S. et al., Transplanted human fetal neural stem cells survive, migrate, and differentiate in ischemic rat cerebral cortex. *Proceedings of the National Academy of Sciences of the United States of America*, 2004. **101**(32): 11839–11844.

70. Zhong, J. et al., Hydrogel matrix to support stem cell survival after brain transplantation in stroke. *Neurorehabilitation and Neural Repair*, 2010. **24**(7): 636–644.

71. Ballios, B.G. et al., A hydrogel-based stem cell delivery system to treat retinal degenerative diseases. *Biomaterials*, 2010. **31**(9): 2555–2564.

72. Algvere, P.V. et al., Transplantation of fetal retinal pigment epithelium in age-related macular degeneration with subfoveal neovascularization. *Graefe's Arch. Clin. Exp. Ophthalmol.*, 1994. **232**(12): 707–716.

73. Das, T.P. et al., The transplantation of human fetal neuroretinal cells in advanced retinitis pigmentosa patients: results of a long-term safety study. *Experimental Neurology*, 1999. **157**(1): 58–68.

74. Algvere, P.V., P. Gouras, and E.D. Kopp, Long-term outcome of RPE allografts in non-immunosuppressed patients with AMD. *European Journal of Ophthalmology*, 1999. **9**(3): 217–230.

75. Eiraku, M. et al., Self-organizing optic-cup morphogenesis in three-dimensional culture. *Nature*, 2011. **472**(7341): 51–56.

76. Gilbert, P.M. et al., Substrate elasticity regulates skeletal muscle stem cell self-renewal in culture. *Science*, 2010. **329**(5995): 1078–1081.

77. Herberts, C.A., M.S. Kwa, and H.P. Hermsen, Risk factors in the development of stem cell therapy. *Journal of Translational Medicine*, 2011. **9**: 29.

3

Fabrication and Application of Gradient Hydrogels in Cell and Tissue Engineering

Azadeh Seidi, Serge Ostrovidov, and Murugan Ramalingam

CONTENTS

3.1 Introduction

The ultimate goal of tissue engineering is to generate physiologically functional tissues or organs in vitro to aid in repairing or replacing diseased or damaged tissues or organs of the body by using synthetic functional components called scaffolding materials, culturing them with appropriate cells that are harvested from patient or donor, and then reimplanting the engineered constructs in the patient's body where the tissue regeneration is required [1,2]. Thus, scaffolding materials play a critical role in tissue engineering by serving as a synthetic extracellular matrix (ECM) to provide temporary mechanical support for the cells and to subject them to conditions highly mimicking the native microenvironments that lead to tissue formation. The ECM is a highly hydrated, viscoelastic network of proteins and carbohydrates that holds cells together. Components of ECM network are comprised of five classes of macromolecules, which can exist in different proportions for different functions: collagens, elastic fibers, proteoglycans, hyaluronan, adhesive glycoproteins, and soluble macromolecules such as growth factors, chemokines, and cytokines [3]. The ECM fibers and soluble macromolecules provide biophysical and biochemical cues to guide cell behavior within a tissue, which often exist in the form of gradients. For example, invading cells create a gradient of ECM around them, with more ECM ahead of the cells in the direction of invasion and less behind. The cells tend to move up gradients of ECM (called haptotaxis) and thus invade the surrounding tissue [4]. Experimental data also show that the gradients in mechanical properties of ECM can induce cellular behaviors such as motility [5] and differentiation [6], while molecular concentration gradients of soluble factors in the ECM regulate biological phenomena such as chemotaxis [7], morphogenesis [8], and wound healing [3]. In addition, chemical and physical gradients exist in vivo in a wide range of length scales and contribute to important biological phenomena such as embryogenesis [9,10], mitosis [11,12], axonal growth [13], immune response [14], interface tissue (i.e., hard-to-soft tissue) formation [15], and heterogeneous structural composition of teeth and bones [16]. Having these points in mind, tissue engineering approaches should impart the key features of natural ECM into biomaterial scaffolds in vitro, which include similar composition and the spatially tunable physical and chemical properties.

Various scaffolding materials have been developed to mimic native ECM, among which hydrogels are important for tissue engineering due to their structural and compositional similarities to natural ECM. Hydrogels are a class of highly hydrated polymer materials composed of hydrophilic chains, which can be from either natural or synthetic origin. The most commonly used materials for fabricating hydrogels are summarized in Table 3.1 [17–40]. Hydrogels possess ECM-like viscoelastic and diffusive transport characteristics; therefore, they mimic the native ECM to some extent. However, in certain cases, their hydrophilic surfaces have to be modified so that they can be used to support better cell attachment and proliferation. The chemical and physical properties of hydrogels are also tunable which makes them suitable for producing tailored 3D cellular microenvironments.

Hydrogels are networks of hydrophilic polymers which serve as excellent candidates to create synthetic environment for cells to grow. Since hydrogels provide unique 3D cellular microenvironments mimicking the ECM, imparting gradient features such as physical, mechanical, chemical, and biological into hydrogel systems has been an attractive tool to engineer functional tissue constructs, closely mimicking the native tissue (see Figure 3.1). Hydrogels with immobilized or soluble gradients of biological agents such as growth factors [41] and adhesion peptides [42] as well as graded physical properties such as stiffness [43] and porosity [44] have been developed to mimic the graded features of ECM for the purpose of high-throughput screening [45], directed cell migration [46],

TABLE 3.1

The Commonly Used Polymers for Making Hydrogels

Origin	Features	Examples	Applications	References
Naturally derived polymers	They are biocompatible, biodegradable, and closely mimic the cellular components. They have macromolecular properties similar to natural ECM	Collagen	Reconstruction of liver, skin, blood vessel, small intestine	[17–19]
		Gelatin	Delivery of growth factors and therapeutic molecules	[21]
		Hyaluronate	Artificial skin, facial intradermal, implants, wound healing	[22]
		Silk	Ligament tissue engineering	[24]
		Fibrin	Engineering tissues with skeletal muscle cells, SMCs, chondrocytes	[25,26]
		Alginate	Injectable cell delivery vehicle, wound dressing, transplantation of chondrocytes	[27,28,30]
		Agarose	Tissue scaffolds used in neurite growth	[31]
		Chitosan	Neural and bone tissue engineering	[32,33]
Synthetic polymers	Their chemistry and properties are easily controllable and reproducible as compared with naturally derived polymers. They are either biodegradable or nonbiodegradable. Surface modifiable with selective functional groups	Polyacrylic acid	Temperature controllable cell sheets	[34,102]
		Polyethylene oxide	Surface modification, drug delivery, artificial articular cartilage	[35,36]
		Polyvinyl alcohol	Artificial articular cartilage	[37]
		Polyphosphazene	Controlled delivery of protein drugs, skeletal tissue engineering	[38]
		Polyethylene glycol	Scaffolds for tissue engineering providing a blank slate for cell adhesion	[39,70,77]
		Polypeptides	pH-responsive silk-like polypeptide hydrogels	[40]

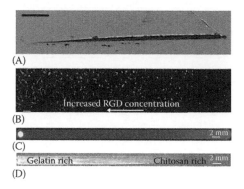

FIGURE 3.1

Examples of hydrogels with gradient in mechanical, biological, and compositional properties. (A) SEM micrographs of cross sections of dried hydrogels fabricated with a gradient of 10 wt.% PEG4000DA (left) to 50 wt.% PEG1000DA (right); bar 100 μm. (From Burdick, J.A. et al., *Langmuir*, 20(13), 5153, 2004.) (B) Phase contrast images of ECs on RGDS-gradient hydrogel; bar 200 mm. (From He, J. et al., *Adv. Funct. Mater.*, 20(1), 131, 2010.) (C) Visualization of a cross-gradient containing green-fluorescent DyLight 549-labeled collagen and red-fluorescent DyLight 649-labeled collagen. (From He, J. et al., *Biotechnol. Bioeng.*, 108, 175, 2010.) (D) Composite material film generated by air drying a gelatin/chitosan concentration cross-gradient. (From He, J. et al., *Biotechnol. Bioeng.*, 108, 175, 2010.)

axonal guidance [47], graded cell differentiation [48], and engineering of hard-to-soft interface tissues [49].

Considering aforementioned issues, in this chapter, the authors focused their attention on the design strategies of hydrogels with structural, mechanical, compositional, chemical, and biological gradients due to their tremendous potential in controlling behavior of multiple cell types in a single experiment and engineering tissues in a biologically mimicking environment. The applications of those gradient hydrogels in basic cell studies and tissue engineering are also discussed. Finally, this chapter concludes with current challenges and future directions in the development of gradient hydrogels for cell and tissue engineering applications.

3.2 Fabrication of Gradient Hydrogels

Gradient hydrogels are often generated by a two-step process involving the formation of concentration gradients of prepolymer solutions followed by appropriate polymer cross-linking to stabilize the gradient. In addition to hydrogels with concentration gradients of a chemical of interest, hydrogels with gradients of various properties that influence the cellular behavior have been developed. In the following sections, we describe basics of the methods of fabricating hydrogels with gradients in structural, mechanical, compositional, chemical, and biological properties and their technological advancements.

3.2.1 Hydrogels with Structural Gradients

The pore size of a scaffolding system is one of the important structural design parameters as it influences cellular infiltration, spreading, intracellular communication, and transport of nutrients and metabolites [50]. While a higher porosity increases the maximum cell accommodation in a scaffold, it greatly decreases mechanical properties of the scaffold as well [51]. Significant attention has been paid to fabricate scaffolds with an interconnected pore network and with controlled pore shapes to minimize the dead volume, facilitate mass transfer, and guide cellular organization [52–54]. The optimal pore size of a scaffold is designed based on the type of cells or tissues of interest to mimic its native ECM, which results in a biomimetic cell alignment and clustering. Thus, the trend in designing scaffolds is changing from isotropic to anisotropic scaffolds with a gradient in pore size that provide a rapid screening of optimal scaffold pore size for the suitable interaction between cell/tissue and the scaffold. In addition to pore-size screening, graded porous scaffolds have been used in the regeneration of interface tissues with zonal properties. For example, Woodfield et al. utilized a 3D fiber deposition technique to fabricate an interconnecting porous poly(ethylene glycol)-terephthalate–poly(butylenes terephthalate) (PEGT/PBT) copolymer scaffolds with pore-size gradients (pore-size diameter range: ~200–1650 μM) [55]. The authors layered patterns of molten copolymer through a nozzle with computer-controlled fiber deposition paths, which resulted in an accurate control on fiber spacing, thus pore size of the fabricated scaffold. Using this technique, scaffolds with a gradient in fiber spacing were generated by initially depositing four layers of copolymer with fiber spacing of 2.0 mm (mimicking the deep zone of articular cartilage with flattened chondrocytes and densely packed collagen fibers), followed by depositing four layers of copolymer with a fiber spacing of 1 mm (mimicking the middle zone of articular cartilage with rounded chondrocytes and randomly oriented collagen fibers), and one layer with fiber spacing of 0.5 mm (mimicking deep zone of articulate cartilage with large spherical chondrocytes and thick collagen fibers). The authors demonstrated that bovine chondrocytes seeded on such scaffolds were distributed in an anisotropic fashion mimicking the superficial, middle, and lower zones of immature bovine articular cartilage. The authors further reported the direct correlation between zonal scaffold volume fraction and DNA, glycosaminoglycan (GAG), and collagen type II content. In the area with 0.5 mm fiber spacing, which had the lowest porosity and highest volume fraction, the concentration of DNA, GAG, and collagen type II was significantly higher than the area with 2.0 mm fiber spacing, indicating an anisotropic distribution of cells in the gradient scaffold [55]. Harley et al. designed a collagen-based scaffold by spinning the collagen suspension in a cylindrical copper mold, followed by freeze drying [56]. The resulting scaffold exhibited a porosity gradient to restrict the migration or ingrowth of cells in the direction of increasing porosity. Such system can find applications in studying path-dependent cell migration processes such as those seen during peripheral nerve regeneration. Orsi et al. developed graded porous scaffolds for use in gene delivery [57]. They generated gene delivering poly(ethylene glycol) (PEG) scaffolds with pore-size gradients by gelatin particles templating, photopolymerization, and gelatin leaching, followed by adsorption of DNA complexes. The fabricated scaffold was loaded with mouse embryo fibroblast NIH3T3 and cultured 16 days; the results indicated that cells exhibited a differential level of scaffold colonization, depending on the local porosity of the scaffold. In another study, by combining phase separation and freeze drying, Vlierberghe et al. generated a gelatin hydrogel [58]. The authors demonstrated that the cooling rate of gelatin solution had an inverse effect on the pore size of the fabricated scaffold and that by creating a temperature gradient during cryogenic treatment, a scaffold with pore-size gradient could be achieved.

A modified freeze-drying procedure has been used by Stokols and Tuszynski to fabricate agarose hydrogels with linear pores [59]. This method utilizes a freezing step that involves exposing only one end of a pillar of agarose to a block of dry ice immersed within a pool of liquid nitrogen. The resulting uniaxial temperature gradient caused ice crystals to form orienting in the direction of the gradient. Upon removal of water by lyophilization, a highly linear network of porous channels was formed with dimensions suitable for cell infiltration. Scaffolds fabricated with this method were able to guide axonal regeneration in a spinal cord injury model.

The experimental studies discussed here demonstrated the potential of hydrogels with structural gradients to influence cellular characteristics such as scaffold colonization, cellular orientation, and migration in an engineered construct.

3.2.2 Hydrogels with Mechanical Gradients

The mechanical properties of substrate, including stiffness, act as a physical signal for cells and influence cellular behaviors such as adhesion, spreading, motility, survival, and differentiation [60–62]. In the studies utilizing fibroblasts, epithelial cells, and smooth muscle cells (SMCs), it has been found that cell contact with the substrate decreases using softer substrates [63,64] and that cells cultured on a substrate with stiffness gradient tend to migrate from softer regions to stiffer regions [43,65–68], a phenomenon termed "durotaxis" or "mechanotaxis." However, different cell types respond to different ranges of substrate stiffness [69], which calls for the use of substrates containing continuous gradients of stiffness in a wide range, to study the cell response to local substrate stiffness and find the optimal substrate properties critical to the target cellular behavior.

Hydrogels with continuous stiffness gradients have been fabricated by controlling photopolymerization processes such as controlling photoexposure [43] or by generating the cross-linker concentration gradient using a microfluidic device [66]. Both methods result in a gradient in the level of hydrogel cross-linking, and thus the stiffness of the fabricated hydrogel. For example, Zaari et al. integrated microfluidics and photopolymerization to generate polyacrylamide (PAAM) substrates with controlled gradient profiles by incorporating a gradient of bis-acrylamide (bis) cross-linker to the gel precursor solution [66]. The authors demonstrated a high spreading, adherence, and defined actin cytoskeleton of vascular smooth muscle cells (VSMCs) on the stiffer regions of the substrate compared to softer side.

Considering different possible approaches, Burdick et al. generated gradients in cross-linking levels along a hydrogel by using two different PEG-diacrylate (PEG-DA) (10 and 50 wt.%) macromers in a microfluidic system [70]. After photopolymerization, the region of hydrogel incorporating mostly 10 wt.% macromer solution produced a thin gel with large mesh, while the other side incorporating mostly 50 wt.% macromer solution produced a thick gel with small mesh. The hydrogel exhibited a thickness gradient from 10 to 50 μm. The cross-linking density gradient was maintained after hydro gel swelling 20 min in phosphate buffer saline (PBS) solution, resulting in a graded release of the embedded rhodamine, which correlated to the gradient in mesh size along the gel.

In addition to microfluidic-based methods, long-range gradient makers have been developed, which consist of two chambers for pouring hydrogel precursors. The chambers contain high and low concentrations of the target material and are mixed before being casted to a mold, incorporating higher portions of the solution of interest with time into the outlet flow and generating a gradient of this solution in the mold, which is then UV-photopolymerized for stabilization of the gradient. Such a gradient maker was also used by Chatterjee et al. to make PEG-based hydrogel biomaterials with encapsulated preosteoblasts MC3T3-E1cells [71]. The authors studied the local osteogenesis of MC3T3-E1 cells and observed that the encapsulated MC3T3-E1 cells exhibited a graded osteogenesis in register to the stiffness gradient along the hydrogel.

In addition to photopolymerization, photodegradation of hydrogels has also been utilized to generate hydrogels with graded mechanical properties. For example, Kloxin et al. fabricated a PEG film with an elastic modulus gradient from 7 to 32 kPa [72]. Loaded with valvular interstitial cells (VICs), the hydrogels induced a differentiation gradient of VIC into myofibroblast, which was the highest at the regions with high elastic modulus. Photodegradation of the hydrogel film caused a reversed differentiation from myofibroblast to VIC.

The experimental examples discussed in this section clearly show the potential of biomaterials, particularly hydrogels with gradient in mechanical properties to influence cellular differentiation.

3.2.3 Hydrogels with Compositional Gradients

Fabricating composite materials that combine the desired properties of two or more constituents is an effective approach to the discovery of new materials with tailored properties to meet specific needs of a tissue to be regenerated. The combination ratios of multiple constituents of a composite material can dramatically influence its chemical and mechanical properties, and thus the corresponding cellular response (e.g., cell morphology, migration, function, and differentiation) [63,73–75]. Therefore, there is a need for new high-throughput combinatorial material synthesis approaches to optimize the concentrations and blending ratios of the target materials in a sensitive fashion. In addition to material discovery, generation of gradient in material composition, in combination with gradient in scaffold pore size, has been utilized to regenerate heterogeneous organization of articular cartilage tissue and reduce delamination problem, which is commonly faced in biphasic scaffold regions [76].

In a typical example, Du et al. used alternated forward/backward flows of hyaluronic acid (HA) and gelatin from the two ports of a straight channel, with 30 s rest between each sequence to allow lateral mixing by diffusion [77]. In this system, convection stretches the fluid along the channel axis while diffusion acts laterally and tends to suppress hydrodynamic stretching. The use of high-speed (in the order of mm/s) flows improves the hydrodynamic stretching and allows to generate long-range gradient of molecules, microbeads, and cells. With this dispersion-based technique, the authors generated an HA–gelatin composite hydrogel with 2–3 cm cross-gradient of HA, which is a cell repellent, and gelatin which is bioactive. Loaded with SMCs and cultured during 24 h, the composite hydrogel accommodated a gradient of cell density matching the gelatin concentration gradient.

In another study, the same group utilized a similar microfluidic platform to fabricate gelatin/chitosan cross-gradient materials [45]. The authors created 3D porous structures of the gelatin/chitosan materials by in situ lyophilization and investigated SMC–material interactions. SMCs showed an elongated morphology on the gelatin-rich side, which gradually transitioned to round morphology on the chitosan-rich side. Furthermore, the authors observed that the adhesion and proliferation rates of the SMCs at the gelatin-rich side of the cross-gradient were higher than the chitosan-rich side.

The methods discussed in this section provide a sensitive platform for screening composite materials with optimal combination ratios that result in the desired cell response.

3.2.4 Hydrogels with Chemical Gradients

As previously discussed in this chapter, hydrogels are important for tissue engineering due to their structural and compositional similarities to natural

ECM. Particularly hydrogels from synthetic origin are appealing for tissue engineering because their chemistry and properties are controllable and reproducible. For example, synthetic polymers can be reproducibly produced with specific molecular weights, block structures, degradable linkages, and cross-linking modes. These properties, in turn, determine gel formation dynamics, cross-linking density, and material's mechanical and degradation properties. However, due to their hydrophilicity, such hydrogels are devoid of biological interactions to cells, and the surface chemistry of hydrogels from synthetic origin has to be modified in order to make them suitable for cell culture. For this purpose, adhesion peptides and natural ECM proteins such as collagen or fibronectin have been commonly immobilized in those hydrogels. The molecules have been incorporated to the gels both before and after polymerization in homogeneous or gradient fashions to generate biomimetic cellular microenvironment and also to study the response of cells to graded adhesion cues.

In order to create a gradient of the adhesion peptide, Arg-Gly-Asp-Ser (RGDS), in a PEG hydrogel, Burdick et al. synthesized Poly(ethylene glycol)-4000 diacrylate (PEG4000DA) and acryloyl-poly(ethylene glycol)-RGDS (Acr-PEG-RGDS) molecules, followed by creating a concentration gradient of Acr-PEG-RGDS in PEG4000DA using a microfluidic-based gradient generator, based on the serial dilutions of Acr-PEG-RGDS, which was stabilized upon UV polymerization [70]. Human umbilical vein endothelial cells (HUVECs) cultured on such a hydrogel exhibited a cell density gradient matching the RGDS gradient as well as a better spreading toward high concentration region after 3 h from seeding.

Taking another microfluidic-based approach, He et al. applied a diffusion strategy based on a passive pump-induced forward flow and an evaporation-induced backward flow in a straight microfluidic channel to generate a centimeter long Acr-PEG-RGDS concentration gradient in a PEG-DA hydrogel [39]. The gradient hydrogel was cultured with HUVECs, which resulted in a cell density gradient as well as a cell morphology gradient (from round shape to well spread) following the RGDS concentration gradient.

In addition to microfluidic-based gradient generators, commercial long-range gradient generators have also been used to generate longer gradients on the order of several centimeters. DeLong et al. used such a gradient maker to immobilize a gradient concentration of RGDS PEG-based hydrogels and achieved the differential attachment of human dermal fibroblasts on these substrates [42].

The mentioned methods of functionalizing hydrogels relied on the incorporation of the target molecules (e.g., adhesion peptides) into the hydrogel prepolymer solution and gradient stabilization by UV polymerization. However, taking a different approach, polymerized hydrogels have been subjected to diffusion of chemicals through the gelled polymers, which results in the formation of graded modifications of hydrogels with functional groups. Those functional groups have then been used for the incorporation of ECM proteins that facilitated a gradient cellular attachment on the hydrogels.

Yamamoto et al. generated a carboxyl group concentration gradient in a PAAM hydrogel by hydrolysis of the amide groups exposed to the diffusion of a sodium hydroxide concentration gradient [78]. Then, they coupled the carboxyl groups to the amino groups of type I collagen to form a collagen concentration gradient hydrogel. The authors loaded the gradient hydrogel with L929 mouse fibroblasts and cultured during 1 day, and they observed a cell density gradient in register to the collagen gradient.

In the same context, Wu et al. fabricated a poly(L-lactic acid) (PLLA) scaffold with a gradient degree of functionalization with gelatin [79]. The authors placed the scaffold vertically in a beaker and aminolyzed it in a gradient manner by wetting at controlled speed from bottom to top with 1,6-hexanediamine/propanol solution. Gelatin was then immobilized on the amino groups via glutaraldehyde coupling agent to form a gelatin gradient.

The experimental examples which were discussed in this section demonstrated the ability of hydrogels with chemical gradients in precisely controlling of cell–hydrogel interaction which is useful in engineering of tissue constructs.

3.2.5 Hydrogels with Biological Gradients

A plethora of studies have demonstrated the critical role of concentration gradients of bioactive signaling molecules (biological gradients) in cellular processes such as morphogenesis [8], wound healing [3], immune response [14], embryogenesis [9,10], and axonal guidance [13]. In the physiological cellular microenvironment, cellular migration and differentiation are precisely controlled by gradients of biochemical signals in the native ECM [80–83]. A successful strategy toward tissue regeneration calls for a proper recreation of the spatial patterns of those signals in vitro. Hydrogel-based approaches have been utilized to generate soluble or immobilized biological gradients in both 2D and 3D. This section contains a brief discussion of the techniques applied to generate biological gradients in hydrogels.

At a 2D level, long-range gradient makers have been used to generate hydrogels with biological gradients. For example, Kapur and Shoichet generated a gradient of nerve growth factor (NGF) along a poly(2-hydroxyethylmethacrylate) [p(HEMA)] hydrogel to culture PC12 cell [47]. The authors demonstrated the guidance of PC12 neurites toward the regions with higher NGF incorporation in the gradient hydrogel [47]. DeLong et al. used a similar approach to generate PEG-based hydrogels with immobilized gradient concentrations of basic fibroblast growth factor (bFGF) to study the migration of human aortic smooth muscle cells (HASMCs) [41].

A different approach toward generation of biological 2D gradients involves the usage of printing techniques. Ilkhanizadeh et al. printed a gradient of ciliary neurotrophic factor (CNTF) on premade PAAM-hydrogel-coated slides [48]. The authors replaced ink cassettes of an inkjet printer with pipette tips, which were fitted over each color orifice. This system was used to print extrinsic factors such as CNTF on hydrogels. The neural stem cells (NSCs) cultured on those gels, with a gradient of CNTF, differentiated into astrocytes, expressing glial fibrillary acidic protein (GFAP). The authors observed a GFAP-positive cell density gradient decreasing from 14% to 6% reflecting the CNTF printed gradient.

In the same context, Phillippi et al. generated a gradient of printed bone morphogenic protein-2 (BMP-2) on fibrin films and studied the differentiation of mouse muscle-derived stem cells (MDSC) in the presence of printed gradients [84]. The printed protein preserved its activity after printing as confirmed by fluorescence tests. As with the NSC studies, MSDC differentiation could be controlled by printing appropriate patterns and gradients of protein. Under myogenic conditions, MSDC seeded on the BMP-2 patterned surfaces differentiated toward the osteogenic lineage, whereas cells off pattern differentiated toward the myogenic lineage.

Weibel et al. printed, with an agarose hydrogel stamp, patterns of *Escherichia coli* K12 bacteria containing a plasmid conferring ampicillin resistance and indigo biosynthesis on an LB-agar layer with ampicillin gradient [85]. In presence of ampicillin, the colonies turned dark indigo, while in absence of ampicillin, bacteria lost their plasmid and the colonies turned white. After 4 day culture at 25°C, they observed that bacterial colonies formed a color gradient from dark indigo to white which matches the ampicillin gradient.

At a 3D level, a few studies have demonstrated the response of cells to biological gradients. Dodla et al. generated a gradient of laminin-1(LN-1) in a cell-laden agarose hydrogel by diffusion. The authors entrapped chicken dorsal root ganglion neurons (DRG) in agarose hydrogel, and after gelation, they placed the cell-laden hydrogel between two sources of cell culture media, containing high and low concentrations of LN-1 conjugated to the Sulfosuccinimidyl-6-[4'-azido-2'-nitrophenylamino] hexanoate (SANPAH). The diffusion of LN-1-SANPAH through the agarose created a concentration gradient of this molecule along the hydrogel, which was stabilized by photoimmobilization. The DRG cells in agarose gel containing a concentration gradient of LN-1 exhibited a higher neurite extension compared with cells cultured in agarose gel containing a homogenous concentration of LN-1.

In a recent study, Aizawa et al. generated a gradient of vascular endothelial growth factor 165 (VEGF165) in a 3D system of agarose-sulfide hydrogel to study the endothelial cell (EC) guidance [86]. The authors immobilized a series of concentration gradients of VEGF165, together with homogeneous concentrations of cell adhesion peptide Gly-Arg-Gly-Asp-Ser (GRGDS), by using coumarin chemistry and patterning the hydrogel photochemically with multiphoton lasers. Using this system, the ECs cultured on those hydrogels guided to form tubule-like structures in 3D hydrogels.

The modification of hydrogels with gradients of bioactive molecules enables the researchers with tremendous tools to mimic complex biological phenomena in vitro and to generate functional tissues with similar characteristics to the tissue to be regenerated.

3.3 Applications of Gradient Hydrogels

The gradient generation methods outlined in the previous section are now being used to recreate cellular microenvironments to answer fundamental questions regarding cellular behavior in response to gradient features and to control cellular behaviors to direct tissue regeneration. In particular, chemical and physical features are being embedded into hydrogels in a gradient way to study and control a range of cellular processes. Based on the available literatures, the gradient hydrogel systems have been used at least in two different categories of applications: basic cell studies and tissue engineering, which are discussed in this section with experimental examples.

3.3.1 Basic Cell Studies

Gradient hydrogels have been used for in vitro study of cellular response to physical, chemical, and biological gradient cues as occur in vivo. These studies have revealed useful information about important biological phenomena such as cell anchorage, migration, proliferation, differentiation, and outgrowth.

The process of cell migration, which is an essential part of morphogenesis [87], inflammation [88,89], wound healing, and tumor metastasis [90], is triggered by the formation of a gradient of chemical or physical cues in the cellular microenvironment. Gradient hydrogels have been employed as promising tools to study cell migration as a response to various cues, which have been hypothesized to affect this process in vivo as well. For example, the role of bFGF in HASMC proliferation and migration was studied by DeLong et al. [41]. In this study, an immobilized concentration of bFGF, in Arg-Gly-Asp-Ser (RGDS) PEG hydrogel scaffolds, resulted in the increase in HASMC proliferation by 41% and migration by 15% and alignment of cells on hydrogels in the direction of increasing tethered bFGF concentration as early as 24h after seeding. A change in the alignment of cells was also observed when human dermal fibroblasts were cultured on a PEG-based hydrogel system with gradient distribution of the immobilized RGDS [42]. Cells cultured on such a hydrogel aligned and migrated in the direction of increasing RGDS concentration. After 24h, 46% of fibroblasts were aligned with the RGDS-gradient axis. In another study, Guarnieri et al. observed same phenomenon when using NIH3T3 cells [46]. In addition to immobilized concentration gradients of molecules discussed earlier, gradients of soluble biomolecular gradients have also had a noticeable effect on cell orientation and migration. In one study, Knapp et al. confined two fibrin matrices in a two-chamber system and separated them by a Teflon plate [91]. Upon removal of the Teflon plate, soluble Gly-Arg-Gly-Asp-Ser-Pro (GRGDSP) peptides diffused from one of the hydrogel matrices into the nonpeptide containing hydrogel created gradients of soluble GRGDSP in the second hydrogel. The formation of concentration gradients of soluble peptides in the 3D matrices has changed the alignment of fibroblasts embedded in the 3D matrices, and the cell migration was induced toward high concentration of the soluble peptides.

Cell migration can be triggered by chemical (chemotaxis) or mechanical (mechanotaxis) gradients. Hydrogels with gradients in mechanical properties have been used to study the effect of mechanical properties of cell culture substrate on cellular movement. Lo et al. formed an acrylamide hydrogel with a gradient in elastic modulus by generating a gradient concentration of bis-acrylamide cross-linker in the hydrogel prepolymer and showed that fibroblasts seeded on these gels moved toward the stiffer regions of the gels than the softer side [65]. In another study, similar phenomenon was observed using VSMCs, which were seeded on PAAM hydrogels with gradients in elastic modulus [43]. In the same context, Marklein and Burdick exploited the human mesenchymal stem cell (hMSC) spreading and proliferation on an HA substrate with gradients in mechanical properties [92]. In this study, the amount of cross-linking agent and UV exposure time was manipulated to create hydrogels with a wide range of mechanical gradients (from 3 to 100 kPa). The hMSCs cultured on these patterned gels became more spread and proliferative on substrates of higher stiffness. In a different approach, photodegradable hydrogels were designed for altering the mechanical properties of cellular microenvironment and change cytoskeletal organization, cell differentiation, and signaling. In one study, VICs, the most prevalent cell type of cardiac valve leaflet, were cultured on such a photodegradable PEG hydrogel exhibiting elasticity gradient in the range of 7–32 kPa [72]. By the third day of culture, αSMC stress fibers were observed, indicating myofibroblast differentiation of VICs in the direction of higher elasticity modulus. However, 5 days after reducing the elastic modulus of the scaffolds to 7 kPa by irradiation, the cells did not exhibit any noticeable αSMC-like phenotype in their cytoskeleton [72]. These studies have supported the previous findings, highlighting the importance of elastic modulus in dictating various cell behaviors.

Another important cellular behavior that has been reported to be influenced by gradient cues in cellular microenvironment is differentiation. This is because cell differentiation is often affected by the substrate properties, which was also examined by Ilkhanizadeh et al. by using stem cells [48]. They reported inkjet printing of fibroblast growth factor-2 (FGF2) and NGF on hydrogels in a homogenous or gradient manner. They found out that NSCs cultured on hydrogels printed with CNTF displayed a rapid induction

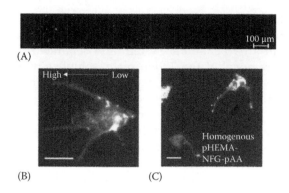

FIGURE 3.2
Examples of cellular response to culture on various gradient hydrogels. (A) The gradient of GFAP-positive cells reflects the underlying gradient of printed CNTF. (From Ilkhanizadeh, S. et al., *Biomaterials*, 28(27), 3936, 2007.) (B) Fluorescent optical microscope images of primed PC12 cells extending neurites on p(HEMA)-NGF-pAA immobilized gradient gels, and on (C) p(HEMA)-NGF-pAA immobilized homogeneous gels; bar 20 μm. (From Kapur, T.A. and Shoichet, M.S., *J. Biomed. Mater. Res. A*, 68(2), 235, 2004.)

of markers for astrocytic (GFAP) and that NSCs cultured on a printed gradient of increasing levels of CNTF showed a linear increase in numbers of cells expressing GFAP, demonstrating a functional gradient of CNTF (see Figure 3.2A).

Another cell phenotype that has been observed to be affected by culture substrate properties has been the regulation of axonal guidance. Immobilized NGF within p(HEMA) microporous gels has been shown to guide the PC12 cell neuritis to culture PC12 cells, up the gradient [47]. Similar phenomenon was observed when DRG cells were cultured on a 3D anisotropic agarose hydrogel scaffolds with photoimmobilized gradients of a growth-promoting glycoprotein, LN-1 (see Figure 3.2B and C) [93].

All these experimental studies, and others, clearly suggest that cells can be manipulated by controlling the physical, chemical, and biological cues in a gradient way.

3.3.2 Tissue Engineering

Major goal of tissue engineering is to engineer physiologically functional tissues under laboratory environment that is capable of repairing or regenerating the defective tissues into normal healthy tissues in the bodily environment. Many critical parameters influence the tissue development in vitro. As discussed in the previous section, basic biological phenomena that occur during the process of tissue development, such as targeted cellular alignment, migration, and differentiation, are influenced and triggered by various gradient cues in vivo. Therefore, applying gradient biomaterials in the form of hydrogels can be of tremendous benefit to engineer a functional tissue with properties close to the native tissue to be regenerated. In this section, the application of gradient hydrogel biomaterials in tissue engineering will be further discussed.

One of the challenges of developing tissues in vitro and in vivo is imparting vascularization into these tissues in order to maintain physiological growth of the cells into particular lineages, as well as to synthesize other cellular components required for the tissue development [94–96]. Chung et al. stimulated angiogenesis in vitro in 3D collagen scaffolds by a gradient concentration of angiogenic vascular endothelial growth factor (VEGF) in the scaffold by using a microfluidic device (see Figure 3.3A) [97]. This study resulted in the formation of sprouting structures of ECs into the collagen scaffold after several days of culture, while those in a control scaffold without a VEGF gradient did not exhibit such a phenotype. Dodla et al. generated 3D agarose hydrogels with gradients of LN-1 and observed neurite extension to the direction of higher concentration of the gradient [93]. Moreover, LN-1 and NGF gradient matrices together with LN-1 homogenous matrices were used in vivo to study rat sciatic nerve regeneration [98]. Gradient LN-1 matrices were placed within 20 mm

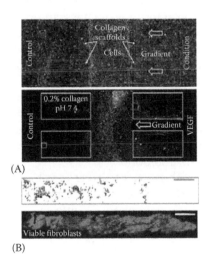

(A)

(B)

FIGURE 3.3
Example applications of gradient hydrogels in tissue engineering. (A) (Upper panel) Growth factor–induced EC migration 1 day after seeding a confluent monolayer in the cell channel. (Lower panel) Migration results of ECs into collagen gel scaffold. White dotted lines indicate the outlines of gel scaffold, and small rectangles in the scaffold region indicate the PDMS posts of 150 µm × 150 µm. (From Chung, S. et al., *Lab Chip*, 9(2), 269, 2009.) (B) (Upper panel) MicroCT image analysis demonstrated spatial patterning of both mineral deposition and non-mineralized, fibroblastic ECM within constructs containing a graded distribution of Runx2 retrovirus after 42 days in vitro culture. (Lower panel) Confocal microscopy image of fluorescently stained cells showing uniform fibroblast distribution across the periphery of a Runx2-retrovirus-coated sample; bar 2 mm. (From Phillips, J.E. et al., *Proc. Natl Acad. Sci. USA*, 105(34), 12170, 2008.)

sciatic nerve gaps. The axon density, myelination, and regain of function of the nerves were analyzed after 4 months in vivo. LN-1 matrices in combination with a gradient of NGF promoted much better nerve regeneration as compared to hydrogel matrices containing homogeneous concentrations of LN-1 or NGF or LN-1 or NGF gradients alone combined with a homogeneous concentration of the respective other molecule.

A new area in the field of tissue engineering which has benefited from utilization of gradient biomaterials is the field of interface tissue engineering (ITE), which aims to regenerate tissues with functional properties of zones at the interface of different tissue types (also called "interface tissues"). Notable examples of the interface tissues in the human body include ligament-to-bone, tendon-to-bone, and cartilage-to-bone. Such tissues are characterized by heterogeneous properties from one end of the tissue to the other in terms of cell type, compositional, mechanical, and biological properties. Engineering interface tissues is a complex process, which is not feasible using conventional biomaterials with homogeneous properties, and monolithic biomaterials are not perfectly suitable for hosting a heterogeneous population of cells to regenerate the functional and native tissue-integrating grafts. For this purpose, a triphasic scaffold was developed by Spalazzi et al. for the regeneration of the anterior cruciate ligament (ACL)-to-bone interface [99,100]. Their scaffold consisted of three phases, which offered three distinct levels of mechanical properties suitable for culturing fibroblasts on the softer phase, osteoblasts on the stiffer phase, and for the formation of a fibrocartilage in the intermediate phase.

The authors implanted such a triphasic scaffold, cultured with fibroblasts, chondrocytes, and osteoblasts, in the posterior dorsum of rats with subcutaneous pouches, for an in vivo evaluation of ACL-to-bone formation [100]. Their reported biomimetic system is an example of regeneration of interface between soft tissue grafts and bone.

Biomaterials with smooth gradients in composition, structure, mechanical, and biomolecular properties were later developed to engineer biomimicking cellular microenvironments at tissue interfaces, which would provide a suitable integration of heterogeneous cell types. Chatterjee et al. encapsulated preosteoblast cell line MC3T3-E1 in a PEG hydrogels with a gradient in stiffness and observed cellular response to this gradient [71]. The authors showed that the stiffness gradient in the scaffold directed osteogenesis and the generation of mineralized tissue happened to occur at the stiffer parts of the scaffold. By using this method, graded osteogenesis and mineralization can be achieved, which is hallmark of soft-to-hard interface tissue, with only varying compressive modulus and without any other modification of the substrate. Other studies have reported the use of gradient hydrogels with biologically active materials such as DNA and growth factors for triggering a graded differentiation of cells. Phillips et al. transfected fibroblasts cultured in a 3D collagen hydrogel with gradient concentration of Runx2/Cbfa1 transcription factor encoding gene in order to generate tendon-to-bone transition zone, which resulted in a graded level of differentiation of fibroblasts into osteoblasts as evidenced by a graded level of mineral deposition by cells along the hydrogel (see Figure 3.3B) [49]. Wang et al. cultured hMSCs on alginate and silk microsphere scaffolds releasing recombinant rhBMP-2 and insulin-like growth factor (rhIGF-I), which were distributed in a gradient way [101]. The authors reported that hMSCs cultured on silk scaffolds exhibited osteogenic and chondrogenic differentiation along concentration gradients of rhBMP-2 and gradient of rhBMP-2/rhIGF-I.

The experimental examples described in this section, and others, indicate that gradient hydrogels have promising potentials to engineer tissue-engineered products with complex structure and function mimicking native interface tissues, which can aid/promote tissue regenerative medicine.

3.4 Concluding Remarks

Gradient hydrogels are unique form of biomaterials and recent addition to bioengineering field, in particular cell and tissue engineering. They offer 2D and 3D platforms for studying the basics of cell behavior in a unique environment and also provide synthetic functional components to assist in regenerating damaged tissues with complex structures. The experimental examples summarized in this chapter represented some of the developments of gradient hydrogels from a variety of micro- and nanoscale approaches. Hydrogels can be loaded with immobilized or soluble concentration gradients of molecules such as growth factors and chemoattractants in a 2D or 3D to elucidate cellular behaviors such as adhesion, migration, differentiation, and angiogenesis. Hydrogels with gradient mechanical properties have proven to greatly influence cellular migration and differentiation. While generation of 2D surfaces of gradient hydrogels provides a useful platform for studying cell–material interactions, biomaterial substrates are typically used in 3D format. Therefore, future direction of using gradient hydrogels should tackle the behavior of cells encapsulated in 3D matrices, which mimic the native cellular microenvironment by developing novel techniques for the fabrication of hydrogels in 3D with functional features quite similar to complex native ECM that could support the physiological and metabolical growth of cells, tissues, and organs for use in regenerative medicine.

Improved techniques of generating stable gradients must be developed and integrated into tissue engineering applications so that the resulting gradient mimics the complex environment in native tissues. To this end, hydrogels with a combination of two distinct properties such as chemical and physical gradients have to be developed in order to truly mimic the functional features of native ECM. Hydrogels incorporating relevant gradient cues need to be created for the regeneration of complex tissues such as interfacial zones to create functional grafts with convenient clinical applicability. This is an exciting time to be involved in gradient hydrogel biomaterials, both micro- and nanoscale dimensions, in order to formulate them as a promising synthetic ECM for in vitro and in vivo cell studies.

Acknowledgments

The first author would like to thank Professor Murugan Ramalingam for the opportunity of contributing a chapter in this book. This study was supported in part by the intramural grant from the World Premier International Research Center–Advanced Institute for Material Research (WPI-AIMR), Tohoku University, Japan.

References

1. Langer, R. and J.P. Vacanti, Tissue engineering. *Science*, 1993. **260**(5110): 920–926.
2. Murugan, R. and S. Ramakrishna, Nano-featured scaffolds for tissue engineering: A review of spinning methodologies. *Tissue Eng*, 2006. **12**(3): 435–447.
3. Lanza, R., R. Langer, and J.P. Vacanti, *Principles of Tissue Engineering*, 2007, New York: Academic Press.
4. Marchant, B.P., J. Norbury, and J.A. Sherratt, Travelling wave solutions to a haptotaxis-dominated model of malignant invasion. *Nonlinearity*, 2001. **14**(6): 1653–1671.
5. Ulrich, T.A., E.M. de Juan Pardo, and S. Kumar, The mechanical rigidity of the extracellular matrix regulates the structure, motility, and proliferation of glioma cells. *Cancer Res*, 2009. **69**(10): 4167–4174.
6. Lam, W.A. et al., Extracellular matrix rigidity modulates neuroblastoma cell differentiation and N-myc expression. *Mol Cancer*, 2010. **9**: 35.
7. Li Jeon, N. et al., Neutrophil chemotaxis in linear and complex gradients of interleukin-8 formed in a microfabricated device. *Nat Biotechnol*, 2002. **20**(8): 826–830.
8. Makarenkova, H.P. et al., Differential interactions of FGFs with heparan sulfate control gradient formation and branching morphogenesis. *Sci Signal*, 2009. **2**(88): ra55.
9. Swartz, M.A., Signaling in morphogenesis: Transport cues in morphogenesis. *Curr Opin Biotechnol*, 2003. **14**(5): 547–550.
10. Ashe, H.L. and J. Briscoe, The interpretation of morphogen gradients. *Development*, 2006. **133**(3): 385–394.
11. Caudron, M. et al., Spatial coordination of spindle assembly by chromosome-mediated signaling gradients. *Science*, 2005. **309**(5739): 1373–1376.

12. Clarke, P.R., Cell biology. A gradient signal orchestrates the mitotic spindle. *Science*, 2005. **309**(5739): 1334–1335.
13. Isbister, C.M. et al., Gradient steepness influences the pathfinding decisions of neuronal growth cones in vivo. *J Neurosci*, 2003. **23**(1): 193–202.
14. Wang, F., The signaling mechanisms underlying cell polarity and chemotaxis. *Cold Spring Harb Perspect Biol*, 2009. **1**(4): a002980.
15. Mikos, A.G. et al., Engineering complex tissues. *Tissue Eng*, 2006. **12**(12): 3307–3339.
16. Ho, S.P. et al., The tooth attachment mechanism defined by structure, chemical composition and mechanical properties of collagen fibers in the periodontium. *Biomaterials*, 2007. **28**(35): 5238–5245.
17. Kaufmann, P.M. et al., Highly porous polymer matrices as a three-dimensional culture system for hepatocytes. *Cell Transplant*, 1997. **6**(5): 463–468.
18. Auger, F.A. et al., Tissue-engineered human skin substitutes developed from collagen-populated hydrated gels: Clinical and fundamental applications. *Med Biol Eng Comput*, 1998. **36**(6): 801–812.
19. Seliktar, D. et al., Dynamic mechanical conditioning of collagen-gel blood vessel constructs induces remodeling in vitro. *Ann Biomed Eng*, 2000. **28**(4): 351–362.
20. Voytik-Harbin, S. et al., Small intestinal submucosa: A tissue-derived extracellular matrix that promotes tissue-specific growth and differentiation of cells in vitro. *Tissue Eng*, 1998. **4**: 157–174.
21. Yamamoto, M., Y. Tabata, and Y. Ikada, Growth factor release from gelatin hydrogel for tissue engineering. *J Bioact Compat Polym*, 1999. **14**: 474–489.
22. Duranti, F. et al., Injectable hyaluronic acid gel for soft tissue augmentation. A clinical and histological study. *Dermatol Surg*, 1998. **24**(12): 1317–1325.
23. Radomsky, M.L. et al., Novel formulation of fibroblast growth factor-2 in a hyaluronan gel accelerates fracture healing in nonhuman primates. *J Orthop Res*, 1999. **17**(4): 607–614.
24. Liu, H. et al., Silk-based scaffold for ligament tissue engineering. *IFMBE Proc*, 2008. **20**: 34–37.
25. Ye, Q. et al., Fibrin gel as a three dimensional matrix in cardiovascular tissue engineering. *Eur J Cardiothorac Surg*, 2000. **17**(5): 587–591.
26. Meinhart, J., M. Fussenegger, and W. Hobling, Stabilization of fibrin-chondrocyte constructs for cartilage reconstruction. *Ann Plast Surg*, 1999. **42**(6): 673–678.
27. Atala, A. et al., Endoscopic treatment of vesicoureteral reflux with a chondrocyte-alginate suspension. *J Urol*, 1994. **152**(2 Pt 2): 641–643; discussion 644.
28. Klock, G. et al., Biocompatibility of mannuronic acid-rich alginates. *Biomaterials*, 1997. **18**(10): 707–713.
29. Wee, S. and W.R. Gombotz, Protein release from alginate matrices. *Adv Drug Deliv Rev*, 1998. **31**(3): 267–285.
30. Gregory, K.E. et al., Abnormal collagen assembly, though normal phenotype, in alginate bead cultures of chick embryo chondrocytes. *Exp Cell Res*, 1999. **246**(1): 98–107.
31. Putnam, A.J. and D.J. Mooney, Tissue engineering using synthetic extracellular matrices. *Nat Med*, 1996. **2**(7): 824–826.
32. Eser Elcin, A., Y.M. Elcin, and G.D. Pappas, Neural tissue engineering: Adrenal chromaffin cell attachment and viability on chitosan scaffolds. *Neurol Res*, 1998. **20**(7): 648–654.

33. Muzzarelli, R.A. et al., Osteoconductive properties of methylpyrrolidinone chitosan in an animal model. *Biomaterials*, 1993. **14**(12): 925–929.
34. Stile, R., W. Burghardt, and K. Healy, Synthesis and characterization of injectable poly(N-isopropylacrylamide)-based hydrogels that support tissue formation in vitro. *Macromolecules*, 1999. **32**: 7370–7379.
35. Sofia, S.J., V.V. Premnath, and E.W. Merrill, Poly(ethylene oxide) grafted to silicon surfaces: Grafting density and protein adsorption. *Macromolecules*, 1998. **31**(15): 5059–5070.
36. Harada, A. and K. Kataoka, Chain length recognition: Core-shell supramolecular assembly from oppositely charged block copolymers. *Science*, 1999. **283**(5398): 65–67.
37. Gu, Z.Q., J.M. Xiao, and X.H. Zhang, The development of artificial articular cartilage—PVA-hydrogel. *Biomed Mater Eng*, 1998. **8**(2): 75–81.
38. Cohen, S. et al., Ionically crosslinkable polyphosphazene: A novel polymer for microencapsulation. *J Am Chem Soc*, 1990. **112**(21): 7832–7833.
39. He, J. et al., Rapid generation of biologically relevant hydrogels containing long-range chemical gradients. *Adv Funct Mater*, 2010. **20**(1): 131–137.
40. Krejchi, M. et al., Observation of a silk-like crystal structure in a genetically engineered periodic polypeptide. *Macromol Sci Pure Appl Chem*, 1996. **A33**: 1389–1398.
41. DeLong, S., J. Moon, and J. West, Covalently immobilized gradients of bFGF on hydrogel scaffolds for directed cell migration. *Biomaterials*, 2004. **26**: 3227–3234.
42. DeLong, S.A., A.S. Gobin, and J.L. West, Covalent immobilization of RGDS on hydrogel surfaces to direct cell alignment and migration. *J Control Release*, 2005. **109**(1–3): 139–148.
43. Wong, J. et al., Directed movement of vascular smooth muscle cells on gradient-compliant hydrogels. *Langmuir*, 2003. **19**: 1908–1913.
44. Dubruel, P. et al., Porous gelatin hydrogels: 2. In vitro cell interaction study. *Biomacromolecules*, 2007. **8**(2): 338–344.
45. He, J. et al., Microfluidic synthesis of composite cross-gradient materials for investigating cell-biomaterial interactions. *Biotechnol Bioeng*, 2010. **108**: 175–185.
46. Guarnieri, D. et al., Engineering of covalently immobilized gradients of RGD peptides on hydrogel scaffolds: Effect on cell behaviour. *Macromol Symp*, 2008. **266**: 36–40.
47. Kapur, T.A. and M.S. Shoichet, Immobilized concentration gradients of nerve growth factor guide neurite outgrowth. *J Biomed Mater Res A*, 2004. **68**(2): 235–243.
48. Ilkhanizadeh, S., A.I. Teixeira, and O. Hermanson, Inkjet printing of macromolecules on hydrogels to steer neural stem cell differentiation. *Biomaterials*, 2007. **28**(27): 3936–3943.
49. Phillips, J.E. et al., Engineering graded tissue interfaces. *Proc Natl Acad Sci USA*, 2008. **105**(34): 12170–12175.
50. Oh, S.H. et al., In vitro and in vivo characteristics of PCL scaffolds with pore size gradient fabricated by a centrifugation method. *Biomaterials*, 2007. **28**(9): 1664–1671.
51. Karageorgiou, V. and D. Kaplan, Porosity of 3D biomaterial scaffolds and osteogenesis. *Biomaterials*, 2005. **26**(27): 5474–5491.
52. Malda, J. et al., The effect of PEGT/PBT scaffold architecture on oxygen gradients in tissue engineered cartilaginous constructs. *Biomaterials*, 2004. **25**(26): 5773–5780.

53. Ma, P.X. and R. Zhang, Microtubular architecture of biodegradable polymer scaffolds. *J Biomed Mater Res*, 2001. **56**(4): 469–477.
54. Zmora, S., R. Glicklis, and S. Cohen, Tailoring the pore architecture in 3-D alginate scaffolds by controlling the freezing regime during fabrication. *Biomaterials*, 2002. **23**(20): 4087–4094.
55. Woodfield, T.B. et al., Polymer scaffolds fabricated with pore-size gradients as a model for studying the zonal organization within tissue-engineered cartilage constructs. *Tissue Eng*, 2005. **11**: 1297–1311.
56. Harley, B.A. et al., Fabricating tubular scaffolds with a radial pore size gradient by a spinning technique. *Biomaterials*, 2006. **27**(6): 866–874.
57. Orsi, S., D. Guarnieri, and P.A. Netti, Design of novel 3D gene activated PEG scaffolds with ordered pore structure. *J Mater Sci Mater Med*, 2010. **21**(3): 1013–1020.
58. Vlierberghe, S.V. et al., Porous gelatin hydrogels: 1. Cryogenic formation and structure analysis. *Biomacromolecules*, 2007. **8**(2): 331–337.
59. Stokols, S. and M.H. Tuszynski, The fabrication and characterization of linearly oriented nerve guidance scaffolds for spinal cord injury. *Biomaterials*, 2004. **25**(27): 5839–5846.
60. Discher, D.E., P. Janmey, and Y.L. Wang, Tissue cells feel and respond to the stiffness of their substrate. *Science*, 2005. **310**(5751): 1139–1143.
61. Schwarz, U., Soft matters in cell adhesion: Rigidity sensing on soft elastic substrates. *Soft Matter*, 2007. **3**: 263–266.
62. Li, S., J.L. Guan, and S. Chien, Biochemistry and biomechanics of cell motility. *Annu Rev Biomed Eng*, 2005. **7**: 105–150.
63. Pelham, R.J., Jr. and Y. Wang, Cell locomotion and focal adhesions are regulated by substrate flexibility. *Proc Natl Acad Sci USA*, 1997. **94**(25): 13661–13665.
64. Engler, A. et al., Substrate compliance versus ligand density in cell on gel responses. *Biophys J*, 2004. **86**(1 Pt 1): 617–628.
65. Lo, C.M. et al., Cell movement is guided by the rigidity of the substrate. *Biophys J*, 2000. **79**(1): 144–152.
66. Zaari, N. et al., Photopolymerization in microfluidic gradient generators: Microscale control of substrate compliance to manipulate cell response. *Adv Mater*, 2004. **16**: 2133–2137.
67. Gray, D.S., J. Tien, and C.S. Chen, Repositioning of cells by mechanotaxis on surfaces with micropatterned Young's modulus. *J Biomed Mater Res A*, 2003. **66**(3): 605–614.
68. Kidoaki, S. and T. Matsuda, Microelastic gradient gelatinous gels to induce cellular mechanotaxis. *J Biotechnol*, 2008. **133**(2): 225–230.
69. Georges, P.C. and P.A. Janmey, Cell type-specific response to growth on soft materials. *J Appl Physiol*, 2005. **98**(4): 1547–1553.
70. Burdick, J.A., A. Khademhosseini, and R. Langer, Fabrication of gradient hydrogels using a microfluidics/photopolymerization process. *Langmuir*, 2004. **20**(13): 5153–5156.
71. Chatterjee, K. et al., The effect of 3D hydrogel scaffold modulus on osteoblast differentiation and mineralization revealed by combinatorial screening. *Biomaterials*, 2010. **31**(19): 5051–5062.
72. Kloxin, A.M., J.A. Benton, and K.S. Anseth, In situ elasticity modulation with dynamic substrates to direct cell phenotype. *Biomaterials*, 2010. **31**(1): 1–8.
73. Meredith, J.C. et al., Combinatorial characterization of cell interactions with polymer surfaces. *J Biomed Mater Res A*, 2003. **66**(3): 483–490.

74. Engler, A.J. et al., Matrix elasticity directs stem cell lineage specification. *Cell,* 2006. **126**(4): 677–689.
75. Liu, C., Z. Han, and J.T. Czernuszka, Gradient collagen/nanohydroxyapatite composite scaffold: Development and characterization. *Acta Biomater,* 2009. **5**(2): 661–669.
76. Sherwood, J.K. et al., A three-dimensional osteochondral composite scaffold for articular cartilage repair. *Biomaterials,* 2002. **23**(24): 4739–4751.
77. Du, Y. et al., Convection-driven generation of long-range material gradients. *Biomaterials,* 2010. **31**(9): 2686–2694.
78. Yamamoto, M., K. Yanase, and Y. Tabata, Generation of type I collagen gradient in polyacrylamide hydrogels by a simple diffusion-controlled hydrolysis of amide groups. *Materials,* 2010. **3**: 2393–2404.
79. Wu, J.D. et al., Covalently immobilized gelatin gradients within three-dimensional porous scaffolds. *Chin Sci Bull,* 2009. **54**: 3174–3180.
80. Gurdon, J.B. and P.Y. Bourillot, Morphogen gradient interpretation. *Nature,* 2001. **413**(6858): 797–803.
81. Eichmann, A. et al., Guidance of vascular and neural network formation. *Curr Opin Neurobiol,* 2005. **15**(1): 108–115.
82. Tessier-Lavigne, M. and C.S. Goodman, The molecular biology of axon guidance. *Science,* 1996. **274**(5290): 1123–1133.
83. Parent, C.A. and P.N. Devreotes, A cell's sense of direction. *Science,* 1999. **284**(5415): 765–770.
84. Phillippi, J.A. et al., Microenvironments engineered by inkjet bioprinting spatially direct adult stem cells toward muscle- and bone-like subpopulations. *Stem Cells,* 2008. **26**(1): 127–134.
85. Weibel, D.B. et al., Bacterial printing press that regenerates its ink: Contact-printing bacteria using hydrogel stamps. *Langmuir,* 2005. **21**(14): 6436–6442.
86. Aizawa, Y., R. Wylie, and M. Shoichet, Endothelial cell guidance in 3D patterned scaffolds. *Adv Mater,* 2010. **22**(43): 4831–4835.
87. Juliano, R.L. and S. Haskill, Signal transduction from the extracellular matrix. *J Cell Biol,* 1993. **120**(3): 577–585.
88. Parente, L. et al., Studies on cell motility in inflammation. I. The chemotactic activity of experimental, immunological and non-immunological, inflammatory exudates. *Agents Actions,* 1979. **9**(2): 190–195.
89. Parente, L. et al., Studies on cell motility in inflammation. II. The in vivo effect of anti-inflammatory and anti-rheumatic drugs on chemotaxis in vitro. *Agents Actions,* 1979. **9**(2): 196–200.
90. Martin, P., Wound healing—Aiming for perfect skin regeneration. *Science,* 1997. **276**(5309): 75–81.
91. Knapp, D.M., E.F. Helou, and R.T. Tranquillo, A fibrin or collagen gel assay for tissue cell chemotaxis: Assessment of fibroblast chemotaxis to GRGDSP. *Exp Cell Res,* 1999. **247**(2): 543–553.
92. Marklein, R.A. and J.A. Burdick, Spatially controlled hydrogel mechanics to modulate stem cell interactions. *Soft Matter,* 2010. **6**: 136–143.
93. Dodla, M.C. and R.V. Bellamkonda, Anisotropic scaffolds facilitate enhanced neurite extension in vitro. *J Biomed Mater Res A,* 2006. **78**(2): 213–221.
94. Khademhosseini, A., J. Vacanti, and R. Langer, Tissue engineering: Next generation tissue constructs and challenges to clinical practice. *Sci Am,* 2009. **300**: 64–71.

95. Hosseinkhani, H. et al., Enhanced angiogenesis through controlled release of basic fibroblast growth factor from peptide amphiphile for tissue regeneration. *Biomaterials*, 2006. **27**(34): 5836–5844.

96. Du, Y. et al., Microfluidic systems for engineering vascularized tissue constructs, *Microfluidics for Biological Applications*, 2008, Berlin, Germany: Springer.

97. Chung, S. et al., Cell migration into scaffolds under co-culture conditions in a microfluidic platform. *Lab Chip*, 2009. **9**(2): 269–275.

98. Dodla, M.C. and R.V. Bellamkonda, Differences between the effect of anisotropic and isotropic laminin and nerve growth factor presenting scaffolds on nerve regeneration across long peripheral nerve gaps. *Biomaterials*, 2008. **29**(1): 33–46.

99. Spalazzi, J.P. et al., Development of controlled matrix heterogeneity on a triphasic scaffold for orthopedic interface tissue engineering. *Tissue Eng*, 2006. **12**(12): 3497–3508.

100. Spalazzi, J.P. et al., In vivo evaluation of a multiphased scaffold designed for orthopaedic interface tissue engineering and soft tissue-to-bone integration. *J Biomed Mater Res A*, 2008. **86**(1): 1–12.

101. Wang, X. et al., Growth factor gradients via microsphere delivery in biopolymer scaffolds for osteochondral tissue engineering. *J Control Release*, 2009. **134**(2): 81–90.

102. Heskins, M. and J. Guillet, Solution properties of poly(N-isopropylacrylamide). *J Macromol Sci Chem Ed*, 1968. **A2**: 1441–1455.

4

Smart Biomaterial Scaffold for In Situ Tissue Regeneration

Jaehyun Kim, Sunyoung Joo, In Kap Ko,
Anthony Atala, James J. Yoo, and Sang Jin Lee

CONTENTS

4.1 Introduction

A major consideration in tissue engineering is the pursuit of scaffolds that provide an architecture on which seeded cells are directed to proliferate and differentiate to form new tissues and organs. Such tissue-engineered constructs composed of scaffolds preseeded with cells are one of the most promising approaches to generate functional replacement tissues. Such strategies are dictated by the primary purpose of the implant and the availability of a suitable cell source. These tissue engineering approaches allow the production of new extracellular matrix (ECM) resembling that of the native tissue to replace or regenerate the injured tissues or organs. Using this concept, various preclinical and clinical studies in different tissue systems have been shown to be effective in tissue repair (El-Kassaby et al. 2003; Ossendorf et al. 2007; Shin'oka et al. 2005).

Despite considerable advancements, such cell-based approaches have limitations such as availability of donor tissue. Harvesting autologous tissue is constrained by anatomic limitations and associated with significant donor site morbidity (Burg et al. 2000; Younger and Chapman 1989). For allografts and xenografts, there is the potential risk of inducing an immunologic response due to genetic differences and the risk of both bacterial and viral transmission from the donor to the host. In addition, this approach requires cell isolation and expansion, which involves labor-intensive cell manipulation (Langer 2000). An alternative cell source is the use of stem and progenitor cells; however, this approach also necessitates ex vivo procedures such as cell isolation, expansion, and/or differentiation of desired cell lineages and may result in loss of cellular function during expansion (Guillot et al. 2007). Although cell-based therapies have limitations and challenges, there have been clinical successes.

Recent progress in tissue engineering has introduced the concept of recruiting host stem cell/progenitor cells for in situ tissue regeneration. The principle of this concept is to take advantage of the body's own regenerating capacity by using the host's ability to mobilize endogenous stem cells to the site of injury. This process would provide a more efficient means of therapy by eliminating ex vivo cell manipulation. Figure 4.1 shows such a concept of in situ tissue regeneration. When scaffolds encapsulated with bioactive molecules are implanted, sustained release of these bioactive molecules unlocks the body's own regenerating capacity. In turn, this induces the recruitment of tissue-specific stem/progenitor cells, drives differentiation of these cells into the targeted cell types, and regenerates functional tissues. This chapter reviews the recent development of strategies for in situ tissue regeneration in terms of mechanism of recruitment, cell sources, cellular and molecular roles in cell differentiation, navigational cues and niche signals,

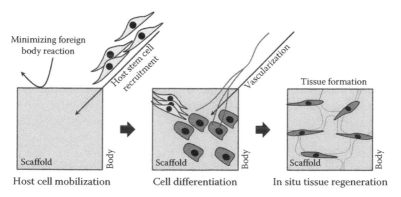

FIGURE 4.1
Illustration of strategy of concept for in situ tissue regeneration.

and a tissue-specific scaffolding system from the perspective of regenerative medicine and tissue engineering.

4.2 Strategy: In Situ Tissue Regeneration

4.2.1 Tissue-Specific Stem/Progenitor Cells

Stem cells, whether derived from embryos, fetuses, or adults, possess a great capacity for the next frontier of regenerative medicine and tissue engineering. This is due to their remarkable potential to develop into many different tissues in the body with specialized function, such as muscle, bone, or blood. Given their unique regenerative abilities, stem cells can also offer new potentials for treating a broad range of diseases, such as diabetes and heart disease. Because of the significant role of stem cells in the regenerative process, a readily available population of stem cells that are highly renewable and that have an extensive ability to differentiate is critical for clinical success.

Recently, the increasing ethical concern for using embryonic stem cells has shifted attention to adult stem cells. An adult stem cell is an undifferentiated cell found in various tissues or organs, which can self-renew and differentiate into tissue-specific cell types. Their primary role is to maintain and repair the tissue in which they are found. The stem cells reside in a specialized microenvironment called a stem cell niche. Tissue-specific stem cells remain quiescent for relatively long periods of time until they are activated by a need for tissue maintenance or by disease or tissue injury. The presence of an underlying regenerative mechanism in the form of tissue-specific stem and progenitor cells suggests that there may be a potential opportunity to bias the host response toward repair and replacement of large tissue defects. It is widely accepted that almost every tissue in the body contains some type of stem or progenitor cells, including brain, liver, circulating blood, heart, skin, fat, kidney, and muscle (Asahara et al. 1997; Bartsch et al. 2005; De Ugarte et al. 2003; Deasy and Huard 2002; Gage 2000; Pfister et al. 2005; Zhang et al. 2003). It would seem that these cells are part of underlying regenerative machinery that is responsible for daily maintenance activities, including repair of normal tissue wear and tear, as well as small, non-life-threatening types of injuries. However, when extensive tissue damage occurs and large tissue defects are present, the regenerative response is overwhelmed, and an immune-based reparative response takes over to maintain some level of function. While the immediate problem may be mitigated by these reparative processes, responses such as inflammation which results in uncontrolled collagen deposition and fibrosis are undesirable because they can lead to further complications and severe deficits in tissue and organ functionality. Therefore, current research efforts have focused on the improvement of the regenerative capacity by controlling the host microenvironment.

4.2.2 Host Cell Mobilization

The concept of in situ tissue regeneration occurs via the recruitment of host stem cells into an injured tissue or other target niche. Currently, various biomaterial scaffolds have been used for the reconstruction of a large tissue defect with functional recovery. From a biomaterial's perspective, placing a biomaterial in the in vivo microenvironment requires injection, insertion, or surgical implantation, all of which injure the tissues or organs involved. In such instances, various reconstructive measures are necessary to restore functionality of the affected tissues or organs. However, it is well known that a biomaterial implant will become populated with host cells that ultimately result in scar tissue. The host cell infiltrate has been assumed to be inflammatory and fibroblastic, as indirect evidence (i.e., the presence of collagen) has suggested that fibroblasts are the predominant cell population present after the initial inflammation has subsided.

Even though inflammatory response and foreign body reaction have been well identified, the cell types that infiltrate the biomaterial have not been identified. Therefore, the possibility of utilizing the body's biologic and environmental resources in situ for tissue regeneration has been investigated. As an initial step, the recruitment of host stem/progenitor cells into an implanted scaffold through the tissue-repairing process has been examined. In our previous study (Lee et al. 2008), poly(glycolic acid) (PGA) nonwoven scaffold, a widely employed biocompatible, biodegradable, and implantable biomaterial, was used in a simple approach to address this dogma. The implant is highly porous and is designed to increase diffusion and accommodate host cell infiltrates. The results showed that the number of host cells continued to increase up to 3 weeks after implantation and began to decrease thereafter as collagen accumulates and fills the pores of the implanted scaffold. This is consistent with a normal inflammatory response seen in many tissue systems. However, we found that a small proportion of infiltrated host cells within the biomaterial implants have multilineage potential (Figure 4.2). These results indicate that some of the host stem cells that can be mobilized into a biomaterial are multipotent, and given an optimal environment, they can differentiate into specific cell types needed for functional regeneration at the implant site.

Based on these findings, a desirable paradigm in which a tissue-specific biomaterial scaffold can be universally applied, without the need for ex vivo cell manipulation, may be attainable. Ideally, the patient's body would supply both the source of cells and the environment for terminal differentiation, provided the appropriate cues can be mediated through the biomaterial scaffold. Therefore, in contrast to current modalities that focus on in vitro manipulation of cells, it may be possible to control tissue morphogenesis

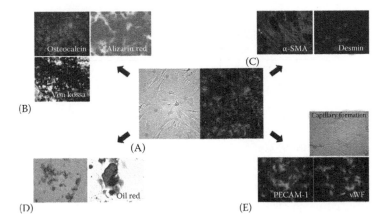

FIGURE 4.2
The infiltrating cells that were induced into different cell lineages under specific conditions demonstrated the expression of their phenotypic and functional characteristics: (A) Infiltrating host cells (Sca-1+), (B) osteogenic differentiation, (C) myogenic differentiation, (D) adipogenic differentiation, and (E) endothelial differentiation as confirmed by specific marker expression.

in vivo by providing the appropriate cues to the infiltrating multipotent cells, leading to the production of functional tissues in situ.

4.3 Design Considerations of Smart Biomaterial Scaffolds

4.3.1 Bioactive Molecules

Bioactive molecules for in situ tissue regeneration play an important role in the control of the microenvironment in vivo. Chemotactic signals from bioactive molecules are responsible for this directed cell migration. An anatomic destination is identified according to certain concentration gradient of chemicals produced in the injured sites in their microenvironment. Bioactive molecules such as growth factors and cytokines can regulate host stem cell migration and proliferation as well as differentiation. In situ tissue regeneration requires activation of stem cells and/or progenitor cells and the recruitment of the cells into a target injury site. Subsequently, the cells must undergo proliferation and differentiation to form a new tissue. Therefore, it is important to utilize a host tissue microenvironment using an implanted biomaterial scaffold for in situ tissue regeneration. Table 4.1 lists bioactive molecules that have been used in implanted scaffolds to help in the recruitment, proliferation, and differentiation of host stem cells.

As an initial step, a sufficient number of stem cells must be recruited into the implanted scaffold for effective tissue regeneration. However, the number

TABLE 4.1

Bioactive Molecules That Are Used in In Situ
Tissue Engineering

Stem cell recruiting factor
 SDF-1
 HGF
 MCP-3
 Granulocyte colony-stimulating factor (G-CSF)
 MMP-2
 Galanin

Collagen synthase inhibitors
 Metalloproteinase inhibitor
 Propyl hydroxylase
 C-proteinase inhibitor
 Halofuginone

Tissue-enhancing factors
 TGF-βs
 Insulin-like growth factors (IGFs)
 Fibroblast growth factor-1 (FGF-1)
 Epidermal growth factor (EGF)

Angiogenic factors
 VEGF
 Fibroblast growth factor-2 (FGF-2)
 PDGF-BB
 TGF-βs
 Angiogenin
 Ang-1
 Ang-2
 Delta-like ligand 4 (Dll4)

Innervation factors
 Brain-derived neurotropic factor (BDNF)
 Glial cell line–derived neurotropic factor (GDNF)
 Nerve growth factor (NGF)
 Agrin

of stem cells in the human body is limited. To overcome this limitation, chemotaxis of mesenchymal stem cells (MSCs) using various chemokines has been evaluated. These cells expressed CCR1, CCR7, CCR9, CXCR4, CXCR5, and CXCR6. The interaction with chemokine and chemokine receptor induces a cellular reaction in response to a specific chemokine and β-actin filament rearrangement (CXCL12). The CXC chemokine, stromal cell-derived factor-1α [SDF-1α (CXCL12)], shows a close relationship with cell survival, migration, proliferation, and differentiation of various host stem cells, including hematopoietic stem cells (HSCs), tissue-specific progenitor cells, and MSC. In addition, it is reported that the MSCs respond in vivo to other bioactive molecules such as hepatocyte growth factor (HGF), matrix metalloproteinase-2 (MMP-2) (Ries et al. 2007), galanin (Louridas et al. 2009), and monocyte chemotactic protein-3 (MCP-3) (Schenk et al. 2007).

More directly, the sustained delivery of chemotactic factors such as SDF-1α using advanced release technology to drive stem cell recruitment to a tissue defect represents a potentially novel approach to regeneration, for example, the encapsulation of these putative bioactive molecules into biodegradable polymeric scaffolds with sustained release kinetics. Several studies have demonstrated that the delivery of exogenous SDF-1α to the myocardium prolongs the presence of SDF-1α after infarction, augmenting stem cell recruitment and improving cardiac function (Chen et al. 2009; Sasaki et al. 2007; Zaruba et al. 2009; Zhang et al. 2007).

4.3.2 Biomaterial Scaffolds

Creation of engineered tissue requires a scaffold that serves as a cell carrier, which would provide structural support until native tissue forms in vivo. Even though the requirements for scaffolds may be different depending on the target applications, a general function of scaffolds that need to be fulfilled is biodegradability, biocompatibility, and temporal structural integrity. In addition, the scaffold's internal architecture should enhance the permeability of nutrients and neovascularization. The latter is particularly important as this porous structure not only can provide space for the recruitment of cells to reside but also can encapsulate bioactive molecules and provide cues that enhance cell migration, proliferation, and differentiation, producing a biofunctional stem cell niche. Principally, to design a tissue-specific scaffolding system for in situ tissue regeneration, it should (1) reduce inflammation and fibrosis, (2) utilize host microenvironment for recruiting host stem/progenitor cells, and (3) control tissue-specific cell differentiation within the scaffold.

One of the significant criteria for developing biomaterial scaffolds especially for the in situ tissue regeneration purposes is to deliver bioactive molecules and regulatory signals in a precise temporal and spatial manner. Development of such scaffolds will serve as a powerful artificial extracellular milieu functioning as an ideal recruiting device (Lutolf et al. 2009). Initially, it is critical to efficiently induce and direct recruitment of a sufficient number of host stem cells to the targeted sites. To achieve this, identifying and understanding the roles of bioactive molecules that initiate the recruiting response of the cells is required. Effective delivery of bioactive molecules requires sustained release to maintain effective concentrations in a local environment. This can be accomplished by incorporation within a scaffold material. The delivery method depends on the characteristics of the scaffold material. Releases of bioactive molecules are controlled by temperature, pH, and material biodegradability.

Especially, the scaffold for in situ tissue regeneration should provide an appropriate microenvironment that could recruit stem and progenitor cells into the implant and support the proliferation and differentiation of the recruited cells to form a desired tissue or a functional organ. For this purpose, it seems that multiple factors should be delivered to a target application due to the complexity of the microenvironment. Mooney and colleagues suggested a multiple protein delivery system for accelerating vascularization and tissue formation because the development of tissues and organs is typically driven by the action of a number of growth factors (Richardson et al. 2001). They reported a new polymeric system that allows for the tissue-specific delivery of two or more growth factors, with controlled dose and rate of delivery. Controlling sustained release of bioactive molecules with different release kinetics enables effective tissue regeneration. Likewise, a recent study shows various methods of sustained release of bioactive molecules over time (Figure 4.3) (Chen et al. 2011). Multiple sustained release mimics actual in vivo tissue regeneration, and it contributes to effective and rapid tissue regeneration. In a recent study, a gelatin-based scaffold was delivered in vivo with chemical conjugations of four different bioactive molecules, vascular endothelial growth factor (VEGF), angiopoietin-1 (Ang-1), keratinocyte growth factor (KGF), and platelet-derived growth factor-BB (PDGF-BB). This combined delivery of multiple bioactive molecules resulted in an increase in angiogenesis with a potential for enhanced tissue regeneration (Elia et al. 2010). Another study in skeletal muscle regeneration shows effective and functional skeletal muscle regeneration using alginate which simultaneously released insulin-like growth factor-1 (IGF-1) and VEGF (Borselli et al. 2010). This study is important because in addition to angiogenesis, a functional skeletal muscle tissue was created with activation of muscle satellite cells.

The mechanical and molecular information coded within the extracellular milieu is guiding the development of a new generation of biomaterials for future tissue regeneration. To this end, ECM-mimicking biomaterials may not only provide structural components for supporting cells but also contain a reservoir of cell signaling motifs and sequestered growth factors that guide cellular anchorage and behavior, inspiring multiple examples of biomimetic design for biomaterial scaffolds. In vascular research, for example, the presence of endothelium-derived macromolecules or their cell interacting domains onto vascular grafts can mimic features of the ECM and thereby assist specific cell adhesion and promote endothelialization (Lutolf and Hubbell 2005).

Another consideration for in situ regeneration is to provide cell attachment sites on the scaffold in conjunction with the release of bioactive molecules involved with target cell recruitment. One of the ECM peptide sequences

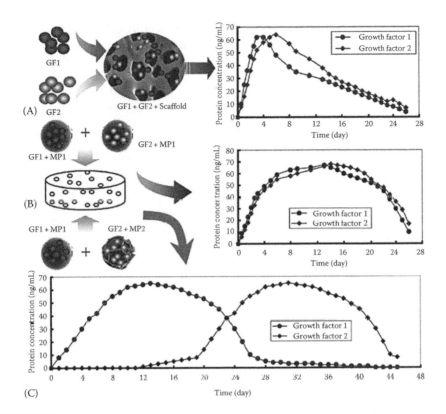

FIGURE 4.3
Schematic illustration of different release profiles of two growth factors (GF1 and GF2) result-ing from different delivery strategies. Dual growth factor combinations are directly intro-duced into the carrier, leading to the simultaneous delivery of two growth factors (A), or else two growth factors are preloaded separately into particulate systems (MP1 or MP2) followed by introduction to the carrier (B,C). The particulate systems for each growth factor can be either the same, leading to the simultaneous sustained delivery of the two growth factors (B), or different, leading to a sequential delivery of the two growth factors (C) (illustration is not to scale). (Data from Chen, F.M. et al., *Biomaterials*, 31(24), 6279, 2010a).

that influence the cell adhesion behavior is the integrin-binding arginine-glycine-aspartic acid (RGD) sequence (Bagno et al. 2007; Benoit and Anseth 2005; Fittkau et al. 2005; Huntley et al. 2006; Kurihara and Nagamune 2005; Marletta et al. 2005; Martino et al. 2009; Oharazawa et al. 2005; Sagnella et al. 2005; Salinas and Anseth 2009). Together with the integrins, the cell surface receptors that recognize the sequence of various proteins, RGD constitutes a major recognition system for cell adhesion. The RGD motif may also enhance the recruitment and activation of endoneurial phagocytes (i.e., phagocytes residing in *Lymnaea*'s nerves) in the injury response of the nervous system of the pond snail *Lymnaea stagnalis* and affect nerve regeneration (Hermann et al. 2008). In an in vitro study, a designer self-assembling peptide scaffold developed by Horii et al. significantly stimulated cell migration into the three-dimensional scaffold, suggesting that it would be possible to apply suitable and active biological scaffolds to stimulate and promote host stem cell recruitment, differentiation, and regeneration of tissues without introducing any foreign cells (Horii et al. 2007).

Cell biomaterial interactions continue to be a principal source of inspiration for biomaterial functionalization and thus play a very important role in future cell recruiting device design (Dutta and Dutta 2009). This is a dynamic and rapidly evolving field that has gained considerable attention as a means of increasing scaffold potency and to improve their biological functionality (Place et al. 2009). Current research has identified many peptides/proteins, various factors, and numerous techniques that could be used for the functionalization of biomaterials. New biomaterials can adapt to the surrounding microenvironment and orchestrate the transport of ions and bioactive molecules and information transfer between cells and their microenvironment (Galler et al. 2008; Horii et al. 2007; Kim et al. 2009; Re'em et al. 2010; Zhang et al. 2009). However, less is known about how such biomaterials can exactly influence and control cell function, how much extrinsic physiochemical information is required to mobilize host stem cells into regenerating a complex tissue, and specifically, taking a combination of clinical performance, marketing, and cost-effectiveness into consideration, what minimum level of biomaterial complexity is required for a given task (Dutta and Dutta 2009). There is an ever-increasing demand for biomaterials that can match both the mechanical and biological properties of real tissue matrix, support vascularization, and recreate nanoscale topographical and cell-specific biochemical cues (Chen et al. 2010a,b).

4.4 Applications of In Situ Tissue Regeneration

4.4.1 Bone

The effect of in situ tissue regeneration has been well studied and documented on regeneration of bone. The required properties of biomaterial scaffolds to ensure successful treatment of bone defects are the temporary mechanical load bearing within the tissue defects. Moreover, it should minimize the immune and/or inflammatory response. The type of biomaterials widely used for this purpose is calcium phosphate, calcium sulfate, and hydroxyapatite. Since bone tissue consists of large amounts of such materials, they have been considered as the major component of scaffold material for bone tissue regeneration. This is due to their close chemical and crystal resemblance to the mineral phase of bone, demonstrating excellent biocompatibility and osteoconductivity (Jarcho et al. 1977). However, as bioactive molecules, bone morphogenetic protein-2 (BMP-2), transforming growth factor-β (TGF-β), basic fibroblast growth factor (bFGF), and VEGF are common but vital growth factors that are introduced into the scaffolds due to their osteoinductive and vascularization properties (Ginebra et al. 2006; Jansen et al. 2005; Seeherman and Wozney 2005). Scaffolds made of natural polymers such as alginate, fibrin, or gelatin and synthetic biodegradable polymers such as polylactide (PLA), poly(lactide-*co*-glycolide) (PLGA) incorporated with factors alone, or in combination have been shown to be osteoinductive. As such, these scaffolds have shown an ability to stimulate and induce neighboring bone marrow stromal cells and enhance bone tissue formation.

4.4.2 Cartilage

When injured, the successful rate of cartilage regeneration is low compared to other types of tissues, which can lead to joint problems such as severe arthritis. In early studies, in vitro cartilage production from chondrocytes and specialized scaffolds has shown success. However, when the engineered cartilage was implanted, serious compatibility issue was noted in vivo. Recently, Erggelet et al. demonstrated the regeneration of cartilage using cell-free biomaterial scaffolds. In this study, this biomaterial constructs composed of biodegradable PLGA scaffold incorporated with plasma and hyaluronic acid were fabricated and implanted into a cartilage microfracture injury site (Erggelet et al. 2009). This result showed that the implanted constructs induced the migration of bone-marrow-derived stem cells and the formation of neocartilage tissue. Mao et al. also demonstrated that the entire articular surface of the synovial joint can regenerate without cell transplantation, using three-dimensional poly(ε-caprolactone) (PCL) and hydroxyapatite composites fabricated by solid free-form technique. These fabricated scaffolds were coated with TGF-β3 and implanted. This was shown to be effective in regenerating cartilage tissue by recruiting host stem cells to the site of implants (Lee et al. 2010).

4.4.3 Skeletal Muscle

Skeletal muscle is composed of bundles of myofibers that are contracted by motor nerve stimulation. Loss of a large amount of muscle mass often results in incomplete recovery, with the development of nonfunctional scar tissue. Minor muscle injury due to exercise and weight-lifting is easily repaired by natural regenerative processes. However, when there is a severe injury or a large defect in skeletal muscle, not only muscle but also surrounding nerves and vessels are destroyed. If the injury is not properly treated, it causes skeletal muscle weakness and atrophy (Huard et al. 2002).

Cell-based approaches have offered new opportunities for repairing such large muscular injuries. As available cell sources, muscle satellite cells (Le Grand and Rudnicki 2007; Sacco et al. 2008) primarily play significant roles in muscle regeneration owing to their self-renewal capabilities and muscle-specific differentiation. Besides muscle satellite cells, several populations of other stem cells, such as muscle-derived stem cells (MDSCs) (Qu-Petersen et al. 2002), pericytes (Crisan et al. 2008), muscle resident macrophages (Polesskaya et al. 2003; Sun et al. 2009), endothelial progenitor cells (EPCs) (Jin et al. 2006), and bone-marrow-derived MSCs (LaBarge and Blau 2002), have been used in engineering muscle tissue and are closely involved in the muscle regeneration process. The roles of these cell populations are critical for efficient muscle regeneration, by maturing blood vessels, secreting trophic factors, and reducing fibrotic formation (Meirelles Lda et al. 2009; Sun et al. 2009).

A novel approach that relies on the body's ability to repair itself has been developed utilizing host stem cell recruitment and control of cell fate. The strategy of this approach is based on the release of tissue-specific stem cell stimulating factors to utilize host stem cells, which is followed by effective tissue regeneration. There have also been a few trials in muscle regeneration (Borselli et al. 2010). In particular, Mooney and his colleagues have developed an injectable system based on alginate material that is able to deliver dual growth factors, IGF-1 and VEGF, for the enhancement of functional muscle regeneration. IGF-1 induces satellite cell mobilization to injured muscle tissue to proliferate and differentiate, and VEGF is a primary proangiogenic factor that recruits vessel-forming stem or progenitor cells. In another study, Kin et al. implanted a collagen scaffold into a rabbit hind limb muscle injury. Twenty-four weeks post-transplantation, the control group (without scaffold) showed severe scar tissue and muscle contraction, whereas the collagen-based scaffold group showed focal tissue adhesion and new muscle tissue formation (Kin et al. 2007).

Table 4.2 lists recent therapeutic applications of cell-free biomaterials for in situ tissue regeneration. Various biomaterial scaffolds have been used for in situ tissue regeneration in the form of injection or implantation. While many technologies are at the early experimental stage, several technologies have been successfully performed in preclinical animal model studies with satisfactory outcomes.

4.5 Conclusion and Future Directions

In situ tissue regeneration is a promising approach to a more simplifying and efficient means of developing functional tissue constructs, by eliminating a tissue biopsy and in vitro cell manipulation; therefore, it holds great potential to provide new therapeutic options for functional tissue regeneration. For this approach to be successful, stem cells need to be directed to the site of injury, and proliferated and differentiated within a local microenvironment provided by biomaterial scaffolds. However,

TABLE 4.2

Recent Therapeutic Applications of Cell-Free Biomaterials for In Situ Tissue Regeneration

	Biomaterials	Bioactive Factors	Animal Model	References
Bone	PLA	FGF-1/spraying	Rat calvarial bone defect	Gomez et al. (2006)
	Fibrin/PLGA	BMP-2/		
	Alginate	encapsulation	Rat calvarial bone defect	Chung et al. (2007)
	Fibrin	BMP-2/		
	P(HEMA-VP) gel	immobilization	Rat muscle	Suzuki et al. (2000)
		Heparin sulfate	Rat cranial defect	
	Gelatin	FGF-2/	Rabbit femoral defect	Woodruff et al. (2007)
	Fibrin/HA	immobilization		
		FGF-2	Mouse maxillae	Mabilleau et al. (2008)
		BMP-2	Mouse calvarial bone defect	Kodama et al. (2009)
				Osathanon et al. (2008)
Bile duct	Collagen/PP mesh		Canine circumferential biliary defect	Nakashima et al. (2007)
Heart and vessel	PGA/PLA/ collagen		Porcine descending aorta, porcine pulmonary arterial trunk, canine ventricular outflow tract	Iwai et al. (2005)
	PGA/PLA/ collagen		Canine carotid arteries	Yokota et al. (2008)
	Porcine SIS/ collagen		Rabbit arterial bypass model	Huynh et al. (1999)
	PEUU		Rat myocardial infarction model	Fujimoto et al. (2007)
	Alginate		Rat myocardial infarction model	Landa et al. (2008)
Cartilage	PGA	Autologous serum/ HA/microfracture	Sheep full-thickness cartilage defect	Erggelet et al. (2009)
	Collagen		Rabbit articular cartilage	Kubo et al. (2007)
Esophagus	UBM		Rat abdominal esophagus	Dahms et al. (1998)
	Rat gastric acellular matrix		Rat abdominal esophagus	Urita et al. (2007)
Fat pad	Collagen/PP	FGF-2/gelatin microsphere	Rat	Hiraoka et al. (2006)
Muscle	Collagen		Rabbit muscle (vastus lateralis)	Kin et al. (2007)

(continued)

TABLE 4.2 (continued)

Recent Therapeutic Applications of Cell-Free Biomaterials for In Situ Tissue Regeneration

	Biomaterials	Bioactive Factors	Animal Model	References
Periodontal tissue	Collagen	FGF-2/gelatin microsphere	Canine periodontal	Nakahara et al. (2003)
	PLGA	GDF-5	Canine periodontal	Herberg et al. (2008)
Skin	Chitosan		Porcine burned skin	Boucard et al. (2007)
Spine	PGA/HA	Blood serum	Rabbit disk defect	Abbushi et al. (2008)
Stomach	Collagen/PGA		Canine stomach	Hori et al. (2001)

PLA, poly(lactic acid); PLGA, poly(lactide-*co*-glycolide); PGA, poly(glycolic acid); PP, polypropylene; PEUU, polyester urethane urea; HA, hyaluronic acid; SIS, small intestine submucosa; UBM, urinary bladder matrix; FGF, fibroblast growth factor; BMP, bone morphogenic protein; GDF, growth differentiation factor.

traditional biomaterial scaffolds, while providing structural support for a new tissue formation, do not adequately mimic the complex interactions between host stem and progenitor cells and the ECM that promote functional tissue regeneration. Therefore, future advances in tissue engineering and regenerative medicine will depend on the development of smart biomaterials that actively participate in the formation of functional tissues. Recent advances in biomaterial processing and fabrication technologies enabled to incorporate bioactive molecules with the aim to accelerate stem cell/progenitor cell mobilization. Incorporating bioactive molecules during scaffold processing, however, is not simple as bioactive molecules are sensitive to elevated temperatures and extreme chemical conditions; thus improvement in the design and structure of the scaffolds achieving optimal therapeutic dose and duration of the bioactive molecules released from an implanted biomaterial should enhance sufficient cell recruitment and robust tissue regeneration. Even though a few molecules were found to be involved in stem cell trafficking as previously reviewed, identifying more molecules and understanding their roles at the molecular level also would facilitate stem cell mobilization.

Lastly, we would like to introduce one more powerful tool which has been currently discovered. In the last decade, RNA interference (RNAi) has provided a great potential for treating various disease states or helping to repair damaged tissue. RNAi is the mechanism in which gene expression is silenced by binding the target mRNAs leading to translational repression or destruction of specific mRNA molecules (Fire et al. 1998). Using this mechanism, expression of specific proteins that are involved in blocking a desired function of proteins of interest for the purpose of targeted tissue regeneration can be suppressed. For example, Nakasa et al. (2010) demonstrated that a local injection of micro-RNAs (miRNAs) that interferes with the regulation of muscle development and homeostasis could accelerate muscle regeneration in a rat skeletal muscle injury model. Administration of exogenous miR-NAs induced expression of myogenic markers, MyoD1, myogenin, and Pax7. Not only for their function in muscle regeneration, they also showed that local injection prevented fibrosis and downregulated myostatin. Even with its enormous potential, the use of RNAi in in situ tissue regeneration is still in the early phase. For a more efficient therapeutic outcome, we need a better understanding of the complex interactions and pathways of the molecules that are involved in the targeted tissue regeneration.

Acknowledgments

We would like to thank Dr. John Jackson for editorial assistance. This work was supported by the Orthopedic Trauma Research Program (USAMRAA OTRP07-07128091) and Armed Forces Institute for Regenerative Medicine (W81XWH-08-2-0032) of Department of Defense.

References

Abbushi, A., M. Endres, M. Cabraja, S. N. Kroppenstedt, U. W. Thomale, M. Sittinger, A. A. Hegewald et al. 2008. Regeneration of intervertebral disc tissue by resorbable cell-free polyglycolic acid-based implants in a rabbit model of disc degeneration. *Spine* 33 (14):32.

Asahara, T., T. Murohara, A. Sullivan, M. Silver, R. van der Zee, T. Li, B. Witzenbichler, G. Schatteman, and J. M. Isner. 1997. Isolation of putative progenitor endothelial cells for angiogenesis. *Science* 275 (5302):964–967.

Bagno, A., A. Piovan, M. Dettin, A. Chiarion, P. Brun, R. Gambaretto, G. Fontana, C. Di Bello, G. Palu, and I. Castagliuolo. 2007. Human osteoblast-like cell adhesion on titanium substrates covalently functionalized with synthetic peptides. *Bone* 40 (3):693–699.

Bartsch, G., J. J. Yoo, P. De Coppi, M. M. Siddiqui, G. Schuch, H. G. Pohl, J. Fuhr, L. Perin, S. Soker, and A. Atala. 2005. Propagation, expansion, and multilineage differentiation of human somatic stem cells from dermal progenitors. *Stem Cells Dev* 14 (3):337–348.

Benoit, D. S. and K. S. Anseth. 2005. The effect on osteoblast function of colocalized RGD and PHSRN epitopes on PEG surfaces. *Biomaterials* 26 (25):5209–5220.

Borselli, C., H. Storrie, F. Benesch-Lee, D. Shvartsman, C. Cezar, J. W. Lichtman, H. H. Vandenburgh, and D. J. Mooney. 2010. Functional muscle regeneration with combined delivery of angiogenesis and myogenesis factors. *Proc Natl Acad Sci USA* 107 (8):3287–3292.

Boucard, N., C. Viton, D. Agay, E. Mari, T. Roger, Y. Chancerelle, and A. Domard. 2007. The use of physical hydrogels of chitosan for skin regeneration following third-degree burns. *Biomaterials* 28 (24):3478–3488.

Burg, K. J., S. Porter, and J. F. Kellam. 2000. Biomaterial developments for bone tissue engineering. *Biomaterials* 21 (23):2347–2359.

Chen, F. M., Y. An, R. Zhang, and M. Zhang. 2011. New insights into and novel applications of release technology for periodontal reconstructive therapies. *J Control Release* 149 (2):92–110.

Chen, F. M., Z. W. Ma, Q. T. Wang, and Z. F. Wu. 2009. Gene delivery for periodontal tissue engineering: Current knowledge—Future possibilities. *Curr Gene Ther* 9 (4):248–266.

Chen, F. M., M. Zhang, and Z. F. Wu. 2010a. Toward delivery of multiple growth factors in tissue engineering. *Biomaterials* 31 (24):6279–6308.

Chen, F. M., J. Zhang, M. Zhang, Y. An, F. Chen, and Z. F. Wu. 2010b. A review on endogenous regenerative technology in periodontal regenerative medicine. *Biomaterials* 31 (31):7892–7927.

Chung, Y. I., K. M. Ahn, S. H. Jeon, S. Y. Lee, J. H. Lee, and G. Tae. 2007. Enhanced bone regeneration with BMP-2 loaded functional nanoparticle-hydrogel complex. *J Control Release* 121 (1–2):91–99.

Crisan, M., S. Yap, L. Casteilla, C. W. Chen, M. Corselli, T. S. Park, G. Andriolo et al. 2008. A perivascular origin for mesenchymal stem cells in multiple human organs. *Cell Stem Cell* 3 (3):301–313.

Dahms, S. E., H. J. Piechota, R. Dahiya, C. A. Gleason, M. Hohenfellner, and E. A. Tanagho. 1998. Bladder acellular matrix graft in rats: Its neurophysiologic properties and mRNA expression of growth factors TGF-alpha and TGF-beta. *Neurourol Urodyn* 17 (1):37–54.

De Ugarte, D. A., P. H. Ashjian, A. Elbarbary, and M. H. Hedrick. 2003. Future of fat as raw material for tissue regeneration. *Ann Plast Surg* 50 (2):215–219.

Deasy, B. M. and J. Huard. 2002. Gene therapy and tissue engineering based on muscle-derived stem cells. *Curr Opin Mol Ther* 4 (4):382–389.

Dutta, R. C. and A. K. Dutta. 2009. Cell-interactive 3D-scaffold; advances and applications. *Biotechnol Adv* 27 (4):334–339.

El-Kassaby, A. W., A. B. Retik, J. J. Yoo, and A. Atala. 2003. Urethral stricture repair with an off-the-shelf collagen matrix. *J Urol* 169 (1):170–173; discussion 173.

Elia, R., P. W. Fuegy, A. VanDelden, M. A. Firpo, G. D. Prestwich, and R. A. Peattie. 2010. Stimulation of in vivo angiogenesis by in situ crosslinked, dual growth factor-loaded, glycosaminoglycan hydrogels. *Biomaterials* 31 (17):4630–4638.

Erggelet, C., M. Endres, K. Neumann, L. Morawietz, J. Ringe, K. Haberstroh, M. Sittinger, and C. Kaps. 2009. Formation of cartilage repair tissue in articular cartilage defects pretreated with microfracture and covered with cell-free polymer-based implants. *J Orthop Res* 27 (10):1353–1360.

Fire, A., S. Xu, M. K. Montgomery, S. A. Kostas, S. E. Driver, and C. C. Mello. 1998. Potent and specific genetic interference by double-stranded RNA in *Caenorhabditis elegans*. *Nature* 391 (6669):806–811.

Fittkau, M. H., P. Zilla, D. Bezuidenhout, M. P. Lutolf, P. Human, J. A. Hubbell, and N. Davies. 2005. The selective modulation of endothelial cell mobility on RGD peptide containing surfaces by YIGSR peptides. *Biomaterials* 26 (2):167–174.

Fujimoto, K. L., J. Guan, H. Oshima, T. Sakai, and W. R. Wagner. 2007. In vivo evaluation of a porous, elastic, biodegradable patch for reconstructive cardiac procedures. *Ann Thorac Surg* 83 (2):648–654.

Gage, F. H. 2000. Mammalian neural stem cells. *Science* 287 (5457):1433–1438.

Galler, K. M., A. Cavender, V. Yuwono, H. Dong, S. Shi, G. Schmalz, J. D. Hartgerink, and R. N. D'Souza. 2008. Self-assembling peptide amphiphile nanofibers as a scaffold for dental stem cells. *Tissue Eng Part A* 14 (12):2051–2058.

Ginebra, M. P., T. Traykova, and J. A. Planell. 2006. Calcium phosphate cements as bone drug delivery systems: A review. *J Control Release* 113 (2):102–110.

Gomez, G., S. Korkiakoski, M. M. Gonzalez, S. Lansman, V. Ella, T. Salo, M. Kellomaki, N. Ashammakhi, and E. Arnaud. 2006. Effect of FGF and polylactide scaffolds on calvarial bone healing with growth factor on biodegradable polymer scaffolds. *J Craniofac Surg* 17 (5):935–942.

Guillot, P. V., W. Cui, N. M. Fisk, and D. J. Polak. 2007. Stem cell differentiation and expansion for clinical applications of tissue engineering. *J Cell Mol Med* 11 (5):935–944.

Herberg, S., M. Siedler, S. Pippig, A. Schuetz, C. Dony, C.-K. Kim, and U. M. E. Wikesjo. 2008. Development of an injectable composite as a carrier for growth factor-enhanced periodontal regeneration. *J Clin Periodontol* 35:976–984.

Hermann, P. M., J. J. Nicol, A. G. Bulloch, and W. C. Wildering. 2008. RGD-dependent mechanisms in the endoneurial phagocyte response and axonal regeneration in the nervous system of the snail *Lymnaea stagnalis*. *J Exp Biol* 211 (Pt 4):491–501.

Hiraoka, Y., H. Yamashiro, K. Yasuda, Y. Kimura, T. Inamoto, and Y. Tabata. 2006. In situ regeneration of adipose tissue in rat fat pad by combining a collagen scaffold with gelatin microspheres containing basic fibroblast growth factor. *Tissue Eng* 12 (6):1475–1487.

Hori, Y., T. Nakamura, K. Matsumoto, Y. Kurokawa, S. Satomi, and Y. Shimizu. 2001. Experimental study on in situ tissue engineering of the stomach by an acellular collagen sponge scaffold graft. *ASAIO J* 47 (3):206–210.

Horii, A., X. Wang, F. Gelain, and S. Zhang. 2007. Biological designer self-assembling peptide nanofiber scaffolds significantly enhance osteoblast proliferation, differentiation and 3-D migration. *PloS One* 2 (2):e190.

Huard, J., Y. Li, and F. H. Fu. 2002. Muscle injuries and repair: Current trends in research. *J Bone Joint Surg Am* 84-A (5):822–832.

Huntley, B. K., S. M. Sandberg, J. A. Noser, A. Cataliotti, M. M. Redfield, Y. Matsuda, and J. C. Burnett, Jr. 2006. BNP-induced activation of cGMP in human cardiac fibroblasts: Interactions with fibronectin and natriuretic peptide receptors. *J Cell Physiol* 209 (3):943–949.

Huynh, T., G. Abraham, J. Murray, K. Brockbank, P. O. Hagen, and S. Sullivan. 1999. Remodeling of an acellular collagen graft into a physiologically responsive neovessel. *Nat Biotechnol* 17 (11):1083–1086.

Iwai, S., Y. Sawa, S. Taketani, K. Torikai, K. Hirakawa, and H. Matsuda. 2005. Novel tissue-engineered biodegradable material for reconstruction of vascular wall. *Ann Thorac Surg* 80 (5):1821–1827.

Jansen, J. A., J. W. Vehof, P. Q. Ruhe, H. Kroeze-Deutman, Y. Kuboki, H. Takita, E. L. Hedberg, and A. G. Mikos. 2005. Growth factor-loaded scaffolds for bone engineering. *J Control Release* 101 (1–3):127–136.

Jarcho, M., J. F. Kay, K. I. Gumaer, R. H. Doremus, and H. P. Drobeck. 1977. Tissue, cellular and subcellular events at a bone-ceramic hydroxylapatite interface. *J Bioeng* 1 (2):79–92.

Jin, D. K., K. Shido, H. G. Kopp, I. Petit, S. V. Shmelkov, L. M. Young, A. T. Hooper et al. 2006. Cytokine-mediated deployment of SDF-1 induces revascularization through recruitment of CXCR4+ hemangiocytes. *Nat Med* 12 (5):557–567.

Kim, K. L., D. K. Han, K. Park, S. H. Song, J. Y. Kim, J. M. Kim, H. Y. Ki et al. 2009. Enhanced dermal wound neovascularization by targeted delivery of endothelial progenitor cells using an RGD-g-PLLA scaffold. *Biomaterials* 30 (22):3742–3748.

Kin, S., A. Hagiwara, Y. Nakase, Y. Kuriu, S. Nakashima, T. Yoshikawa, C. Sakakura, E. Otsuji, T. Nakamura, and H. Yamagishi. 2007. Regeneration of skeletal muscle using in situ tissue engineering on an acellular collagen sponge scaffold in a rabbit model. *ASAIO J* 53 (4):506–513.

Kodama, N., M. Nagata, Y. Tabata, M. Ozeki, T. Ninomiya, and R. Takagi. 2009. A local bone anabolic effect of rhFGF2-impregnated gelatin hydrogel by promoting cell proliferation and coordinating osteoblastic differentiation. *Bone* 44 (4):699–707.

Kubo, M., S. Imai, M. Fujimiya, E. Isoya, K. Ando, T. Mimura, and Y. Matsusue. 2007. Exogenous collagen-enhanced recruitment of mesenchymal stem cells during rabbit articular cartilage repair. *Acta Orthop* 78 (6):845–855.

Kurihara, H. and T. Nagamune. 2005. Cell adhesion ability of artificial extracellular matrix proteins containing a long repetitive Arg-Gly-Asp sequence. *J Biosci Bioeng* 100 (1):82–87.

LaBarge, M. A. and H. M. Blau. 2002. Biological progression from adult bone marrow to mononucleate muscle stem cell to multinucleate muscle fiber in response to injury. *Cell* 111 (4):589–601.

Landa, N., L. Miller, M. S. Feinberg, R. Holbova, M. Shachar, I. Freeman, S. Cohen, and J. Leor. 2008. Effect of injectable alginate implant on cardiac remodeling and function after recent and old infarcts in rat. *Circulation* 117 (11):1388–1396.

Langer, R. 2000. Tissue engineering. *Mol Ther* 1 (1):12–15.

Le Grand, F. and M. A. Rudnicki. 2007. Skeletal muscle satellite cells and adult myogenesis. *Curr Opin Cell Biol* 19 (6):628–633.

Lee, C. H., J. L. Cook, A. Mendelson, E. K. Moioli, H. Yao, and J. J. Mao. 2010. Regeneration of the articular surface of the rabbit synovial joint by cell homing: A proof of concept study. *Lancet* 376 (9739):440–448.

Lee, S. J., M. Van Dyke, A. Atala, and J. J. Yoo. 2008. Host cell mobilization for in situ tissue regeneration. *Rejuvenation Res* 11 (4):747–756.

Louridas, M., S. Letourneau, M. E. Lautatzis, and M. Vrontakis. 2009. Galanin is highly expressed in bone marrow mesenchymal stem cells and facilitates migration of cells both in vitro and in vivo. *Biochem Biophys Res Commun* 390 (3):867–871.

Lutolf, M. P., P. M. Gilbert, and H. M. Blau. 2009. Designing materials to direct stem-cell fate. *Nature* 462 (7272):433–441.

Lutolf, M. P. and J. A. Hubbell. 2005. Synthetic biomaterials as instructive extracellular microenvironments for morphogenesis in tissue engineering. *Nat Biotechnol* 23 (1):47–55.

Mabilleau, G., E. Aguado, I. C. Stancu, C. Cincu, M. F. Basle, and D. Chappard. 2008. Effects of FGF-2 release from a hydrogel polymer on bone mass and microarchitecture. *Biomaterials* 29 (11):1593–1600.

Marletta, G., G. Ciapetti, C. Satriano, S. Pagani, and N. Baldini. 2005. The effect of irradiation modification and RGD sequence adsorption on the response of human osteoblasts to polycaprolactone. *Biomaterials* 26 (23):4793–4804.

Martino, M. M., M. Mochizuki, D. A. Rothenfluh, S. A. Rempel, J. A. Hubbell, and T. H. Barker. 2009. Controlling integrin specificity and stem cell differentiation in 2D and 3D environments through regulation of fibronectin domain stability. *Biomaterials* 30 (6):1089–1097.

Meirelles Lda, S., A. M. Fontes, D. T. Covas, and A. I. Caplan. 2009. Mechanisms involved in the therapeutic properties of mesenchymal stem cells. *Cytokine Growth Factor Rev* 20 (5–6):419–427.

Nakahara, T., T. Nakamura, E. Kobayashi, M. Inoue, K. Shigeno, Y. Tabata, K. Eto, and Y. Shimizu. 2003. Novel approach to regeneration of periodontal tissues based on in situ tissue engineering: Effects of controlled release of basic fibroblast growth factor from a sandwich membrane. *Tissue Eng* 9 (1):153–162.

Nakasa, T., M. Ishikawa, M. Shi, H. Shibuya, N. Adachi, and M. Ochi. 2010. Acceleration of muscle regeneration by local injection of muscle-specific microRNAs in rat skeletal muscle injury model. *J Cell Mol Med* 14 (10):2495–2505.

Nakashima, S., T. Nakamura, L. Han, K. Miyagawa, T. Yoshikawa, C. Sakakura, A. Hagiwara, and E. Otsuji. 2007. Experimental biliary reconstruction with an artificial bile duct using in situ tissue engineering technique. *Inflamm Regen* 27 (6):579–585.

Oharazawa, H., N. Ibaraki, K. Ohara, and V. N. Reddy. 2005. Inhibitory effects of Arg-Gly-Asp (RGD) peptide on cell attachment and migration in a human lens epithelial cell line. *Ophthalmic Res* 37 (4):191–196.

Osathanon, T., M. L. Linnes, R. M. Rajachar, B. D. Ratner, M. J. Somerman, and C. M. Giachelli. 2008. Microporous nanofibrous fibrin-based scaffolds for bone tissue engineering. *Biomaterials* 29 (30):4091–4099.

Ossendorf, C., C. Kaps, P. C. Kreuz, G. R. Burmester, M. Sittinger, and C. Erggelet. 2007. Treatment of posttraumatic and focal osteoarthritic cartilage defects of the knee with autologous polymer-based three-dimensional chondrocyte grafts: 2-year clinical results. *Arthritis Res Ther* 9 (2):R41.

Pfister, O., F. Mouquet, M. Jain, R. Summer, M. Helmes, A. Fine, W. S. Colucci, and R. Liao. 2005. CD31− but Not CD31+ cardiac side population cells exhibit functional cardiomyogenic differentiation. *Circ Res* 97 (1):52–61.

Place, E. S., N. D. Evans, and M. M. Stevens. 2009. Complexity in biomaterials for tissue engineering. *Nat Mater* 8 (6):457–470.

Polesskaya, A., P. Seale, and M. A. Rudnicki. 2003. Wnt signaling induces the myogenic specification of resident CD45+ adult stem cells during muscle regeneration. *Cell* 113 (7):841–852.

Qu-Petersen, Z., B. Deasy, R. Jankowski, M. Ikezawa, J. Cummins, R. Pruchnic, J. Mytinger et al. 2002. Identification of a novel population of muscle stem cells in mice: Potential for muscle regeneration. *J Cell Biol* 157 (5):851–864.

Re'em, T., O. Tsur-Gang, and S. Cohen. 2010. The effect of immobilized RGD peptide in macroporous alginate scaffolds on TGFbeta1-induced chondrogenesis of human mesenchymal stem cells. *Biomaterials* 31 (26):6746–6755.

Richardson, T. P., M. C. Peters, A. B. Ennett, and D. J. Mooney. 2001. Polymeric system for dual growth factor delivery. *Nat Biotechnol* 19 (11):1029–1034.

Ries, C., V. Egea, M. Karow, H. Kolb, M. Jochum, and P. Neth. 2007. MMP-2, MT1-MMP, and TIMP-2 are essential for the invasive capacity of human mesenchymal stem cells: Differential regulation by inflammatory cytokines. *Blood* 109 (9):4055–4063.

Sacco, A., R. Doyonnas, P. Kraft, S. Vitorovic, and H. M. Blau. 2008. Self-renewal and expansion of single transplanted muscle stem cells. *Nature* 456 (7221):502–506.

Sagnella, S., E. Anderson, N. Sanabria, R. E. Marchant, and K. Kottke-Marchant. 2005. Human endothelial cell interaction with biomimetic surfactant polymers containing peptide ligands from the heparin binding domain of fibronectin. *Tissue Eng* 11 (1–2):226–236.

Salinas, C. N. and K. S. Anseth. 2009. Decorin moieties tethered into PEG networks induce chondrogenesis of human mesenchymal stem cells. *J Biomed Mater Res A* 90 (2):456–464.

Sasaki, T., R. Fukazawa, S. Ogawa, S. Kanno, T. Nitta, M. Ochi, and K. Shimizu. 2007. Stromal cell-derived factor-1alpha improves infarcted heart function through angiogenesis in mice. *Pediatr Int* 49 (6):966–971.

Schenk, S., N. Mal, A. Finan, M. Zhang, M. Kiedrowski, Z. Popovic, P. M. McCarthy, and M. S. Penn. 2007. Monocyte chemotactic protein-3 is a myocardial mesenchymal stem cell homing factor. *Stem Cells* 25 (1):245–251.

Seeherman, H. and J. M. Wozney. 2005. Delivery of bone morphogenetic proteins for orthopedic tissue regeneration. *Cytokine Growth Factor Rev* 16 (3):329–345.

Shin'oka, T., G. Matsumura, N. Hibino, Y. Naito, M. Watanabe, T. Konuma, T. Sakamoto, M. Nagatsu, and H. Kurosawa. 2005. Midterm clinical result of tissue-engineered vascular autografts seeded with autologous bone marrow cells. *J Thorac Cardiovasc Surg* 129 (6):1330–1338.

Sun, D., C. O. Martinez, O. Ochoa, L. Ruiz-Willhite, J. R. Bonilla, V. E. Centonze, L. L. Waite, J. E. Michalek, L. M. McManus, and P. K. Shireman. 2009. Bone marrow-derived cell regulation of skeletal muscle regeneration. *FASEB J* 23 (2):382–395.

Suzuki, Y., M. Tanihara, K. Suzuki, A. Saitou, W. Sufan, and Y. Nishimura. 2000. Alginate hydrogel linked with synthetic oligopeptide derived from BMP-2 allows ectopic osteoinduction in vivo. *J Biomed Mater Res* 50 (3):405–409.

Urita, Y., H. Komuro, G. Chen, M. Shinya, S. Kaneko, M. Kaneko, and T. Ushida. 2007. Regeneration of the esophagus using gastric acellular matrix: An experimental study in a rat model. *Pediatr Surg Int* 23 (1):21–26.

Woodruff, M. A., S. N. Rath, E. Susanto, L. M. Haupt, D. W. Hutmacher, V. Nurcombe, and S. M. Cool. 2007. Sustained release and osteogenic potential of heparan sulfate-doped fibrin glue scaffolds within a rat cranial model. *J Mol Histol* 38 (5):425–433.

Yokota, T., H. Ichikawa, G. Matsumiya, T. Kuratani, T. Sakaguchi, S. Iwai, Y. Shirakawa et al. 2008. In situ tissue regeneration using a novel tissue-engineered, small-caliber vascular graft without cell seeding. *J Thorac Cardiovasc Surg* 136 (4):900–907.

Younger, E. M. and M. W. Chapman. 1989. Morbidity at bone graft donor sites. *J Orthop Trauma* 3 (3):192–195.

Zaruba, M. M., H. D. Theiss, M. Vallaster, U. Mehl, S. Brunner, R. David, R. Fischer et al. 2009. Synergy between CD26/DPP-IV inhibition and G-CSF improves cardiac function after acute myocardial infarction. *Cell Stem Cell* 4 (4):313–323.

Zhang, Y., X. F. Bai, and C. X. Huang. 2003. Hepatic stem cells: Existence and origin. *World J Gastroenterol* 9 (2):201–204.

Zhang, G., Y. Nakamura, X. Wang, Q. Hu, L. J. Suggs, and J. Zhang. 2007. Controlled release of stromal cell-derived factor-1 alpha in situ increases c-kit+ cell homing to the infarcted heart. *Tissue Eng* 13 (8):2063–2071.

Zhang, L., F. Rakotondradany, A. J. Myles, H. Fenniri, and T. J. Webster. 2009. Arginine-glycine-aspartic acid modified rosette nanotube-hydrogel composites for bone tissue engineering. *Biomaterials* 30 (7):1309–1320.

5

Fabrication of 3D Scaffolds and Organ Printing for Tissue Regeneration

Ferdous Khan and Sheikh R. Ahmad

CONTENTS

5.1 Introduction

The generation of artificial tissues and the creation of complex 3D structures of biological organs had, until recently, been a fanciful myth and a dream. In recent years, an extensive research in this field has made this dream come true. Tissue engineering technology has now advanced to a stage where generation of new tissue to replace the damaged ones or, for that matter, growing artificial organs for use in organ transplantation is now undergoing extensive clinical trials in many countries (Cima et al. 1991; Yang et al. 2001; Griffith and Naughton 2002; Vacanti 2003; Weinand et al. 2006; Marcacci et al. 2007; Iwasa et al. 2009).

Examples of a list of scaffolds used in preclinical animal and human clinical trials are presented in Table 5.1.

Biological tissues are extremely complex structures having concomitant complicated mechanical functions and mass transport characteristics.

TABLE 5.1

Examples of a List of Scaffolds Used in Preclinical Animal and Human
Clinical Trials

| | | Application | | |
| | Fabrication | Preclinical | Human | |
Materials Types	Technique	Animal Trial	Clinical Trial	References
Hydroxyapatite	Sintering	Ectopic bone formation in rats	—	Kuboki et al. (2001), Kuboki et al. (2002), Jin et al. (2000)
Hydroxyapatite/ tricalcium phosphate	Sintering	Femoral defect/bone in dogs	—	Yuan et al. (2001)
Glass-ceramics	Phase transformation	Femoral defects/bone in rabbits	—	El-Ghannam (2004)
Titanium fiber	Sintering	Cranial defects/bone in rats	—	Sikavitsas et al. (2003)
Collagen	Freeze drying	Tibia defects/ bone in rats	—	Rocha et al. (2002)
Collagen/ hyaluronate	Cross-linking	Cranial defects/bone in rats	—	Liu et al. (1999)
Poly(lactide-*co*-glycolide)/ poly(vinyl alcohol)	Leaching	Cranial defects/bone in rabbits	—	Oh et al. (2003)
Poly(propylene glycol-*co*-fumaric acid)	Gas foaming with effervescent reaction (in vivo)	Cortical defects bone in rats	—	Trantolo et al. (2003)
Poly(caprolactone)/ poly(L-lactide)	Solvent evaporation	Trabecular bone in mice	—	Khan et al. (2010)
Poly(L-lactide), poly(L-lactide-*co*-glycolide)	Leaching	Cartilage in mice	—	Tanaka et al. (2010)
Poly(L-lactide-*co*-glycolide)/ collagen	Freeze drying	Cartilage in mice	—	Dai et al. (2010)
Poly(glycolic acid), poly(caprolactone), poly(hydroxyl butyrate)	Solvent casting/ leaching	Neocartilage in mice	—	Shieh et al. (2004)
Gelatin	Leaching	Artificial skin in mice	—	Lee et al. (2005)

TABLE 5.1 (continued)

Examples of a List of Scaffolds Used in Preclinical Animal and Human
Clinical Trials

Materials Types	Fabrication Technique	Application Preclinical Animal Trial	Human Clinical Trial	References
Poly(ethylene glycol), cystamine and poly(caprolactone)	Cross-linking	Connective tissue in rats	—	Hamid et al. (2010)
Poly(ethylene glycol)-poly(butyl terephthalate) copolymer	Robotic deposition	Cartilage in mice	—	Woodfield et al. (2004)
Silk	Freeze drying	Ligament in pig	—	Fan et al. (2009)
Hyaluronan based	—	—	Chondrocyte transplantation for cartilage tissue	Kon et al. (2011)
Collagen, hyaluronan, alginate, poly(lactic acid)	—	—	Cartilage tissue	Iwasa et al. (2009)
Hyaluronan based	—	—	Chondrocyte implantation, transplantation	Marcacci et al. (2007), Marcacci et al. (2002)

Tissue engineering seeks to recapitulate such structures and functions using
biomaterial scaffolds, delivering therapeutic biologics such as cells, proteins,
and genes for tissue reconstruction. In tissue engineering, 3D scaffold structure plays a critical role in tissue regeneration, both in vitro and in vivo,
to improve the clinical potential of regenerative medicine. The surface and
structural properties of the scaffolds directly influence the cell viability,
migration, proliferation, and differentiation.

The majority of tissue engineering techniques currently under investigation utilize a 3D scaffold seeded with cells. Scaffolds are often designed for
specific applications and fabricated from a variety of materials such as biopolymers, synthetic polymers, ceramics, metals, etc. Although the materials
of the scaffolds can be from different origins, all scaffolds need to have the
following essential characteristics:

1. High porosity with appropriate pore size and pore interconnectivity.

2. High surface area.

3. Good mechanical properties to maintain the predesigned tissue structure.

4. Biocompatibility.

5. High adhesion capability, growth, and migration rates for cells within the scaffolds.

6. High qualities of differentiation function allowing the scaffolds perform efficient nutrient transportation. It is noted that, in the normal in vivo condition, vasculature provides most of the nutrients essential for cells to function.

Techniques used for the fabrication of 3D scaffolds are classified into two categories. The first one is the conventional scaffold fabrication technique which includes fiber meshes and bonding, gas foaming, phase separation, freeze drying, and particulate leaching, among others. Fibrous nonwoven, woven, or knitted scaffolds can be fabricated from polymeric fibers manufactured with standard textile technologies (Moroni et al. 2008). These scaffolds, however, lack structural stability, and consequently, these may undergo high deformations due to cell contractility and motility (Smilenov et al. 1999). The second category is produced by rapid prototyping techniques, which include the 3D printing, selective laser sintering (SLS), and laser ablation (LA). The 3D printing is the first rapid prototyping devices developed for tissue engineering applications and regenerative medicine. Rapid prototyping techniques can process a wide number of biomaterials in a custom-made shape and with desired mechanical properties for specific applications. Products of this technology have finely controllable porosity, pore size, and shape and have a completely interconnected pore network, allowing better cell migration and nutrient perfusion than those provided by the 3D scaffolds fabricated with conventional techniques (Malda et al. 2004). This chapter focuses on the recent advancement in 3D printing techniques for scaffold fabrication in bio-related and medicinal research, particularly in the study of cells and tissue engineering applications. Additionally, this chapter also discusses the organ printing, an alternative to the 3D scaffold-based tissue regeneration.

5.2 3D Printing Techniques for Scaffold Fabrication

Recently, 3D printing techniques have attracted the attention of both academic and industrial communities (Hollister 2005; Théry et al. 2005). Various methods have been developed for the use of polymeric biomaterials and hydrogels as integral components in micro-devices. Recent advances in patterning of polymer using micro-contact printing (μCP), inkjet printing, robotic deposition, dip-pen lithography, and nanoimprinting and their application in biomedical sciences are discussed in the following.

5.2.1 Micro-Contact Printing

μCP is a process that transfers molecules from a patterned stamp to a substrate through the formation of covalent bonds. In μCP, a rigid or elastic stamp with a relief is allowed to transfer an "inked" material on to a substrate (Figure 5.1). The conformal contact between the substrate and the raised regions of the stamp provides high fidelity when transferring an

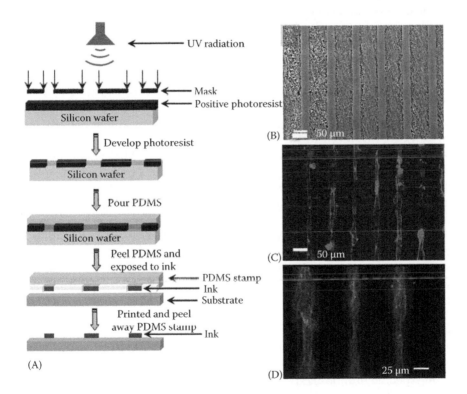

FIGURE 5.1
(A) Schematic representation of μCP. (B) Optical micrograph of OEGMA/MA patterned on chitosan film. (C) Cytoskeletal alignment in microvascular endothelial cells cultured on 20 μm lines (magnification × 200). (D) Alignment of cytoskeleton and nuclei in NIH 3T3 fibroblasts cultured on 30 μm wide lines of PLGA substrates after 24 h. Actin microfilaments were visualized by Alexa 488–labeled phalloidin. Cell nuclei were visualized by DAPI. (B and C: Reprinted with permission from Kumar, G. et al., *Langmuir*, 19, 10550, 2003; D: Lin, C.C. et al., *Biomaterials*, 26, 3655, 2005.)

"ink" to the surface. High-quality μCP patterns can be generated under optimized conditions, avoiding contamination and deformation of stamps and the lateral diffusion of the ink (Quist et al. 2005). The molecules transfer a pattern defined by the topography of the stamp, which means the regions of atomic-level contact between the stamp and substrate exist. The flexibility of the stamp and the ability to achieve conformal, atomic-level contact between the stamp and the substrate are both advantageous for printing over large areas (>50 cm^2) and on curved surfaces.

In recent time, μCP has drawn much interest from the surface science community, engineers, biologists, and tissue engineers for 3D surface/scaffold fabrication. This is primarily due to the possibility of achieving lower spatial resolution down to the nanometer range. At such low resolutions, μCP process offers much improvement, particularly for the introduction of new technical innovations and design to suit new applications (Gates et al. 2005). The technique, primarily developed for PDMS stamps, has recently demonstrated the feasibility of effective and efficient surface modifications of the stamps and other variety of polymeric materials to achieve 3D structure. This allowed implementation of precise control for cell adhesion and the immobilization of biological molecules on various substrates. Studies on μCP method for patterning proteins and cells on chitosan substrates of copolymers of oligoethyleneglycol methacrylate and methacrylic acid (poly(OEGMA-*co*-MA)) have shown to resist protein and cell adsorption (Kumar et al. 2003). In this method, poly(OEGMA-*co*-MA) was transferred from the plateaus of a PDMS stamp onto chitosan. Poly(OEGMA-*co*-MA)-printed areas were found to resist protein adsorption and cell attachment, while the unprinted bare chitosan regions allowed proteins and cells to adsorb naturally. Other studies have also expanded the μCP techniques to pattern proteins or cells by physical or chemical attachment of cell-adhesive molecules such as extracellular matrix proteins and peptides or protein-resistant polymers (Mrksich et al. 1997; Scholl et al. 2000; Liu et al. 2002; Hyun et al. 2003; Khademhosseini et al. 2003). The 3D cell patterning techniques have been reported to be used on a variety of substrates, including silicon-based substrates (Lin et al. 2005), polystyrene (PS) (Liu et al. 2002), and biodegradable polymeric materials (Kumar et al. 2003; Lin et al. 2005) which are suitable for tissue engineering applications. PEG-methacrylate-based polymers, used for coating surface to create protein-resistant areas for site-specific immobilization of NIH 3T3 fibroblasts cells (Lin et al. 2005), are also potential substrates. Examples of cell patterning are shown in Figure 5.1C and D, illustrating the feasibility of employing poly-OEGMA and poly(OEGMA-*co*-MA) for spatially controlled protein adsorption and cell attachments on PLGA and PLA substrates. Aligned patterns were stable for up to 2 weeks and were able to confine endothelial and 3T3 fibroblast cells effectively in serum-containing media. The anisotropic cell-adhesive micro-patterns controlled the spatial distribution of the extracellular matrix secreted by individual cells for guiding and orientation of cell polarity (Théry et al. 2005).

5.2.2 Inkjet Fabrication

Inkjet printing is a versatile tool for various industrial manufacturing processes to deposit minute quantities of materials. Inkjet printing of a material can be performed using the solution either in "continuous" mode or in "drop-by-drop" mode. The continuous mode of inkjet printing is mainly used for high-speed graphical applications while the drop-by-drop mode is used for the applications where high placement accuracy is needed.

Inkjet printing is one of the key technologies in the field of "defined polymer deposition" for a variety of applications, including the manufacturing of multicolor polymer light-emitting diode displays, polymer electronics, and 3D structure patterning for a broad range of biotechnological applications (Gans et al. 2004; Kim et al. 2010). In this section, the focus is placed on the use of inkjet printing for biomedical applications, in particular, to the 3D tissue engineering scaffolds and cell-based drug screening.

The 3D pattern of polymeric biomaterials can be fabricated via inkjet printing of preformed polymer solution onto a substrate. In this method, a jet of a polymer solution breaks up into droplets and gets deposited on to the substrate to form a 3D pattern following evaporation of the solvent (Figure 5.2A). Polymer solutions used for efficient patterning should possess appropriate (Newtonian) rheological properties as well as high surface tension and solvent volatility (Christanti and Walker 2001). For polymer solutions with non-Newtonian properties (e.g., viscosity), complications may arise in the breakup of jets. Nonlinear effects, such as strain hardening, may occur due to high extensional stresses with the consequent suppression of the breakup of jets. It was noted that the viscosity of polymer solutions will depend on the nature of polymer architectures, molecular weights and the concentration of polymer.

Inkjet printing technologies can be used to fabricate persistent biomimetic 3D structures that can be used to study the underlying biology of tissue regeneration. Such structures may, ultimately, find applications in clinical therapies. However, recapitulating nature, even at the most primitive levels such that printed cells, extracellular matrices and hormones become integrated into hierarchical and spatially organized 3D tissue structures with appropriate functionality, which remains a significant challenge to be addressed. This remains a significant challenge to be addressed. Inkjet printing has also been used for positioning cells onto target materials (Roth et al. 2004; Yamazoe and Tanabe 2009) and for the fabrication of organized 3D structures (Xu et al. 2006a,b; Yamazoe and Tanabe 2009) (Figure 5.2B through F). Xu et al. (2006a,b) developed 3D cellular structures as sheets of neural cells using layer-by-layer process by alternate inkjet printing of NT2 cells

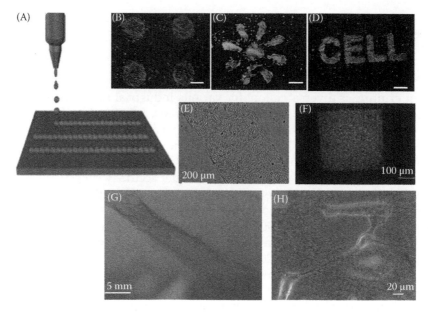

FIGURE 5.2

(A) Schematic representation of inkjet printing drop-by-drop mode. (B–D) Fluorescence micro-scope images of hASCs on PLGA-patterned PS substrates after 5 days of culture, (B) dot pattern, (C) flower pattern, and (D) "CELL" letter pattern. (E) The morphology of a printed collagen line pattern seeded with smooth muscle cells (SMCs) after culture times of 4 days. (F) Cell viability of patterned cells; PEI was printed onto a cross-linked albumin substrate with a 3D structure and L929 cells were seeded onto the substrate and fluorescence image shows the cell viability stained with calcein-AM. (G, H) 3D neural sheets were fabricated by alternately printing fibrin gels and NT2 cells. (G) Photography of a printed neural 3D sheet in the culture medium at day 1. The gross dimension of the 3D neural sheet was 25 mm × 5 mm × 1 mm. (H) Microscopy images of NT2 cells within the printed neural 3D sheet (phase contrast). NT2 cells within the fibrin gel developed their processes out after 12 days of culture. (B through D: Reprinted with permission from Kim, J.D. et al., *Polymer*, 51, 2147, 2010; E: Roth, E.A. et al., *Biomaterials*, 25, 3707, 2004; F: Yamazoe, H. and Tanabe, T., *J. Biomed. Mater. Res.*, 91A, 1202, 2009; G and H: Xu, T. et al., *Biomaterials*, 27, 3580, 2006b.)

and fibrin gels (Figure 5.2G and H). This technique is rapidly evolving into a tissue fabrication method to build functional neural tissue regeneration.

5.2.3 Robotic Deposition

Robotic deposition technique offers new opportunities for 3D complex structure of materials at finer length scales, which rely on the quality of the filament of the ink material which is continuously extruded from a nozzle and deposited onto a substrate to yield complex structures in a layer-by-layer build sequence (Figure 5.3A). This technique had been used for fabricating 3D scaffolds with a variety of materials, including reactive prepolymers, conductive polymers and polyelectrolytes, either in melts or in solution

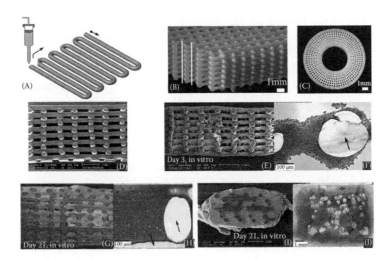

FIGURE 5.3
(A) One layer scaffold produced by 3D plotting. Optical images illustrating (B) a 3D periodic structure with a simple tetragonal symmetry reveal the high-integrity interfaces formed between layers and (C) a 3D radial structure comprised of alternating layers deposited using radial and concentric fill patterns. (D) SEM section of 3D-deposited scaffolds with homogeneous 1 mm fiber spacing showing typical fiber diameter and pore geometries × 20. SEM (E, G, I) and safranin-O stained (F, H, J) of 3D-deposited 300/55/45 scaffolds following (E, F) 3 days dynamic seeding of bovine articular chondrocytes; (G, H) 21 days dynamic culture in vitro; (I, J) 21 days subcutaneous implantation in nude mice; (arrows indicate PEGT/PBT fiber, * indicates fibrous capsule). (A: Reprinted with permission from Landers, R. et al., *Biomaterials*, 23, 4437, 2002; B and C: Smay, J.E. et al., *Langmuir*, 18, 5429, 2002; D through J Woodfield, T.B.F. et al., *Biomaterials*, 25, 4149, 2004.)

systems. The process requires the optimization of viscosity and viscoelasticity of the ink and the hardening of the ink after extrusion from the nozzle. High resolution of complex 3D scaffold structures is achievable with feature sizes from a few hundred microns to submicron scale. By controlling ink rheology, a complex 3D scaffold consisting of a continuous solid, high porosity, and high pore interconnectivity had been constructed (Smay et al. 2002; Vozzi et al. 2002, 2003). Smay et al. (2002) have generated 3D periodic lattices and radial arrays (Figure 5.3C) using robotic deposition of gels.

Gels are excellent candidate inks for building a complex 3D structure due to their suitable viscoelastic properties allowing drastic improvement in the efficiency of flow through nozzles and produce patterned filaments that maintain their shape. Ang et al. (2002) have fabricated 3D chitosan-hydroxyapatite gel-based scaffolds by robotic dispensing system and demonstrated both the patterned materials and pore channels are interconnected in all three dimensions. The interconnected pores allowed uniform cell distribution, proliferation, and migration throughout the 3D network within the scaffold.

Many other research groups have produced complex 3D scaffolds using a variety of biocompatible polymers, including poly(L-lactic acid), poly(caprolactone) (Vozzi et al. 2002), poly(D-L-lactide-*co*-glycolide) (Vozzi et al. 2003), poly(ethylene glycol terephthalate-*b*-butylene terephthalate) (Woodfield et al. 2004), agarose, gelatin (Landers et al. 2002), and polyelectrolytes (Gratson et al. 2004), for tissue engineering applications via robotic deposition and achieved precisely controlled geometries. The 3D polymer scaffold generated by robotic deposition, having well-defined geometry, porosity, good mechanical properties, and appropriate biological cues, was used for the organization of cells over the length scales that were required for tissue function (Hutmacher 2000; Therriault et al. 2003; Hollister 2005; Gratson et al. 2006; Xu et al. 2006a,b). Poly(ethylene glycol terephthalate-*b*-butylene terephthalate) scaffolds have been used for the regeneration of cartilage and bone tissues by seeding and culturing bovine articular chondrocytes (Landers et al. 2002) which had provided a homogenous distribution of cells and supported the formation of the cartilage-like tissue. Cell attachment and proliferation of expanded human articular chondrocytes on the scaffolds led to the filling of pores with high percentage of living cells (Figure 5.3). A study of cell growth on the scaffold, fabricated in PCL and PCL–hydroxyapatite composites, showed that human bone marrow–derived osteoprogenitor confirmed the developed cells along the osteogenic lineage (Endres et al. 2003). This technique offers an efficient approach for generating complex 3D architectures that are not accessible by conventional lithographic methods. The technique provides an alternative to stereolithography, as it does not require polymer photocuring and causes no damage to light-sensitive materials.

5.2.4 Dip-Pen Lithography

Dip-pen nanolithography (DPN) (Piner and Mirkin 1997; Mirkin 2000; Jang et al. 2002; Lee et al. 2002; Noy et al. 2002; Shin et al. 2010; Hernandez-Santana et al. 2011) uses chemically coated atomic force microscope (AFM) tips to deposit nanoscale chemical patterns on a substrate in a direct write process in an arbitrary configuration (Figure 5.4A). Multicomponent nanostructures can easily be prepared with near-perfect alignment using DPN technique. One can write an initial nanostructure (an array of dots) (Figure 5.4B) with octadecanethiol and then write within that initial structure another nanostructure with a different type of ink. DPN has successfully been employed to fabricate biomolecular nanoarrays to demonstrate specific adhesion of cells and virus particles (Shin et al. 2010). These studies relied on fabricating nanoarrays with proteins or inks using DPN that have affinity for the cell membranes or virus capsids.

DPN has also been applied to construct arrays of proteins with features of dimension in the range of 100–350 nm (Lee et al. 2002). This has been reported to exhibit almost nondetectable nonspecific binding affinity of proteins to their passivated portions even in complex mixtures of proteins. It can, therefore, be expected that the technique provides the opportunity to study a

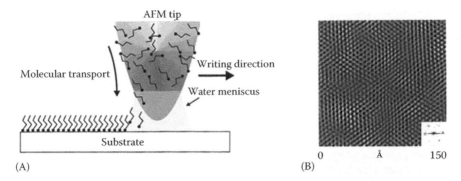

FIGURE 5.4
(A) Scheme of dip-pen lithography and (B) image of octadecanethiol nanostructure deposited via DPN on Au(111). (Reprinted with permission from Mirkin, C.A., *Inorg. Chem.*, 39, 2258, 2000.)

variety of surface-mediated biological recognition processes. Lee et al. (2002) demonstrated cell adhesion studies using a protein nanoarray composed of a cell binding protein called retronectin. Cell adhesion on retronectin and subsequent spreading of cell occur by interaction with the extracellular matrix protein on cell membranes by focal adhesion. A typical protein array was fabricated by patterning 1,6-mercaptohexadecanoic acid (MHA) into a square array of dots 200 nm in diameter and separated by 700 nm using DPN. AFM images showed that on subsequent treatment retronectin was adsorbed almost exclusively to the MHA nanopatterns. The protein nanoarray was then incubated with bovine calf serum containing fibroblast cells followed by rinsing with PBS and studied under optical microscope. The cells were found to attach only to the patterned regions of the substrate and spread, but not completely, into a more flattened morphology than that in their unbound state.

5.2.5 Nanoimprinting Lithography

Nanoimprint lithography (NIL) is a high-throughput, low-cost and non-conventional method of fabricating high-resolution nanoscale patterns. Patterns are created by mechanical deformation of imprint resist, typically a monomer or polymer formulation that is cured either by heat or UV radiation. In this technique, mold is being pressed against a softened thermoplastic polymer or a liquid polymer precursor and trapping the pattern in the solid state by either changing temperature, or by UV radiation curing the polymer precursor, as demonstrated in Figure 5.5. The 3D patterning materials can be spun-cast, dispensed as droplets or allowed to fill the space between the mold and substrate by capillary forces. After embossing, the resulting pattern can be further processed to etch away the thin regions of the pattern, to etch the exposed substrate material or to selectively deposit other materials.

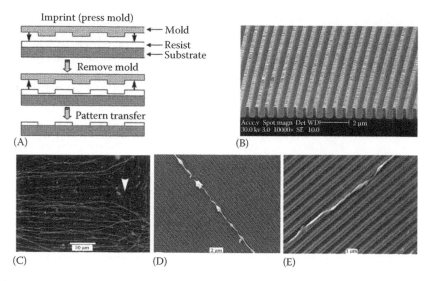

FIGURE 5.5
(A) Scheme of nanoimprinting. (B) NIL results using a 250 nm linewidth PDMS-*b*-PS grating. (C) Axons aligning along a horizontally imprinted pattern (200 nm width and 400 nm pitch). The arrow indicates the border of the pattern. The more random growth of axons outside the pattern is clearly visible. Note that the larger axons are not as well guided as the thinner ones. (D) 100 nm width and 500 nm pitch and (E) 400 nm width and 800 nm pitch. SEM images show the axons grow on the ridge edges, and not in the grooves. (B: Reprinted with permission from Guo, L.J., *Adv. Mater.*, 19, 495, 2007; C through E: Johansson, F. et al., *Biomaterials*, 27, 1251, 2006.)

In NIL, the selection of resist materials is critically important for correct pattern replication without fracture defects during mild release that are not tolerable in many applications. In majority of thermal NIL methods, homopolymers, for example, PMMA and PS, are used as resists. However, such materials are prone to fracture defects. In this respect, the polymer should satisfy the requirements of having a low surface energy for easy de-molding while maintaining sufficient adhesion to the substrate. Materials that possess dual surface properties are needed to address these critical issues. Specific interest materials are the PDMS organic block or graft copolymers. A 250 nm linewidth grating imprinted in a block copolymer of PDMS and PS (Figure 5.5) (Guo 2007) shows much superior mold releasing capabilities to those offered by PMMA and PS homopolymers. In this technique, the most widely used system is the UV-curing formulations. This is based on highly reactive free radical polymerization of acrylic and methacrylic monomers. However, such material systems suffer oxygen sensitivity problems, in which oxygen affects the polymerization reaction at the resist surface. To address this issue, a UV-curable epoxysilicone material, based on the cationic cross-linking of cycloaliphatic epoxides, has been developed (Cheng et al. 2005).

Sophisticated chemistries can be introduced for NIL patterning for biological applications such as cell patterning (Kane et al. 1999), tissue engineering (Lenhert et al. 2004; Johansson et al. 2006), and enhancing genetic response of cells on the patterned surfaces (Dalby et al. 2003). Surface topography plays an important role as almost all types of cells are reactive even to small differences in the scale of topography (Curtis 2004). For example, patterned groove surfaces having dimension of 100–400 nm widths and 300 nm depths showed guided cell growth along the top of the patterns, and the axion displayed contact guidance on all patterns (Figure 5.5C through E) (Johansson et al. 2006). Most cells follow the discontinuities of grooves and ridges and form an elongated shape due to surface-induced rearrangements of the cytoskeleton (Johansson et al. 2006), and this surface also induces transcript up and down the regulation of genes (Dalby et al. 2003). Osteoblasts extracted from bovine periosteum tissue, cultured on groove surfaces with a depth of 150 nm, were aligned with the grooves (Lenhert et al. 2004). However, the depth of the grooves has significant influence on the cell alignment due to the surface wetting properties (Lenhert et al. 2004). Future applications of this technique in biotechnological field depend mainly on the careful control of moulds geometries and process parameter yields. For highly reproducible patterns, the development of new photochemically sensitive materials, not susceptible to oxygen inhibition, not prone to shrinkage, and are easily removable from moulds with high aspect ratios is critically important. However, recent attempts using rapid prototyping technologies to design solid synthetic scaffolds (Landers et al. 2002; Sodian et al. 2002; Yang et al. 2002; Zein et al. 2002) suffer from the inability to precisely place cells or cell aggregates into a printed scaffold. Therefore, tissue engineering researchers have recently developed an alternative approach called "organ printing" as discussed in the following section.

5.3 Organ Printing

5.3.1 Principle of the Process

Organ printing, introduced in 1999 (Vladimir et al. 2003), is an alternative technology to the classical scaffold-based tissue engineering. It is based on the principle of rapid prototyping technology of layer-by-layer deposition of cells or matrices and has shown potential for engineering of new tissues or organs. The aim of the organ printing procedure is to fabricate 3D vascularized functional living human organs, suitable for clinical implantation. The concept of solid scaffold-free tissue engineering is based on the assumption that tissues and organs are self-organizing systems, and cells, especially micro-tissues, can undergo biological self-assembly and self-organization without any external influence of solid scaffolds. In this section, the applicability of developmental biology to organ printing and essential steps and various elements of this novel technology are described. The challenging technological barriers, possible strategies to overcome them, and estimation of overall feasibility of printing 3D human tissues and organs are also discussed in the following section.

5.3.2 Technology of Organ Printing

In organ printing technology, a biomedically relevant variant of rapid prototyping technology, based on tissue fluidity, is used. In this application of micro-fluidic design for cells and cell aggregates by triggering biologically relevant phenomena such as fusion, computer-assisted printing of cells or matrix is performed. The process uses one layer at a time until a particular 3D form is achieved which has the potential for surpassing traditional solid scaffold-based tissue engineering (Vladimir et al. 2003; Jakab et al. 2004; Mironov et al. 2008; Mironov et al. 2009).

Organ printing procedure is divided into three sequential steps: (1) preprocessing, (2) processing, and (3) postprocessing. Preprocessing step primarily deals with the development of a computer-aided design (CAD), that is, creation of a blue print of a specific organ. The design can be derived from digitized image reconstruction of a natural organ or tissue. Imaging data can be derived from various modalities, including noninvasive scanning of the human body (e.g., MRI or micro-CT) or a detailed 3D reconstruction of serial sections of specific organs (Sun and Lal 2002). Processing usually refers to actual computer-aided printing or layer-by-layer placement of cells or cell aggregates into a 3D environment using CAD or blueprints. Finally, the postprocessing is concerned with the perfusion of printed organs and their biomechanical conditioning to both direct and accelerate organ maturation.

It has been hypothesized that the printing of cells, one layer at a time, tends to aggregate and fuse to form a complete disc or tube of tissue, as illustrated in Figure 5.6. Organ printing includes many different printer designs and components of the deposition process such as jet-based cell printers, cell dispensers, bioplotters, and different types of 3D hydrogels and varying cell types. The development of a printer which can print cells and/or aggregates; demonstration of a procedure for the "layer-by-layer," sequential deposition and solidification of a thermo-reversible gel or matrix; and demonstration of fusion from closely placed cell aggregates into ring-like or tube-like structure within the gel are the areas of active research and development at present. Thermo-reversible gel (Mironov et al. 2009) was printed one layer at a time each on the other with individual layer thickness comparable to the diameter of the cell aggregates (Figure 5.6) (Sun and Lal 2002). In accordance with mathematical predictions (An et al. 2001), it was shown that closely placed cell aggregates and embryonic heart mesenchymal (cushion tissue) fragments could fuse into ring- and tube-like structures in 3D gel (Figure 5.6B through G).

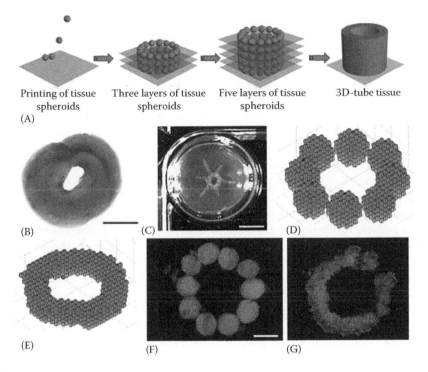

FIGURE 5.6
(A) Schematic representation of principle of organ printing technology. Bioassembly of tubular tissue construct using bioprinting of self-assembled tissue spheroids illustrating sequential steps of layer-by-layer tissue spheroid deposition in solidifying gel and tissue fusion process into 3D tube. (B) Printed bagel-like ring that consists of several layers of sequentially (layer-by-layer) deposited collagen type I gel. (C) Manually printed living tube with radial branches from the chick 27 stage HH embryonic heart cushion tissue placed in 3D collagen type I gel. Tube was formed as a result of fusion of three sequential rings, and each ring consists of 16–18 closed placed and fused embryonic cushion tissue explants. Image was taken after 24 h incubation in M199 medium with 10% of chicken serum plus ITS (insulin transferrin-selenium). (D, E) Mathematical model shows cell aggregate behavior after implantation in a 3D model gel. (D) Eight aggregates each containing 123 cells before fusion. (E) Fused disc. (F, G) Fusion of aggregates of Chinese hamster ovary (CHO) cells implanted into RGD containing thermo-reversible gel and genetically labeled with green fluorescent protein. (F) Ten aggregates (containing 5000 cells) before fusion. (G) Final disc-like configuration after fusion. (Reprinted with permission from Vladimir, M.V. et al., *Trends Biotechnol.*, 21, 157, 2003.)

5.4 Conclusions and Future Prospect

The μCP has been shown to have an important role in the fabrication of 3D surface structures, especially in combination with photolithography, dip-pen lithography or the self-assembly of copolymers. However, it has several limitations in the production of multilayer and multicomponent system and routinely generated micrometer-size features. In contrast to the lithographic methods, the inkjet printing is a solid "free-form" fabrication process and a solution-based direct-writing technique. This technique has many advantages, such as the generation of micro-patterned structures with flexibility, simplicity and arbitrary geometry. In comparison to printing methods, the µCP and dip-pen lithography have advantages in high-resolution patterning; inkjet and µCP printing are more effective in high-throughput large-area patterning. Additionally, the DPN and inkjet printing allow easy patterning of multicomponent polymer patterns with good control over feature position. The ultimate technique of choice depends on the selected polymer, the substrate and the intended application. The robotic deposition is the most efficient technique for complex 3D architecture scaffold fabrication which is accessible using other conventional techniques.

Some of the techniques are still in an early stage of development, and researchers working in these fields are trying to acquire knowledge and tools for constructing structures with best performance. Two major concerns are in the improvement of 3D patterning efficiency and cost-effectiveness, without compromising pattern performance. The challenge is to achieve a balance between high-speed and high-resolution, and a low-cost patterning process. For example, inkjet printing has great advantages in the straightforward low-cost patterning of polymers; however, enhanced patterning resolution of this method has to be traded off by the increased cost of printing. 3D patterning of polymers for cell deposition has to provide an environment that mimics natural in vivo conditions. Spatial localization of different cells and tailoring of growth factors would allow control of the behavior of cells. Currently, the fabrication of complex 3D structures with spatial distribution of chemical functionalities remains a challenge.

To achieve cost-effective 3D patterning and/or to achieve patterns in multiple length scales, a combination of different patterning techniques will be necessary. For example, the use of both photolithography and printing methods together would allow the fabrication of low-cost patterns with enhanced performance.

Organ printing or computer-aided layer-by-layer assembly of biological tissues and organs is currently technically feasible and predicted to be major technologies in future clinical tissue engineering. Organ printing uses the principle of cellular self-assembly into tissues (Woodfield et al. 2004), similar to the way embryonic-like tissues sort and fuse into functional forms as dictated by the rules laid out in developmental biology. Besides, their possible application in organ transplantation, 3D perfused, vascularized, and printed human tissues (or structural functional units of human organs) could become popular screening assays for drug discovery and testing, therefore allowing further biomedical research activities. Many technical difficulties need to be solved before noble organ construct, ready for implantation, can be achieved. Despite the technical difficulties, it is predicted that in the twenty-first century, cell and organ printers will be widely used as biomedical research tools.

References

An, Y. H., Webb, D., Gutowska, A., Mironov, V. A., Friedman, R. J. 2001. Regaining chondrocyte phenotype in thermosensitive gel culture. *Anat Rec* 263 (4): 336–341.

Ang, T. H., Sultana, F. S. A., Hutmacher, D. W. et al. 2002. Fabrication of 3D chitosan–hydroxyapatite scaffolds using a robotic dispensing system. *Mater Sci Eng C* 20: 35–42.

Cheng, X., Guo, L. J., Fu, P. F. 2005. Room-temperature, low-pressure nanoimprinting based on cationic photopolymerization of novel epoxysilicone monomers. *Adv Mater* 17: 1419–1424.

Christanti, Y., Walker, L. M. 2001. Surface tension driven jet break up of strain-hardening polymer solutions. *J Non-Newtonian Fluid Mech* 100: 9–26.

Cima, L. G., Vacanti, J. P., Vacanti, C., Ingber, D., Mooney, D., Langer, R. 1991. Tissue engineering by cell transplantation using degradable polymer substrates. *J Biomech Eng* 113: 143–151.

Curtis, A. S. G. 2004. Small is beautiful but smaller is the aim: Review of a life of research. *Eur Cells Mater* 8: 27–36.

Dai, W., Kawazoe, N., Lin, X., Dong, J., Chen, G. 2010. The influence of structural design of PLGA/collagen hybrid scaffolds in cartilage tissue engineering. *Biomaterials* 31: 2141–2152.

Dalby, M. J., Riehle, M. O., Yarwood, S. J., Wilkinson, C. D. W., Curtis, A. S. G. 2003. Nucleus alignment and cell signaling in fibroblasts: Response to a micro-grooved topography. *Exp Cell Res* 284: 274–282.

de Gans, B. J., Duineveld, P. C., Schubert, U. S. 2004. Inkjet printing of polymers: State of the art and future developments. *Adv Mater* 16: 203–213.

El-Ghannam, A. R. 2004. Advanced bioceramic composite for bone tissue engineering: Design principles and structure–bioactivity relationship. *J Biomed Mater Res A* 69(3): 490–501.

Endres, M., Hutmacher, D. W., Salgado, A. J. et al. 2003. Osteogenic induction of human bone marrow-derived mesenchymal progenitor cells in novel synthetic polymer hydrogel matrices. *Tissue Eng* 9: 689–702.

Fan, H., Liu, H., Siew L. T., Goh, J. C. H. 2009. Anterior cruciate ligament regeneration using mesenchymal stem cells and silk scaffold in large animal model. *Biomaterials* 30: 4967–4977.

Gates, B. D., Xu, Q., Stewart, M., Ryan, D., Willson, C. G., Whitesides, G. M. 2005. New approaches to nanofabrication: Moulding, printing, and other techniques. *Chem Rev* 105: 1171–1196.

Gratson, G. M., Garcia-Santamaria, F., Lousse, V. et al. 2006. Direct-write assembly of three-dimensional photonic crystals: Conversion of polymer scaffolds to silicon hollow-woodpile structures. *Adv Mater* 18: 461–465.

Gratson, G. M., Xu, M. J., Lewis, J. A. 2004. Microperiodic structures: Direct writing of three-dimensional webs. *Nature* 428: 386–386.

Griffith, L. G., Naughton, G. 2002. Tissue engineering—Current challenges and expanding opportunities. *Science* 295: 1009–1010.

Guo, L. J. 2007. Nanoimprint lithography: Methods and material requirements. *Adv Mater* 19: 495–513.

Hamid, Z. A. A., Blencowe, A., Ozcelik, B. et al. 2010. Epoxy-amine synthesised hydrogel scaffolds for soft-tissue engineering. *Biomaterials* 31: 6454–6467.

Hernandez-Santana, A., Irvine, E., Faulds K., Graham, D. 2011. Rapid prototyping of poly(dimethoxysiloxane) dot arrays by dip-pen nanolithography. *Chem Sci* 2: 211–215.

Hollister, S. J. 2005. Porous scaffold design for tissue engineering. *Nat Mater* 4: 518–524.

Hutmacher, D. W. 2000. Scaffolds in tissue engineering bone and cartilage. *Biomaterials* 21: 2529–2543.

Hyun, J. H., Ma, H. W., Zhang, Z. P., Beebe, T. P., Chilkoti, A. 2003. Universal route to cell micropatterning using an amphiphilic comb polymer. *Adv Mater* 15: 576–579.

Iwasa, J., Engebretsen, L., Shima, Y., Ochi, M. 2009. Clinical application of scaffolds for cartilage tissue engineering. *Knee Surg Sports Traumatol Arthrosc* 17: 561–577.

Jakab, K., Neagu, A., Mironov, V., Forgacs, G. 2004. Organ printing: Fiction or science. *Biorheology* 41: 371–375.

Jang, J. Y., Schatz, G. C., Ratner, M. A. 2002. Liquid meniscus condensation in dip-pen nanolithography. *J Chem Phys* 116(9): 3875–3886.

Jin, Q., M., Takita, H., Kohgo, T., Atsumi, K., Itoh, H., Kuboki, Y. 2000. Effects of geometry of hydroxyapatite as a cell substratum in BMP-induced ectopic bone formation. *J Biomed Mater Res* 51(3): 491–499.

Johansson, F., Carlberg, P., Danielsen, N., Montelius, L., Kanje, M. 2006. Axonal outgrowth on nano-imprinted patterns. *Biomaterials* 27: 1251–1258.

Kane, R. S, Takayama, S., Ostuni, E., Ingber, D. E., Whitesides, G. M. 1999. Patterning proteins and cells using soft lithography. *Biomaterials* 20: 2363–2376.

Khademhosseini, A., Jon, S., Suh, K. Y. et al. 2003. Direct patterning of protein- and cell-resistant polymeric monolayers and microstructures. *Adv Mater* 15: 1995–2000.

Khan, F., Tare, R. S., Kanczler, J. M., Oreffo, R. O. C., Bradley, M. 2010. Strategies for cell manipulation and skeletal tissue engineering using high-throughput polymer blend formulation and microarray techniques. *Biomaterials* 31(8): 2216–2228.

Kim, J. D., Choi, J. S., Kim, B. S., Choi, Y. C., Cho, Y. W. 2010. Piezoelectric inkjet printing of polymers: Stem cell patterning on polymer substrates. *Polymer* 51: 2147–2154.

Kon, E., Di Martino, A., Filardo, G. et al. 2011. Second-generation autologous chondrocyte transplantation: MRI findings and clinical correlations at a minimum 5-year follow-up. *Eur J Radiol* 79: 382–388.

Kuboki, Y., Jin, Q., Kikuchi, M., Mamood, J., Takita, H. 2002. Geometry of artificial ECM: Sizes of pores controlling phenotype expression in BMP-induced osteogenesis and chondrogenesis. *Connect Tissue Res* 43(2–3): 529–534.

Kuboki, Y., Jin, Q., Takita, H. 2001. Geometry of carriers controlling phenotypic expression in BMP-induced osteogenesis and chondrogenesis. *J Bone Joint Surg Am* 83A (Suppl. 1(Pt 2)): S105–S115.

Kumar, G., Wang, Y. C., Co, C., Ho, C. C. 2003. Spatially controlled cell engineering on biomaterials using polyelectrolytes. *Langmuir* 19: 10550–10556.

Landers, R., Hubner, U., Schmelzeisen, R., Mulhaupt, R. 2002. Rapid prototyping of scaffolds derived from thermoreversible hydrogels and tailored for applications in tissue engineering. *Biomaterials* 23: 4437–4447.

Lee, S. B., Kim, Y. H., Chong, M. S., Hong, S. H., Lee, Y. M. 2005. Study of gelatin-containing artificial skin V: Fabrication of gelatin scaffolds using salt-leaching method. *Biomaterials* 26: 1961–1968.

Lee, K. B., Park, S. J., Mirkin, C. A., Smith, J. C., Mrksich, M. 2002. Protein nanoarrays generated by dip-pen nanolithography. *Science* 295: 1702–1705.

Lenhert, S., Zhang, L., Mueller, J. et al. 2004. Self-organized complex patterning: Langmuir-Blodgett lithography. *Adv Mater* 16: 619–624.

Lin, C. C., Co, C. C., Ho, C. C. 2005. Micropatterning proteins and cells on poly(lactic acid) and poly(lactide-*co*-glycolide). *Biomaterials* 26: 3655–3662.

Liu, V. A., Jastromb, W. E., Bhatia, S. N. 2002. Engineering protein and cell adhesivity using PEO-terminated triblock polymers. *J Biomed Mater Res* 60: 126–134.

Liu, L. S., Thompson, A. Y., Heidaran, M. A., Poser, J. W., Spiro, R. C. 1999. An osteo-conductive collagen/hyaluronate matrix for bone regeneration. *Biomaterials* 20(12): 1097–1108.

Malda, J., Woodfield, T. B., van der Vloodt, F. et al. 2004. The effect of PEGT/PBT scaffold architecture on oxygen gradients in tissue engineered cartilaginous constructs. *Biomaterials* 25: 5773–5780.

Marcacci, M., Elizaveta, K., Kon, E. et al. 2007. Arthroscopic second generation autologous chondrocyte implantation. *Knee Surg Sports Traumatol Arthrosc* 15: 610–619.

Marcacci, M., Zaffagnini, S., Kon, E., Visani, A., Iacono, F., Loreti, I. 2002. Arthroscopic autologous chondrocyte transplantation: Technical note. *Knee Surg Sports Traumatol Arthrosc* 10: 154–159.

Mirkin, C. A. 2000. Programming the assembly of two- and three-dimensional architectures with DNA and nanoscale inorganic building blocks. *Inorg Chem* 39: 2258–2272.

Mironov, V., Kasyanov, V., Drake, C., Markwald, R. R. 2008. Organ printing: Promises and challenges. *Regen Med* 3: 93–103.

Mironov, V., Visconti, R. P., Kasyanov, V. et al. 2009. Organ printing: Tissue spheroids as building blocks. *Biomaterials* 30: 2164–2174.

Moroni, L., DeWijn, J. R., Van Blitterswijk, C. A. 2008. Integrating novel technologies to fabricate smart scaffolds. *J Biomater Sci Polymer Edn* 19(5): 543–572.

Mrksich, M., Dike, L. E., Tien, J., Ingber, D. E., Whitesides, G. M. 1997. Using microcontact printing to pattern the attachment of mammalian cells to self-assembled monolayers of alkanethiolates on transparent films of gold and silver. *Exp Cell Res* 235: 305–313.

Noy, A., Miller, A. E., Klare, J. E., Weeks, B. L., Woods, B. W., DeYoreo, J. J. 2002. Fabrication of luminescent nanostructures and polymer nanowires using dip-pen nanolithography. *Nano Lett* 2: 109–112.

Oh, S. H., Kang, S. G., Kim, E. S., Cho, S. H., Lee, J. H. 2003. Fabrication and characterization of hydrophilic poly(lactic-*co*-glycolic acid)/poly(vinyl alcohol) blend cell scaffolds by melt-molding particulate-leaching method. *Biomaterials* 24(22): 4011–4021.

Piner, R. D., Mirkin C. A. 1997. Effect of water on lateral force microscopy in air. *Langmuir* 13: 6864–6868.

Quist, A. P., Pavlovic, E., Oscarsson, S. 2005. Recent advances in microcontact printing. *Anal Bioanal Chem* 381: 591–600.

Rocha, L. B., Goissis, G., Rossi, M. A. 2002. Biocompatibility of anionic collagen matrix as scaffold for bone healing. *Biomaterials* 23(2): 449–456.

Roth, E. A., Xu, T., Das, M., Gregory, C., Hickman, J. J., Boland T. 2004. Inkjet printing for high-throughput cell patterning. *Biomaterials* 25: 3707–3715.

Scholl, M., Sprossler, C., Denyer, M. et al. 2000. Ordered networks of rat hippocampal neurons attached to silicon oxide surfaces. *J Neurosci Methods* 104: 65–75.

Shieh, S. J., Terada, S., Vacanti, J. P. 2004. Tissue engineering auricular reconstruction: In vitro and in vivo studies. *Biomaterials* 25: 1545–1557.

Shin, Y.-H., Yun, S.-H., Pyo, S.-H. et al. 2010. Polymer-coated tips for patterning of viruses by dip-pen nanolithography. *Angew Chem Int Ed* 49: 9689–9692.

Sikavitsas, V. I., van den Dolder, J., Bancroft, G. N., Jansen, J. A., Mikos, A. G. 2003. Influence of the in vitro culture periodon the in vivo performance of cell/titanium bone tissue-engineered constructs using a rat cranial critical size defect model. *J Biomed Mater Res A* 67(3): 944–951.

Smay, J. E., Cesarano, J., Lewis, J. A. 2002. Colloidal inks for directed assembly of 3-D periodic structures. *Langmuir* 18: 5429–5437.

Smilenov, L. B., Mikhailov, A., Pelham, R. J., Marcantonio, E. E., Gundersen, G. G. 1999. Focal adhesion motility revealed in stationary fibroblasts. *Science* 286(5442): 1172–1174.

Sodian, R., Loebe, M., Hein, A. et al. 2002. Application of stereolithography for scaffold fabrication for tissue engineered heart valves. *ASAIO J* 48: 12–16.

Sun, W., Lal, P. 2002. Recent development on computer aided tissue engineering—A review. *Comput Methods Programs Biomed* 67: 85–103.

Tanaka, Y., Yamaoka, H., Nishizawa, S. et al. 2010. The optimization of porous polymeric scaffolds for chondrocyte/atelocollagen based tissue-engineered cartilage. *Biomaterials* 31: 4506–4516.

Therriault, D., White, S. R., Lewis, J. A. 2003. Chaotic mixing in three-dimensional microvascular networks fabricated by direct-write assembly. *Nat Mater* 2: 265–271.

Théry, M., Racine, V., Pépin, A. et al. 2005. The extracellular matrix guides the orientation of the cell division axis. *Nat Cell Biol* 7: 947–953.

Trantolo, D. J., Sonis, S. T., Thompson, B. M., Wise, D. L., Lewandrowski, K. U., Hile, D. D. 2003. Evaluation of a porous, biodegradable biopolymer scaffold for mandibular reconstruction. *Int J Oral Maxillofac Implants* 18(2): 182–188.

Vacanti, J. P. 2003. Tissue and organ engineering: Can we build intestine and vital organs? *J Gastrointest Surg* 7: 831–835.

Vladimir, M. V., Thomas, B. T., Thomas, T. T., Gabor, F. G., Markwald, R. R. 2003. Organ printing: Computer-aided jet-based 3D tissue engineering. *Trends Biotechnol* 21: 157–161.

Vozzi, G., Flaim, C., Ahluwalia, A., Bhatia, S. 2003. Fabrication of PLGA scaffolds using soft lithography and microsyringe deposition. *Biomaterials* 24: 2533–2540.

Vozzi, G., Previti, A., De Rossi, D., Ahluwalia, A. 2002. Microsyringe-based deposition of two-dimensional and three-dimensional polymer scaffolds with a well-defined geometry for application to tissue engineering. *Tissue Eng* 8: 1089–1098.

Weinand, C., Pomerantseva, I., Neville, C. M. et al. 2006. Hydrogel-β-TCP scaffolds and stem cells for tissue engineering bone. *Bone* 38: 555–563.

Woodfield, T. B. F., Malda, J., de Wijn, J., Peters, F., Riesle, J., van Blitterswijk, C. A. 2004. Design of porous scaffolds for cartilage tissue engineering using a three-dimensional fiber-deposition technique. *Biomaterials* 25: 4149–4161.

Xu, M. J., Gratson, G. M., Duoss, E. B., Shepherd, R. F., Lewis, J. A. 2006a. Biomimetic silicification of 3D polyamine-rich scaffolds assembled by direct ink writing. *Soft Matter* 2: 205–209.

Xu, T., Gregory, C. A., Molnar, P. et al. 2006b. Viability and electrophysiology of neural cell structures generated by the inkjet printing method. *Biomaterials* 27: 3580–3588.

Yamazoe, H., Tanabe, T. 2009. Cell micropatterning on an albumin-based substrate using an inkjet printing technique. *J Biomed Mater Res* 91A: 1202–1209.

Yang, S., Leong, K.-F., Du, Z., Chua, C.-K. 2001. The design of scaffolds for use in tissue engineering. Part I. Traditional factors. *Tissue Eng* 7: 679–689.

Yang, S., Leong, K.-F., Du, Z., Chua, C.-K. 2002. The design of scaffolds for use in tissue engineering Part II. Rapid prototyping techniques. *Tissue Eng* 8: 1–11.

Yuan, H., Yang, Z., de Bruijn, J. D., de Groot, K., Zhang, X. 2001. Material-dependent bone induction by calcium phosphate ceramics: A 2.5-year study. *Biomaterials* 22: 2617–2623.

Zein, I., Hutmacher, D. W., Tan, K. C., Teoh, S. H. 2002. Fused deposition modeling of novel scaffold architectures for tissue engineering applications. *Biomaterials* 23: 1169–1185.

6

Natural Membranes as Scaffold for Biocompatible Aortic Valve Leaflets: Perspectives from Pericardium

Maria Cristina Vinci, Francesca Prandi,
Barbara Micheli, Giulio Tessitore, Anna Guarino,
Luca Dainese, Gianluca Polvani, and Maurizio Pesce

CONTENTS

6.1 Introduction

The pericardium (from the Greek περι, "around," and κάρδιον, "heart,"/perikardion/) is the sac surrounding the heart. Its primary function is to provide a natural barrier to infections of the myocardium and prevent adhesion to surrounding tissues. In addition, it mechanically protects the heart from overdilation, it maintains its correct anatomical position, and regulates the pressure/volume ratio in the left ventricle during diastole (Ishihara et al. 1980). For its structure and resistance to mechanical stress, pericardium has been for long proposed as a suitable material to construct bioprosthesis for various medical applications such as vascular grafts (Matsagas et al. 2006; Schmidt and Baier 2000), patches for abdominal and vaginal wall repara-tion (Lazarou et al. 2005; Limpert et al. 2009), brain surgery (Filippi et al. 2001), and heart valves (Flanagan and Pandit 2003; Schoen and Levy 1999; Vesely 2005).

Derivation of fully engineered aortic valves (AOVs) with optimal biologi-cal compatibility is an alternative to the use of animal-derived bioprosthetic valves, donor-derived homografts, or mechanical valves. This possibility is particularly attractive for pediatric AOV surgery. In fact, pediatric patients need to be transplanted with valve tissues able to grow in concert with the natural growth of the heart, thereby generating durable grafts that do not need further substitution during the life span.

Decellularization is the main way to ensure biocompatibility of peri-cardial tissue for surgical implantation. In fact, this procedure avoids the host versus graft immune reaction when animal or human pericardium is used. Various protocols to remove pericardium cellular components have been devised. These include aldehyde fixation, enzymatic treatments of the fresh tissue, and incubation with mild hypotonic agents able to remove donor cells.

Mechanical resistance of human pericardium is compatible with that of AOV leaflets. It has been therefore hypothesized that pericardial tissue might be a suitable and abundant source of cell-free scaffold that may be combined to valve-extrinsic or valve-resident (stem) cells to construct fully compatible engineered leaflets having the necessary resistance to mechani-cal strain occurring during the cardiac cycle.

This chapter will describe anatomy, structure, and biological properties of human pericardium with reference to applications in cardiac surgery. This will be discussed in the light of possible future applications in AOV engi-neering using stem cells.

6.2 Structure, Composition, and Mechanical Properties of Pericardium

The pericardium is normally described as a unique tissue. However, it con-sists of two sacs intimately connected together. The outer sac, known as the fibrous/parietal pericardium, consists of fibrous tissue, while the inner sac, or serous/visceral pericardium, is a thin membrane juxtaposed between the

Parietal

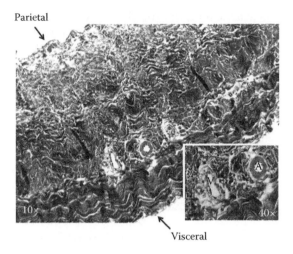

Visceral

FIGURE 6.1
Masson trichrome staining of a transverse section of human pericardium. The orientation of the tissue is shown with the parietal portion at the top of the picture and the visceral portion at the bottom. Note the arrangement of collagen bundles and (inset) venules (V) and arterioles (A).

fibrous pericardium and the heart. Between the visceral and parietal pericardium is present the pericardial space, which is filled with a small volume of fluid.

The tissue covering the left ventricle has the highest resistance to stress; for this reason, it is the most commonly used in bioprosthesis derivation (Hiester and Sacks 1998; Simionescu et al. 1993) (Figure 6.1). Mechanical properties of the fibrous pericardium are ensured by the unique multilayered composition of collagen bundles, interwoven with elastin fibers and embedded into an amorphous matrix, mainly composed of free glycosaminoglycans (GAGs) and proteoglycans (GAGs linked to protein cores; dermatan sulfate, chondroitin sulfate, hyaluronan; Simionescu et al. 1989). The most represented collagen fibril species in pericardial tissue are made of type I and III collagen (95%); these fibrils are hierarchically organized in fibers, fiber bundles, and laminates (Allen and Didio 1984). Arrangement of collagen structures does not follow a precise orientation. In fact, observation under polarized light reveals a relatively high disarrangement of bundle orientation with consequences for zonal anisotropy of resistance to mechanical stress (Zioupos and Barbenel 1994a,b; Zioupos et al. 1994).

Pericardium has been introduced in the clinical practice as a repair biomaterial since early days of cardiovascular surgery; however, only during the past 20 years, a specific focus has been made to clarify the relationships between mechanical resistance and fiber orientation and whether protocols used to increase in vivo biocompatibility of this tissue affect its mechanical resistance. The basic mechanical properties of the pericardium such as elastic deformation (elastic module), maximum withstanding stress (ultimate tensile strength), or the limit at which the tissue is not able to return to the original shape after strain removal (yield strength) are calculated by uniaxial tensile tests (Crofts and Trowbridge 1989). Other protocols that can be used to test pericardium mechanical resistance are (1) the shear test, which allows calculation of the tissue deformation in response to tangentially applied forces (Boughner et al. 2000), and (2) the compressive buckling, which is used to measure resistance to deformation under compressive forces (Vesely and Mako 1998).

6.3 Structure, Composition, and Mechanical Properties of AOV Leaflets

The structure of the AOV leaflets is shown in Figure 6.2; in a cross-section view, the leaflet is composed of three main components: (1) the fibrosa, facing the aorta, (2) the spongiosa present in the middle portion, and (3) the ventricularis, facing the ventricular cavity.

In analogy with pericardium, the main structural components of AOV leaflets are represented by collagen I, III, and IV bundles, elastin, and GAGs.

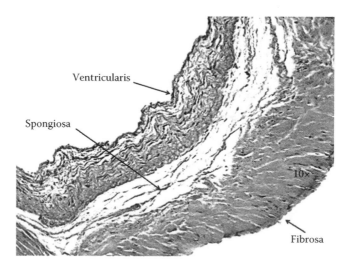

FIGURE 6.2
Hematoxylin–eosin staining of an AOV leaflet in transverse section. The orientation is shown with the ventricularis at the top of the panel and the fibrosa at the bottom. Note the presence of elastin bundles in the ventricularis, necessary for valve closure, and the transverse arrangement of collagen bundles in the fibrosa necessary to provide resistance and stiffness during diastole, thereby preventing blood backflow.

The presence of these components ensures an optimal resistance to strain/stretch cycles during the transition from systolic (valve opening) to diastolic phases (valve closing). The spatial arrangement of structural components is, however, different from that in the pericardium. In fact, elastic fibers are not interwoven with collagen bundles but are mostly abundant in the ventricularis where they are radially aligned; this arrangement ensures the necessary elasticity for the cusp contraction against the wall when the valve is open and provides the necessary tension to close the valve at the end of the systolic phase. Collagen fibers are mostly abundant in the fibrosa, where they are aligned circumferentially to provide the necessary resistance and stiffness to each cusp to maintain the curvature, thereby preventing blood backflow. Finally, the loose arrangement of collagen fibers and GAGs in the spongiosa is necessary to balance changes in cusp orientation throughout the cardiac cycle, to dissipate shear stresses caused by the differential movement of the layers, and to maintain leaflet resistance to compression forces (Sutton et al. 1995).

The anisotropic arrangement of leaflet structural components is associated to anisotropic mechanical properties. In fact, cusps have to be at the same time extremely soft and flexible when unloaded during systolic phase, but need to resist strong back pressures insisting onto the edges during diastole. Mechanical resistance of leaflets to stretch and strain forces associated to cardiac cycle can be reproduced, as in the case of pericardium, by mechanical testing of the cusps (May-Newman and Yin 1995). In a typical stress/strain curve, the diastolic deformation under peak diastolic load of AOV is reproduced (Schoen and Levy 1999, page 442, Figure 1. Section C). In the figure, it is possible to distinguish three distinct phases of stress/strain behavior of valve tissue. In the first phase (elastin phase), the leaflet offers little resistance to elongation since force transmission and load bearing are provided mainly by the elastin fibers. During this phase, the collagen layer in the fibrosa has a minimal contribution to force transmission and the tissue behaves almost as a Hookean elastic solid with the stress increasing linearly with increasing strain. In the second phase (transition and collagen phase), collagen fibers become aligned and uncoiled, thus contributing to force transmission. The slope of the stress–strain curve for this phase is steep and almost constant, reflecting the material properties of the collagen fibers which allow limited elongation to fracture. The collagen phase of a leaflet usually continues beyond physiological range, corresponding to its reserve strength. At the end of the collagen phase, the slope of the curve starts decreasing, and a further increase in the load will finally result in tissue rupture.

6.3.1 Role of the Valve Interstitial Cells and Valve Endothelial Cells

Different cellular types are present in the three layers in AOV leaflets: endothelial cells, interstitial cells, and neuronal cells (Flanagan and Pandit 2003; Liu et al. 2007). The endothelial component is represented by the so-called valve endothelial cells (VECs), covering both sides of the leaflet surface; interstitial cells are represented by mesenchymal-like cells (the so-called valve interstitial cells [VICs]) present in all three layers of the valve. Finally, neuronal elements arising from ventricular endocardial plexuses (Marron et al. 1996) and from the aortic adventitial wall penetrate in the leaflets and predominantly localize in the ventricularis.

A particular role in the homeostasis and pathology of the AOV is represented by VICs. These cells closely resemble myofibroblasts and express a variety of mesenchymal markers. It has been hypothesized that VICs function as sensors translating mechanical forces into molecular responses involved in maintenance of leaflet structural integrity through interactions with ECM and promoting ECM remodeling. In fact, VICs are able to synthesize new ECM components as well as matrix degrading enzymes that repair functional damages normally occurring to matrix components during physiological load conditions. VIC physiologic activity can turn into pathologic phenotypes that can be at the base of AOV stenosis. In fact, in healthy valves, VICs have a quiescent fibroblast-like phenotype morphologically resembling fibroblasts. However, following injury or during progression of valve pathology, they change phenotype (Schoen 2008).

Overall, at least five different VIC phenotypes have been recognized; some of them are only relevant for valve formation during embryogenesis, while others play a specific role in valve homeostasis and during progression or stenotic valve pathology. During ontogeny, a complex process called endothelial/mesenchymal transition (EMT) occurs in a subset of endothelial cells overlying the endocardial cushion and gives rise to embryonic VICs (eVICs). This EMT generates mesenchymal progenitor cells that contribute to valvuloseptal structures as well as to adult VICs. Initially, the mesenchymal cells of the endocardial cushions are highly proliferative; however, during tissue valve formation and remodeling, cell proliferation decreases and there is little to no proliferation of adult VICs (Hinton and Yutzey 2011; Liu et al. 2007). At postdevelopmental stages, VICs (pVICs) can also derive from migratory waves of bone-marrow-derived circulating progenitors. These cells, characterized by stem cell marker expression of CD34, CD133, and/or S100 (an intracellular calcium-binding protein), are recruited from circulation in response to injury and contribute to activated VICs (aVICs) for heart valve repair. In physiological state, valve structure and function are maintained by quiescent VICs (qVICs) that, under pathological conditions, become aVICs. These cells express alpha-smooth muscle actin (α-SMA), proliferate, migrate, and induce profound matrix remodeling in response to injuries promoted by risk-factor-related (e.g., hypercholesterolemia, hypertension) inflammation, infections, or abnormal hemodynamic/mechanical forces. A variety of factors (cytokines and chemokines), secreted by the same aVICs and neighboring cells, regulate this process that in a chronic loop leads to AOV stenosis. Finally, "osteoblastic" VICs (oVICs) are cells directly playing a role in the calcification of AOV leaflets with secretion of alkaline phosphatase, osteocalcin, osteopontin, and bone sialoprotein; this condition is normally related to bone and cartilage formation in the valve tissue (Liu et al. 2007). Another relevant cell type for the valve leaflet homeostasis is represented by VECs. Although morphologically indistinguishable from EC cells obtained from other vascular sources, VECs may have specific features. In fact, it has been found that the transcriptional gene expression profile of VECs is different from that of aortic EC cells when these cells are exposed to the same mechanical and chemical stimuli. Recent studies have suggested an interplay between VECs and VICs in maintenance of VICs phenotype and valve homeostasis (Butcher and Nerem 2006).

6.4 Bioprosthetic AOVs: Modalities of Preparation, Durability, and Main Failure Causes

Valve replacement still represents the treatment of choice for severe cardiac valve dysfunction. There are two major valve replacement options: mechanical devices and biological/bioprosthetic valves. Modern mechanical valves are an evolution of the pioneer ball-and-cage valve first implanted in 1952 (DeWall et al. 2000). A satisfactory progress has been reached in the 1970s, with the evolution of the St. Jude valve, with an excellent durability and good hemodynamic. These valves are, still today, the most commonly implanted (e.g., in the United States); however, they require lifetime treatment with anticoagulants to reduce thromboembolic complications with inherent increase of hemorrhage risk (Wang 1989). From the early 1960s, the term "bioprosthesis" was introduced to describe valves which are chemically treated with or without being mounted onto a supporting stent. Bioprosthetic valves can be classified in two major groups: animal tissue valves and human tissue valves. Compared to mechanical valves, they have a closer physiology to that of natural valves, do not require long-term treatment with anticoagulants, and have better hemodynamic performances. However, like all biological implants, bioprosthetic valves have a limited life span, and their use is indicated only for the elderly, in whom life expectancy does not exceed implant durability.

6.4.1 Types of Bioprosthetic Valves for Human Implantation

The most commonly used AOV bioprosthetic valves in cardiac surgery are (1) human-derived valves (allografts/homografts) or (2) xenograft valves.

Human-derived valves are obtained from organ and tissue donors; they are cryopreserved together with the aortic root and trimmed to size and shape at the time of implantation in recipient. Homograft valves ensure good hemodynamic and durability. However, their major limitation consists in the lack of wide donor availability and their postimplantation longevity, confined to 10–20 years postimplantation, due to structure deterioration (Dohmen et al. 2002; Senthilnathan et al. 1999).

Animal-derived tissue valves are often referred to as xenograft; these are of porcine or calf origin. Porcine xenografts consist of intact pig AOVs, preserved in low-concentration fixation solution, while bovine pericardial valves are fabricated using three separate pieces of calf pericardium treated with low concentrations of fixative, affixed to a supporting stent and sewing cuff.

Animal-derived AOVs have a durability of 10–15 years and, like homografts, have excellent hemodynamic performance. Besides limiting the immune reaction associated to implantation of xenograft tissues, cross-linking of animal-derived tissues with fixing solutions is necessary to ensure long-term stabilization of collagen bundles, thereby maintaining long-term performance of these grafts (Dohmen et al. 2002).

6.4.2 Fixation Protocols That Ensure Prosthetic AOV Biocompatibility

Production of AOV xenografts requires the use of protocols that ensure tissue preservation and grant their biocompatibility. Aldehyde fixation, and in particular, the use of glutaraldehyde (GA), is, since its primary introduction in 1970, the method of choice. In fact, aldehyde treatment leads to formation of Schiff base bonds with the primary amines of amino acid residues lysine and hydroxylysine on collagen, which render the tissue more stable against mechanical loading. Aldehyde-based fixation also decreases tissue antigenicity; makes the tissue thromboresistant, thus eliminating the need for anticoagulation treatments (Paez and Jorge-Herrero 1999); and protects tissues against rapid enzymatic and chemical degradation.

GA treatment is not optimal. In fact, this stabilizing agent does not preclude tissues from undergoing progressive structural deterioration, calcification, and early mechanical failure. This is likely due to the absence of an active cellular machinery necessary for remodeling and renewing extracellular matrix components and scavenging cell debris. In addition, aldehyde residues may persist in the tissues also after extensive washing; these may have cytotoxic effects that are considered a primary cause for calcification induction and ultimate failure of the graft.

Common GA-based fixation procedure is performed using 0.2% GA solution in 0.2 M phosphate buffer (pH = 6.5). These conditions allow formation of a maximal number of cross-links and maintain the highest thermal stability of collagen membranes (Jastrzebska et al. 2008).

6.4.3 Failure of Bioprosthetic Valves

The majority of bioprosthetic valves fail due to calcification and structural deterioration. The determinants of bioprosthetic valve mineralization include several factors related to (1) host metabolism, (2) implant structure and chemistry, and (3) mechanical factors. Mineralization of the prosthetic tissues begins early after graft implantation with formation of small hydroxyapatite deposits. This occurs especially in young-age recipients, and it is correlated to high mechanical stress. The fixed cells in the bioprosthetic tissues are the predominant site of mineralization (Schoe and Levy 1999).

Calcification process in porcine AOVs and bovine pericardium grafts is qualitatively, quantitatively, and mechanistically similar. As deposits grow in the inner part of the tissue, calcific nodules reach the tissue surface, inducing further structural damage (Schoen et al. 1986). The damages of calcification and structural deterioration may be cooperative; in fact, mechanical stress caused by loading and unloading of cusps may result in further increased calcification (Berbacca et al. 1992); this occurs especially in areas subject to high motion and hemodynamic stress such as the commissural regions (Schoen et al. 1986, 1988).

At a cellular level, pathologic mineralization of the prosthetic valve tissue can be reconducted to physiologic bone calcification. In fact, calcium deposition typically occurs because nonviable cells of the prosthetic valve do not maintain a lower intracellular calcium concentration. As a consequence, calcium binds to cell and organelle phospholipid membranes, creating mineralization nuclei. Collagen, elastin, and GAGs; cellular debris; and remaining traces of fixative can also function as additional nuclei for calcium deposition. Recent studies suggest that the mechanisms responsible for pathologic calcification are not passive but regulated similarly to physiologic mineralization of the bone. During normal bone calcification, the growth of apatite crystals is regulated by several noncollagen matrix proteins such as osteopontin, an acidic calcium-binding phosphoprotein with high affinity to hydroxyapatite, or osteonectin, secreted by osteoblasts during bone formation and responsible for initiating mineralization and promoting mineral crystal formation (McKee and Nanci 1996a,b). All these noncollagen matrix proteins have been found in calcified bioprosthetic valves, suggesting that they play a regulatory role in these forms of bioprosthesis calcification in patients (Schoen et al. 1986). How these regulatory proteins are correlated to mineralization of the fixed tissues is still an active area of investigation; however, it cannot be excluded as an important contribution by the recipient's circulating cells.

6.4.4 Approaches to Prevent AOV Xenograft Calcification

Bioengineering is an approach that may be suitable to prevent calcification in AOV bioprostheses. Different strategies include (1) treatment of the tissue at postfixation stages, (2) development of fixative-free decellularization procedures, and, finally, (3) combining artificial matrix and stem cell technology. While some of these approaches are already in clinical use, others are still a matter of preclinical investigation.

Anticalcification treatments for GA-fixed tissue have been devised with the aim to reduce calcium nucleation. These include, for example, the use of diphosphonates, drugs generally used to treat malign hypercalcemia and osteoporosis. It has been found that cusp pretreatment with these drugs or their systemic administration significantly reduces bioprosthetic valve calcification (Levy et al. 1985). Other substances such as metal ions (aluminum and iron) (Levy et al. 1991) or amino oleic acid (Chen et al. 1994a,b) remove calcium via complex formation or its flux through bioprosthetic cusps. Surfactants such as sodium dodecyl sulfate (SDS) (Hirsch et al. 1993) and preincubation with ethanol (Vyavahare et al. 1997) remove acidic phospholipids, thus reducing the initial cell-membrane-oriented calcification. All these treatments, alone or in combination, are clinically approved in Europe for porcine valves.

Since the cellular components of homografts and bioprosthetic xenografts contribute to calcification or immune reactions, another strategy to overcome this issue is to use decellularization protocols that remove cells without the adverse effects consequent to fixation. In this way, biological materials used to derive prosthetic AOVs are likely to maintain a native structure that may serve as natural scaffolds for homing of host-derived cells after implantation or for recellularization protocols using various types of (stem) cells before transplanting them into patients.

The recent refinement of decellularization techniques has led to outstanding examples of whole bioartificial organ derivation such as artificial lung, liver, and heart with maintenance of fundamental physiologic activities (Ott et al. 2008). Different methods to remove cells from aortic and pulmonary valves (both human and porcine) have been investigated, with various degrees of success in terms of cell removal, maintenance of matrix structure/function, and susceptibility to cell seeding (Tudorache et al. 2007). These methods include detergent cell extraction, using Triton or SDS, enzymatic digestion such as trypsin (Kasimir et al. 2003; Kim et al. 2002), freeze-thawing or freeze-drying (Curtil et al. 1997), or incubation in hypotonic/hypertonic buffers (Meyer et al. 2005). Trypsin and the detergents Triton X-100 and SDS are the most promising agents for maintenance of matrix architecture and biomechanical properties both in porcine valve (Kim et al. 2002) and bovine pericardium, even if the latter requires subsequent cross-linking to provide the suitable tissue stiffness (Courtman et al. 1994). An example of typical decellularization protocol is the following: first wash in DI water for 30 min, incubation with agitation in 1% SDS in phosphate-buffered saline (PBS) for 24 h, and overnight rinse in DI water (Seif-Naraghi et al. 2010).

Preclinical results demonstrated that decellularized valve tissues have good biocompatibility both in vitro (after cell seeding) and in vivo. However, preliminary results obtained in a clinical study in children have disclosed adverse effects of these tissues, such as fibrous overgrowth and strong inflammatory response. Research in this field is at the very beginning, and future experimental studies are needed to minimize these effects (Simon et al. 2003).

6.5 Future Perspectives for Pericardium-Based Heart Valve Tissue Engineering

The next step in heart valve tissue engineering is to generate living heart valve replacements composed of healthy cells able to repair stress-associated ECM damage and grow in concert with heart growth in young recipients. Innovative work toward this objective has been performed in various laboratories, where trileaflet valves have been produced by seeding mesenchymal stem cells (MSCs) of bone marrow and cord blood origin onto synthetic, biodegradable, polymer scaffolds fabricated in the shape of trileaflet valves, followed by incubation into a bioreactor, controlling metabolic and mechanical environment (Sacks et al. 2005). This approach has led to derivation of functional tissue-engineered heart valve (TEHV), which showed intact, mobile, and pliable leaflets and functional competence during valve closure even under supraphysiological flow and pressure conditions. Histological analysis of TEHV leaflets revealed that biodegradable polyglycolic acid mesh scaffold was gradually replaced by a viable tissue organized in a layered fashion with heart valve ECM proteins even if without the typical three-layered structural composition of native valve leaflets (Hoerstrup et al. 2002).

An alternative to the use of synthetic scaffolds for derivation of fully biocompatible AOV leaflets is represented by employment of decellularized natural membranes to seed stem cells. In this respect, decellularized human pericardium obtained from multi organ donors may be suitable to achieve this goal. In fact, while the pericardium structure is fundamentally different from that of valve leaflets, resistance of these two tissues to mechanical stress is comparable (Liao et al. 1992; Mirsadraee et al. 2006). In addition, the collagen pericardium structure represents, at least potentially, a good microenvironment for migratory activity and proliferation of seeded cells and for progenitor cell-based endothelialization and matrix deposition. Finally, pericardium decellularization abolishes immunogenic response derived from cells and collagen matrix without compromising its mechanical resistance, as demonstrated by Mirsadraee et al. (2007).

What are the ideal cells that may be used in this context? As in the case of artificial scaffold seeding, the use of pericardial MSC seeding is particularly promising. The advantage of these cells compared with other types is that they are natural components of the stroma in several tissues, where they are committed to synthesize and remodel ECM. Another interesting MSC feature is that these cells do not express class II MHC; this makes MSCs immune, privileged, and suitable for heterologous transplantation in humans (Liechty et al. 2000; Prockop 1997). Recent characterizations of the heart-derived stem cell population have shown that both c-kit+ cardiac progenitors and stromal cells express markers in common with MSCs (Gambini et al. 2011; Rossini et al. 2011). Heart-derived MSCs appear very intriguing for valve-like tissue engineering. In fact, compared with BM-derived MSCs, they have less propensity to give rise to adipogenic or osteogenic lineages, at least in vitro, but maintain a higher performance to repair the damaged heart. This latter feature may be very important to prevent unwanted tissue calcification in bioengineered leaflets.

Another cell type that may well adapt to pericardium environment is represented by VICs. These cells are the most physiologic candidates to reconstitute living leaflet tissues. However, since these cells are turned into pro-osteogenic cells as a response to cardiovascular risk-factor-associated inflammation, it will have to be assessed whether these cells maintain a noncalcific phenotype also following rounds of ex vivo amplification prior to seeding them into pericardium.

6.6 Pericardium Medical Applications and International Regulatory Requirements for Human and Animal Tissue Use

Human and animal pericardium is widely used in production of bioprosthetic implants for various medical applications (see Table 6.1). The use of this material is subject to various legislation which may profoundly differ from country to country. In the European Union, for example, regulatory guidelines are established for ensuring safety and quality of allografts and xenografts. In this legislation, animal-derived pericardium, as a nonliving tissue of animal origin, is likened to medical devices, and its use is therefore under the Council Directive 93/42/EEC by Article 1.5.g. Different is the policy for human origin tissues, including pericardium, whose employment for banking and transplantation is still regulated independently by each European country, thus causing problems in free exchange and circulation of these products and hampering the welfare of patients and the progress of research. Recently, a European guideline of recommendations (Commission Directive 2004/23/EC; Commission Directive 2006/17/EC and 2006/86/EC) for tissue banking has been proposed; this framework is structured into four parts: (1) quality systems that apply to tissue banking and general quality system requirements, (2) regulatory framework in Europe, (3) standards available, and (4) recommendations of the fundamental quality and safety

TABLE 6.1

Additional Medical Applications of Pericardium

Pericardium	Medical Applications	References
Human and bovine pericardium	*Ear, nose, and throat repair*	Wheeler et al. (2007)
Human and bovine pericardium	*Ophthalmology*	Khanna and Mokhtar (2008)
		Gupta et al. (2007)
Human and bovine pericardium	*Urology*	Moon et al. (2011)
		Palese and Burnett (2001)
Bovine pericardium	*Gynecology*	Guerette et al. (2009)
Human and bovine pericardium	*Dental implantation*	Steigmann (2006)
Human and bovine pericardium	*Brain surgery*	Anson and Marchand (1996)
Human and bovine pericardium	*Abdominal wall reconstruction and hernia*	Bellows et al. (2006) Cavallaro et al. (2010)

key points. For more details, visit the website http://ec.europa.eu/health/blood_tissues_organs/policy/index_en.htm.

In the United States, human tissue products and tissue banks that supply allografts are regulated by the FDA. In addition, some tissue banks are voluntarily accredited by the American Association of Tissue Banks (AATB). The AATB recommends standards with regard to retrieval, processing, storage, and/or distribution. Specific FDA regulations for allografts can be found at www.fda.gov/cber/tiss.htm, and information about the AATB can be found at www.aatb.org. Tissues of animal origin for xenografts are also regulated by the FDA, and like in Europe, these products are regulated as medical devices. Specific guidelines in the use of animal tissues for xenotransplantation are found at www.fda.gov/cber/gdlns/clinxeno.htm.

6.7 Conclusions

Research on stem-cell-based tissue engineering has just begun in the cardiovascular area. While technology for bioprosthetic valve production has already delivered good devices for treating heart valve diseases, the derivation of a fully biocompatible AOVs is still to come. In our view, combination of stem cell technology with research on native and artificial scaffolds, such as pericardium, will provide important answers to control unwanted cellular responses and prevent mechanical failure that, today, limit graft durability in patients with AOV disease.

References

Allen, DJ and Didio, LJA. 1984. The structure of native human, bovine and porcine parietal pericardium. *Anatom Record* 208: 7A.

Anson, JA and Marchand, EP. 1996. Bovine pericardium for dural grafts: Clinical results in 35 patients. *Neurosurgery* 39: 764–768.

Bellows, CF, Alder, A, and Helton, WS. 2006. Abdominal wall reconstruction using biological tissue grafts: Present status and future opportunities. *Expert Rev Med Devices* 3: 657–675.

Berbacca, M, Fisher, AC, Wilkinson, R, Mackay, TG, and Wheatley, DJ. 1992. Calcification and stress distribution in bovine pericardial heart valves. *J Biomed Mater Res* 26: 959–966.

Boughner, DR, Haldenby, M, Hui, AJ, Dunmore-Buyze, J, Talman, EA, and Wan, WK. 2000. The pericardial bioprosthesis: Altered tissue shear properties following glutaraldehyde fixation. *J Heart Valve Dis* 9: 752–760.

Butcher, JT and Nerem, RM. 2006. Valvular endothelial cells regulate the phenotype of interstitial cells in co-culture: Effects of steady shear stress. *Tissue Eng* 12: 905–915.

Cavallaro, A, Lo Menzo, E, Di Vita, M et al. 2010. Use of biological meshes for abdominal wall reconstruction in highly contaminated fields. *World J Gastroenterol* 16: 1928–1933.

Chen, W, Kim, JD, Schoen, FJ, and Levy RJ. 1994a. Effect of 2-amino oleic acid exposure conditions on the inhibition of calcification of glutaraldehyde cross-linked porcine aortic valves. *J Biomed Mater Res* 28: 1485–1495.

Chen, W, Schoen, FJ, and Levy RJ. 1994b. Mechanism of efficacy of 2-amino oleic acid for inhibition of calcification of glutaraldehyde-pretreated porcine bioprosthetic heart valves. *Circulation* 90: 323–329.

Commission Directive 2004/23/EC of the European Parliament and of the Council setting standards of quality and safety for the donation, procurement, testing, processing, preservation, storage and distribution of human tissues and cells. *Official Journal of the European Union*, Luxembourg, March 31, 2004.

Commission Directive 2006/17/EC and 2006/86/CE implementing Directive 2004/23/EC as regards certain technical requirements for the donation, procurement and testing of human tissues and cells. *Official Journal of the European Union*, Luxembourg, February 8, 2006.

Courtman, DW, Pereira, CA, Kashef, V, McComb, D, Lee, JM, and Wilson GJ. 1994. Development of a pericardial acellular matrix biomaterial: Biochemical and mechanical effects of cell extraction. *J Biomed Mater Res* 28: 655–666.

Crofts, CE and Trowbridge, EA. 1989. Local variation in the tearing strength of chemically modified pericardium. *Biomaterials* 10: 230–234.

Curtil, A, Pegg, DE, and Wilson, A. 1997. Freeze drying of cardiac valves in preparation for cellular repopulation. *Cryobiology* 34: 13–22.

DeWall, RA, Qasim, N, and Carr, L. 2000. Evolution of mechanical heart valves. *Ann Thorac Surg* 69: 1612–1621.

Dohmen, PM, Scheckel, M, Stein-Konertz, M, Erdbruegger, W, Affeld, K, and Konertz, W. 2002. In vitro hydrodynamics of a decellularized pulmonary porcine valve, compared with a glutaraldehyde and polyurethane heart valve. *Int J Artif Organs* 25: 1089–1094.

Filippi, R, Schwarz M, Voth D et al. 2001. Bovine pericardium for duraplasty: Clinical results in 32 patients. *Neurosurg Rev* 24: 103–107.

Flanagan, TC and Pandit, A. 2003. Living artificial heart valve alternatives: A review. *Eur Cells Mater* 6: 28–45.

Gambini, E, Pompilio, G, Biondi, A et al. 2011. C-kit+ cardiac progenitors exhibit mesenchymal markers and preferential cardiovascular commitment. *Cardiovasc Res* 89: 362–373.

Guerette, NL, Peterson, TV, Aguirre, OA, Vandrie, DM, Biller, DH, and Davila, GW. 2009. Anterior repair with or without collagen matrix reinforcement: A randomized controlled trial. *Obstet Gynecol* 114: 59–65.

Gupta, M, Lyon, F, Singh, AD, Rundle, PA, and Rennie, IG. 2007. Bovine pericardium (Tutopatch) wrap for hydroxyapatite implants. *Eye* 21: 476–479.

Hiester, ED and Sacks, MS. 1998. Optimal bovine pericardial tissue selection sites. I. Fiber architecture and tissue thickness measurements. *J Biomed Mater Res* 39: 207–214.

Hinton, RB and Yutzey, KE. 2011. Heart valve structure and function in development and disease. *Annu Rev Physiol* 73: 29–46.

Hirsch, D, Drader, J, Thomas, TJ, Schoen, FJ, Levy, JT, and Levy, RJ. 1993. Inhibition of calcification of glutaraldehyde pretreated porcine aortic valve cusps with sodium dodecyl sulfate: Preincubation and controlled release studies. *J Biomed Mater Res* 27: 1477–1484.

Hoerstrup, SP, Kadner, A, Melnitchouk, S et al. 2002. Tissue engineering of functional trileaflet heart valves from human marrow stromal cells. *Circulation* 106(Suppl I): I-143–I-150.

Ishihara, T, Ferrans, VJ, Jones, M, Boyce, SW, Kawanami, O, and Roberts, WC. 1980. Histologic and ultrastructural features of normal human parietal pericardium. *Am J Cardiol* 46: 744–753.

Jastrzebska, M, Mroz, I, Barwinski, B, Zalewska-Rejdak, J, Artur Turek, A, and Cwalina, B. 2008. Supramolecular structure of human aortic valve and pericardial xenograft material: Atomic force microscopy study. *J Mater Sci: Mater Med* 19: 249–256.

Kasimir, MT, Rieder, E, Seebacher, G et al. 2003. Comparison of different decellularization procedures of porcine heart valves. *Int J Artif Organs* 26: 421–427.

Khanna, RK and Mokhtar, E. 2008. Bovine pericardium in treating large corneal perforation secondary to alkali injury: A case report Indian. *J Ophthalmol* 56: 429–430.

Kim, WG, Park, JK, and Lee, WY. 2002. Tissue engineered heart valve leaflets: An effective method of obtaining acellularized valve xenografts. *Int J Artif Organs* 25: 791–797.

Lazarou, G, Powers, K, Pena, C, Bruck, L, and Mikhail, MS. 2005. Inflammatory reaction following bovine pericardium graft augmentation for posterior vaginal wall defect repair. *Int Urogynecol J* 16: 242–244.

Levy, RJ, Schoen, FJ, Flowers, WB and Staelin, ST. 1991. Initiation of mineralization in bioprosthetic heart valves: Studies of alkaline phosphatase activity and its inhibition by AlCl3 or FeCl3 preincubations. *J Biomed Mater Res* 25: 905–935.

Levy, RJ, Wolfrum, J, Schoen, FJ, Hawley, MA, Lund, SA, and Langer, R. 1985. Inhibition of calcification of bioprosthetic heart valves by local controlled-release diphosphonate. *Science* 228: 190–192.

Liao, K, Seifter, E, Hoffman, D., Yellin, EL, and Frater, WM. 1992. Bovine pericardium versus porcine aortic valve: Comparison of tissue biological properties as prosthetic valves. *Artif Organs* 16: 361–365.

Liechty, KW, MacKenzie, TC, Shaaban, AF et al. 2000. Human mesenchymal stem cells engraft and demonstrate site-specific differentiation after in utero transplantation in sheep. *Nat Med* 6(11): 1282–1286.

Limpert, JN, Desai, AR, Kumpf, AL, Fallucco, MA, and Aridge, DL. 2009. Repair of abdominal wall defects with bovine pericardium. *Am J Surg* 198: e60–e65.

Liu, AC, Joag, VR, and Gotlieb, AI. 2007. The emerging role of valve interstitial cell phenotypes in regulating heart valve pathobiology. *Am J Pathol* 171: 1407–1418.

Marron, K, Yacoub, MH, Polak, JM et al. 1996. Innervation of human atrioventricular and arterial valves. *Circulation* 94: 368–375.

Matsagas, MI, Bali, C, Arnaoutoglou, E et al. 2006. Carotid endarterectomy with bovine pericardium patch angioplasty: Mid-term results. *Ann Vasc Surg* 20: 614–619.

May-Newman, K and Yin, FC. 1995. Biaxial mechanical behavior of excised porcine mitral valve leaflets. *Am J Physiol* 269: 1319–1327.

McKee, MD and Nanci, A. 1996a. Osteopontin deposition in remodeling bone: An osteoblast mediated event. *J Bone Miner Res* 11: 873–875.

McKee, MD and Nanci, A. 1996b. Osteopontin: An interfacial extracellular matrix protein in mineralized tissues. *Connect Tissue Res* 35: 197–205.

Meyer, SR, Nagendran, J, Desai, LS et al. 2005. Decellularization reduces the immune response to aortic valve allografts in the rat. *J Thorac Cardiovasc Surg* 130: 469–476.

Mirsadraee, SD, Wilcox, HE, Korossis, SA et al. 2006. Development and characterization of an acellular human pericardial matrix for tissue engineering. *Tissue Eng* 12: 763–773.

Mirsadraee, SD, Wilcox, HE, Watterson, KG et al. 2007. Biocompatibility of acellular human pericardium. *J Surg Res* 143: 407–414.

Moon, SJ, Kim, DH, Jo, JK et al. 2011. Bladder reconstruction using bovine pericardium in a case of enterovesical fistula. *Korean J Urol* 52: 150–153.

Ott, HC, Matthiesen, TS, Goh, SK et al. 2008. Perfusion-decellularized matrix: Using nature's platform to engineer a bioartificial heart. *Nat Med* 14: 213–221.

Paez, JMG and Jorge-Herrero, E. 1999. Assessment of pericardium in cardiac bioprostheses: A review. *J Biomater Appl* 13: 351–388.

Palese, MA and Burnett, AL. 2001. Corporoplasty using pericardium allograft (tutoplast) with complex penile prosthesis surgery. *Urology* 58: 1049–1052.

Prockop, DJ. 1997. Marrow stromal cells as stem cells for nonhematopoietic tissues. *Science* 276: 71–74.

Rossini, A, Frati, C, Lagrasta, C et al. 2011. Human cardiac and bone marrow stromal cells exhibit distinctive properties related to their origin. *Cardiovasc Res* 89: 650–660.

Sacks, MS, Engelmayr, GC, and Hildebrand, D. 2005. Bioreactors for heart valve tissue engineering. *Bioreactors for Tissue Engineering*. Springer, Dordrecht, the Netherlands, pp. 235–267.

Schmidt, CE and Baier, JM. 2000. Acellular vascular tissues: Natural biomaterials for tissue repair and tissue engineering. *Biomaterials* 21: 2215–2231.

Schoen, FJ. 2008. Valvular heart disease: Changing concepts in disease management. *Circulation* 118: 1864–1880.

Schoen, FJ, Kujovich, JL, Levy, RJ, and Sutton, MS. 1988. Bioprosthetic valve failure. *Cardiovasc Clin* 18: 289–317.

Schoen, FJ and Levy, RJ. 1999. Tissue heart valves: Current challenges and future research perspectives. *J Biomed Mater Res* 47: 439–475.

Schoen, FJ, Tsao, JW, and Levy, RJ. 1986. Calcification of bovine pericardium used in cardiac valve bioprostheses. Implications for the mechanisms of bioprosthetic tissue mineralization. *Am J Pathol* 123: 134–145.

Seif-Naraghi, SB, Salvatore, MA, Schup-Magoffin, PJ, Hu, DP, and Christman, KL. 2010. Design and characterization of an injectable pericardial matrix gel: A potentially autologous scaffold for cardiac tissue engineering. *Tissue Eng* 16(Part A): 2017–2027.

Senthilnathan, V, Treasure, T, Grunkemeier, G, and Starr, A. 1999. Heart valves: Which is the best choice? *Cardiovasc Surg* 7: 393–397.

Simionescu, D, Iozzo, RV, and Kefalides, NA. 1989. Bovine pericardial proteoglycan: Biochemical, immunochemical and ultrastructural studies. *Matrix* 9: 301–310.

Simionescu, D, Simionescu, A, and Deac, R. 1993. Mapping of glutaraldehyde-treated bovine pericardium and tissue selection for bioprosthetic heart valves. *J Biomed Mater Res* 27: 697–704.

Simon, P, Kasimir, MT, Seebacher, G et al. 2003. Early failure of the tissue engineered porcine heart valve SYNERGRAFT in pediatric patients. *Eur J Cardiothorac Surg* 23: 1002–1006.

Steigmann, M. 2006. Pericardium membrane and xenograft particulate grafting materials for horizontal alveolar ridge defects. *Implant Dent* 15:186–191.

Sutton, JP, Ho, SY, and Anderson, RH. 1995. The forgotten interleaflet triangles: A review of the surgical anatomy of the aortic valve. *Ann Thorac Surg* 59: 419–427.

Tudorache, I, Cebotari, S, Sturz, G et al. 2007. Tissue engineering of heart valves: Biomechanical and morphological properties of decellularized heart valves. *J Heart Valve Dis* 16: 567–574.

Vesely, I. 2005. Heart valve tissue engineering. *Circ Res* 97: 743–755.

Vesely, I and Mako, WJ. 1998. Comparison of the compressive buckling of porcine aortic valve cusps and bovine pericardium. *J Heart Valve Dis* 7: 34–39.

Vyavahare, N, Hirsch, D, Lerner, E et al. 1997. Prevention of bioprosthetic heart valve calcification by ethanol preincubation. Efficacy and mechanisms. *Circulation* 95: 479–488.

Wang, JH. 1989. The design simplicity and clinical elegance of the St Jude medical heart valve. *Ann Thorac Surg* 48: S55–S56.

Wheeler, DS, Wong, HR, and Shanley, TP. 2007. *Pediatric Critical Care Medicine: Basic Science and Clinical Evidence*. Springer, London, U.K.

Zioupos, P and Barbenel, JC. 1994a. Mechanics of native bovine pericardium. I. The multiangular behaviour of strength and stiffness of the tissue. *Biomaterials* 15: 366–373.

Zioupos, P and Barbenel, JC. 1994b. Mechanics of native bovine pericardium. II. A structure based model for the anisotropic mechanical behaviour of the tissue. *Biomaterials* 15: 374–382.

Zioupos, P, Barbenel, JC, and Fisher, J. 1994. Anisotropic elasticity and strength of glutaraldehyde fixed bovine pericardium for use in pericardial bioprosthetic valves. *J Biomed Mater Res* 28: 49–57.

7

Spatially Designed Nanofibrous Membranes for Periodontal Tissue Regeneration

Marco C. Bottino, Yogesh K. Vohra, and Vinoy Thomas

CONTENTS

7.1 Introduction

Data from the United States National Institute of Dental and Craniofacial Research (NIDCR) indicate that more than 86% of adult populations over 70 years old have at least moderate level of periodontal-related diseases and over a quarter have lost their teeth (Nakashima and Reddi 2003). Periodontitis and gingivitis are the two most common periodontal-related inflammatory diseases seen in adults. The progressive destruction of the tooth-support apparatus, including gingiva, alveolar bone, and periodontal ligament (PDL), and root cementum, if no clinical intervention, may ultimately lead to tooth loss (Haffajee and Socransky 1994; Pihlstrom et al. 2005; Nanci and Bosshardt 2006). Research findings have suggested an association of periodontitis with various systemic disorders, such as diabetes, cardiovascular diseases, and respiratory diseases, thus considered as a chronic condition (Pihlstrom et al. 2005; Southerland et al. 2006; Nishimura et al. 2007).

In the last decade, various regenerative surgical modalities have been proposed, tested, and evaluated for regenerating periodontium, the tooth-supporting apparatus (McClain and Schallhorn 2000; Hanes 2007; Kao et al. 2009). This has included the use of a wide range of surgical procedures, the use of occlusive barrier membranes, and the use of a variety of bone grafts and other osteoconductive/inductive biomaterials or proteins, exogenous growth factors, cell-based technology, and recombinant-gene technology (McClain and Schallhorn 2000; Hanes 2007; Kao et al. 2009). Among them, the regeneration of damaged periodontal structures with various bone graft materials and guided tissue regeneration (GTR) strategies has achieved some success in certain ideal clinical scenarios. Nonetheless, outcomes are variable, depending upon multiple factors such as defect size and type, patient age, and genetics, among others (McClain and Schallhorn 2000; Hämmerle and Jung 2003; Adriaens and Adriaens 2004; Ishikawa and Baehni 2004; Trombelli 2005; Zeichner-David 2006). In the pioneering work, Langer and Vacanti (1993) proposed tissue engineering as a possible route to regenerate lost tissue and ultimately to restore various human tissues and organs. Since then, tissue engineering has been regarded as "the Holy Grail" in regenerative medicine and dentistry. The principles of tissue engineering for the regeneration of damaged tissues have involved a combination and interplay of mainly three elements, scaffolds or membranes as constructive component, regenerative cells or stem cells as productive component, and cell-signaling molecules or growth factors as inductive component, traditionally known as "tissue engineering triad." The use of the principles of tissue engineering offers promise for regenerating or developing de novo periodontal tissues or hybrid tissues, in a rationale manner as with other tissues. However, the complete and functional regeneration of large periodontal defects is still in its early stages, owing to its complex physiology with multiple cells/matrix structures/functions and limited regenerative capacity of adult periodontium (Nakahara 2006; Ripamonti 2007; Chen and Jin 2010; Izumi et al. 2011).

A literature survey using keywords "periodontal regeneration" and "scaffolds" based on a "Web of Science" search presented in Figure 7.1 clearly demonstrates the surging interest in membrane-/scaffolds-based periodontal tissue regeneration. With the advances in nanoscience and nanotechnology, there is an increased enthusiasm toward the nanotechnology-enabled approaches such as electrospinning and self-assembly for the development of nanostructured biomimetic multifunctional growth-enhancing regenerative membranes for periodontal tissue engineering. Electrospinning is a simple and facile polymer-processing technique to produce nanofibrous polymers, blends, and composites (Ramakrishna et al. 2005; Thomas et al. 2006a; Renekar and Yarin 2008). Electrospun nanofibrous membranes mimic more closely the nanoscaled morphologies of the extracellular matrix (ECM) proteins (fibers with diameters ranging from 50 to 500 nm). Multifunctional electrospun scaffolds/membranes with therapeutic delivery of the necessary

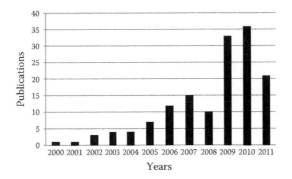

FIGURE 7.1
Publications involving the use of exogenous growth factors, cell based technology, bone grafting materials, and scaffolds for periodontal tissue regeneration during the last 10 years until September 2011. (From ISI Web of Knowledge, Thomson Reuters, New York, USA.)

growth factors and other cell-signaling molecules further enhance the functional tissue regeneration. Therefore, this chapter focuses mainly on spatially designed membranes for periodontal tissue engineering with a special emphasis on functionally graded membrane (FGM) fabricated by electrospinning and in vitro antibacterial-/cell-related studies.

7.2 Periodontium and Pathological Conditions

Periodontium is an extremely complex tissue comprised of root cementum, PDL, bone lining the tooth socket (alveolar bone), and the part of the gingival tissue that faces the tooth (dentogingival junction) (Figure 7.2). The periodontium functions to anchor the teeth to the jaws (mandible and maxilla) and preserve their positions, as well as to provide nourishment to the teeth. In addition, it assures proper function and the dissipation of forces, preventing any injury to teeth and the mandibular and maxillary bones. The cementum is an avascular (not innervated) calcified connective tissue that covers the root dentin and anchors PDL fibers. The PDL is composed of collagen fibers and responsible to absorb and distribute functional forces to the alveolar bone. The PDL is connected to the root cementum on one side and to the alveolar bone on the other (Gillett et al. 1990; Nanci and Bosshardt 2006). The alveolar bone supports the teeth and the gingival tissues. It distributes and absorbs the forces of mastication via the PDL. Proper functioning of the periodontium is only achieved through structural integrity and interactions between its components.

Periodontitis affects the integrity of the periodontium. Loss of alveolar bone results in the formation of a periodontal pocket around the tooth which in turn acts as a reservoir favoring the growth of anaerobic

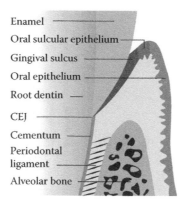

Enamel

Oral sulcular epithelium

Gingival sulcus

Oral epithelium

Root dentin

CEJ

Cementum

Periodontal ligament

Alveolar bone

FIGURE 7.2
Schematic illustration showing a longitudinal section through dento-gingival part of a tooth and its periodontal tissue apparatus.

bacteria (e.g., *Porphyromonas gingivalis* and *Prevotella intermedia*) (Haffajee and Socransky 1994). The progression of periodontitis leads to the destruction of the periodontium, ultimately resulting into tooth loss. Another periodontal problem that affects millions of people worldwide is gingival recession, a multifactorial condition (i.e., periodontal disease, gender, trauma, tobacco consumption, etc.) found primarily in adult patients regardless of oral hygiene (Tugnait and Clerehugh 2001). It is a result of apical migration of the gingival margin that leads to the root surface exposure (Santos et al. 2005). Gingival recession is not only an aesthetic concern; it may lead to root sensitivity and cervical root caries (Kassab and Cohen 2003).

Different treatment modalities have been proposed to promote periodontal regeneration in cases of both gingival recession and periodontitis. In order to treat cases of gingival recession, traditional surgical procedures use transplanted autogenous tissue (i.e., palatal area). The employment of tissue-derived collagen-based membranes has started to replace this procedure, which avoids the associated pain and morbidity due to multiple surgeries (Santos et al. 2005; Felipe et al. 2007). As mentioned earlier, the strategy of using either synthetic or tissue-derived membranes as barriers with or without calcium-phosphate-based bone graft materials has received great attention for restoring and regenerating the functions of damaged or pathological periodontium (Chen et al. 2010; Izumi et al. 2011).

7.3 Periodontal Tissue Regeneration

Therapeutic approaches for correcting the different diseases that affect the periodontium and ultimately lead to periodontal defects include flap debridement/flap curettage, resective procedures, and periodontal regenerative therapy (Chen et al. 2010). Overall, the different strategies utilized in periodontal therapy aim (1) to reduce and/or eliminate inflamed periodontal tissues caused by bacterial plaque, (2) to correct defects or anatomical problems provoked by the disease, and (3) to regenerate new periodontal tissues (Wang 2005). Most importantly, periodontal regeneration with the development of new connective tissue attachment, that is, new cementum with inserting PDL fibers and alveolar bone, is the main goal (Nyman et al. 1982, 1987; Karring et al. 1993; Linde et al. 1993; Piattelli et al. 1996; Karring 2000; Taba et al. 2005; Polimeni et al. 2006, 2008; Behring et al. 2008; Geurs et al. 2008; Sculean et al. 2008).

A critical challenge for regenerative periodontal treatment is to restore the attachment of the PDL and gingival collagen fibers to both the cementum and alveolar bone, that is, to obtain full regeneration of the periodontium (Nanci and Bosshardt 2006; Chen and Jin 2010). Two surgical techniques have been increasingly used to restore the periodontium: GTR and guided bone regeneration (GBR) (Nyman et al. 1982, 1987; Karring et al. 1993; Linde et al. 1993; Piattelli et al. 1996; Karring 2000; Taba et al. 2005; Polimeni et al. 2006, 2008; Behring et al. 2008; Geurs et al. 2008; Sculean et al. 2008). The use of an occlusive membrane barrier between the gingiva and the tooth to promote formation of new periodontium is called GTR. Briefly, the utilization of an occlusive periodontal membrane will act as a barrier when placed into the surgical site, preventing epithelial and connective tissue downgrowth into the defect. Therefore, progenitor cells located in the remaining PDL in the adjacent alveolar bone or in the blood will be able to recolonize the root area and differentiate into a new periodontal supporting apparatus with the regeneration of new bone, PDL, and cementum. Another important application of the guided regeneration concept concerns the restoration of deficient alveolar sites (e.g., extraction site and deficient alveolar ridge) for posterior implant placement. This process has been named GBR. In general, barrier membranes have been successfully used for the treatment of furcations and intrabony defects, as well as for the correction of marginal tissue recession defects (Wang 2005; Retzepi and Donos 2010). A meta-analysis comparing GTR with open-flap debridement reported greater clinical attachment level (of 1.22 mm) and greater probing depth reduction (of 1.21 mm) in GTR-treated sites (Needleman et al. 2006). Indeed, a systematic review by Murphy and Gunsolley (2003) highlighted important clinical advantages promoted by guided regeneration procedures compared to conventional open flap debridement for the treatment of intrabony and furcations defects. Interestingly, it has been shown that even though the utilization of grafting materials in conjunction with physical membranes enhances the regeneration of furcation defects, which does not seem to have a similar effect in the treatment of intrabony defects (Wang 2005).

7.4 GTR and GBR Membranes

The barrier strategy to isolate the periodontal defect with a material (i.e., resorbable or nonresorbable) in a mat-like form that will function as a physical barrier to avoid the downgrowth of gingival cells into the bone defect led to the development of GTR membranes. It was shown that conditions enabling periodontal regeneration arise when epithelial cells or gingival fibroblasts are excluded from the defect, and the PDL cells and other bone-forming cells (osteoblasts) are allowed to migrate, proliferate, and grow to the defect. Generally, these GTR/GBR membranes need to exhibit (1) biocompatibility, to allow its integration with the host tissues without eliciting inflammatory responses; (2) proper degradation rates, to match those of new tissues formation; and (3) sufficient mechanical and physical properties, to allow its placement in vivo and also to sustain proper strength avoiding the membrane collapse and the performance of their barrier characteristics (Magnusson et al. 1988; Linde et al. 1993; Piattelli et al. 1996; Karring 2000; Polimeni et al. 2006, 2008; Behring et al. 2008; Geurs et al. 2008). According to their degradability capacity, GTR/GBR membranes can be divided into two groups: nonresorbable and resorbable membranes.

7.4.1 Nonresorbable versus Resorbable Membranes

The "gold standard" periodontal membrane for GTR/GBR procedures is fabricated from expanded polytetrafluoroethylene (e-PTFE). e-PTFE membranes are biocompatible, act as a cellular barrier, provide space for tissue regeneration, and allow tissue integration. It has been suggested that there is a positive relationship between space maintenance and the amount of bone regenerated (Polimeni et al. 2005). Titanium wire reinforcement of e-PTFE membranes has allowed the restoration of larger defects. However, the great disadvantage of nonresorbable membranes is the necessity of performing a second surgical procedure for removal. In order to eliminate the second surgical procedure, resorbable barrier membranes have been developed (Behring et al. 2008).

The majority of resorbable membranes for periodontal regeneration currently on the market are either based on polyesters (e.g., poly(lactic acid) (PLA), poly(glycolic acid) (PGA), poly(lactic-*co*-glycolic acid) (PLGA), poly(ε-caprolactone) (PCL), and their copolymers) (Piattelli et al. 1996; Milella et al. 2001; Geurs et al. 2008; Gielkens et al. 2008) or tissue-derived collagens (Felipe et al. 2007; Behring et al. 2008). The polyester-based membranes are biocompatible and biodegradable, allow tissue integration, and are easier to handle surgically compared to e-PTFE membranes. Their resorption rate is important since these membranes must function for at least 4–6 weeks to achieve successful regeneration of the periodontal system (Becker et al. 1987; Piattelli et al. 1996; Liao et al. 2007). The process of biodegradation involves nonenzymatic cleavage of PGA and PLA polymers into pyruvic and lactic acids, respectively, common end products of carbohydrate digestion. These biodegradable polymeric membranes are long-term tissue compatible and predictable in terms of resorption (i.e., membrane thickness can be controlled). Milella et al. (2001) have evaluated both the morphological and mechanical characteristics of commercially available polyester-based membranes (i.e., Resolut® LT and Biofix®). Although the membranes demonstrated initially high strength results (~12–14 MPa), they completely lost their structural and mechanical properties within 4 weeks of incubation in culture medium. The maximum stress after 7 days of exposure decreased to approximately 3 MPa. After 14 days, the results were far below 1 MPa. Unfortunately, in addition to the lack of mechanical stability and unpredictable degradation profiles (Milella et al. 2001), the poor cell response also limits their use in GTR/GBR applications (Kikuchi et al. 2004; Behring et al. 2008).

Tissue-derived collagen-based membranes from human (i.e., skin), bovine (i.e., Achilles tendon), or porcine (i.e., intestinal submucosa) are an important alternative to synthetic polymers in GTR/GBR procedures due to their excellent cell affinity and biocompatibility (Santos et al. 2005; Felipe et al. 2007; Behring et al. 2008). For example, AlloDerm® (LifeCell, Branchburg, NJ) is an acellular freeze-dried dermal matrix graft derived from human cadaveric skin after the removal of both epidermis and cellular components (Livesey et al. 1995). This process does not cause any damage to the critical biochemical structural cues needed to maintain the tissue's natural regenerative properties and leaves behind an extracellular collagenous matrix that provides the basis for tissue structure and guides cellular functions (Livesey et al. 1995; Gouk et al. 2008; Bottino et al. 2009). Collagen-based membranes have shown very poor performance in vivo as the membrane starts to degrade early (Behring et al. 2008). Improvements in collagen stability and mechanical properties can be achieved via physical or chemical cross-linking, using methods such as ultraviolet radiation (UV), dehydrothermal (DHT), glutaraldehyde (Glu), 1-ethyl-3-(3-dimethylaminopropyl) carbodiimide hydrochloride (EDC), and genipin (Gp), among others. It has been reported that exogenous cross-linking agents (e.g., Gp) not only significantly increases the stability of collagen-based tissues but also reduce its antigenicity. In fact, the formations of supplementary inter- or intramolecular cross-links within the collagen fibers enhances mechanical properties of biological tissues (Nimni et al. 1988; Huang et al. 1998; Sung et al. 1999; Thomas et al. 2007; Bhrany et al. 2008; Sundararaghavan et al. 2008). Bottino et al. studied how the incorporation of a natural cross-linking agent, Gp, into a collagen-based GTR membrane (i.e., AlloDerm) rehydration protocol affects the biomechanical properties and the stability of the collagenous matrix (Bottino et al. 2010). A significant enhancement in tensile strength was found when Gp exposure time was increased from 30 min to 6 h (Table 7.1). Additionally, differential scanning calorimetry analyses indicated an important shift in the denaturation temperature for cross-linked samples, which coincides with the increase on the enthalpy of denaturation (Bottino et al. 2010). This finding agrees with prior investigations on the use of a cross-linking agent to enhance the stabilization of collagenous matrices derived

TABLE 7.1

Biomechanical Properties Obtained from
Tensile Experiment

Experimental Groups	σ_{UTS} (MPa)	E (MPa)
G1—Control (30 min)	21.16 ± 2.98	70.92 ± 6.33
G2—1 wt.% Gp (30 min)	25.52 ± 0.21*	71.93 ± 3.61
G3—1 wt.% Glu (30 min)	23.44 ± 1.97	61.78 ± 4.91
G4—1 wt.% Gp (6 h)	28.50 ± 2.47**	78.86 ± 1.56**

source: (From Bottino, M.C., Thomas, V., Jose, M.V., Dean, D.R., and Janowski, G.M.: Acellular dermal matrix graft: Synergistic effect of rehydration and natural crosslinking on mechanical properties. *J. Biomed. Mater. Res. Appl. Biomater.* 2010.95. 276–282. Copyright Wiley-VCH Verlag GmbH & Co. KGaA. Reprinted with permission.)
Data are mean ± SD ($n=4$).
* $P < 0.05$ compared with control group (saline rehydration, 30 min).
** $P < 0.05$ compared with G2 (Gp cross-linking, 30 min).

from different biological tissues (Nimni et al. 1988; Huang et al. 1998, Sung et al. 1999; Thomas et al. 2007; Bhrany et al. 2008; Sundararaghavan et al. 2008).

It has been suggested that there is a positive relationship between space maintenance and the amount of bone regenerated since membrane collapse into the bone defect reduces the space available for new bone formation (Polimeni et al. 2005). An in vivo study compared the biodegradation of cross-linked and non-cross-linked collagen membranes (CLM and NCLM, respectively) after exposure to the oral environment (Tal et al. 2008). Even though both membranes resisted to tissue degradation, none of the membranes resist degradation when exposed to the oral environment (Tal et al. 2008).

7.4.2 Spatially Designed Bioactive Membranes

The disadvantages of both e-PTFE (e.g., second surgery) and collagen-based resorbable membranes (e.g., insufficient mechanical properties) make it clear that the "ideal" periodontal membrane has yet to be found. In the past decade, numerous researchers have attempted to develop GTR/GBR periodontal membranes with the requisite properties and features from blends of natural and synthetic polymers. These studies ranged from the preparation via film casting (Park et al. 2000; Milella et al. 2001; Liao et al. 2005; Liao et al. 2007; Li et al. 2009), dynamic filtration (Teng et al. 2009), and electrospinning (Fujihara et al. 2005; Yang et al. 2009) of synthetic (e.g., PLA) and/or natural (e.g., collagen, chitosan) polymer GTR/GBR membranes. The membranes have been prepared with or without therapeutic drugs (Park et al. 2000), growth factors (Nie et al. 2008; Srouji et al. 2011), and/or calcium phosphate particles (Kikuchi et al. 2004;

Fujihara et al. 2005; Liao et al. 2005, 2007; Thomas et al. 2006a; Li et al. 2009; Yang et al. 2009; Jose et al. 2010; Bottino et al. 2011). A material with the necessary mechanical, degradation, and biological properties is still needed to guarantee GTR/GBR its best in vivo performance. The majority of these methods result in membranes with very low clinical potential due to high density (i.e., difficulty in handling) and heterogeneity (i.e., nonuniform degradation rate). On the other hand, the electrospinning technique has demonstrated great potential for processing membranes for periodontal regeneration (Fujihara et al. 2005; Yang et al. 2009; Zamani et al. 2010; Bottino et al. 2011). Electrospinning is a particularly promising technique for synthesizing biomimetic nanomatrices. With this approach, scaffolds can be fabricated with nanoscale fibers that mimic the size and arrangement of native collagen fibers (50–500 nm). Additionally, electrospun scaffolds have a high surface to volume ratio and interconnecting pores, which facilitates cell adhesion and formation of cell-cell junctions (Jose et al. 2010; Zhang et al. 2010a; Phipps et al. 2011). Li et al. have shown the ability of nanofiber structure in supporting cell attachment and proliferation by culturing fibroblasts, cartilage cells, or mesenchymal stem cells on electrospun nanofibers of PLGA or PCL (Li et al. 2002, 2006a, 2006b). Because of the inherent very high surface area, surface functional groups, interconnected pores, and nanoscaled size, nanofiber-based scaffolds are more favorable than microfibers or any other morphological forms. Nanofibrous scaffolds induce favorable cell–ECM interactions, increase the proliferation rate, maintain cell phenotype, support differentiation of stem cells, and promote in vivo-like three-dimensional (3D) matrix adhesion and activate cell-signaling pathways by providing the chemical and physical stimuli to cells (Li et al. 2006a; Venugopal et al. 2008b). Systematic reviews on the electrospinning process and applications of the these nanofibers in tissue engineering can be seen elsewhere (Ramakrishna et al. 2005; Murugan and Ramakrishna 2006; Pham et al. 2006; Thomas et al. 2006a; Greiner and Wendorff 2007; Sell et al. 2007; Reneker and Yarin 2008; Sill and von Recum 2008; Xie et al. 2008; Hong and Madihally 2011).

Usually, sequential spinning or multilayering is employed to fabricate scaffolds or membranes with different layer structures. Sequential electrospinning of different polymers, both synthetic polymers alone and in combination with natural proteins, enhances the mechanical integrity and dimensional stability of electrospun meshes (Kidoaki et al. 2005; Vaz et al. 2005; Thomas et al. 2007, 2009; Zhang et al. 2009). Synthetic/protein layer-by-layer membranes can also be fabricated by electrospinning in combination with freeze-drying (Jeong et al. 2007) or phase separation process (Soletti et al. 2010).

7.4.3 Functionally Graded Nanofibrous Membranes

GTR/GBR membranes for periodontal regeneration can be considered as an interface implant, which interface with gingival epithelial tissue and alveolar bone tissue. An interface implant needs to utilize a graded structure with compositional gradients and components that meet the local functional requirements (Bottino et al. 2011). Therefore, the use of homogeneous biomaterials (either compositionally or structurally) may not be sufficient to engineer periodontium by enhancing bone growth while preventing the gingival tissue downgrowth. Toward this end, fabrication of a functionally graded three-layered membrane from PLGA, collagen, and nanohydroxyapatite (n-HAp) by a layer-by-layer casting method was reported earlier (Liao et al. 2005, 2007). The membrane was designed with one face of 8% nanocarbonated hydroxyapatite (nCHAC)/collagen/PLGA acid (nCHAC/PLGA) porous membrane allowing cell adhesion, and the opposite face with a smooth PLGA nonporous film. The transitional middle layer was 4% nCHAC/PLGA, which inhibits cell attachment. Very recently, using electrospinning technique, functionally graded nanofibrous tubular scaffolds with distinct chemical compositions as well as tuned mechanical and degradation characteristics have been reported (Thomas et al. 2007b, 2009; McClure et al. 2010). The design and fabrication of a FGM via electrospinning holds promise as an interface implant between alveolar bone and epithelial tissue. The rationale of having a periodontal membrane with a graded structure again relies on the idea that one can tailor the properties of the different layers to design a membrane that will retain its structural, dimensional, and mechanical integrity long enough to enhance periodontal regeneration. Accordingly, a novel FGM was designed and fabricated via sequential multilayering electrospinning (Bottino et al. 2011) (Figure 7.3). The FGM consists of a core layer

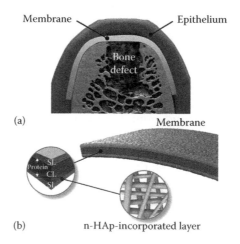

FIGURE 7.3
Schematic illustration of the spatially designed and FGM. (a) Membrane placed in a GBR scenario; (b) details of the CL and the functional SL interfacing bone (n-HAp) and epithelial (MET) tissues. Note the chemical composition stepwise grading from the CL to SLs, that is, polymer content decreased and protein content increased. (Reprinted from *Acta Biomater.*, 7, Bottino, M.C., Thomas, V., and Janowski, G.M., A novel spatially designed and functionally graded electrospun membrane for periodontal regeneration, 216–224, 2011, Copyright 2011, with permission from Elsevier.)

(CL) and two functional surface layers (SL) interfacing bone (n-HAp) and epithelial (metronidazole, MET) tissues. The CL comprises a neat PLCL layer surrounded by two composite layers composed of a gelatin/polymer ternary blend (PLCL:PLA:GEL). Poly(D,L-lactide-*co*-caprolactone) (PLCL) is a copolymer of PCL and PLA. SEM images showing the morphological characterization of the FGM layers are given in Figure 7.4. While most biodegradable polymers are rigid and brittle, PLCL possesses exceptional elastic properties, which is critical for periodontal membranes during in vivo placement. Electrospun P(LLA-CL) tubular grafts implanted into the epigastric vein of

FIGURE 7.4
SEM micrographs of the individual electrospun layers used in the FGM fabrication. (a,b) PLA:GEL + 25%MET. (c,d) PLA:GEL + 10%-n-HAp. The arrows in (d) and in the inset evidence the presence of the n-HAp particles embedded on the fibers surface. (e,f) Ternary PLCL:PLA:GEL blend. (g,h) Neat PLCL fibers. (Reprinted from *Acta Biomater.*, 7, Bottino, M.C., Thomas, V., and Janowski, G.M., A novel spatially designed and functionally graded electrospun membrane for periodontal regeneration, 216–224, 2011, Copyright 2011, with permission from Elsevier.)

a rabbit have remained stable up to 7 weeks (He et al. 2009b). The degradation process, mainly by hydrolysis of ester bonds, involves the formation of lactic and hydroxy hexanoic acids. PLA is a semicrystalline polymer, which has been used as sutures, bone screws and plates, drug delivery systems, and many other biomedical applications (Dong et al. 2009).

In GTR/GBR regeneration, mechanical properties of the membranes are of utmost importance for the clinical success of the therapy (Milella et al. 2001; Fujihara et al. 2005; Liao et al. 2005; Yang et al. 2009). The mechanical properties of the CL are a major contributor to and, therefore, predictor of the in vivo mechanical performance of the present FGM for the intended duration. Under hydrated conditions, CL presented high tensile strength (8.7 MPa) and tensile modulus (156 MPa) with strain at failure of 375%. No delamination was observed in the CL membrane, indicating that the compositionally graded layers remained intact under physiological conditions (Bottino et al. 2011). Cross-section SEM images are given in Figure 7.5. Addition of two SLs rich in protein has decreased the mechanical tensile properties in both dry and hydrated conditions. The FGM exhibited a tensile strength of 3.5 MPa and a tensile modulus of 80 MPa with a strain at break equal to 297%. During testing, the n-HAp-containing SL failed first (lower strain), but the CL- (responsible to provide the overall mechanical integrity the membrane in vivo) and MET-incorporated SL remained intact. The decrease in elastic modulus from 156 to 80 MPa for the FGM can be attributed to the hydration of considerably amounts of uncross-linked

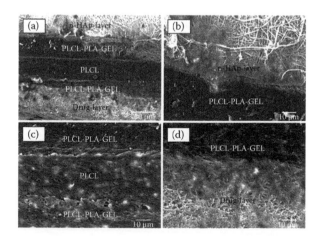

FIGURE 7.5
Cross-section SEM micrographs of the FGM processed via multilayering electrospinning: (a) general view of the FGM, (b) interface n-HAp-containing layer/PLCL:PLA:GEL, (c) CL structure, and (d) interface MET-loaded layer/PLCL:PLA:GEL. (Reprinted from *Acta Biomater.*, 7, Bottino, M.C., Thomas, V., and Janowski, G.M., A novel spatially designed and functionally graded electrospun membrane for periodontal regeneration, 216–224, 2011, Copyright 2011, with permission from Elsevier.)

hydrophilic gelatin in the SLs. The mechanical properties of FGM under dry and wet conditions are given in Table 7.2. The mechanical properties of the five-layered FGM could be improved by protein (i.e., gelatin) cross-linking. Overall, it is extremely important, in terms of membrane design, to bear in mind that an ideal periodontal membrane should balance two important material properties (i.e., stiffness and elasticity) to sustain mechanical loading without membrane collapse and good manageability, respectively.

7.4.4 Design of Bioactive Interfacing Functional Layers

The "ideal" periodontal membrane must be biocompatible and biodegradable as well as sufficiently strong to maintain the defect space and, at the same time, flexible and clinically manageable. Moreover, membranes intended for GTR/GBR applications capable of promoting faster bone growth as well as impeding the infiltration of epithelial tissue into the defect and bacteria colonization would be unique and highly effective when implanted in vivo. The functional SLs of membrane layers must be designed based upon the tissue with which each would interface. In our new FGM, the layer that is supposed to interface with the bone defect, PLA:GEL + n-HAp, was designed to mimic the collagen–HA matrix of bone. In contrast, the layer that interfaces with the epithelial tissue (PLA:GEL + MET) was loaded with 25 wt.% of MET. From the CL to SLs, the chemical composition showed a stepwise grading, namely, the polymer content decreased and protein content increased (Figure 7.3). Mature bone matrix is composed of 65% mineral and 35% protein. The mineral phase is a calcium phosphate mixture that is predominantly hydroxyapatite (HA). The organic phase consists of 90% collagen I fibers, and the remaining 10% is composed of various proteoglycans and other proteins (Karsenty 1999). The use of collagen in tissue engineering has grown substantially due to its distinctive biological characteristics, such as a hemostatic effect, low antigenicity, and great biocompatibility and biodegradability (Behring et al. 2008; Kasaj et al. 2008; Tal et al. 2008).

Coupled with its osteoconductivity and osteoinductivity, HA $[Ca_{10}(PO_4)_6(OH)_2]$ has been recognized as an important component for tissue-engineered bone substitutes and scaffolds (El-Ghannam 2005). In an open pore bone scaffold, HA enables tissue penetration by attracting the cells toward it, resulting in better biointegration as well as mechanical/dimensional integrity of the scaffold (Wang et al. 2002; Thomas et al. 2006a). Techniques such as coelectrospinning of HA nanoparticles (Ito et al. 2005; Kim et al. 2005, 2006; Bishop et al. 2006; Thomas et al. 2006b, 2007a; Deng et al. 2007; Erisken et al. 2008; Jose et al. 2010), electrospraying of HA suspension (Gupta et al. 2009) and biomimetic growth of HA from simulated body fluids (SBF) (Cui et al. 2010; Liu et al. 2011) have been frequently employed for fabricating fibrous composites of polymer and HA with enhanced bioactivity, osteoinductivity, adhesion, and proliferation of osteoblasts or mesenchymal

TABLE 7.2

Mechanical Properties of the Individual Electrospun Layers, CL, and FGM

Electrospun Fibers	Tensile Modulus (MPa)		Yield Strength (MPa)		UTS (MPa)		Strain (%)	
	Dry	Hydrated	Dry	Hydrated	Dry	Hydrated	Dry	Hydrated
NEAT PLCL	52.7±6.6	21.7±13.4	2.3±0.3	1.5±0.5	6.7±0.6	8.6±1.7	416.1±48.8	466.6±62.4
PLCL:PLA:GEL	133.7±18.1	157.6±67.1	3.0±0.5	6.6±1.5	3.9±0.3	10.6±0.6	94.0±9.1	465.0±59.0
PLA:GEL:10%nHAp	60.5±18.9	4.6±1.1	1.3±0.3	n/a	1.7±0.2	1.0±0.2	46.2±8.2	51.5±10.2
PLA:GEL:25%MET	56.6±13.3	95.4±2.2	1.7±0.4	3.8±0.3	2.1±0.3	3.9±0.3	121.3±20.4	42.5±4.7
CL	147.7±17.4	156.2±34.1	4.1±0.5	5.3±0.8	4.1±0.6	8.7±0.8	73.1±22.0	375.0±55.0
FGM	37.8±10.5	80.6±25.8	1.7±0.1	3.0±0.7	2.9±0.6	3.5±0.5	234.0±27.1	296.9±60.6

Source: Reprinted from *Acta Biomater.*, 7, Bottino, M.C., Thomas, V., and Janowski, G.M., A novel spatially designed and functionally graded electrospun membrane for periodontal regeneration, 216–224, 2011, Copyright 2011, with permission from Elsevier.

n/a, not available.

stem cells. Coelectrospinning of HA and collagen with polymers has facilitated the formation of membranes with improved mechanical properties (Thomas et al. 2006b; Yang et al. 2009). Alternatively, we have employed electrophoretic deposition (EPD) also as an efficient method for depositing HA coating on the nanofibrous scaffold for therapeutic delivery of bone morphogenetic protein-2 (Deshpande et al. 2011). During EPD, the charged colloidal-sized particles in suspension migrate toward the countercharged electrode and get discharged and flocculated on the substrate electrode.

In cases of periodontitis, the loss of alveolar bone results in the formation of a periodontal pocket around the tooth that acts as a reservoir favoring the growth of anaerobic bacteria (Haffajee and Socransky 1994). Indeed, the presence of periodontopathogens such as *P. gingivalis* and *P. intermedia* may negatively affect the success of periodontal regeneration; therefore, it is of extreme importance to control and/or reduce the bacterial contamination of the periodontal defect in order to enhance periodontal regeneration (Haffajee and Socransky 1994; Slots et al. 1999). MET is a drug commonly used in the treatment of periodontal infections (Martin et al. 1991; Freeman et al. 1997). MET acts against anaerobic Gram-negative (e.g., *P. gingivalis*) and anaerobic spore forming Gram-positive bacilli (El-Kamel et al. 2007). Clinically, the oral administration of antibiotics to fight periodontal diseases has been helpful in reducing and/or eliminating the presence of pathogens in the gingival fluid (Al-Mubarak et al. 2000). Prolonged usage of systemic antimicrobials has significant issues (e.g., bacterial strain resistance), which has led investigators to design materials capable of localized drug delivery in a controlled fashion to secure its effects for an appropriate period of time. Different investigators have successfully incorporated tetracycline hydrochloride (TCH) and MET into different polymeric solutions aiming to develop a material with therapeutic properties (Kenawy et al. 2002; He et al. 2009a; Zamani et al. 2010).

It is well known that the successful regeneration of the periodontal apparatus depends on a precise combination of key elements such as growth factors, cells, blood supply, and a scaffold material (Taba et al. 2005; Nakahara 2006; Chen et al. 2010). The local delivery of a wide selection of growth factors (e.g., platelet-rich plasma [PRP], bone morphogenetic proteins [BMPs], platelet-derived growth factor [PDGF]) has demonstrated to enhance both periodontal healing and regeneration by modulating the cellular activity and providing stimuli to cells to differentiate and synthesize ECM to develop new tissue (Giannobile and Somerman 2003; Jung et al. 2003; Nevins et al. 2005; Chen et al. 2010). To date, the use of bone grafting materials, exogenous growth factors, and cell-based approaches have contributed to the advancement of the treatment of periodontitis and for the overall promotion of periodontal health. These issues, although important, are beyond the scope of this review paper but reviewed in detail by others (Nakashima and Reddi 2003; Nakahara 2006; Ripamonti 2007; Chen et al. 2009, 2010).

7.5 Antibacterial and Bone Regeneration Studies

Wound stability is a key factor for a successful regenerative periodontal surgery (Wikesjo et al. 1995). The 3D nanofibrous structure of multifunctional scaffolds with a high surface area of improved hydrophilicity and wettability is able to regulate cell functions, leading to the formation of new bone into the defect (Thomas et al. 2006a; Venugopal et al. 2008b; Phipps et al. 2011). In vitro release studies on TCH incorporated various electrospun scaffolds, namely, PLA, poly(ethylene-co-vinyl acetate) (PEVA), and a 50:50 PLA/PEVA exhibited a highest drug release rate by PEVA scaffold, followed by the PLA/PEVA blend (Kenawy et al. 2002). In the same line of reasoning, TCH was incorporated into poly(L-lactic acid) (PLLA) fibers either via blend or coaxial electrospinning (He et al. 2009a). Two relevant conclusions were drawn based on the in vitro drug release evaluation. First, it was demonstrated that threads processed from the core-shell fibers (coaxial electrospinning) had a lower initial burst and a more sustained release. Second, the threads processed from the blend fibers resulted in a large initial burst release, which can be of great value in avoiding bacterial infection. More recently, Zamani et al. (2010) fabricated PCL nanofibrous scaffolds for periodontal regeneration containing distinct amounts of MET using different dichloromethane (DCM)/N,N-dimethyl formamide (DMF) solvent ratios. According to drug release data, both the different solvent ratios and the amount of drug incorporated into the PCL fibers played a role on drug release rate. More importantly, the burst release was low and sustained for at least 19 days, which may be important for the treatment and regeneration of new periodontal tissues.

As previously mentioned, GBR procedures may fail when resorbable and nonresorbable membranes become exposed to the oral cavity, for example, in cases of soft tissue dehiscences commonly seen after immediate implant placement in fresh extraction sockets and bone dehiscences adjacent to implants (Tal et al. 2008). In these situations, the membrane may be colonized by oral bacteria that not only will eventually result in inflammation and bone resorption (Mombelli et al. 1993; Chen et al. 1997) but also will impair the attachment of the PDL cells (Hung et al. 2002). Chou et al. (2007) reported on the antibacterial effects of two commercially available biodegradable membranes (i.e., a polyester-based membrane, Resolut Adapt LT [Gore-Tex; W.L. Gore & Associates, Inc., Flagstaff, AZ] and a type I bovine collagen, BioMend Extend [Zimmer Dental, Carlsbad, CA]), after mineralization with zinc phosphate. Regardless of the membrane composition, zinc phosphate mineralization led to a significant decrease in *Actinobacillus actinomycetemcomitans* activity when compared to nonmineralized membranes (Chou et al. 2007). Park et al. (2000) fabricated porous poly(L-lactide) films loaded with tetracycline on poly(glycolide) meshes. Tetracycline-loaded membranes induced better cell attachment compared to unloaded membranes. In addition, drug-loaded membranes had a positive effect on new bone formation and stimulated bony reunion after 2 weeks of implantation in rat calvarial defects (Park et al. 2000). In a recent study, a newly developed GBR membrane based on a silver–HA–titania/polyamide nanocomposite (nAg–HA–TiO_2/PA) revealed higher osteogenic activity when compared to e-PTFE as per the higher levels of alkaline phosphatase. Overall, based on both radiographic and histomorphometric data demonstrated a fully closed cranial defect for both nAg–HA–TiO_2/PA and e-PTFE groups after 8 weeks in vivo (Zhang et al. 2010b).

Several in vitro studies have demonstrated that GTR/GBR membrane's surface topography, chemical composition (e.g., collagen-based versus non-resorbable membranes), and pore structure may influence proliferation of different cell types including human gingival fibroblasts (HGFs), PDL fibroblasts (PDLFs), and human osteoblast-like (HOBs) cells (Behring et al. 2008; Kasaj et al. 2008). The use of collagen as a membrane in GTR and GBR has grown substantially due to its distinctive biological characteristics, such as a hemostatic effect where the ability to aggregate platelets may aid clot formation and wound stabilization (Behring et al. 2008; Tal et al. 2008). According to Kasaj et al. (2008), the rate of HGFs, PDLFs, and HOBs cells proliferation was greater when using collagen-based membranes. Worth to mention, collagen type I may have limitations on its use due to the high cost and poor definition of its commercial sources (e.g., bovine, porcine, human), which may make it difficult to control degradation and mechanical properties. More importantly, non-cross-linked collagen-based GTR/GBR membranes have shown rapid resorbability in vivo (Behring et al. 2008). A recently published systematic review addressed the effects of collagen origin and cross-linking method on different cells attachment, proliferation, and migration processes. Overall, the findings revealed that distinct fibroblast cells and bone-forming cells attach to collagen membranes regardless of collagen origin or processing method (Behring et al. 2008).

Many research groups have studied the effect of nanosized HA particles in electrospun matrices for bone tissue regeneration in vitro (Wutticharoenmongkol et al. 2006; Venugopal et al. 2008a; Ngiam et al. 2009; Prabhakaran et al. 2009; Jose et al. 2010). Studies on the three-layered membrane prepared by Liao et al. (2005), having a porous side (to allow cell ingrowth) containing nCHAC/collagen/PLGA, the opposite side being a pure PLGA nonporous membrane (to discourage cell adhesion), and a transitional middle layer consisting of nCHAC/PLGA, have demonstrated that the addition of nCHAC improved both the biocompatibility and the osteoconductivity of the membrane. In a recent study (Yang et al. 2009), electrospun composite nanofibrous membranes based on nanoapatite (nAp) and PCL, that is, n0 (nAp:PCL = 0:100), n25 (nAp:PCL = 25:100), and n50 (nAp:PCL = 50:100), were fabricated. The incorporation of nAp played a significant role in terms of improving membrane bioactivity and facilitating early cell differentiation. We have studied the effect of collagen and HA on aligned electrospun multifunctional scaffolds based on PLGA/collagen/HA by seeding of

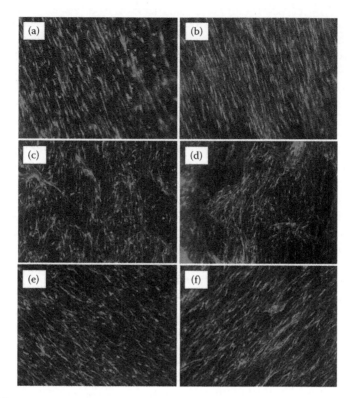

FIGURE 7.6
Fluorescent microscopic images of hMSCs seeded (1×10^5) on nanofibrous scaffolds (a,d) Neat PLGA, (b,e) 80/20, and (c,f) 80/20/0.5 for a duration of 1 day (a–c) and 7 days (d–f). The magnification is 6.3×. (Jose, M.V., Thomas, V., Xu, Y.Y., Bellis, S., Nyairo, E., Dean, D: Aligned bioactive multi-component nanofibrous nanocomposite scaffolds for bone tissue engineering. *Macromol. Biosci.* 2010. 10. 433–444. Copyright Wiley-VCH Verlag GmbH & Co. KGaA. Reprinted with permission.)

human mesenchymal stem cells on the scaffolds (Figure 7.6). The fluorescent microscopic images showed the scaffolds supported cell adhesion and proliferation and the aligned morphology of the nanofibers resulted in a highly oriented cell morphology (Jose et al. 2010). The proliferation and increase in the population density of cells on scaffold surface indicated the cytocompatibility of the multicomponent scaffolds with time. Western blot data and microBCA assay showed protein adsorption was enhanced with the addition of collagen and nano-HA into the polymer matrix (Figure 7.7).

Chitosan, a known carbohydrate polymer, is not toxic to the living tissues; presents good biocompatibility, biodegradability, and wound healing properties; and has been recognized for its antibacterial characteristics (Zhang and Zhang 2004; Teng et al. 2008, 2009). Based on that, nanofibrous chitosan membranes were recently fabricated by electrospinning. The nanofibrous chitosan membranes demonstrated good in vitro cytocompatibility

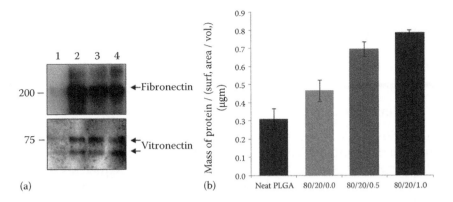

FIGURE 7.7
(a) Western blot analysis for fibronectin and vitronectin in the multicomponent system: 1, Neat PLGA; 2, 80.20; 3, 80/20/0.5; 4, 80/20/1.0. (b) Total protein adsorption on scaffold calculated from microBCA assay. (Jose, M.V., Thomas, V., Xu, Y.Y., Bellis, S., Nyairo, E., Dean, D: Aligned bioactive multi-component nanofibrous nanocomposite scaffolds for bone tissue engineering. *Macromol. Biosci.* 2010. 10. 433–444. Copyright Wiley-VCH Verlag GmbH & Co. KGaA. Reprinted with permission.)

(i.e., osteosarcoma cell line MG63) as well as new bone formation in rabbit critical-sized calvarial defects after 4 weeks (Shin et al. 2005). Further, chitosan/HAp membranes were prepared by Teng et al. (2009) based on the coprecipitation method followed by dynamic filtration and freeze-drying and the cell-related assays indicated that the highest alkaline phosphatase level associated with membranes doped with 30% of Hap. Recently, we have shown that a tricomponent biocomposite fibrous electrospun membrane comprised of PCL, collagen, and n-HAp (50/30/20 weight ratio) adsorbed a substantially greater quantity of the adhesive proteins, fibronectin, and vitronectin than PCL or PCL/HA following in vitro exposure to serum or placement into rat tibiae (Phipps et al. 2011). The tricomponent substrate exhibited a rapid spreading and significantly greater proliferation of cultured mesenchymal stem cells than PCL/HA, PCL, and collagen scaffolds. In addition, cells seeded onto PCL/col/HA scaffolds showed markedly increased levels of phosphorylated FAK, a marker of integrin activation known to be involved in osteoblastic differentiation signaling cascades (Phipps et al. 2011).

The bone regeneration capacity of calcium-alginate-based films (CAF) was compared to collagen membranes when applied to a rabbit mandible defect model (He et al. 2008). Overall, it was demonstrated that the CAF membrane promoted early bone growth and seemed more effective than the collagen membrane counterpart (He et al. 2008). Park et al. (2009) reported a novel approach to synthesize PLGA grafted hyaluronic acid (HA-PLGA) structures with predictable degradation properties. In vitro studies revealed that the decomposition of the HA-PLGA/PLGA film was much slower than that of PLGA film and HA-coated PLGA film. In terms of bone regeneration capability, the newly developed HA-PLGA/PLGA film exhibited a significant amount of new bone filling the defect.

7.6 Summary and Future Outlook

Periodontitis is a very aggressive pathology that can lead to the destruction of the periodontium, which may eventually result in tooth loss. During the past few decades, several therapeutic approaches, including flap debridement and/or flap curettage and periodontal regenerative therapy, have been utilized to treat the different diseases that affect the periodontium. Surgical regenerative approaches utilize an occlusive membrane based on bioresorbable polymers and/or processed-collagen for the regeneration of both the structure and function of damaged periodontal tissues. It is well known that these membranes act as a barrier to prevent epithelial and connective tissue downgrowth into the defect, enabling periodontal regeneration. However, these conventional membranes possess many structural, mechanical, and biofunctional drawbacks. Based on graded-biomaterials approach, we have hypothesized that a spatially designed and FGM that mimics closely the native ECM could potentially be the next generation of GTR/GBR membranes for periodontal tissue engineering (Bottino et al. 2011). The rationale of having a periodontal membrane with a functionally graded structure again relies on the idea that one can tailor the properties of the different layers to design a membrane that will retain its structural, dimensional, and mechanical integrity long enough to enhance periodontal regeneration.

In the context of regenerative dentistry, various stem cells existing in human dental tissues such as dental pulp, exfoliated deciduous teeth, and PDL are being isolated (Gronthos et al. 2000; Miura et al. 2003; Lutolf and Hubbell 2005), the translational research necessary to apply a growth-enhancing nanomatrix and stem cells hybrid system to regenerative medicine is still in its early stages. An ideal system for clinical use will be a simple procedure that provides one-step delivery of gene/protein of interest with minimal manipulation (Nakahara 2006). To accelerate the bench-to-bedside application of newly emerged periodontal nanotherapies, future studies must require multidisciplinary collaboration among material scientists and stem cell biologists, as well as basic and clinical dental researchers.

Acknowledgments

We acknowledge support by Award Number R01AR056665 from the National Institute of Arthritis and Musculoskeletal and Skin Diseases. The content is solely the responsibility of the authors and does not necessarily represent the official views of the National Institute of Arthritis and Musculoskeletal and Skin Diseases or the National Institutes of Health. We thank Drs. Dean DR, Janowski GM, Jose MV, and Bellis SL for their collaborations rendered during the cell-scaffolds interactions and mechano-morphological characterizations of membranes presented in this chapter. The authors are also thankful to Mr. Mark A. Dirlam (IUSD, Indianapolis, IN) for helping with the illustration of the periodontal system. We apologize to those authors whose work may have been relevant to this chapter but was not cited due to perceived lack of fit or due to the space limitations.

References

Adriaens, P. A., Adriaens, L. M. 2004. Effects of nonsurgical periodontal therapy on hard and soft tissues. *Periodontol* 2000(36):121–145.

Al-Mubarak, S. A., Karring, T., Ho, A. 2000. Clinical evaluation of subgingival application of metronidazole 25%, and adjunctive therapy. *J Int Acad Periodontol* 2:64–70.

Becker, W., Becker, B. E., Prichard, J. F., Caffesse, R., Rosenberg, E., Gian-Grasso, J. 1987. Root isolation for new attachment procedures. A surgical and suturing method: Three case reports. *J Periodontol* 58:819–826.

Behring, J., Junker, R., Walboomers, X. F., Chessnut, B., Jansen, J. A. 2008. Toward guided tissue and bone regeneration: Morphology, attachment, proliferation, and migration of cells cultured on collagen barrier membranes. A systematic review. *Odontology* 96:1–11.

Bhrany, A. D., Lien, C. J., Beckstead, B. L. et al. 2008. Crosslinking of an oesophagus acellular matrix tissue scaffold. *J Tissue Eng Regen Med* 2(6):365–372.

Bishop, A., Balázsi, C., Yang, J. H. C., Gouma, P. I. 2006. Biopolymer-hydroxyapatite composite coatings prepared by electrospinning. *Polym Adv Technol* 17:902–906.

Bottino, M. C., Jose, M. V., Thomas, V., Dean, D. R., Janowski, G. M. 2009. Freeze-dried acellular dermal matrix graft: Effects of rehydration on physical, chemical, and mechanical properties. *Dent Mater* 25:1109–1115.

Bottino, M. C., Thomas, V., Janowski, G. M. 2011. A novel spatially designed and functionally graded electrospun membrane for periodontal regeneration. *Acta Biomater* 7:216–224.

Bottino, M. C., Thomas, V., Jose, M. V., Dean, D. R., Janowski, G. M. 2010. Acellular dermal matrix graft: Synergistic effect of rehydration and natural crosslinking on mechanical properties. *J Biomed Mater Res Appl Biomater* 95B:276–282.

Chen, F.-M., Jin, Y. 2010. Periodontal tissue engineering and regeneration: Current approaches and expanding opportunities *Tissue Eng Part B: Rev* 16(2):219–255.

Chen, F. M., Shelton, R. M., Jin, Y., Chapple, I. L. C. 2009. Localized delivery of growth factors for periodontal tissue regeneration: Role, strategies, and perspectives. *Med Res Rev* 29:472–413.

Chen, Y. T., Wang, H. L., Lopatin, D. E., Oneal, R., MacNeil, R. L. 1997. Bacterial adherence to guided tissue regeneration barrier membranes exposed to the oral environment. *J Periodontol* 68:172–179.

Chen, F. M., Zhang, J., Zhang, M., An, Y., Chen, F., Wu, Z. F. 2010. A review on endogenous regenerative technology in periodontal regenerative medicine. *Biomaterials* 31:7892–7827.

Chou, A. H. K., LeGeros, R. Z., Chen, Z., Li, Y. H. 2007. Antibacterial effect of zinc phosphate mineralized guided bone regeneration membranes. *Implant Dent* 16:89–100.

Cui, W. G., Li, X. H., Xie, C. Y., Zhuang, H. H., Zhou, S. B., Weng, J. 2010. Hydroxyapatite nucleation and growth mechanism on electrospun fibers functionalized with different chemical groups and their combinations. *Biomaterials* 31:4620–4629.

Deng, X.-L., Sui, G., Zhao, M.-L., Chen, G.-Q., Yang, X.-P. 2007. Poly(L-lactic acid)/hydroxyapatite hybrid nanofibrous scaffolds prepared by electrospinning. *J Biomater Sci Polym Ed* 18:117–130.

Deshpande, H., Shindler, C., Dean, D. R. et al. 2011. Nanocomposite scaffolds based on electrospun polycaprolactone/modified carbon nanofiber/nanohydroxyapatite by electrophoretic deposition. *J Biomater Tissue Eng* 1:177–184.

Dong, Y., Liao, S., Ngiam, M. et al. 2009. Degradation behaviors of electrospun resorbable polyester nanofibers. *Tissue Eng Part B Rev* 15:333–351.

El-Ghannam, A. 2005. Bone reconstruction: From bioceramics to tissue engineering. *Expert Rev Med Devices* 2:87–101.

El-Kamel, A. H., Ashri, L. Y., Alsarra, I. A. 2007. Micromatricial metronidazole benzoate film as a local mucpadhesive delivery system for treatment of periodontal diseases. *AAPS Pharm Sci Tech* 8(3):E_1–E_{11}.

Erisken, C., Kalyon, D. M., Wang, H. 2008. Functionally graded electrospun polycaprolactone and β-tricalcium phosphate nanocomposites for tissue engineering applications. *Biomaterials* 29:4065–4073.

Felipe, M. E., Andrade, P. F., Grisi, M. F. et al. 2007. Comparison of two surgical procedures for use of the acellular dermal matrix graft in the treatment of gingival recession: A randomized controlled clinical study. *J Periodontol* 78:1209–1217.

Freeman, C. D., Klutman, N. E., Lamp, K. C. 1997. A therapeutic review and update. *Drugs* 54(5):679–708.

Fujihara, K., Kotaki, M., Ramakrishna, S. 2005. Guided bone regeneration membrane made of polycaprolactone/calcium carbonate composite nano-fibers. *Biomaterials* 26:4139–4147.

Geurs, N. C., Korostoff, J. M., Vassilopoulos, P. J. et al. 2008. Clinical and histologic assessment of lateral alveolar ridge augmentation using a synthetic long-term bioabsorbable membrane and an allograft. *J Periodontol* 79:1133–1140.

Giannobile, W. V., Somerman, M. J. 2003. Growth and amelogenin-like factors in periodontal wound healing. A systematic review. *Ann Periodontol* 8:193–204.

Gielkens, P. F., Schortinghuis, J., de Jong, J. R. et al. 2008. The influence of barrier membranes on autologous bone grafts. *J Dent Res* 87:1048–1052.

Gillett, I. R., Johnson, N. W., Curtis, M. A. et al. 1990. The role of histopathology in the diagnosis and prognosis of periodontal diseases. *J Clin Periodontol* 17:673–684.

Gouk, S. S., Lim, T. M., Teoh, S. H. et al. 2008. Alterations of human acellular tissue matrix by gamma irradiation: Histology, biomechanical property, stability, *in vitro* cell repopulation, and remodeling. *J Biomed Mater Res Appl Biomater* 84B:205–217.

Greiner, A., Wendorff, J. H. 2007. Electrospinning: A fascinating method for the preparation of ultrathin fibers. *Angew Chem Int Ed* 46:5670–5703.

Gronthos, S., Mankani, M., Brahim, J. et al. 2000. Postnatal human dental pulp stem cells (DPSC) in vitro and in vivo. *Proc Natl Acad Sci USA* 97:13625–13630.

Gupta, D., Venugopal, J., Mitra, S., Giri Dev, V. R., Ramakrishna, S. 2009. Nanostructured biocomposite substrates by electrospinning and electrospraying for the mineralization of osteoblasts. *Biomaterials* 30:2085–2094.

Haffajee, A. D., Socransky, S. S. 1994. Microbial etiological agents of destructive periodontal diseases. *Periodontol 2000* 5:78–111.

Hämmerle, C. H., Jung, R. E. 2003. Bone augmentation by means of barrier membranes. *Periodontol 2000* 33:36–53.

Hanes, P. J. 2007. Bone replacement grafts for the treatment of periodontal intrabony defects. *Oral Maxillofac Surg Clin North Am* 19:499–512.

He, C. L., Huang, Z. M., Han, X. J. 2009a. Fabrication of drug-loaded electrospun aligned fibrous threads for suture applications. *J Biomed Mater Res* 89A:80–95.

He, W., Ma, Z., Teo, W. E. et al. 2009b. Tubular nanofiber scaffolds for tissue engineered small-diameter vascular grafts. *J Biomed Mater Res* 90A:205–216.

He, H., Yan, W. Q., Chen, G. F., Lu, Z. H. 2008. Acceleration of de novo bone formation with a novel bioabsorbable film: A histomorphometric study in vivo. *J Oral Pathol Med* 37:378–382.

Hong, J. K., Madihally, S. V. 2011. Next generation of electrosprayed fibers for tissue regeneration. *Tissue Eng* 17B:125–142.

Huang, L. L., Sung, H. W., Tsai, C. C., Huang, D. M. 1998. Biocompatibility study of a biological tissue fixed with a naturally occurring crosslinking reagent. *J Biomed Mater Res* 42:568–576.

Hung, S. L., Lin, Y. W., Wang, Y. H., Chen, Y. T., Su, C. Y., Ling, L. J. 2002. Permeability of *Streptococcus mutans* and *Actinobacillus actinomycetemcomitans* through guided tissue regeneration membranes and their effects on attachment of periodontal ligament cells. *J Periodontol* 73:843–851.

Ishikawa, I., Baehni, P. 2004. Nonsurgical periodontal therapy-where do we stand now? *Periodontol* 2000(36):9–13.

Ito, Y., Hasuda, H., Kamitakahara, M. et al. 2005. A composite of hydroxyapatite with electrospun biodegradable nanofibers as a tissue engineering material. *J Biosci Bioeng* 100:43–49.

Izumi, Y., Aoki, A., Yamada Y. et al. 2011. Current and future periodontal tissue engineering. *Periodontol* 2000(56):166–187.

Jeong, S.I., Kim, S. Y., Cho, S. K. et al. 2007. Tissue-engineered vascular grafts composed of marine collagen and PLGA fibers using pulsatile perfusion bioreactors. *Biomaterials* 28:1115–1122.

Jose, M. V., Thomas, V., Xu, Y. Y., Bellis, S., Nyairo, E., Dean, D. 2010. Aligned bioactive multi-component nanofibrous nanocomposite scaffolds for bone tissue engineering. *Macromol Biosci* 10:433–444.

Jung, R. E., Glauser, R., Scharer, P., Hammerle, C. H. F., Sailer, H. F., Weber, F. E. 2003. Effect of rhBMP-2 on guided bone regeneration in humans—A randomized, controlled clinical and histomorphometric study. *Clinical Oral Impl Res* 14:556–568.

Kao, R. T., Murakami, S., Beirne, O. R. 2009. The use of biologic mediators and tissue engineering in dentistry. *Periodontol* 2000(50):127–153.

Karring, T. 2000. Regenerative periodontal therapy. *J Int Acad Periodontol* 2:101–109.

Karring, T., Nyman, S., Gottlow, J., Laurell, L. 1993. Development of the biological concept of guided tissue regeneration—Animal and human studies. *Periodontol* 2000(1):26–35.

Karsenty, G. 1999. The genetic transformation of bone biology. *Genes Dev* 13:3037–3051.

Kasaj, A., Reichert, C., Gotz, H., Rohrig, B., Smeets, R., Willershausen, B. 2008. In vitro evaluation of various bioabsorbable and nonresorbable barrier membranes for guided tissue regeneration. *Head Face Med* 4:22 (October) http://www.head-face-med.com/content/4/22

Kassab, M. M., Cohen, R. E. 2003. The etiology and prevalence of gingival recession. *J Am Dent Assoc* 134:220–225.

Kenawy, el-R., Bowlin, G. L., Mansfield, K. et al. 2002. Release of tetracycline hydrochloride from electrospun poly(ethylene-co-vinylacetate), poly(lactic acid), and a blend. *J Control Release* 81:57–64.

Kidoaki, S., Kwon, K., Matsuda, T. 2005. Mesoscopic spatial designs of nano- and microfiber meshes for tissue-engineering matrix and scaffold based on newly devised multilayering and mixing electrospinning techniques. *Biomaterials* 26:36–47.

Kikuchi, M., Koyama, Y., Yamada, T. et al. 2004. Development of guided bone regeneration membrane composed of beta-tricalcium phosphate and poly (L-lactide-co-glycolide-epsilon-caprolactone) composites. *Biomaterials* 25:5979–5986.

Kim, H. W., Lee, H. H., Knowles, J. C. 2006. Electrospinning biomedical nanocomposite fibers of hydroxyapatite/poly (lactic acid) for bone regeneration *J Biomed Mater Res* 79A:643–649.

Kim, H. W., Song, J. H., Kim, H. E. 2005. Nanofiber generation of gelatin-hydroxyapatite biomimetics for guided tissue regeneration. *Adv Funct Mater* 15:1988–1994.

Langer, R., Vacanti, J. P. 1993. Tissue engineering. *Science* 260(5110):920–926.

Li, W. J., Cooper, J. A., Mauck, R. L., Tuan, R. S. 2006a. Fabrication and characterization of six electrospun poly (hydroxyl ester)-based fibrous scaffolds for tissue engineering applications. *Acta Biomater* 2:377–385.

Li, W. J., Laurencin, C. T., Caterson, E. J., Tuan, R. S., Ko, F. K. 2002. Electrospun nanofibrous structure: A novel scaffold for tissue engineering. *J Biomed Mater Res* 60:613–621.

Li, W. J., Shanti, R. M., Tuan, R. S. 2006b. Nanotechnologies for the life sciences. In: *Tissue, Cell and Organ Engineering*, Vol. 9. eds. S. S. Challa, R. Kumar, p. 135. Weinheim, Germany: Wiley-VCH Verlag GmbH & Co.

Li, J. D., Zuo, Y., Cheng, X. M., Yang, W. H., Wang, H. N., Li, Y. B. 2009. Preparation and characterization of nano-hydroxyapatite/polyamide 66 composite GBR membrane with asymmetric porous structure. *J Mater Sci Mater Med* 20:1031–1038.

Liao, S., Wang, W., Uo, M. et al. 2005. A three-layered nano-carbonated hydroxyapatite-collagen-PLGA composite membrane for guided tissue regeneration. *Biomaterials* 26:7564–7571.

Liao, S., Watari, F., Zhu, Y. et al. 2007. The degradation of the three layered nano-carbonated hydroxyapatite/collagen/PLGA composite membrane in vitro. *Dent Mater* 23:1120–1128.

Linde, A., Alberius, P., Dahlin, C., Bjurstam, K., Sundin, Y. 1993. Osteopromotion: A soft-tissue exclusion principle using a membrane for bone healing and bone neogenesis. *J Periodontol* 64:1116–1128.

Liu, W., Yeh, Y.C., Lipner, L. et al. 2011. Enhancing the stiffness of electrospun nanofiber scaffolds with controlled surface coating and mineralization. *Langmuir* 27(15):9088–9093, doi: 10.1021/la2018105.

Livesey, S. A., Herndon, D. N., Hollyoak, M. A. et al. 1995. Transplanted acellular allograft dermal matrix-potential as a template for the reconstruction of viable dermis. *Transplantation* 60:1–9.

Lutolf, M. P., Hubbell, J. A. 2005. Synthetic biomaterials as instructive extracellular microenvironments for morphogenesis in tissue engineering. *Nat Biotechnol* 23:47–55.

Magnusson, I., Batich, C., Collins, B. R. 1988. New attachment formation following controlled tissue regeneration using biodegradable membranes. *J Periodontol* 59:1–6.

Martin, A. R. 1991. Anti-infective agents. In: *Textbook of Organic, Medicinal and Pharmaceutical Chemistry*,10th edn. eds. J. N. Delgado, W. A. Remers, pp. 173–219. Philadelphia, PA: Lippincott-Ravan Pub.

McClain, P. K., Schallhorn, R. G. 2000. Focus on furcation defects-guided tissue regeneration in combination with bone grafting. *Periodontol* 2000(22):190–212.

McClure, M. J., Sell, S. A., Simpson, D. G., Walpoth, B. H., Bowlin, G. L. 2010. A three-layered electrospun matrix to mimic native arterial architecture using polycaprolactone, elastin, and collagen: A preliminary study. *Acta Biomater* 6:2422–2433.

Milella, E., Ramires, P. A., Brescia E., La Sala, G., Di Paola, L., Bruno, V. 2001. Physicochemical, mechanical, and biological properties of commercial membranes for GTR. *J Biomed Mater Res* 58:427–435.

Miura, M., Gronthos, S., Zhao, M. et al. 2003. Stem cells from human exfoliated deciduous teeth. *Proc Natl Acad Sci USA* 100:5807–5812.

Mombelli, A., Lang, N. P., Nyman, S. 1993. Isolation of periodontal species after guided tissue regeneration. *J Periodontol* 64:1171–1175.

Murphy, K. G., Gunsolley, J. C. 2003. Guided tissue regeneration for the treatment of periodontal intrabony and furcation defects. A systematic review. *Ann Periodontol* 8:266–302.

Murugan, R., Ramakrishna, S. 2006. Nano-featured scaffolds for tissue engineering: A review of spinning methodologies. *Tissue Eng* 12:435–447.

Nakahara, T. 2006. A review of new developments in tissue engineering therapy for periodontitis. *Dent Clin N Am* 50:265–276.

Nakashima, M., Reddi, H. A. 2003. The application of bone morphogenetic proteins to dental tissue engineering. *Nat Biotechnol* 21(9):1025–1032.

Nanci, A., Bosshardt, D. D. 2006. Structure of periodontal tissues in health and disease. *Periodontol* 2000(40):11–28.

Needleman, I. G., Worthington, H. V., Giedrys-Leeper, E., Tucker, R. J. 2006. Guided tissue regeneration for periodontal infra-bony defects. *Cochrane Database Syst Rev* 19:CD001724.

Nevins, M., Giannobile, W. V., McGuire, M. K. et al. 2005. Platelet-derived growth factor stimulates bone fill and rate of attachment level gain: Results of a large multicenter randomized controlled trial. *J Periodontol* 76:2205–2215.

Ngiam, M., Liao, S. S., Patil, A. J. et al. 2009. The fabrication of nanohydroxyapatite on PLGA and PLGA/collagen nanofibrous composite scaffolds and their effects in osteoblastic behavior for bone tissue engineering. *Bone* 45:4–16.

Nie, H., Soh, B. W., Fu, Y. C., Wang, C. H. 2008. Three-dimensional fibrous PLGA/HAp composite scaffold for BMP-2 delivery. *Biotechnol Bioeng* 99:223–234.

Nimni, M. E., Cheung, D., Strates, B., Kodama, M., Sheikh, K. 1988. Bioprosthesis derived from cross-linked and chemically modified collagenous tissues. In: *Collagen*, Vol. III. ed. M. E. Nimni, pp. 1–38. Boca Raton, FL: CRC Press.

Nishimura, F., Iwamoto, Y., Soga, Y. 2007. The periodontal host response with diabetes. *Periodontol* 2000(43):245–253.

Nyman, S., Gottlow, J., Karring, T., Lindhe, J. 1982. The regenerative potential of the periodontal ligament. An experimental study in the monkey. *J Clin Periodontol* 9:257–265.

Nyman, S., Gottlow, J., Lindhe, J., Karring, T., Wennstrom, J. 1987. New attachment formation by guided tissue regeneration. *J Periodont Res* 22:252–254.

Park, Y. J., Lee, Y. M., Park, S. N. et al. 2000. Enhanced guided bone regeneration by controlled tetracycline release from poly(L-lactide) barrier membranes. *J Biomed Mater Res* 51:391–397.

Park, J. K., Yeom, J., Oh, E. J. et al. 2009. Guided bone regeneration by poly(lactic-co-glycolic acid) grafted hyaluronic acid bi-layer films for periodontal barrier applications. *Acta Biomater* 5:3394–3403.

Pham, Q. P., Sharma, U., Mikos, A. G. 2006. Electrospinning of polymeric nanofibers for tissue engineering applications: A review. *Tissue Eng* 12:1197–1211.

Phipps, M. C., Clem, W. C., Catledge, S. A. et al. 2011. Mesenchymal stem cell responses to bone-mimetic electrospun matrices composed of polycaprolactone, collagen I and nanoparticulate hydroxyapatite. *PlosOne* 6(2):e16813.

Piattelli, A., Scarano, A., Russo, P., Matarasso, S. 1996. Evaluation of guided bone regeneration in rabbit tibia using bioresorbable and non-resorbable membranes. *Biomaterials* 17:791–796.

Pihlstrom, B. L., Michalowicz, B. S., Johnson, N. W. 2005. Periodontal diseases. *Lancet* 366:1809–1820.

Polimeni, G., Albandar, J. M., Wikesjö, U. M. 2005. Prognostic factors for alveolar regeneration: Effect of space provision. *J Clin Periodontol* 32:951–954.

Polimeni, G., Koo, K. T., Pringle, G. A. et al. 2008. Histopathological observations of a polylactic acid-based device intended for guided bone/tissue regeneration. *Clin Implant Dent Relat Res* 10:99–105.

Polimeni, G., Xiropaidis, A. V., Wikesjoe, U. M. E. 2006. Biology and principles of periodontal wound healing/regeneration. *Periodontol* 2000(41):30–47.

Prabhakaran, M. P., Venugopal, J., Ramakrishna, S. 2009. Electrospun nanostructured scaffolds for bone tissue engineering. *Acta Biomater* 5:2884–2893.

Ramakrishna, S., Fujihara, K., Teo, W. E., Lim, T. C., Ma, Z. 2005. *Introduction to Electrospinning and Nanofibers*. Singapore: World Scientific Publishing Company, Inc.

Reneker, D. H., Yarin, A. L. 2008. Electrospinning jets and polymer nanofibers. *Polymer* 49:2387–2425.

Retzepi, M., Donos, N. 2010. Guided bone regeneration: Biological principle and therapeutic applications. *Clin Oral Implants Res* 21:567–576.

Ripamonti, U. 2007. Recapitulating development: A template for periodontal tissue engineering. *Tissue Eng* 13:51–71.

Santos, A., Goumenos, G., Pascual, A. 2005. Management of gingival recession by the use of an acellular dermal graft material: A 12-case series. *J Periodontol* 76:1982–1990.

Sculean, A., Nikolidakis, D., Schwarz, F. 2008. Regeneration of periodontal tissues: Combinations of barrier membranes and grafting materials—Biological foundation and preclinical evidence: A systematic review. *J Clin Periodontol* 35:106–116.

Sell, A., Barnes, C., Smith, M. et al. 2007. Extracellular matrix regenerated: Tissue engineering via electrospun biomimetic nanofibers. *Polym Int* 56:1349–1360.

Shin, S. Y., Park, H. N., Kim, K. H. et al. 2005. Biological evaluation of chitosan nanofiber membrane for guided bone regeneration. *J Periodontol* 76:1778–1784.

Sill, T. J., von Recum, H. A. 2008. Electro spinning: Applications in drug delivery and tissue engineering. *Biomaterials* 29:1989–2006.

Slots, J., MacDonald, E. S., Nowzari, H. 1999. Infectious aspects of periodontal regeneration. *Periodontol* 2000(19):164–172.

Soletti, L., Hong, Y., Guan, J. et al. 2010. A bilayered elastomeric scaffold for tissue engineering of small diameter vascular grafts. *Acta Biomater* 6:110–122.

Southerland, J. H., Taylor, G. W., Moss, K., Beck, J. D., Offenbacher, S. 2006. Commonality in chronic inflammatory diseases: Periodontitis, diabetes, and coronary artery disease. *Periodontol* 2000(40):130–143.

Srouji, S., Ben-David, D., Lotan, R., Livne, E., Avrahami, R., Zussman, E. 2011. Slow-release human recombinant bone morphogenetic protein-2 embedded within electrospun scaffolds for regeneration of bone defect: In vitro and in vivo evaluation. *Tissue Eng* 17:269–277.

Sundararaghavan, H. G., Monteiro, G. A., Lapin, N. A., Chabal, Y. J., Miksan, J. R., Shreiber, D. I. 2008. Genipin-induced changes in collagen gels: Correlation of mechanical properties to fluorescence. *J Biomed Mater Res* 87A:308–320.

Sung, H. W., Chang, Y., Chiu, C. T., Chen, C. N., Liang, H. C. 1999. Mechanical properties of a porcine aortic valve fixed with a naturally occurring crosslinking agent. *Biomaterials* 20:1759–1772.

Taba, M., Jr., Jin, Q., Sugai, J. V., Giannobile, W. V. 2005. Current concepts in periodontal bioengineering. *Orthod Craniofac Res* 8:292–302.

Tal, H., Kozlovsky, A., Artzi, Z., Nemcovsky, C. E., Moses, O. 2008. Cross-linked and non-cross-linked collagen barrier membranes disintegrate following surgical exposure to the oral environment: A histological study in the cat. *Clin Oral Implants Res* 19:760–766.

Teng, S. H., Lee, E. J., Wang, P., Shin, D. S., Kim, H. E. 2008. Three-layered membranes of collagen/hydroxyapatite and chitosan for guided bone regeneration. *J Biomed Mater Res Appl Biomater* 87B:132–138.

Teng, S. H., Lee, E. J., Yoon, B. H., Shin, D. S., Kim, H. E., Oh, J. S. 2009. Chitosan/nanohydroxyapatite composite membranes via dynamic filtration for guided bone regeneration. *J Biomed Mater Res* 88A:569–580.

Thomas, V., Dean, D. R., Jose, M. V., Mathew, B., Chowdhury, S., Vohra, Y. K. 2007a. Nanostructured biocomposite scaffolds based on collagen co-electrospun with nanohydroxyapatite. *Biomacromolecules* 8:631–637.

Thomas, V., Dean, D. R., Vohra, Y. K. 2006a. Nanostructured biomaterials for regenerative medicine. *Curr Nanosci* 2:155–177.

Thomas, V., Jagani, S., Johnson, K. et al. 2006b. Electrospun bioactive nanocomposite scaffolds of polycaprolactone and nanohydroxyapatite for bone tissue engineering. *J Nanosci Nanotechnol* 6:487–493.

Thomas, V., Zhang, X., Catledge, S. A., Vohra, Y. K. 2007b. Functionally graded electrospun scaffolds with tunable mechanical properties for vascular tissue regeneration. *Biomed Mater* 2:224–232.

Thomas, V., Zhang, X., Vohra, Y. K. 2009. A biomimetic tubular scaffold with spatially designed nanofibers of protein/PDS bio-blends. *Biotechnol Bioeng* 104:1025–1033.

Trombelli, L. 2005. Which reconstructive procedures are effective for treating the periodontal intraosseous defect? *Periodontol* 2000(37):88–105.

Tugnait, A., Clerehugh, V. 2001. Gingival recession-its significance and management. *J Dent* 29:381–394.

Vaz, C. M., van Tuijl, S., Bouten, C. V., Baaijens, F. P. 2005. Design of scaffolds for blood vessel tissue engineering using a multi-layering electrospinning technique. *Acta Biomater* 1(5):575–582.

Venugopal, J., Low, S., Choon, A. T., Kumar, T. S. S., Ramakrishna, S. 2008a. Mineralization of osteoblasts with electrospun collagen/hydroxyapatite nanofibers. *J Mater Sci Mater Med* 19:2039–2046.

Venugopal, J., Low, S., Choon, A. T., Ramakrishna, S. 2008b. Interaction of cells and nanofiber scaffolds in tissue engineering *J Biomed Mater Res Appl Biomater* 84B:34–48.

Wang, H. L. 2005. Periodontal regeneration. *J Periodontol* 76:1601–1622.

Wang, C., Cheng, W., Zhang, R., Ma, J. 2002. Thick hydroxyapatite coatings by electrophoretic deposition. *Mater Lett* 57:99–105.

Wikesjo, U. M., Sigurdsson, T. J., Lee, M. B., Tatakis, D. N., Selvig, K. A. 1995. Dynamics of wound healing in periodontal regenerative therapy. *J Calif Dent Assoc* 23:30–35.

Wutticharoenmongkol, P., Sanchavanakit, N., Pavasant, P., Supaphol, P. 2006. Preparation and characterization of novel bone scaffolds based on electrospun polycaprolactone fibers filled with nanoparticles. *Macromol Biosci* 6:70–77.

Xie, J., Li, X., Xia, Y. 2008 Putting electrospun nanofibers to work for biomedical research. *Macromol Rapid Commun* 29(22):1775–1792.

Yang, F., Both, S. K., Yang, X. C., Walboomers, X. F., Jansen, J. A. 2009. Development of an electrospun nano-apatite/PCL composite membrane for GTR/GBR application. *Acta Biomater* 5:3295–3304.

Zamani, M., Morshed, M., Varshosaz, J., Jannesari, M. 2010. Controlled release of metronidazole benzoate from poly epsilon-caprolactone electrospun nanofibers for periodontal diseases. *Eur J Pharm Biopharm* 75:179–185.

Zeichner-David, M. 2006. Regeneration of periodontal tissues: Cementogenesis revisited. *Periodontol* 2000(41):196–217.

Zhang, X., Thomas, V., Vohra, Y. K. 2009. In vitro biodegradation of designed tubular scaffolds of electrospun protein/polyglyconate blend fibers. *J Biomed Mater Res Appl Biomater* 89B:135–147.

Zhang, X., Thomas, V., Xu, Y., Bellis, S. L., Vohra, Y. K. 2010a. An in vitro regenerated functional human endothelium on a nanofibrous electrospun scaffold. *Biomaterials* 31:4376–4381.

Zhang, J., Xu, Q., Huang, C., Mo, A., Li, J., Zuo, Y. 2010b. Biological properties of an anti-bacterial membrane for guided bone regeneration: An experimental study in rats. *Clin Oral Implants Res* 21:321–327.

Zhang, Y., Zhang, M. Q. 2004. Cell growth and function on calcium phosphate reinforced chitosan scaffolds. *J Mater Sci Mater Med* 15:255–260.

8

Autoinductive Scaffolds for Osteogenic Differentiation of Mesenchymal Stem Cells

Esmaiel Jabbari and Murugan Ramalingam

CONTENTS

8.1 Introduction

Growth and differentiation factors play a central role in modulation and control of cell migration, differentiation and maturation, and morphogenesis [1]. The proteins in the collagenous phase of the bone matrix play a central role in controlling the function of osteoblasts and progenitor bone mesenchymal stem (BMS) cells [2]. For example, extracellular matrix (ECM) proteins including collagen, fibronectin, sialoprotein, osteopontin (OP), and small integrin-binding ligands (SIBLINGS) contain the short RGD peptide, which is critical for cell adhesion to the matrix through integrin-binding cell surface receptors [3,4]. The RGD peptide has been conjugated to poly(ε-caprolactone) [5], poly(L-lactide) (PLA)[6], poly(lactide-co-glycolide) (PLGA) [7], and polyethylene glycol [8,9] to improve cell adhesion and differentiation. The previous studies have shown that conjugation of the RGD peptide to tissue-engineered

(TE) scaffolds promotes focal-point adhesion of BMS cells and their osteo-genic differentiation by increasing alkaline phosphatase (ALPase) activity, osteocalcin (OC) and OP expression, and mineralization in a dose-dependent manner.

In bone regeneration, BMS cells migrate from the bone marrow or periosteum to the regeneration site as a result of expression of bone morphogenetic proteins (BMPs) [10]. BMPs are a group of proteins that play important roles in tissue regeneration and remodeling processes after injuries [11]. BMP signaling is highly regulated by pro-BMP domains [12], binding peptides [13], and other extracellular factors [14] that synergistically enhance the effect of BMPs. Therefore, doses 4–5 orders of magnitude higher than the amount found endogenously [11] have to be loaded in the graft. In particular, BMP-2 plays a major role in initiating the cascade of chemotaxis, differentiation of BMS cells, and bone regeneration [15,16], and recombinant human BMP-2 (rhBMP-2) is used clinically in spinal fusion [11,17]. High doses coupled with diffusion of rhBMP-2 away from the intended site of regeneration [18] cause adverse effects such as bone overgrowth and immunological reactions [19]. After loading rhBMP-2 in the TE scaffold, a large fraction of the protein is lost in the process of irrigating the wound and by the action of antibiotics. Furthermore, the carrier should retain rhBMP-2 at the regeneration site for a sufficient time and concentration to chemotactically recruit osteoprogenitor cells and allow their differentiation toward osteoblast phenotype [20]. Encapsulation in micro-/nanoparticles (NPs) has been used to reduce diffusion of rhBMP-2 away from the regeneration site and to reduce its enzymatic degradation [21–25]. Although encapsulated rhBMP-2 has been shown to enhance mineralization and bone formation [26], a large fraction of the protein is deactivated in the encapsulation process [27]. Therefore, relatively high doses have to be loaded in micro-/nanoparticles, affecting the safety profile of rhBMP-2 for clinical applications [19,28].

Active sequences from the active domain of proteins have been shown to promote differentiation of progenitor cells to a specific lineage. For example, the amino acid sequence KIPKA SSVPT ELSAI STLYL (hereafter referred to as the BMP peptide), corresponding to residues 73–92 of the knuckle epitope of rhBMP-2, promotes matrix mineralization and bone healing [29–31]. The BMP peptide has a significantly higher ALPase activity than other active sequences of rhBMP-2 (amino acid sequences 68–87, 68–92, 78–97, 44–58), inhibits the binding of rhBMP-2 to both BMP receptors type IA and type II, and promotes the expression of OC mRNA [29]. In another study, a multi-domain peptide consisting of a BMP receptor targeting sequence, a hydrophobic spacer, and a heparin-binding sequence promoted cell adhesion, and it was a positive modulator of rhBMP-2 [14]. The previous works coupled with the fact that the combination of a collagen network and soluble BMP is required for inducing mineralization [32] indicate that multiple peptides from the insoluble and soluble fractions of the bone ECM contribute to the cascade of osteogenesis and differentiation of BMS cells.

Here, we present two strategies to develop autoinductive scaffolds to support osteogenic differentiation of progenitor BMS cells and bone formation. In the first strategy, combination of two peptides, the RGD amino acid sequence from the insoluble collagenous network of the bone ECM and the KIPKA SSVPT ELSAI STLYL sequence derived from BMP-2 in the soluble fraction of the bone ECM, is grafted to the scaffold to produce autoinductive osteogenic scaffolds. In the second strategy, degradable NPs with large surface area for grafting are used as a delivery platform for rhBMP-2. In this approach, the scaffold is dipped in a hydrogel precursor solution containing rhBMP-2 grafted NPs. The grafted NPs suspended in the scaffold serve as a platform for sustained delivery of rhBMP-2 to recruit and differentiate osteoprogenitor cells as well as reduce diffusion of protein away from the regeneration site.

8.2 Peptide-Grafted Scaffolds

The BMP peptide sequence KIPKA SSVPT ELSAI STLYL, corresponding to residues 73–92 of the knuckle epitope of rhBMP-2, promotes matrix mineralization [29–31]. The BMP peptide significantly inhibits the binding of rhBMP-2 to both BMP receptors type I and type II and promotes the expression of OC mRNA [29]. Furthermore, the combination of a collagen network and soluble BMP is required for inducing mineralized matrix formation [32], which indicates that synthetic peptides with cell adhesion domains can potentially modulate the action of BMPs or peptides to enhance osteogenic differentiation of BMS cells. The objective was to determine the effect of RGD integrin-binding cell adhesion peptide and the BMP osteoinductive peptide, grafted to an inert degradable hydrogel, on osteogenic differentiation of bone marrow derived mesenchymal stem (BMS) cells. The inert degradable hydrogel used in this work was based on the poly(lactide-co-ethylene oxide fumarate) (PLEOF) macromer, consisting of low-molecular-weight poly(L-lactide) (LMW PLA) and polyethylene glycol (PEG) blocks linked by unsaturated fumarate units [33,34]. The PLA and PEG are FDA approved and fumaric acid occurs naturally in the Kreb's cycle. The water content can be adjusted by the ratio of the hydrophilic PEG to hydrophobic PLA blocks and by the molecular weight of PEG. The degradation rate of the network can be tailored to a particular application by the ratio of PLA to PEG blocks. Acrylated RGD peptide was conjugated to the hydrogel by reaction with unsaturated fumarate units. BMP peptide was grafted to the hydrogel by "click chemistry" [35,36].

8.2.1 Hydrogel Synthesis

LMW PLA was synthesized by ring opening polymerization of LA, as described [37]. The number average molecular weight (M_n) and polydispersity index (PI) of the synthesized LMW PLA were 1.2 kDa and 1.4, respectively, and those of PEG were 3.4 kDa and 1.1. PLEOF macromer was synthesized by condensation polymerization of LMW PLA and PEG with fumaryl chloride (FuCl) [33]. The PEG:LMW PLA weight ratio was 70:30 to produce a hydrophilic water-soluble macromer. The M_n and PI of the PLEOF synthesized from LMW PLA and PEG were 8.1 kDa and 1.6, respectively.

8.2.2 Synthesis of Functionalized Peptides

Acrylated GRGD (Ac-GRGD) was synthesized on the Rink Amide NovaGel™ resin, using a procedure similar to that of Ac-QPQGLAK-Ac peptide [38]. After coupling the last amino acid and selectively decoupling the Fmoc protecting group of the amino acid residues, the GRGD peptide was acrylated directly on the peptidyl resin at the N-terminal by coupling acrylic acid to the amine group of the glycine residue. The HPLC retention time of the Ac-GRGD peptide was 8.04 min. Mass numbers 457 and 479 corresponding to monovalent hydrogen and sodium cations of the Ac-GRGD peptide, respectively, were observed in the ESI-MS spectrum. The BMP peptide was synthesized on the Rink Amide NovaGel resin, using a procedure similar to that for Ac-GRGD with some modifications. To increase solubility of the azide-functionalized BMP peptide in aqueous solution and reduce steric hindrance in the grafting reaction, an Fmoc-protected mini-PEG spacer was inserted between the BMP peptide and functional azide group. After coupling the mini-PEG spacer and deprotecting the Fmoc-protecting group of the amino acid residues, the mPEG-BMP peptide was functionalized with an azide group, directly on the resin, by the reaction of 4-carboxybenzenesulfonazide with the amine end group of the mini-PEG spacer. The HPLC retention time of the peptide was 23.77 min. In the mass spectrum, mass numbers 2445 and 2467 corresponded to the monovalent hydrogen and sodium cations of the peptide, respectively. ALPase activity was used as a marker to determine osteogenic activity of Az-mPEG-BMP peptide with BMS cells at a concentration of 200 ng/mL [14]. ALPase activity increased after 7 days and peaked in 14 days with the addition of Az-mPEG-BMP peptide, but peak activity was significantly higher than the group without BMP peptide. Furthermore, BMS cells cultured with Az-mPEG-BMP for 21 days stained positive for alizarin red and produced mineralized matrix, while those without the peptide did not induce mineralization.

8.2.3 Grafting Peptides to the Hydrogel

RGD sequence of the Ac-GRGD peptide provides integrin-binding sites on the hydrogel surface for adhesion of BMS cells while the BMP peptide initiates the cascade of osteogenesis and mineralization. Schematic diagram for grafting of the peptides to the hydrogel is shown in Figure 8.1. The hydrogel

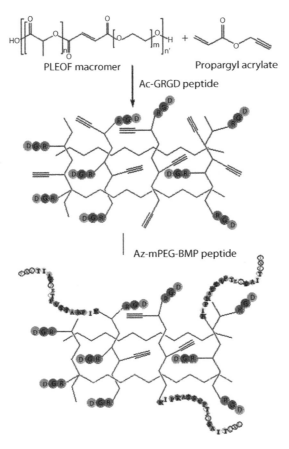

FIGURE 8.1
Schematic diagram showing the steps for the synthesis of PLEOF hydrogel, conjugated with Ac-GRGD peptide and surface-grafted with Az-mPEG-BMP peptide.

was prepared by the polymerization of PLEOF macromer, propargyl acrylate, and Ac-GRGD peptide in aqueous solution using redox initiation under sterile conditions [33,34]. The solution of PLEOF macromer, propargyl acrylate, Ac-GRGD peptide, and initiation system was cross-linked to form an RGD-conjugated hydrogel with propargyl groups. Next, Az-mPEG-BMP peptide was covalently attached to the surface of the hydrogel by click reaction [35,36] between the azide group of the peptide and propargyl group of the hydrogel [39]. It has been shown that BMS cells exhibit focal-point adhesion and the highest level of mRNA expression of OP, an early marker for osteogenesis, when seeded on scaffolds conjugated with 1 wt.% Ac-GRGD [40]. Therefore, the hydrogel was conjugated with 1 wt.% Ac-GRGD. The density of Az-mPEG-BMP peptide, measured by Kaiser test [41,42], was $5.2 \pm 0.6 \, \text{nmol/cm}^2$.

8.2.4 Osteogenic Differentiation and Mineralization of the BMS Cells

BMS cells were isolated from the bone marrow of young adult male Wistar rats [33,34]. BMS cells were seeded at a density of 2×10^5 cells/cm^2 on the hydrogel and incubated in osteogenic media from 7 to 21 days. Samples were removed after 7, 14, and 21 days and analyzed for ALPase activity (osteogenic differentiation) and calcium content (mineralization). Experimental groups included hydrogel without peptide grafting (control group), with RGD (RGD), with BMP peptide (BMP), with RGD and BMP peptides (RGD + BMP), and with RGD and mutant BMP peptides (RGD + muBMP). Cell count was highest on surfaces conjugated with integrin-binding RGD ligand (RGD and RGD + BMP groups). The cell count demonstrated that RGD-conjugated hydrogels enhanced adhesion of BMS cells to the substrate, especially in the first week after cell seeding, consistent with previous results [8,43,44]. The ALPase activity of BMS cells seeded on hydrogels conjugated with RGD and grafted with BMP peptide increased fivefold from 7 to 14 days and then returned to the baseline level after 21 days. BMS cells seeded on peptide-modified hydrogels stained positive for alizarin red (Figure 8.2a) and produced a mineralized matrix. The extent of mineralization (calcium content) of the ECM produced by BMS cells, seeded on peptide-modified hydrogels, is shown in Figure 8.2b. BMS cells seeded on hydrogels without peptide modification did not show increase in calcium content with increasing incubation time while those cultured on RGD-conjugated hydrogels showed sevenfold

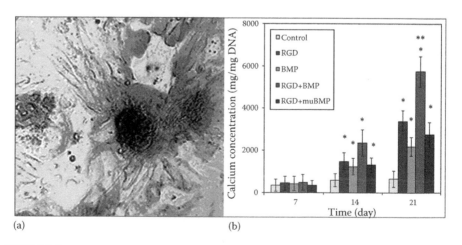

(a)   (b)

FIGURE 8.2
(a) Demonstration of mineralized matrix formation by alizarin red staining of BMS cells seeded on the hydrogel with Ac-GRGD/Az-mPEG-BMP peptide grafting after 3 weeks incubation in osteogenic media; (b) calcium content of the BMS cells seeded on hydrogels and incubated in osteogenic media. Experimental groups include BMS cells seeded on hydrogels without peptide grafting (control), conjugated with RGD (RGD), grafted with BMP peptide (BMP), conjugated with RGD and grafted with BMP peptide (RGD + BMP), and conjugated with RGD and grafted with mutant BMP peptide (RGD + muBMP).

increase in calcium content from day 14 to 21. Therefore, RGD-conjugation functions as a mild promoter of osteogenic differentiation of BMS cells. BMS cells cultured on BMP peptide grafted hydrogels (without Ac-GRGD) showed fivefold increase in calcium content from day 14 to 21, indicating that osteogenic activity of the BMP peptide is dependent on the presence of RGD for adhesion. BMS cells seeded on PLEOF substrates conjugated with RGD and grafted with the BMP peptide, showed 12-fold increase in calcium content from day 14 to 21, which was significantly higher than the BMP peptide grafted or RGD-conjugated or substrates. These findings clearly demonstrate that RGD and BMP peptides, grafted to a hydrogel substrate, act synergistically to increase osteogenic differentiation of BMS cells.

RGD peptide which is associated with ECM components like collagen type I, fibronectin, sialoprotein, and OP [4] interacts with the BMS cells through integrin cell surface receptors to facilitate spreading and focal-point adhesion of the BMS cells. It has been shown that covalent attachment of RGD peptide to polymer surfaces facilitates spreading and focal-point adhesion of BMS cells and increases the expression of markers for osteogenic differentiation [8,43–46]. In essence, grafting of cell-adhesive RGD and the osteogenic BMP peptide to a hydrogel scaffold synergistically enhances osteogenic differentiation of BMS cells. RGD peptide provides sites for cell attachment to the scaffold, which in turn enhances the interaction of the BMP peptide with type I and type II transmembrane serine/threonine receptor kinase, leading to an increase in osteogenesis and mineralization.

8.3 Protein-Grafted Nanoparticles

The rate at which rhBMP-2 is released from the carrier can affect the efficacy of bone induction [47]. The carrier should retain rhBMP-2 at the site for a sufficient time and concentration to chemotactically recruit osteoprogenitor cells to the site and allow their differentiation to the osteoblast phenotype [20]. Encapsulation in micro-/nanoparticles has been used to reduce diffusion of rhBMP-2 away from the regeneration site and to reduce its enzymatic degradation [21–25]. Although encapsulated rhBMP-2 has been shown to enhance mineralization and bone formation [26], a large fraction of the protein is deactivated in the process of emulsification [27]. Therefore, relatively high doses have to be loaded in the particles, affecting the safety profile of rhBMP-2 for clinical applications [19,28]. Functionalized NPs provide large surface area and reactive groups for grafting proteins. The objective was to investigate the osteogenic activity of rhBMP-2 protein grafted to self-assembled biodegradable NPs. Biodegradable NPs based on poly(lactide-*co*-glycolide fumarate) (PLGF) and poly(lactide-*co*-ethylene oxide fumarate) (PLEOF) macromers were used in this investigation [37].

In the process of self-assembly, PLEOF macromers act as a surfactant to stabilize the PLGF NPs. NPs ranging 50–500 nm in size can be produced by varying the ratio of PLEOF to PLGF in the blend. The macromers can be functionalized with succinimide groups for grafting proteins in aqueous media [48].

8.3.1 Nanoparticle Synthesis

LMW PLA and LMW PLGA were synthesized by ring-opening polymerization of LA and GL monomers, respectively [37]. Number average molecular weight ($\overline{M_n}$), weight average molecular weight ($\overline{M_w}$), and PI of the LMW PLA macromer were 1450 Da, 1730 Da, and 1.2, respectively, and those of LMW PLGA were 1660 Da, 2150 Da, and 1.3. Next, PLAF or PLGF was synthesized by condensation polymerization of LMW PLA or PLGA, respectively, with FuCl [37,40]. Similarly, PLEOF was synthesized by reacting LMW PLA and PEG with FuCl [33,34,49]. $\overline{M_n}$, $\overline{M_w}$, and PI of the synthesized PLAF were 4.5 kDa, 8.6 kDa, and 1.9, respectively; those of PLGF were 5.2 kDa, 10.9 kDa, and 2.1; and those of PLEOF were 9.7 kDa, 15.4 kDa, and 1.6. PLAF and PLGF chains were succinimide-terminated (PLAF-NHS and PLGF-NHS) by the reaction of hydroxyl end groups of the macromers with *N,N*-disuccinimidyl carbonate (DSC) [50]. The succinimide-terminated macromers and PLEOF were self-assembled into NPs by dialysis, as shown in Figure 8.3. The macromers were dissolved in dimethylformamide (DMF) and dialyzed against phosphate buffer saline (PBS) to form NPs. Attachment of succinimide end-groups to PLAF macromer resulted in an increase in number average from 4.5 to 4.7 kDa and a slight increase in weight average from 8.61 to 8.63 kDa (PI decreased slightly from 1.91 to 1.83).

8.3.2 Protein Grafting

The NPs were centrifuged, resuspended in PBS, and rhBMP-2 protein was added to the NPs suspension. The protein was grafted by the reaction of succinimide groups of the macromer on the NPs with the protein's amine

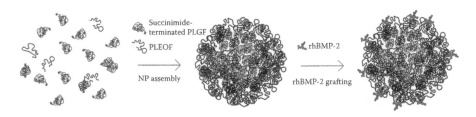

FIGURE 8.3
Schematic diagram for self-assembly of PLGF-NHS/PLEOF macromers to form NPs. After self-assembly, rhBMP-2 is grafted to the NPs by the reaction between amine groups of the protein with succinimide end-groups of the PLGF-NHS macromers of the NPs.

groups. The average number of succinimide end-groups per macromer was 1.4. NPs had spherical geometry. The average size of PLAF-NHS NPs increased from 240 ± 70 to 250 ± 80 nm after grafting with rhBMP-2 while that of PLGF-NHS NPs increased from 195 ± 40 to 200 ± 70 nm, consistent with size of the protein in the native conformation [51]. Grafting efficiency was determined by measuring the concentration of rhBMP-2 in the supernatant after centrifugation of the NPs by enzyme-linked immunosorbent assay (ELISA). The grafting efficiency was $97\% \pm 1\%$ and $98\% \pm 1\%$ for PLAF-NHS and PLGF-NHS NPs, respectively. The stability of the protein with time was tested by incubating rhBMP-2 in PBS with PLAF and PLGF. The relative enzymatic activity after incubation for 1, 15, and 30 days was $100\% \pm 8\%$, $102\% \pm 8\%$, and $103\% \pm 8\%$, respectively, demonstrating that incubation time did not affect stability of the protein. The release characteristics of enzymatically active rhBMP-2 from PLAF-NHS and PLGF-NHS NPs are shown in Figure 8.4a. Release of active rhBMP-2 from PLGF-NHS and PLAF-NHS NPs was linear in the first 15 and 20 days, respectively. Nearly $25\% \pm 2\%$ and $50\% \pm 1\%$ of the grafted rhBMP-2 was released in enzymatically active conformation after complete degradation of the PLGF-NHS and PLAF-NHS NPs, respectively.

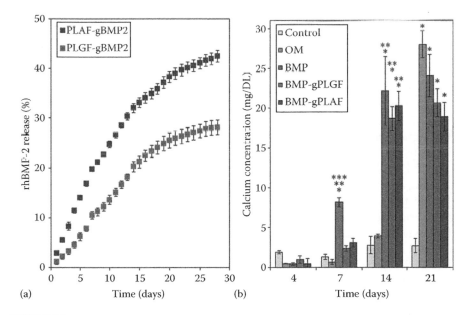

FIGURE 8.4
(a) Release kinetics of bioactive rhBMP-2 protein grafted PLAF and PLGF NPs; (b) the extent of mineralization (calcium content) of BMS cells as a function of incubation time cultured in osteogenic media supplemented with rhBMP-2 grafted NPs. One star in (b) indicates statistically significant difference between test group and control; two stars between test group and OM; and three stars between grafted-BMP groups and BMP.

8.3.3 Osteogenic Activity of Protein-Grafted Nanoparticles

BMS cells were isolated from the bone marrow of young adult male Wistar rats [33,39] and seeded at a density of 5×10^4 cells/mL. After cell attachment, the media was replaced with standard osteogenic media with the addition of 200 ng/mL rhBMP-2 grafted NPs. rhBMP-2 directly added to the BMS cell cultures in osteogenic media was used as the positive control and BMS cell cultured without rhBMP-2 were used as the negative control. Groups included BMS cells in primary media (control), osteogenic media (OM), OM supplemented with 200 ng/mL rhBMP-2 (BMP), rhBMP-2 grafted to PLGF-NHS NPs (PLGF-gBMP), and rhBMP-2 grafted to PLAF-NHS NPs (PLAF-gBMP). DNA content of the control group was significantly higher than the BMP groups consistent with previous results that the addition of rhBMP-2 reduces proliferation and enhances osteogenic differentiation of BMS cells [52,53]. The increase in ALPase activity for PLGF-gBMP and PLAF-gBMP were similar to that of rhBMP-2 added directly to the culture media. The calcium content of the BMS cells with incubation time for the five groups is shown in Figure 8.4b. The calcium content of BMP-gPLAF, BMP-gPLGF, and BMP groups showed a significant increase from day 4 to 7 and 7 to 14. The group with BMP directly added to the culture media showed early increase in calcium content in day 7 but all three groups ultimately showed similar levels mineralization after 21 days. The expression level of osteogenic markers OP and OC with incubation time is shown in Figure 8.5a and b, respectively. At each time point, BMP-gPLAF group had significantly higher expression of OP and OC, followed by BMP-gPLGF and BMP groups. At each time point, OP and OC expression of BMP-gPLAF was significantly higher than that of BMP-gPLGF, consistent with higher cumulative release of rhBMP-2 from BMP-gPLAF NPs. After 21 days, OP expression of control, OM, BMP, BMP-gPLGF, and BMP-gPLAF was 2, 6.5, 6.8, 13, and 29, respectively, and OC expression was 12, 68, 67, 83, and 150. Higher OP and OC expression of rhBMP-2 grafted NP groups may be related to other factors in the cascade of osteogenesis, such as differentiation of the BMS cells to the vasculogenic lineage and formation of a vascularized/mineralized matrix.

While the reduction in entropy of unfolding in the grafted state has a stabilizing effect, the energetic interaction of the grafted protein negatively affects stability. rhBMP-2 protein grafted to the less polar PLAF-NHS NPs has less energetic interaction with the NPs surface, resulting in higher stability of the grafted protein. On the other hand, rhBMP-2 grafted to the relatively more polar PLGF-NHS interacts more strongly with the NPs surface, resulting in lower stability of the grafted protein. The scaffold should retain rhBMP-2 for a sufficient time and concentration to recruit osteoprogenitor cells [20] and reduce diffusion of the protein away from the site of regeneration. Since rhBMP-2 is a potent factor for osteogenic differentiation of progenitor BMS cells, diffusion of rhBMP-2 away from the intended site can cause bone overgrowth [11,54]. Grafting of rhBMP-2 to degradable NPs not

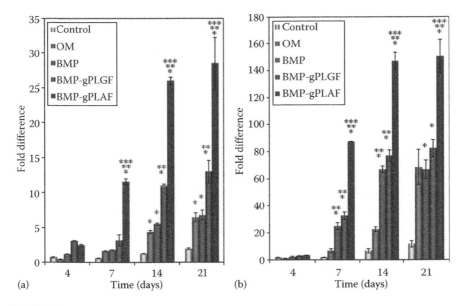

FIGURE 8.5
mRNA expression levels (as fold difference) of OP (a) and OC (b) for BMS cells as a function of incubation time cultured in osteogenic media supplemented with rhBMP-2 grafted NPs. One star in (a and b) indicates statistically significant difference between test group and control; two stars between test group and OM; and three stars between grafted-BMP groups and BMP.

only provides a sustained rhBMP-2 delivery system for recruitment and differentiation of osteoprogenitor cells, it also reduces diffusion of the protein away from the regeneration site. Results demonstrate that rhBMP-2 grafted to PLGF-NHS NPs can be as effective in inducing mineralization as the native rhBMP-2.

8.4 Conclusions

In this chapter, two strategies were presented for synthesizing autoinductive scaffolds to support osteogenic differentiation of the progenitor BMS cells. In the first strategy, combination of two peptides, the RGD amino acid sequence from the insoluble collagenous network of the bone ECM and the KIPKA SSVPT ELSAI STLYL sequence of rhBMP-2 from the soluble fraction of the bone ECM, was grafted to the scaffold to produce an autoinductive bone graft. RGD peptide was coupled to the scaffold by the reaction between the acrylate functional group of the peptide and the fumarate groups of the scaffold. The BMP peptide was grafted to the scaffold by the click reaction between the azide functional group of the peptide and the propargyl groups of the scaffold. The BMP and RGD peptides, grafted to the scaffold, synergistically enhanced osteogenic differentiation and mineralization of the BMS cells. In the second strategy, degradable NPs with large surface area for grafting were used as a delivery platform for rhBMP-2. In this approach, the grafted NPs suspended in the scaffold served as a platform for recruitment and differentiation of the osteoprogenitor BMS cells. rhBMP-2 was grafted to succinimide-functionalized degradable NPs. rhBMP-2 grafted NPs were as effective as the native protein in stimulating osteogenic differentiation of the BMS cells. Furthermore, rhBMP-2 grafted NPs had higher expression of osteogenic markers OP and OC compared to the native protein. Higher OP and OC expression of rhBMP-2 grafted NP groups may be related to other factors in the cascade of osteogenesis, such as differentiation of the BMS cells to the vasculogenic lineage and formation of a vascularized/mineralized matrix.

Acknowledgments

Preparation of this chapter was supported by grants to E. Jabbari from the National Science Foundation under grant nos. CBET0756394, CBET0931998, and DMR1049381; the National Institutes of Health under grant no. DE19180; and the Arbeitsgemeinschaft Fur Osteosynthesefragen (AO) Foundation under grant no. C10–44J. S.N. E.J. likes to thank Dr. Angel Mercado and Dr. Xuezhong He for the preparation of this book chapter.

References

1. Lisi S., Peterkova R., Peterka M., Vonesch J.L., Ruch J.V., Lesot H., Tooth morphogenesis and pattern of odontoblast differentiation. *Connect. Tissue Res.* **2003**, 44, 167–170.
2. Buckwalter J., Cooper R., Bone biology: Part II: Formation, form, modeling, remodeling, and regulation of cell function. *J. Bone Joint Surg.* **1995**, 77A(8), 1276–1289.
3. Wu S.C., Yu J.C., Hsu S.H., Chen D.C.H., Artificial extracellular matrix proteins contain heparin-binding and RGD-containing domains to improve osteoblast-like cell attachment and growth. *J. Biomed. Mater. Res.* **2006**, 79A, 557–565.
4. Ruoslahti E., Pierschbacher M.D., New perspectives in cell-adhesion—RGD and integrins. *Science* **1987**, 238, 491–497.
5. Santiago L.Y., Nowak R.W., Rubin J.P., Marra K.G., Peptide-surface modification of poly(caprolactone) with laminin-derived sequences for adipose-derived stem cell applications. *Biomaterials* **2006**, 27, 2962–2969.

6. Ho M.H., Hou L.T., Tu C.Y., Hsieh H.J., Lai J.Y., Chen W.J., Wang D.M., Promotion of cell affinity of porous PLA scaffolds by immobilization of RGD peptides via plasma treatment. *Macromol. Biosci.* **2006**, 6, 90–98.
7. Hsu S.H., Chang S.H., Yen H.J., Whu S.W., Tsai C.L., Chen D.C., Evaluation of biodegradable polyesters modified by type II collagen and Arg-Gly-Asp as tissue engineering scaffolding materials for cartilage regeneration. *Artif. Organs* **2006**, 30, 42–55.
8. Yang F., Williams C.G., Wang D.A., Lee H., Manson P.N., Elisseeff J., The effect of incorporating RGD adhesive peptide in polyethylene glycol diacrylate hydrogel on osteogenesis of bone marrow stromal cells. *Biomaterials* **2005**, 26, 5991–5998.
9. Deng C., Tian H.Y., Zhang P.B., Sun J., Chen X.S., Jing X.B., Synthesis and characterization of RGD peptide grafted poly(ethylene glycol)-b-poly(L-lactide)-b-poly(L-glutamic acid) triblock copolymer. *Biomacromolecules* **2006**, 7, 590–596.
10. Rosen V., Thies R., Adult skeletal repair. In *The Cellular and Molecular Basis of Bone Formation and Repair.* Springer, New York, 1995, pp. 97–142.
11. McKay B., Sandhu H.S., Use of recombinant human bone morphogenetic protein-2 in spinal fusion applications. *Spine* **2002**, 27, S66–S85.
12. Hillger F., Herr G., Rudolph R., Schwarz E., Biophysical comparison of BMP-2, ProBMP-2, and the free pro-peptide reveals stabilization of the pro-peptide by the mature growth factor. *J. Biol. Chem.* **2005**, 280, 14974–14980.
13. Behnam K., Phillips M.L., Sliva J.D.P., Brochmann E.J., Duarte M.E.L., Murray S.S., BMP binding peptide: A BMP-2 enhancing factor deduced from the sequence of native bovine bone morphogenetic protein/non-collagenous protein. *J. Orthop. Res.* **2005**, 23, 175–180.
14. Lin X.H., Zamora P.O., Albright S., Glass J.D., Pena L.A., Multidomain synthetic peptide B2A2 synergistically enhances BMP-2 in vitro. *J. Bone Miner. Res.* **2005**, 20, 693–703.
15. Ripamonti U., Reddi A.H., Tissue engineering, morphogenesis, and regeneration of the periodontal tissues by bone morphogenetic proteins. *Crit. Rev. Oral Biol. Med.* **1997**, 8, 154–163.
16. Wozney J.M., Overview of bone morphogenetic proteins. *Spine* **2002**, 27, S2–S8.
17. Robinson Y., Heyde C.E., Tschoke S.K., Mont M.A., Seyler T.M., Ulrich S.D., Evidence supporting the use of bone morphogenetic proteins for spinal fusion surgery. *Expert Rev. Med. Dev.* **2008**, 5, 75–84.
18. Lee S.H., Shin H., Matrices and scaffolds for delivery of bioactive molecules in bone and cartilage tissue engineering. *Adv. Drug Deliv. Rev.* **2007**, 59, 339–359.
19. Shields L.B.E., Raque G.H., Glassman S.D., Campbell M., Vitaz T., Harpring J., Shields C.B., Adverse effects associated with high-dose recombinant human bone morphogenetic protein-2 use in anterior cervical spine fusion. *Spine* **2006**, 31, (5), 542–547.
20. Li R.H., Wozney J.M., Delivering on the promise of bone morphogenetic proteins. *Trends Biotechnol.* **2001**, 19, (7), 255–265.
21. Lin H., Zhao Y., Sun W., Chen B., Zhang J., Zhao W., Xiao Z., Dai J., The effect of crosslinking heparin to demineralized bone matrix on mechanical strength and specific binding to human bone morphogenetic protein-2. *Biomaterials* **2008**, 29, 1189–1197.

22. Chen B., Lin H., Wang J., Zhao Y., Wang B., Zhao W., Sun W., Dai J., Homogeneous osteogenesis and bone regeneration by demineralized bone matrix loading with collagen-targeting bone morphogenetic protein-2. *Biomaterials* **2007**, 28, 1027–1035.

23. Wei G., Jin Q., Giannobile W.V., Ma P.X., The enhancement of osteogenesis by nano-fibrous scaffolds incorporating rhBMP-7 nanospheres. *Biomaterials* **2007**, 28, 2087–2096.

24. Fu Y.C., Nie H., Ho M.L., Wang C.K., Wang C.H., Optimized bone regeneration based on sustained release from three-dimensional fibrous PLGA/HAP composite scaffolds loaded with BMP-2. *Biotechnol. Bioeng.* **2008**, 99, 996–1006.

25. Chung Y.I., Ahn K.M., Jeon S.H., Lee S.Y., Lee J.H., Tae G., Enhanced bone regeneration with BMP-2 loaded functional nanoparticle-hydrogel complex. *J. Control Rel.* **2007**, 121, 91–99.

26. Schrier J.A., Fink B.F., Rodgers J.B., Vasconez H.C., DeLuca P.P., Effect of a freeze-dried CMC/PLGA microsphere matrix of rhBMP-2 on bone healing. *AAPS PharmSciTech.* **2001**, 2, E18.

27. Wei Q., Wei W., Lai B., Wang L.Y., Wang Y.X., Su Z.G., Ma G.H., Uniform-sized PLA nanoparticles: Preparation by premix membrane emulsification. *Int. J. Pharm.* **2008**, 359(1–2), 294–297.

28. Meikle M.C., On the transplantation, regeneration and induction of bone: The path to bone morphogenetic proteins and other skeletal growth factors. *Surgeon-J. R. Coll. Surgeons Edinburgh Irel* **2007**, 5, 232–243.

29. Saito A., Suzuki Y., Ogata S., Ohtsuki C., Tanihara M., Activation of osteo-progenitor cells by a novel synthetic peptide derived from the bone morphogenetic protein-2 knuckle epitope. *Biochim. Biophys. Acta* **2003**, 1651, 60–67.

30. Saito A., Suzuki Y., Ogata S., Ohtsuki C., Tanihara M., Prolonged ectopic calcification induced by BMP-2-derived synthetic peptide. *J. Biomed. Mater. Res.* **2004**, 70A, 115–121.

31. Saito A., Suzuki Y., Ogata S., Ohtsuki C., Tanihara M., Accelerated bone repair with the use of a synthetic BMP-2-derived peptide and bone-marrow stromal cells. *J. Biomed. Mater. Res.* **2005**, 72A, (1), 77–82.

32. Urist M.R., Bone—Formation by autoinduction. *Science* **1965**, 150, (3698), 893–899.

33. He X., Jabbari E., Material properties and cytocompatibility of injectable MMP degradable poly(lactide ethylene oxide fumarate) hydrogel as a carrier for marrow stromal cells. *Biomacromolecules* **2007**, 8, 780–792.

34. Sarvestani A., He X., Jabbari E., Viscoelastic characterization and modeling of gelation kinetics of injectable in situ cross-linkable poly(lactide-co-ethylene oxide-co-fumarate) hydrogels. *Biomacromolecules* **2007**, 8, 406–415.

35. Ossipov D.A., Hilborn J., Poly(vinyl alcohol)-based hydrogels formed by "click chemistry." *Macromolecules* **2006**, 39, 1709–1718.

36. Wang Q., Chan T.R., Hilgraf R., Fokin V.V., Sharpless K.B., Finn M.G., Bioconjugation by copper(I)-catalyzed azide-alkyne [3 + 2] cycloaddition. *J. Am. Chem. Soc.* **2003**, 125, 3192–3193.

37. Jabbari E., He X., Synthesis and characterization of bioresorbable in situ cross-linkable ultra low molecular weight poly(lactide) macromer. *J. Mater. Sci. Mater. Med.* **2008**, 19, 311–318.

38. He X., Jabbari E., Solid-phase synthesis of reactive peptide crosslinker by selective deprotection. *Prot. Pept. Lett.* **2006**, 13, 715–718.

39. He X., Ma J., Jabbari E., Effect of grafting RGD and BMP-2 protein-derived peptides to a hydrogel substrate on osteogenic differentiation of marrow stromal cells. *Langmuir* **2008**, 24, 12508–12516.
40. Jabbari E., He X., Valarmathi M., Sarvestani A.S., Xu W., Material properties and bone marrow stromal cells response to in situ crosslinkable RGD-functionlized lactide-co-glycolide scaffolds. *J. Biomed. Mater. Res.* **2009**, 89A, 124–137.
41. Kaiser E., Colescott R.L., Bossinger C.D., Cook P.I., Color test for detection of free terminal amino groups in the solid-phase synthesis of peptides. *Anal. Biochem.* **1970**, 34, 595–598.
42. Sarin V.K., Kent S.B., Tam J.P., Merrifield R.B., Quantitative monitoring of solid-phase peptide synthesis by the ninhydrin reaction. *Anal. Biochem.* **1981**, 117, 147–157.
43. Shin H., Jo S., Mikos A.G., Modulation of marrow stromal osteoblast adhesion on biomimetic oligo[poly(ethylene glycol) fumarate] hydrogels modified with Arg-Gly-Asp peptides and a poly(ethylene glycol) spacer. *J. Biomed. Mater. Res.* **2002**, 61A, 169–179.
44. Chen J.S., Altman G.H., Karageorgiou V., Horan R., Collette A., Volloch V., Colabro T., Kaplan D.L., Human bone marrow stromal cell and ligament fibroblast responses on RGD-modified silk fibers. *J. Biomed. Mater. Res.* **2003**, 67A, 559–570.
45. Shin H., Temenoff J.S., Bowden G.C., Zygourakis K., Farach-Carson M.C., Yaszemski M.J., Mikos A.G., Osteogenic differentiation of rat bone marrow stromal cells cultured on Arg-Gly-Asp modified hydrogels without dexamethasone and beta-glycerol phosphate. *Biomaterials* **2005**, 26, 3645–3654.
46. Dee K., Anderson T., Bizios R., Cell function on substrates containing immobilized bioactive peptides. *Mar. Res. Soc. Symp. Proc.* **1994**, 331, 115–119.
47. Jeon O., Song S.J., Yang H.S., Bhang S.H., Kang S.W., Sung M.A., Lee J.H., Kim B.S., Long-term delivery enhances in vivo osteogenic efficacy of bone morphogenetic protein-2 compared to short-term delivery. *Biochem. Biophys. Res. Commun.* **2008**, 369, 774–780.
48. Mercado A.E., He X., Xu W.J., Jabbari E., The release characteristics of a model protein from self-assembled succinimide-terminated poly(lactide-co-glycolide ethylene oxide fumarate) nanoparticles. *Nanotechnology* **2008**, 19, #325609.
49. Sarvestani A., Xu W., He X., Jabbari E., Gelation and degradation characteristics of in situ photo-crosslinked poly(L-lactid-co-ethylene oxide-co-fumarate) hydrogels. *Polymer* **2007**, 48, 7113–7120.
50. Morpurgo M., Bayer E.A., Wilchek M., N-hydroxysuccinimide carbonates and carbamates are useful reactive reagents for coupling ligands to lysines on proteins. *J. Biochem. Biophys. Methods* **1999**, 38, 17–28.
51. Scheufler C., Sebald W., Hülsmeyer M., Crystal structure of human bone morphogenetic protein-2 at 2.7 Å resolution. *J. Mol. Biol.* **1999**, 287, 103–115.
52. Ter Brugge P.J., Jansen J.A., In vitro osteogenic differentiation of rat bone marrow cells subcultured with and without dexamethasone. *Tissue Eng.* **2002**, 8, 321–331.
53. Porter R.M., Huckle W.R., Goldstein A.S., Effect of dexamethasone withdrawal on osteoblastic differentiation of bone marrow stromal cells. *J. Cell. Biochem.* **2003**, 90, 13–22.
54. Poynton A.R., Lane J.M., Safety profile for the clinical use of bone morphogenetic proteins in the spine. *Spine* **2002**, 27, S40–S48.

9

Ophthalmic Applications of Biomaterials in Regenerative Medicine

Victoria Kearns, Rosalind Stewart, Sharon Mason, Carl Sheridan, and Rachel Williams

CONTENTS

9.1 Introduction

Severe loss of vision is a major disability, resulting in a significant burden on society. Loss of vision can be caused by injury, disease, or aging of a specific tissue in the eye while other parts function as normal. There is, therefore, a role for regenerative medicine strategies to address the replacement of the damaged tissue. This chapter will evaluate the current position on the use of biomaterials and cells to treat conditions including ocular surface disorders, corneal disease, cataract, and age-related macular degeneration involving tissues of the conjunctiva, cornea, lens, and retinal complex, respectively. For each tissue, there is a need to optimize the properties of the biomaterials, the source of the cells, and the appropriate route to generate the replacement tissue. In some situations, it is necessary to culture the cells on a substrate in vitro for subsequent transplantation back into the eye, whereas in others, it may be possible to provide a scaffold for stem cells (SCs) within the eye to grow and repopulate a damaged area. The advantages and disadvantages of these different approaches and the optimization of the materials and cells will be discussed along with the potential for future developments in this area.

9.2 Structure of the Eye

The eye is a complex organ responsible for transmission, refraction, and conversion of light energy into cellular signals for the brain to process as vision (Figure 9.1). It contains cells from three embryonic-derived lineages, namely, the neural ectoderm (e.g., neuronal cells of the retina and its supporting cells), surface ectoderm (e.g., avascular lens and cornea), and the mesoderm (e.g., specialized fibroblasts and vascular sclera and uveal tunic). In addition, the eye consists of a vast array of extracellular proteins which are organized

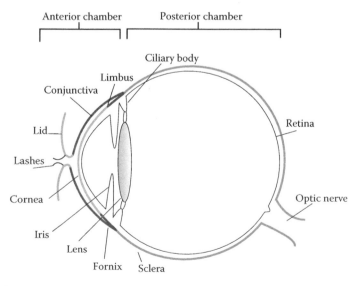

FIGURE 9.1
Schematic diagram of a cross section (front to back) of the eye and eyelids.

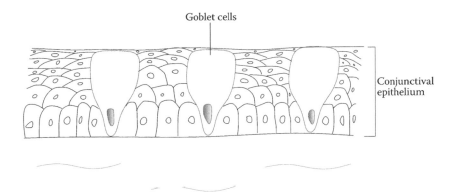

FIGURE 9.2
Schematic diagram of the conjunctival epithelium demonstrating the large mucin-producing goblet cells.

in a structural and functional dependant manner and include collagen-, proteoglycan-, and glycoprotein-rich matrices and basement membranes.

The conjunctiva is a transparent membrane which covers the inner surface of the eyelid from the mucocutaneous junction and over the globe to the limbus. Although it is a continuous tissue, it is described in three areas as the palpebral (covering the inner eyelids), bulbar (covering the sclera of the eye), and forniceal (sac adjoining the two) conjunctiva, as shown in Figure 9.1. It is a stratified non-keratinized epithelium varying from two to nine cells in thickness and from cuboidal to squamous cell shape, with an underlying submucosal lamina propria (Figure 9.2). It plays an important role not only as a barrier protecting the underlying tissues but also as a mucous membrane. Large goblet cells secrete gel-forming mucins which maintain the tear film and in turn protect the corneal epithelium from infection and desiccation.

The cornea and lens are responsible for the refraction of light. These tissues and the aqueous and vitreous humor through which light passes must remain optically clear to maintain good vision. The cornea is a complex structure composed of five layers as shown in Figure 9.3 whose transparency is due to the regularity and smoothness of the outer stratified squamous non-keratinized epithelium, its avascularity, and the regular arrangement of collagenous lamellae and cellular components in the stroma. The endothelium is a simple squamous epithelium which has a critical role in maintaining stromal hydration (and hence transparency) by an ATP-driven ion pump. Although the lens has less refractive power than the cornea, it has the ability to alter its refractive power by changing shape, thus altering the focal distance of the eye to focus on objects at varying distances. It too has regularly arranged lens cells/fibers to maintain transparency. The lens is made up of three structures (Figure 9.4). The outermost layer is the lens capsule which is the basement membrane of the lens epithelial cells. It is a transparent tissue

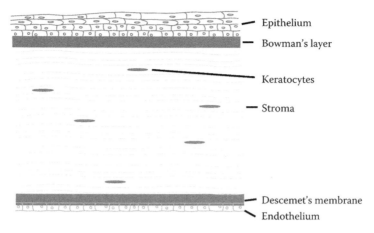

FIGURE 9.3
Schematic diagram of the cornea showing the five layers from the outer epithelium to the inner endothelium.

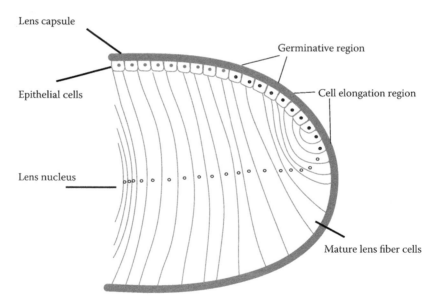

FIGURE 9.4
Schematic diagram illustrating the cells and cell fibers of the mature lens.

composed of collagen in a glycosaminoglycan matrix and is highly elastic. The lens epithelial cells are attached as a monolayer to the anterior portion of the lens capsule. The bulk of the lens is made up of the lens fibers, which are long, thin transparent cells derived from the lens epithelial cells. They are packed tightly in concentric layers; thus, the oldest cells are in the center, and mature lens fibers do not contain organelles or nuclei. Changes in the

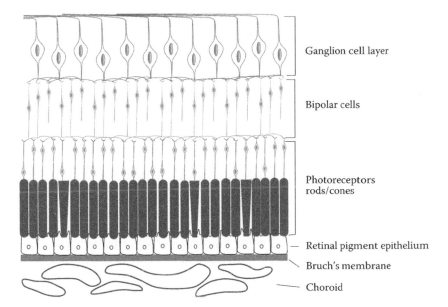

FIGURE 9.5
Schematic diagram of the retinal complex and underlying Bruch's membrane and choroid. The ganglion cell fibers traverse across the retina to form the optic nerve.

structure and composition of the lens can lead to changes in its transparency (formation of cataract). Light is focused on the retina, a multi-layered light-sensitive tissue at the back of the eye (Figure 9.5). A specialized area of the central neural retina known as the macula is responsible for high-resolution central vision. Light is detected by the photoreceptors (rods and cones) in the retina, generating a complex cascade of chemical and electrical events with signals transmitted to the brain via the ganglion cells that comprise the optic nerve. The retinal pigment epithelium (RPE) is a monolayer of cells that sits beneath this neural retina on a thin basement membrane known as Bruch's membrane. The RPE monolayer of cells is the crucial component of the support tissue of the macula and is known to play a key role in maintaining normal functions of the neural retina, particularly the photoreceptors. The retina and RPE are supplied with blood by the central retinal artery and the choroidal blood vessels.

9.3 Ocular Stem Cells

9.3.1 Cornea and Conjunctiva

Like all surface epithelia, the cornea and conjunctiva are renewed constantly, a process for which they rely on the presence of SCs and transient amplifying cells (TACs), which are the only proliferative cells in normal tissue (Barrandon 1993, Lavker et al. 1993, Morrison et al. 1997). SCs can be defined as cells that can give rise to multiple differentiated cell types (multipotency), have the ability to self-renew and resist progression along the line of specialization (Mikkers and Frisen. 2005, Potten and Loeffler 1990). TACs are derived from these SCs, have a high proliferative capacity, and represent the largest group of dividing cells. TACs generate terminally differentiated cells which no longer have the capacity to divide (Barrandon 1993, Lajtha 1979, Lavker et al. 1993). Corneal and conjunctival epithelia are believed to belong to two separate lineages arising from different SC populations (Wei et al. 1996). There is much evidence to suggest corneal epithelial SCs are located in the transitional zone between the cornea and the conjunctiva in an area called the limbus, and hence, they are often referred to as limbal stem cells (LSCs) (Cotsarelis et al. 1989, Davanger and Evensen 1971, Ebato et al. 1988, Huang and Tseng 1991). They are concentrated in the basal layers of limbal epithelial crypts which might offer them a sheltered location (Dua et al. 2005) from where they migrate in a centripetal manner to replenish the central cornea with terminally differentiated cells (Figure 9.6). The assumption that corneal epithelial SCs are concentrated in the limbus has, however, been questioned with evidence from one group presented in favor of the existence of SCs throughout the cornea (Majo et al. 2008).

The location of conjunctival SCs is less clear as there are only a few studies of human tissue. An in vitro study analyzing the clonogenic properties of the ocular surface epithelia indicated that SCs are uniformly distributed in the bulbar and forniceal conjunctiva (Pellegrini et al. 1999). Expression of the LSC marker ABCG2 (Chen et al. 2004) has been demonstrated in clusters of conjunctival epithelia basal cells (Budak et al. 2005). Similarly, a study assessing a battery of candidate and progenitor cell markers concluded the presence of bulbar SCs but did not assess other areas of the conjunctiva

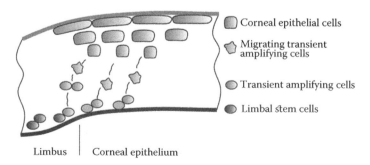

Corneal epithelial cells

Migrating transient amplifying cells

Transient amplifying cells

Limbal stem cells

Limbus | Corneal epithelium

FIGURE 9.6
Schematic diagram illustrating corneal epithelial renewal. SCs are located in the basal layers of the limbus and produce TACs which migrate to the central cornea and to more superficial epithelial layers to form mature corneal epithelial cells.

(Vascotto and Griffith 2006). Further investigation is thus required to clarify their location.

9.3.2 Lens

The lens is the only tissue in the body that grows throughout life, albeit very slowly and in a limited area. The most proliferative area is found in the germinative zone (Figure 9.4), a narrow band of epithelial cells that rings the lens epithelium, and it is protected somewhat from ultraviolet rays by the iris. These proliferative cells (TACs) migrate along the inner capsular surface toward, and posteriorly past, the lens equator. They elongate while they migrate and mature into lens fiber cells which renew the periphery of the posterior lens pushing the older cells toward the center (Piatigorsky 1987). Central lens fiber cells are retained throughout life and are not renewed, while the epithelium which lines the anterior of the lens is renewed very slowly, if at all. Although it is generally agreed that the TACs reside in the germinative zone, the location of the SCs that precede the TACs is more controversial. While most believe they are located in the germinative zone, some believe that they reside in the central lens based on the evidence of DNA labeling studies in mice (Griep 2006, Zhou et al. 2006). Others have postulated that the SCs could not reside in the lens due to the absence of nerves, veins, and immune cells which are generally required to furnish a SC niche (Jones and Wagers 2008) and the fact that tumors do not arise in the lens (Seigel and Kummer 2001). They suggested the SCs may reside in the ciliary body (CB) due to its proximity to the lens and the fact that pigmented cells are able to form lens fiber-like cells called lentoids (Remington and Meyer 2007). As yet there is no clear consensus on where the lens SCs may reside, let alone their potential to be cultured and used therapeutically.

9.3.3 Retinal Complex

Unlike amphibians and fish, humans cannot carry out retinal neurogenesis after birth. However, over the last decade, evidence has accumulated to support the existence of several populations of cells in the eye with progenitor/SC properties and the plasticity to develop into mature cells of the retinal complex in vitro. Due to the limited ability of these cells to proliferate like SCs and differentiate into all cell types, we will call them progenitors rather than SCs. CB pigmented epithelium (Coles et al. 2004, Tropepe et al. 2000), iris pigmented epithelium (Sun et al. 2006), the Müller glia (Fischer and Reh 2001), and the neonatal retina (Akagi et al. 2003, Canola et al. 2007) have all been proposed as sites of retinal progenitor cells (RPCs). It appears that the extracellular environment in the eye causes these cells to remain dormant in vivo but that, provided with the correct signals, they can be released from their dormancy to proliferate and differentiate in vitro.

9.4 Ocular Biomaterials

Biomaterials have played a central part in clinical treatment of eye disorders and vision loss (Colthurst et al. 2000, Lloyd et al. 2001) for many years. Various tissues have been replaced with synthetic materials including the cornea, lens, vitreous, and retina. For example, contact lenses are worn on the corneal surface to improve visual acuity; corneal implants, or keratoprostheses, have been designed to replace diseased or damaged cornea; intraocular lenses (IOLs) are implanted following cataract surgery to replace the opaque crystalline lens; tamponade agents are used to replace the vitreous and treat retinal detachments; and retinal implants are designed to transmit electrical signals to the brain via the retina and the optic nerve.

Biomaterials are exploited for their physical and chemical properties to achieve the therapeutic goals in ophthalmic surgery. For this to be successful, it is necessary to understand the biological requirements and how these can be met by the appropriate choice of biomaterial. There are a number of essential requirements that need to be addressed by ophthalmic biomaterials which include the ability to maintain the light path and control refraction, to integrate with the tissues and to modulate the wound healing response. Over the years, a large number of biomaterials have been investigated ranging from nonreactive synthetic polymers (e.g., acrylics, silicone rubber, and polytetrafluoroethylene) to biological materials (e.g., collagen, lens capsule, and amniotic membrane [AM]). More recently, there have been developments in biomaterial research in which their physical, chemical, and mechanical properties are being tailored to elicit the specific behavior required for particular applications (e.g., surface topography to control cell migration and orientation and surface chemistry to provide biomimetic surfaces).

9.5 Regenerative Medicine Approaches

9.5.1 Conjunctiva

A wide variety of diseases severely affect the ocular surface, causing scarring and secondary mucin deficiency. As a result, the ocular surface becomes keratinized, causing considerable long-term ocular discomfort (Figure 9.7). This in turn leads to corneal desiccation, vascularization, and ulceration with loss of sight. These conditions include mucocutaneous disease (such as mucous membrane pemphigoid and Stevens Johnson syndrome), systemic inflammatory diseases (including Sjogren's syndrome and graft-versus-host disease), infections, drug-induced conjunctival disease (particularly following prolonged use of topical medications), pterygia and glaucoma

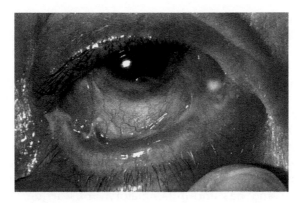

FIGURE 9.7
Clinical photograph of mucous membrane pemphigoid demonstrating significant conjunctival inflammation and scarring.

surgery requiring graft tissue, conjunctival carcinomas, and chemical and physical injuries. Severe conjunctival disease is associated with symbleph- aron (partial or complete adhesion of the conjunctiva covering the eye to that covering the inner eyelid) formation and forniceal shortening and may constitute complete conjunctival SC deficiency. It is extremely difficult to manage clinically, with traditional treatment options limited and usually aimed at symptom control or preventing disease progression, often through systemic immunosuppression. Results are often poor and treatment side effects significant. Although artificial tears may be used to ease discom- fort, many of these patients have concurrent corneal disease, and attempted limbal epithelial or corneal transplantation to improve vision is prone to failure in the presence of significant conjunctival disease (Shortt and Tuft 2011, Tseng et al. 2005).

Conjunctival autografting has been used to cover small conjunctival defects with success (Thoft 1977, Vastine et al. 1982) but would be insuf- ficient to reconstruct the entire fornix as is often required in severe con- junctival disease. Although autologous oral or nasal mucosal grafts offer greater availability, they are not without significant morbidity to the donor site, and this tissue is also often impaired in patients suffering systemic autoimmune cicatrizing disease. The use of biomaterials to enhance wound healing has been investigated. Both a porous collagen-glycosaminogly- can copolymer matrix and a porous collagen and hyaluronic acid modi- fied poly(lactide-*co*-glycolide) scaffold have been demonstrated to reduce conjunctival scarring and shortening when transplanted into rabbit con- junctival wounds (Hsu et al. 2000, Lee et al. 2003). However, neither was transplanted with an epithelium, and each took considerably longer to re- epithelialize than the ungrafted wounds, and as substrates, they lack the elasticity required to successfully reconstruct the fornix. Tissue engineer- ing has aimed to address these issues, generating a functional conjunctival epithelial equivalent. This must comprise epithelial and goblet cells on a suitable flexible, stable, and elastic matrix that is well tolerated in vivo and promotes the self-renewal potential of progenitor cells. In addition, the material must act as a mechanical barrier preventing re-scarring and symblepharon formation in forniceal reconstruction. To this latter aim, additional measures to prevent re-adhesion of the conjunctival surfaces include insertion of symblepharon rings (Patel et al. 1998) or silicone sheet implants (Choy et al. 1977, Ralph 1975).

Like all epithelial cells, growth and differentiation of conjunctival epithelial cells are determined by their underlying matrix. Human AM is currently the material favored for this matrix as it is thin and elastic, promotes rapid epithelialization and has anti-inflammatory, antiscarring, antiangiogenic, and antipain properties (Liu et al. 2010). Histologically it comprises a single epithelial layer, a thick basement membrane and an avascular stroma. It has been used both as a protective onlay and for fornix reconstruction in a variety of disease processes and demonstrated to enhance conjunctival regeneration (Barabino and Rolando 2003, Honavar et al. 2000, Solomon et al. 2003). It is expected to stay in place for 2 weeks (Liu et al. 2010), but its degradation behavior seems to vary, with studies reporting in vivo degradation times of less than 3 weeks and patients in which degradation did not occur by 24 weeks (Vyas and Rathi 2009). It is prone to recurrent shrinkage in the presence of uncontrolled inflammation with recurrent symblepharon formation in 10%–44% (Barabino and Rolando 2003, Honavar et al. 2000) and progressive loss of approximately half of the fornix depth originally obtained within 4 months of surgery (Barabino and Rolando 2003). This variation suggests a role for a synthetic scaffold with reproducible degradation properties.

Despite these limitations, AM has been used as a substrate for a variety of in vitro expanded ocular surface epithelial progenitor cells including conjunctival cells (Ang and Tan 2005, Meller et al. 2002, Ono et al. 2007, Sangwan et al. 2003, Tanioka et al. 2006). Autologous transplantation of cultivated conjunctival epithelial on AM has been reported in conditions with relatively localized conjunctival disease with initial success (Ang and Tan 2005, Ang et al. 2005, Tan et al. 2004). Histologic examination of the transplanted tissue in each case showed a four-to-five-layer-deep stratified squamous epithelial sheet, but no mention of the presence of goblet cells. It may be that there is sufficient mucin production from the remaining healthy conjunctiva to maintain a healthy ocular surface in these cases, but in conditions with extensive conjunctival destruction and SC loss where large grafts would be required, this may no longer hold true. Long-term success rates may also depend upon mucin productivity and subsequent tear film stability.

Although conjunctival epithelial progenitor cells capable of generating goblet cell phenotypes have been cultured in vitro on AM (Budak et al. 2005, Meller et al. 2002), and the presence of the goblet cell mucin MUC5AC mRNA has been demonstrated in cultured conjunctival epithelial cells (Ang et al. 2004), it is noted that non-goblet cell differentiation is either preferentially promoted or results in potential loss of progenitor cells (Meller et al. 2002). It may be that goblet cell differentiation requires a more stringent stromal environment. An initial greater concentration of progenitor cells may aid goblet cell production.

9.5.2 Cornea

Many of the ocular surface diseases described earlier not only severely affect the conjunctiva but can also result in LSC deficiency with resulting painful corneal epithelial breakdown, scarring, and invasion of the conjunctival epithelium over the clear cornea leading to loss of vision. The corneal stroma is also susceptible to scarring and/or thinning/edema from a variety of causes. These conditions include infections (microbial and viral [particularly herpetic] keratitis), corneal dystrophies (most commonly Fuchs' endothelial dystrophy, but also include various epithelial, bowman layer and stromal dystrophies), corneal ectasias (predominantly keratoconus) which cause progressive corneal thinning, peripheral corneal disorders (most notably rheumatoid-associated corneal melt), corneal degeneration, and trauma. These stromal conditions usually require full or partial thickness replacement of the corneal tissue to restore sight which is presently achieved by penetrating (full thickness) or lamellar (partial thickness) corneal grafts using human donor cadaveric tissue. However, good quality tissue is in limited supply, and there is a significant risk of tissue failure or rejection. There are needs, therefore, for tissue engineering approaches to the replacement of both the LSCs (corneal epithelium) and full or partial thickness cornea.

9.5.2.1 Corneal Epithelium

Resurfacing of the corneal epithelium requires the recruitment and expansion of LSCs. Expanding LSCs and their progeny in vitro before they are returned to the patient removes the need to harvest large portions of the limbus from the donor eye, thus decreasing the chance of damage to the healthy eye through LSC deficiency. Although the explant culture is probably more widely used, both explant (Koizumi et al. 2000, Tseng et al. 2002) and suspension culture (Nishida et al. 2004, Yokoo et al. 2008) have been used to successfully produce a sheet of limbal epithelial cells. LSCs for transplantation are often expanded with a feeder layer of mouse 3T3 fibroblasts. This is not, however, ideal due to risk of potential immunological reactions and pathogen transfer, and a wide range of alternative substrates and culture systems have thus been investigated. A system that is feeder cell-free, based on the use of the thermo-responsive polymer poly(N-isopropylacrylamide) (NIPAAm) (Nishida et al. 2004) has been investigated. In this system, an epithelial sheet is cultured and then released from the NIPAAm surface as an intact sheet when subjected to a temperature change. This may avoid problems with graft integration, transparency of carrier substrates and the degradation of resorbable substrates, but the tissue may have poor handling properties. As a consequence, the use of a substrate has been a more popular approach. Early studies (Pellegrini et al. 1997, Schwab 1999) examined various substrates and reported that AM was most successful in vitro and in vivo. A summary of those substrates used clinically can be found in the review by Baylis et al. (2011). These results have strongly influenced clinical practice, with AM being the most commonly used substrate for LSC transplant. However, it is semitransparent, a distinct disadvantage for a tissue which needs to be optically clear, and there is some difference in opinion as to whether the AM should have its epithelial layer removed—a process known as denuding. Denuded AM has been demonstrated to provide better cell attachment and results in more mature tissue construct, but this process may remove components that help maintain SCs and prevent LSCs from differentiating. A recent attempt to clarify this issue confirmed these findings but reported that denuded AM supports a smaller population of undifferentiated cells (Chen et al. 2010).

Substrates other than AM have been investigated, although few have been progressed to the clinical stage. Naturally derived materials are particularly popular, as they often support cell attachment and proliferation. Fibrin in the form of films and gels is one such example and has been used in patients for over 10 years (Rama et al. 2010). In vitro these cells are reported to support the formation of a more differentiated construct than denuded AM (Higa et al. 2007), while maintaining a population of undifferentiated cells (Han et al. 2002, Higa et al. 2007). An animal-free culture system has also been developed (Han et al. 2002), and mechanical properties can be improved by cross-linking (Han et al. 2002). Fibrin gels, however, degrade within a few days (Han et al. 2002, Talbot et al. 2006), which is unlikely to be sufficient time for the cells to lay down a new replacement basement membrane. Mechanical properties of other biologically derived materials, such as cross-linked (Dravida et al. 2008) and compressed collagen gels (Levis et al. 2010, Mi et al. 2010), cross-linked collagen-chitosan hydrogels (Rafat et al. 2008), keratin from human hair (Reichl et al. 2011), and silk fibroin (Bray et al. 2011, Chirila et al. 2008), are reported to be superior to AM in addition to their ability to support corneal epithelial cell attachment. Further studies will be needed to determine how many of these materials perform in vivo, although recent results indicate that fibroin is well tolerated in rabbit corneas and is partially degraded by 6 months (Higa et al. 2011).

9.5.2.2 Corneal Endothelium

Although the majority of studies have concentrated on replacing the corneal epithelium, there is also the need to replace corneal endothelium. Corneal endothelial cells do not normally proliferate in vivo, although they maintain the capacity to do so (Joyce and Zhu 2004). Their number reduces with age, with significant loss leading to corneal edema.

As with corneal epithelial cells, AM has been proposed as a carrier substrate for corneal endothelial cell transplantation. It is reported to support cell growth in vitro and maintain cell function in vivo (Ishino et al. 2004). Further processing to remove the epithelial and stromal layers from AM (i.e., leaving just the basement membrane) has also provided similarly promising in vivo results (Wencan et al. 2007), with the added benefit that this substrate is thinner and may have better optical properties. The substrate-free strategy based on NIPAAm has also been adapted for corneal endothelial cells (Ide et al. 2006, Nitschke et al. 2007), with cell morphology and function being maintained subsequent to detachment from the culture surface, although poor handling resulted in the addition of a degradable gelatin carrier layer to the construct, which is reported to perform well in a rabbit model (Lai et al. 2007). Corneal endothelial cells have also been co-cultured with corneal epithelial and stromal cells for whole cornea replacement strategies—more details can be found in Section 9.5.2.3.

9.5.2.3 Full Thickness Ocular Surface Replacement

The clinical solution to replacement of the full corneal thickness is the transplantation of allogenic cornea from cadavers. However, the limited supply of tissue leads to the need to alternative solutions. Initially the development of keratoprostheses investigated the use to polymeric materials already used in lenses, for example, the polyacrylics, poly(methyl methacrylate) (PMMA) and poly(hydroxyethyl methacrylate) (pHEMA), and polydimethylsiloxane (PDMS) (Nguyen and Yiu 2008). These were constructed with a solid core to act as the optic and allow transport of the light into the eye and a porous skirt to enhance incorporation into the tissue. In general, it was difficult to promote tight tissue ingrowth, and there was poor epithelialization (Princz et al. 2011). Emerging technologies to enhance the tissue response has involved the development of modified surfaces to enhance the tissue response. This has involved the production of biomimetic surfaces either using synthetic analogues (Rimmer et al. 2007) or covalent binding of biological molecules, for example, collagen I to synthetic polymers such as polyethylene glycol/ polyacrylic acid copolymers and NIPAAm (Evans and Sweeney 2010). There have also been investigations into the architectural design of substrates to encourage cell ingrowth and surface epithelial growth using 3D collagen gels and protein hydrogels with an aligned molecular structure to enhance cell orientation, migration, proliferation, and differentiation. Further development of these systems may provide multifunctional materials with the potential to interact with the appropriate SCs to enhance corneal replacement.

Some groups have used multiple cell types to construct a full thickness cornea in vitro. The aims of these studies were to obtain tissue for in vitro cytotoxicity testing of ocular drugs; however, recently the focus has turned toward tissue engineering. One early example of the technique is (Schneider et al. 1999) using porcine corneal epithelial, endothelial, and stromal cells from the same donor. The stromal cells were cultured in a collagen gel, which was then placed on an endothelial monolayer. Epithelial cells were subsequently seeded onto the collagen. This approach resulted in the formation of basement membranes at the interface between the layers, similar to the Descemet's membrane and Bowman's layer found in the cornea (Figure 9.3), but was limited in the number of stromal cells that could be used because they caused contraction of the collagen gel. The first studies using human cells were performed by Germain et al. (2000). They also used collagen but attempted to restrict gel contraction by employing a filter paper anchorage ring. This approach did not result in the synthesis of the intra-layer membranes. A subsequent study by the same group used native stromal cells to produce a basement membrane (avoiding the requirement for a biomaterial scaffold), which was then seeded with corneal endothelial cells on one side and epithelial cells on the other (Proulx et al. 2010). This tissue construct was similar, although not identical to native corneas, in terms of its basement membrane composition.

9.5.3 Lens

Any opacity in the lens capsule or substance is termed a cataract and may cause visual loss. This is most commonly age-related but may be congenital, associated with conditions such as diabetes mellitus or steroid usage, or secondary to trauma. Cataract surgery with implantation of an artificial IOL is now the most commonly performed surgical technique in the United Kingdom, producing excellent results. The most common complication is the subsequent development of posterior capsule opacity (PCO) in which lens epithelial cells migrate and proliferate, causing surface wrinkling of the capsular bag. This complication is presently managed by performing a capsulotomy with nd:YAG laser which is an expensive treatment and therefore may not be available to much of the developing world.

In some species, such as newts and salamanders, complete regeneration of the lens can be achieved. In mammals, lens regeneration is less successful, but the potential for a tissue engineering approach has been reported (Tsonis and Del Rio-Tsonis 2004). In the mammalian eye, successful lens regeneration requires the presence of both the lens capsule and lens epithelial cells. The requirement for a relatively intact lens capsule, which acts both as a source of new lens material and a substrate for regeneration, suggests a potential role for the use of biomaterial substrates to support regeneration. Various studies have used a technique where embryonic skin ectoderm is implanted into the capsule following the removal of the lens. Although the outcomes for these studies are not always clear (reviewed in Gwon 2006), the implanted tissue appears to act as a source of growth factors rather than cells indicating the importance of these for regeneration. Lens regeneration, in particular lens cell proliferation, is affected by the physical properties of lens capsule (particularly taut versus wrinkled) and mechanical forces applied to it (Bito and Harding 1965). A collagen seal and air fill following capsulectomy resulted in faster regeneration of rabbit lenses with better shape and structure (Gwon et al. 1993) than without. The shape of the lens and lens fiber morphology, however, were not as good as normal lenses, possibly because rate of regeneration was not the same throughout the capsular bag leading to distortion. In regenerated lens, cell dedifferentiation occurs (Gwon et al. 1990), suggesting that the environment, including growth factors, needs perfecting. Additionally, the requirement for oriented lens fibers to achieve the correct optical properties suggests a role for orienting stimuli such as topographical cues on substrates and electric fields. Nano-grooved substrates have been demonstrated to cause lens epithelial cells to orient (Rajnicek et al. 2008), although the authors suggested that micron-scale features may be more akin to the natural scale of orientation for the cells. Electrical fields (within a certain range) are also effective in inducing lens epithelial cell orientation and migration (Wang et al. 2003).

9.5.4 Retinal Complex

The retina and underlying RPE (Figure 9.5) are susceptible to a wide variety of pathologies. Age-related macular degeneration is the leading cause of irreversible visual loss in the Western world and is characterized by central visual loss due to disease processes in the photoreceptor-RPE complex (Figure 9.8). Deposition of abnormal lipid material (drusen) in Bruch's membrane with subsequent degeneration of the overlying RPE results in

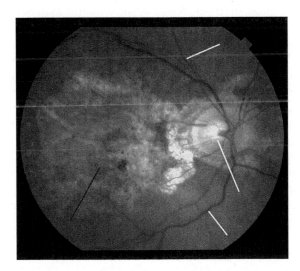

FIGURE 9.8
Clinical photograph of nonvascular (atrophic) age-related macular degeneration. Note the normal optic disc (optic nerve head) and normal retinal vessels (white arrows). The macula (center of picture) has extensive underlying RPE atrophy (pallor) and scarring (black arrow) in comparison to the normal surrounding retina.

nonvascular (atrophic) or neovascular disease. The current anti-vascular endothelial growth factor (VEGF) therapies being employed often stabilize neovascular disease but only offer hope of improvement of vision in a minority of patients. Retinal dystrophies similarly cause atrophy of both the photoreceptor layer and underlying RPE, resulting in irreversible gradual loss of peripheral and/or central vision. Examples include retinitis pigmentosa and Stargardt's disease. The retina may also be affected by degenerative vascular disease: either macrovascular occlusion (such as central retinal artery or vein occlusion) or microvascular occlusion (of the retinal capillaries). Aging, diabetes, and hypertension are predisposing factors to each. Although retinal laser ablation may control complications due to secondary ischemia, there is at present no treatment available to improve visual outcomes. While a number of the diseases mentioned earlier target both the neural retina and supporting tissue, due to the complex nature of the tissue, regenerative therapies have tended to focus on specific cell types within the complex, and so this section will focus on the strategies employed to target the neural retina (Section 9.5.4.1) and its underlying RPE (Section 9.5.4.2).

9.5.4.1 Neural Retina

As mentioned previously, potential sources of progenitors with the capacity for forming neuroretinal cells include the CB pigmented epithelium, Müller glia, and the neonatal retina. Human CB–derived RPCs express markers seen in the developing retina such as Nestin, Pax6, and Chx10, and undirected differentiation of these RPCs results in the expression of markers of all of the mature retinal complex cell types. A high percentage of these cells tend to show markers of rod photoreceptors; therefore, most groups working with RPCs are trying to direct them to a photoreceptor phenotype (Inoue et al. 2010). Rat Müller glia can proliferate and produce new rods and bipolar cells in response to injury, and this differentiation can be influenced by genes and extrinsic factors (Ooto et al. 2004). When glial SCs are transplanted into rats, they can slow the progression of retinal degeneration and result in the generation of new photoreceptors from the transplanted glia (Tian et al. 2011). Iris-derived progenitor cell expresses nestin and shows some ability to develop into mature neuroretinal cells in vitro (Asami et al. 2007). When transplanted over the RPE, they have the ability to integrate into the subretinal space and produce photoreceptor-like cells (Sun et al. 2006). Progenitor-like cells with the capacity to form neuroretinal have also been obtained from embryonic and neonatal retina (Akagi et al. 2003, Canola et al. 2007), which being less mature, have a higher plasticity. In the following text, we discuss the body of work which has been carried out with RPCs in relation to substrates. Most of this work involved obtaining cells from the whole of the postnatal day one neural retina due to the immature nature of these cells. While a relatively large body of work into the various aspects of regeneration of the retina is available, this section will concentrate on investigation into the use of RPCs to produce neural retinal tissue. Reviews of other transplantation strategies can be found elsewhere (Aramant and Seiler 2002, Klassen 2006).

Various studies of implantation of suspensions and sheets of fetal neural retina and adult photoreceptor sheets into the subretinal space in human patients demonstrated that the implants are generally well tolerated and the procedures are relatively safe (Das et al. 1999, Humayun et al. 2000, Kaplan et al. 1997, Radtke et al. 2002), but definite long-term visual improvement is yet to be reported. The low viability of cells implanted in suspension suggests that implantation of a sheet may be preferable (Tomita et al. 2005). There is also a potential advantage of implanting intact retinal sheets rather than simply the photoreceptor layer; the inclusion of Müller cells appears to be crucial to the outcome of the implanted photoreceptors. For example, Silverman et al. (1992) reported that the co-transplantation of the inner retina (containing Müller cells) and photoreceptor layer in a rodent model encouraged better organization of the healed tissue. This study also demonstrated that transplantation of mature adult photoreceptors was as successful as when immature tissue was used, and the authors suggested that the transplanted cells were forming new synapses with the host retina. Such integration of grafted cells into the host tissue is vital to transplant success. The potential to use mature photoreceptors is encouraging for regenerative medicine therapies, as other central nervous system neurons and RPCs must be immature to be transplanted (MacLaren et al. 2006). While the use of fetal tissue has its obvious drawbacks in a translational setting, the use of such tissue does have some advantages in research as it contains pluripotent precursor cells influenced by its surrounding milieu that can allow cells to differentiate down the retinal neuronal and glial lineages (Aramant and Seiler 2002). Some authors have speculated that the neural retina and underlying RPE (see the following section) should be transplanted together; however, this is not well studied and would present some technical difficulties, particularly at the surgical stage. More details can be found in Aramant and Seiler (2004) and Zarbin (2008).

The limited regeneration potential of mature retinal cells means that the many investigators have concentrated on immature RPCs for tissue engineering studies. Scaffolds can be used to induce differentiation of RPCs prior to their subretinal implantation. Various scaffold materials have been investigated. These include nondegradable (PMMA; Tao et al. 2007) and degradable (polycaprolactone [Redenti et al. 2008, Steedman et al. 2010], poly(glycerol-sebacate), PGS [Neeley et al. 2008, Redenti et al. 2009], PLA/poly(DL-lactic-co-glycolic acid) [PLGA] [Lavik et al. 2005, Tomita et al. 2005]) scaffolds. The group using PLA/PLGA highlighted the importance of tailoring the scaffold to influence cell differentiation, noting that their scaffold was able to induce changes in gene expression suggestive of a photoreceptor lineage, although complete cell differentiation to photoreceptor cells was not observed. They also reported the ability of the scaffolds to direct cell orientation and polarization in vitro, which are important for the functioning of photoreceptors that may develop from the progenitors. Scaffolds pre-seeded with progenitor cells supported increased cell survival compared to scaffold-free implantation into a mouse model. The authors speculated that this might be due to adsorption of growth factors by the scaffold, which would then be available to the cells over a sustained period. Implantation into a pig model was also achieved, although cells implanted in the PLGA substrate did not migrate from it, compared to cells implanted in suspension, which integrated into the inner retina. The authors suggest that scaffolds may be useful for delivering spatially organized cells but only in situations where the outer retina is to be regenerated. The in vivo response to the degradation of scaffolds is of particular concern, with PLGA and PGS being reported to cause chronic inflammation when placed in contact with rat sciatic nerves (Sundback et al. 2005). The substrates tested frequently require pre-coating with laminin in order that RPCs can adhere; development of a scaffold that has the appropriate surface chemistry to avoid this is desirable. Further limitations of some of the substrates include the mechanical properties, which in particular do not match the modulus or thickness of the retina. Biodegradable, hyaluroncellulose hydrogels may have more suitable mechanical properties and have been demonstrated to act as a delivery vehicle for cultured progenitor spheres. Although these are well tolerated in vivo, however, the majority of cells integrated into the RPE layer rather than the neural retina (Ballios et al. 2010). RPCs delivered via PLA/PLGA scaffolds, however, are reported to remain within the scaffold rather than integrating with surrounding tissue (Warfvinge et al. 2005).

The arrangement of retinal neurons in a 3D structure suggests that there is a requirement for organized regeneration which is a significant obstacle for in vitro studies. The use of a bioreactor to reproduce the 3D organization has had some promising results, with differentiation of various retinal cell types and primitive layer formation from a retinal precursor cell line being reported (Dutt et al. 2003). As in most areas of tissue engineering, the importance of tailoring the macrostructure and surface topography of biomaterial substrates to elicit the desired cellular response has been investigated. 25μm wide pits in thin PCL films have been reported to replace the requirement for pre-coating with laminin for RPC adherence and increase expression of neural differentiation markers (Steedman et al. 2010). Micron-scale pores in PMMA substrates encouraged adherence, survival, and integration of RPCs into host tissue in a mouse model compared to a nonporous control, despite similar results between the two surfaces in in vitro experiments (Tao et al. 2007). Furthermore, once implanted, the cells migrated toward different layers of the retina and differentiated toward glial or photoreceptor cells, depending on their location. Redenti et al. (2009) reported RPC alignment to nanoscale variations in their porous poly(glycerol-sebacate) substrates. Neurites from rat retinal ganglion cells and neuritis and glia from whole retinal explants can be encouraged to extend along micron-scale zigzag patterns (Leng et al. 2004). Although the focus of that study was to direct retinal cell neurites toward stimulating electrodes in a retinal implant, it does demonstrate the potential use of surface patterning.

9.5.4.2 Retinal Pigment Epithelium

It is known that the function of the macula can be preserved if it is placed on healthy rather than the diseased support tissue characteristic of AMD. Previous studies with animal work, RPE transplantation in humans and macular relocation have all shown that replacing diseased RPE with healthier RPE can rescue photoreceptors, prevent further visual loss or even promote visual improvement (Pertile and Claes 2002, Phillips et al. 2003, Van Meurs et al. 2004). The feasibility of RPE cell replacement has been further demonstrated by autologous transplantation of full thickness RPE/choroid from the periphery to the macula or patch graft transplant (MacLaren et al. 2007, van Meurs and Van Den Biesen 2003, Wong 2000, 2004). The patch grafting technique often results in a number of potentially blinding complications due to two complex surgical sites and moving a potentially diseased RPE-Bruch's complex. It does, however, demonstrate the potential for restoration of sight by implantation of a "healthier" RPE-Bruch's membrane construct. This has led to a plethora of studies examining the potential of cells from various sources to differentiate and perform RPE functions and the selection of an underlying substrate to facilitate surgical success.

The source of cells for transplant and the ability of a monolayer to perform the function of RPE are crucial to the success of the treatment, and an autologous approach to transplantation is attractive in discussing replacement of the RPE monolayer. The use of donated RPE cells is not desirable as the aging eye is not necessarily immunologically privileged (Zhang and Bok 1998), especially in case where the blood-retinal barrier is impaired or diseased. The loss of immune privilege therefore implies that immune rejection becomes a more likely outcome unless an immunosuppressive regime is followed. Unfortunately the extended age of the patients requiring such treatment means that they do not tolerate immunosuppressive drugs well, and the complications can often outweigh the benefits of treating a sight-threatening, but not life-threatening, condition. While the majority of initial studies concentrated on using the patient's own RPE cells, obtaining these cells requires extra surgery and may cause complications at the harvest site, and peripheral RPE cells may exhibit some of the characteristics of diseased cells. Hence, a number of alternative autologous sources have been researched. Iris pigment epithelial cells (Figure 9.1) have the same embryonic origin as RPE and demonstrate many of the same features and functions in vitro (Schraermeyer et al. 1997, Thumann et al. 1998). Although gene expression in RPE and IPE from the same donor is reported to differ significantly (Cai et al. 2006), the appropriate in vitro and in vivo conditions may encourage transdifferentiation of IPE toward RPE. Preliminary clinical studies implanting IPE suspensions have also shown some promise; the cells are well tolerated, although there is little significant improvement in vision (Abe et al. 1999, Aisenbrey et al. 2006, Lappas et al. 2000), probably because a functional monolayer is required.

Cells with RPE-like characteristics have been derived from adult SCs from various sites, but standardized methods for their isolation, expansion and differentiation are not well characterized, and RPE monolayers have yet to be achieved. Induced pluripotent stem cells (iPS) may provide an optimal solution in the future; it has already been demonstrated that these cells can differentiate into RPE-like cells from a potential autologous source (Buchholz et al. 2009). However, despite the huge advances in recent years in this field, and the continued advances in reducing the type of vector and/or oncogene, they are unlikely to be of benefit in the short term until the issues of concerning cellular memory, which may predispose the cells to dedifferentiate into fibroblast-like cells, as well as the ability to form tumors are addressed fully.

As mentioned previously, human CB–derived cells (Coles et al. 2004) can be cultivated and differentiated to express markers seen in the developing retina. Unfortunately with undirected differentiation, the percentage of cells differentiating to RPE-like lineage is extremely low (<1%) (Coles et al. 2004), and a differentiated RPE monolayer has yet to be achieved (Aruta et al. 2011, Marzo et al. 2010, Vossmerbaeumer et al. 2008). Therefore, other progenitor sources are being explored to varying degrees of success with mesenchymal SCs reported to express epithelial cell markers (Vossmerbaeumer et al. 2009) while both Müller cells and neonatal retina have as yet not been shown to differentiate into RPE cells.

Another source of RPE may be utilizing embryonic stem (ES) cells, which have an advantage in their ability to self-renew, thus providing a reservoir of cells compared to the restricted availability of adult progenitor cells. Their use, however, is potentially compromised by serious ethical and safety concerns such as the possibility of teratoma formation (Prokhorova et al. 2009). The first evidence that human ES cells could differentiate into RPE was reported in 2004 (Klimanskaya et al. 2004) when ES cells were overgrown (allowed to become overconfluent) for 6–8 weeks without LIF or bFGF and small patches of pigmented epithelial cells appeared. iPS, being very similar to ES cells, can also spontaneously differentiate into RPE in culture (Buchholz et al. 2009), especially if Wnt and nodal antagonists are used (Hirami et al. 2009). However, the expression profiles of ESE-derived and IPS-derived RPE have been found to differ slightly (Liao et al. 2010). Recently RPE-like cells that had many properties of RPE such as ion transport, membrane potential, and polarized VEGF secretion were produced from iPS by the addition of nicotinamide and activin A to the growth medium (Kokkinaki et al. 2011). Thus, using ES cells or IPS cells has potential for generating enough cells for therapeutic purposes (even tailored to the recipient in the case of IPS cells), but safety worries, especially concerning teratoma formation, have to be addressed first.

It has been demonstrated that injection of a cell suspension gives only limited improvement in vision (Binder et al. 2004, Van Meurs et al. 2004), so in recent years, efforts have concentrated on developing and implanting a functioning monolayer of RPE cells. Temperature-sensitive polymer substrates can be used to culture an RPE monolayer (von Recum et al. 1998); however, there are problems with the fragile handing properties of the cell sheet when they are not supported by a basement membrane (Del Priore et al. 2001). Work has, therefore, focused on a finding a suitable synthetic Bruch's membrane substitute to facilitate the transplant procedure. The use of an artificial membrane to replace the native Bruch's membrane (Figure 9.9) may overcome the handling problems and avoid placement of healthy implanted RPE on aged Bruch's membrane, which is unable to support maintenance of their functional phenotype (Gullapalli et al. 2005, Itaya et al. 2004, Sun et al. 2007). The artificial Bruch's membrane should meet several requirements. It should (1) support a functional layer of RPE cells, (2) be porous in order to allow nutrient flow, (3) have appropriate mechanical properties to facilitate surgical handling and protection of the cells during implantation, and (4) induce no adverse inflammatory or immune response. A wide range of different substrates have been investigated; recent reviews can be found in Binder et al. (2007), Hynes and Lavik (2010), and Sheridan et al. (2007).

Cadaver Bruch's membrane, anterior lens capsule (Hartmann et al. 1999), AM (Stanzel et al. 2005), and Descemet's membrane (Thumann et al. 1997) are all reported to support the formation of RPE monolayers and often have mechanical properties similar to Bruch's membrane (Binder 2011). These tissues, however, may have limited availability, disease transmission may occur, and their surgical handling properties may not be optimal. Collagen has been reported to support RPE attachment and is well tolerated under rabbit retinas (Bhatt et al. 1994, Thumann et al. 2009). In vitro, other extracellular matrix components such as fibronectin and laminin have been demonstrated to encourage a good RPE monolayer (Heth et al. 1987, Wagner et al. 1995),

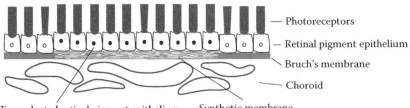

Transplanted retinal pigment epithelium   Synthetic membrane

— Photoreceptors

— Retinal pigment epithelium

Bruch's membrane

— Choroid

FIGURE 9.9
Schematic diagram illustrating an RPE transplant on an artificial membrane which replaces both the diseased RPE and Bruch's membrane.

but again their mechanical and degradation properties may not be suitable for implantation. Oganesian et al. (1996) reported a technique using fibrinogen microspheres as a carrier substrate, from which RPE cells migrated after implantation and gelatin has also been used as a temporary protective carrier (Del Priore et al. 2004). The advantageous properties of these naturally derived materials in terms of their ability to support a monolayer of RPE should be identified in order that they can be exploited in the design of the optimal synthetic substrate.

Synthetic polymers have the potential advantage over naturally derived materials in that they can be tailored to have the desired chemical, mechanical and architectural features. Synthetic polymer substrates can be divided into two categories: degradable and nondegradable. Several degradable polymers have been demonstrated to support the formation of an RPE monolayer in vitro. PLGA is used in many tissue engineering investigations and degrades by hydrolysis to lactic and glycolic acids. PLGA is well tolerated in the subretinal space, at least over a few weeks (Lavik et al. 2005, Tomita et al. 2005), although longer-term studies are required to determine response to the degrading polymer. Various blends of this co-polymer are reported to support cultured RPE monolayers (Giordano et al. 1997, Hadlock et al. 1999, Lu et al. 1998, 2001). Preliminary findings with poly(3-hydroxybutyrate-*co*-3-hydroxyvalerate (PHBV8)) demonstrated RPE attachment and tight junction formation (Tezcaner et al. 2003, Zorlutuna et al. 2006), but more detailed in vitro and in vivo studies to determine how these materials perform in the eye are required.

A wide range of nondegradable polymers have been investigated. Preliminary studies often report RPE monolayer formation on nonporous materials, which must be made porous in order to support nutrient and waste transport in the eye and achieve RPE polarity. Substrates that have been investigated in vitro include plasma-treated or protein-coated PDMS (Krishna et al. 2007, Lim et al. 2004), porous polyester (Stanzel et al. 2012), and nanofibrous polyamide (Thieltges et al. 2011). Polyurethanes have also been shown to support functional RPE monolayers, capable of phagocytosing photoreceptor outer segments, and can be manufactured into thin porous films (Alias et al. 2011, Williams et al. 2005). Surface modified expanded polytetrafluoroethylene gives promising in vitro results (Krishna et al. 2011), and is, at least in the short-term, well tolerated in the subretinal space (Stanzel et al. 2010) of rabbits. Microporous poly(acrylamide-acrylate) hydrogels coated with fibronectin or poly-D-lysine to improve cell attachment (Singh et al. 2001) have been demonstrated to support RPE monolayers in vitro, although caution must be exercised with hydrogels as their swelling properties may lead to variation in size and mechanical properties. Consideration should be given to the surface chemistry patterning and topography of the substrate, as this has been shown to influence RPE behavior (Lim et al. 2004, Lu et al. 1999, 2001, Thieltges et al. 2011).

9.6 Conclusion and Future Directions

The loss of vision is a debilitating condition with enormous social and economic cost. Biomaterials have been used in the treatment of vision loss for many years; however, this has often been by the replacement of tissue with synthetic materials with the aim of providing a clear optical path. More recently, the evidence of ocular tissue SCs in various tissues has led to the development of alternative routes to treat ocular conditions by recruitment of the SCs and their direction to renewal of the diseased tissue. As with all regenerative medicine approaches, the recruited SCs require a substrate on which to grow, and at the present time, most of the studies report the use of biologically derived substrates. There are advantages of this approach since the biologically derived substrates contain many of the necessary biological cues to direct the SC behavior. However, the disadvantage of this approach is the potential limited supply of the biological material, its variability, the ability to process the material to ensure sterility, control immunogenicity and maintain sufficient mechanical properties to allow handling and in some cases its degradation. A significant improvement of clinical treatment could be achieved if synthetic materials could be produced that were tailored to provide the specific biological cues needed by the SCs to stimulate their regenerative potential. These synthetic materials could be produced reliably and reproducibly and in sufficient quantities to be used therapeutically. A cross-disciplinary approach is needed to design and develop these substrates taking into account of the complexity of the in vivo environment, the diseased eye and the surgical and therapeutic options available.

References

Abe, T., H. Tomita, T. Ohashi et al. 1999. Characterization of iris pigment epithelial cell for auto cell transplantation. *Cell Transplantation* 8(5):501–510.

Aisenbrey, S., B.A. Lafaut, P. Szurman et al. 2006. Iris pigment epithelial translocation in the treatment of exudative macular degeneration: A 3-year follow-up. *Archives of Ophthalmology* 124(2):183–188.

Akagi, T., M. Haruta, J. Akita et al. 2003. Different characteristics of rat retinal progenitor cells from different culture periods. *Neuroscience Letters* 341(3):213–216.

Alias, E., V.R. Kearns, I.M. Harrison, C.M. Sheridan, and R.L. Williams. 2011. Comparison of growth and monolayer formation of primary bovine RPE and IPE on polyurethane (PU) membranes. *Histology and Histopathology. Cellular and Molecular Biology* 26(S1):342–342.

Ang, L.P.K. and D.T.H. Tan. 2005. Autologous cultivated conjunctival transplantation for recurrent viral papillomata. *American Journal of Ophthalmology* 140(1):136–138.

Ang, L.P.K., D.T.H. Tan, H. Cajucom-Uy, and R.W. Beuerman. 2005. Autologous cultivated conjunctival transplantation for pterygium surgery. *American Journal of Ophthalmology* 139(4):611–619.

Ang, L.P.K., D.T.H. Tan, T.T. Phan et al. 2004. The in vitro and in vivo proliferative capacity of serum-free cultivated human conjunctival epithelial cells. *Current Eye Research* 28(5):307–317.

Aramant, R.B. and M.J. Seiler. 2002. Retinal transplantation—Advantages of intact fetal sheets. *Progress in Retinal and Eye Research* 21(1):57–73.

Aramant, R. and M.J. Seiler. 2004. Progress in retinal sheet transplantation. *Progress in Retinal and Eye Research* 23(5):475–494.

Aruta, C., F. Giordano, A. De Marzo et al. 2011. In vitro differentiation of retinal pigment epithelium from adult retinal stem cells. *Pigment Cell and Melanoma Research* 24(1):233–240.

Asami, M., G. Sun, M. Yamaguchi, and M. Kosaka. 2007. Multipotent cells from mammalian iris pigment epithelium. *Developmental Biology* 304(1):433–446.

Ballios, B.G., M.J. Cooke, D. van der Kooy, and M.S. Shoichet. 2010. A hydrogel-based stem cell delivery system to treat retinal degenerative diseases. *Biomaterials* 31(9):2555–2564.

Barabino, S. and M. Rolando. 2003. Amniotic membrane transplantation elicits goblet cell repopulation after conjunctival reconstruction in a case of severe ocular cicatricial pemphigoid. *Acta Ophthalmologica Scandinavica* 81(1):68–71.

Barrandon, Y. 1993. The epidermal stem cell: An overview. *Seminars in Developmental Biology* 4(4):209–215.

Baylis, O., F. Figueiredo, C. Henein, M. Lako, and S. Ahmad. 2011. 13 Years of cultured limbal epithelial cell therapy: A review of the outcomes. *Journal of Cellular Biochemistry* 112(4):993–1002.

Bhatt, N.S., D.A. Newsome, T. Fenech et al. 1994. Experimental transplantation of human retinal pigment epithelial cells on collagen substrates. *American Journal of Ophthalmology* 117(2):214–221.

Binder, S. 2011. Scaffolds for retinal pigment epithelium (RPE) replacement therapy. *British Journal of Ophthalmology* 95(4):441–442.

Binder, S., I. Krebs, R.D. Hilgers et al. 2004. Outcome of transplantation of autologous retinal pigment epithelium in age-related macular degeneration: A prospective trial. *Investigative Ophthalmology and Visual Science* 45(11):4151–4160.

Binder, S., B.V. Stanzel, I. Krebs, and C. Glittenberg. 2007. Transplantation of the RPE in AMD. *Progress in Retinal and Eye Research* 26(5):516–554.

Bito, L.Z. and C.V. Harding. 1965. Patterns of cellular organization and cell division in the epithelium of the cultured lens. *Experimental Eye Research* 4(3):146–161.

Bray, L.J., K.A. George, S.L. Ainscough et al. 2011. Human corneal epithelial equivalents constructed on *Bombyx mori* silk fibroin membranes. *Biomaterials* 32(22):5086–5091.

Buchholz, D.E., S.T. Hikita, T.J. Rowland et al. 2009. Derivation of functional retinal pigmented epithelium from induced pluripotent stem cells. *Stem Cells* 27(10):2427–2434.

Budak, M.T., O.S. Alpdogan, M. Zhou et al. 2005. Ocular surface epithelia contain ABCG2-dependent side population cells exhibiting features associated with stem cells. *Journal of Cell Science* 118(8):1715.

Cai, H., M.C. Shin, T.H. Tezel, H.J. Kaplan, and L.V. Del Priore. 2006. Use of iris pigment epithelium to replace retinal pigment epithelium in age-related macular degeneration: A gene expression analysis. *Archives of Ophthalmology* 124(9):1276–1285.

Canola, K., B. Angénieux, M. Tekaya et al. 2007. Retinal stem cells transplanted into models of late stages of retinitis pigmentosa preferentially adopt a glial or a retinal ganglion cell fate. *Investigative Ophthalmology and Visual Science* 48(1):446–454.

Chen, Z., C.S. de Paiva, L. Luo et al. 2004. Characterization of putative stem cell phenotype in human limbal epithelia. *Stem Cells* 22(3):355–366.

Chen, B., S. Mi, B. Wright, and C.J. Connon. 2010. Differentiation status of limbal epithelial cells cultured on intact and denuded amniotic membrane before and after air-lifting. *Tissue Engineering Part A* 16(9):2721–2729.

Chirila, T.V., Z. Barnard, Zainuddin et al. 2008. *Bombyx mori* silk fibroin membranes as potential substrata for epithelial constructs used in the management of ocular surface disorders. *Tissue Engineering Part A* 14(7):1203–1211.

Choy, A.E., R.L. Asbell, and H.B. Taterka. 1977. Symblepharon repair using a silicone sheet implant. *Annals of Ophthalmology* 9(2):197–204.

Coles, B.L.K., B. Angenieux, T. Inoue et al. 2004. Facile isolation and the characterization of human retinal stem cells. *Proceedings of the National Academy of Sciences the United States of America* 101(44):15772–15777.

Colthurst, M.J., R.L. Williams, P.S. Hiscott, and I. Grierson. 2000. Biomaterials used in the posterior segment of the eye. *Biomaterials* 21(7):649–665.

Cotsarelis, G., S.-Z. Cheng, G. Dong, T.-T. Sun, and R.M. Lavker. 1989. Existence of slow-cycling limbal epithelial basal cells that can be preferentially stimulated to proliferate: Implications on epithelial stem cells. *Cell* 57(2):201–209.

Das, T., M. del Cerro, S. Jalali et al. 1999. The transplantation of human fetal neuroretinal cells in advanced retinitis pigmentosa patients: Results of a long-term safety study. *Experimental Neurology* 157(1):58–68.

Davanger, M. and A. Evensen. 1971. Role of the pericorneal papillary structure in renewal of corneal epithelium. *Nature* 229(5286):560–561.

Del Priore, L.V., H.J. Kaplan, T.H. Tezel et al. 2001. Retinal pigment epithelial cell transplantation after subfoveal membranectomy in age-related macular degeneration: Clinicopathologic correlation. *American Journal of Ophthalmology* 131(4):472–480.

Del Priore, L.V., T.H. Tezel, and H.J. Kaplan. 2004. Survival of allogeneic porcine retinal pigment epithelial sheets after subretinal transplantation. *Investigative Ophthalmology and Visual Science* 45(3):985–992.

Dravida, S., S. Gaddipati, M. Griffith et al. 2008. A biomimetic scaffold for culturing limbal stem cells: A promising alternative for clinical transplantation. *Journal of Tissue Engineering and Regenerative Medicine* 2(5):263–271.

Dua, H.S., V.A. Shanmuganathan, A.O. Powell-Richards, P.J. Tighe, and A. Joseph. 2005. Limbal epithelial crypts: A novel anatomical structure and a putative limbal stem cell niche. *British Journal of Ophthalmology* 89(5):529–532.

Dutt, K., S. Harris-Hooker, D. Ellerson et al. 2003. Generation of 3D retina-like structures from a human retinal cell line in a NASA bioreactor. *Cell Transplantation* 12:717–731.

Ebato, B., J. Friend, and R.A. Thoft. 1988. Comparison of limbal and peripheral human corneal epithelium in tissue culture. *Investigative Ophthalmology and Visual Science* 29(10):1533–1537.

Evans, M.D.M. and D.F. Sweeney. 2010. Synthetic corneal implants. In *Biomaterials and Regenerative Medicine in Ophthalmology*, T.V. Chirila (Ed.), pp. 65–133. Cambridge, U.K.: Woodhead Publishing Limited.

Fischer, A.J. and T.A. Reh. 2001. Müller glia are a potential source of neural regeneration in the postnatal chicken retina. *Nature Neuroscience* 4(Journal Article):247–252.

Germain, L., P. Carrier, F.A. Auger, C. Salesse, and S.L. Guérin. 2000. Can we produce a human corneal equivalent by tissue engineering? *Progress in Retinal and Eye Research* 19(5):497–527.

Giordano, G.G., R.C. Thomson, S.L. Ishaug et al. 1997. Retinal pigment epithelium cells cultured on synthetic biodegradable polymers. *Journal of Biomedical Materials Research* 34(1):87–93.

Griep, A.E. 2006. Cell cycle regulation in the developing lens. *Seminars in Cell and Developmental Biology* 17(6):686–697.

Gullapalli, V.K., I.K. Sugino, Y. Van Patten, S. Shah, and M.A. Zarbin. 2005. Impaired RPE survival on aged submacular human Bruch's membrane. *Experimental Eye Research* 80(2):235–248.

Gwon, A. 2006. Lens regeneration in mammals: A review. *Survey of Ophthalmology* 51(1):51–62.

Gwon, A., L.J. Gruber, and C. Mantras. 1993. Restoring lens capsule integrity enhances lens regeneration in New Zealand albino rabbits and cats. *Journal of Cataract and Refractive Surgery* 19(6):735–746.

Gwon, A.E., L.J. Gruber, and K.E. Mundwiler. 1990. A histologic study of lens regeneration in aphakic rabbits. *Investigative Ophthalmology and Visual Science* 31(3):540–547.

Hadlock, T., S. Singh, J.P. Vacanti, and B.J. McLaughlin. 1999. Ocular cell monolayers cultured on biodegradable substrates. *Tissue Engineering* 5(3):187–196.

Han, B., I.R. Schwab, T.K. Madsen, and R.R. Isseroff. 2002. A fibrin-based bioengineered ocular surface with human corneal epithelial stem cells. *Cornea* 21(5):505–510.

Hartmann, U., F. Sistani, and U.H. Steinhorst. 1999. Human and porcine anterior lens capsule as support for growing and grafting retinal pigment epithelium and iris pigment epithelium. *Graefe's Archive for Clinical and Experimental Ophthalmology* 237(11):940–945.

Heth, C.A., M.A. Yankauckas, M. Adamian, and R.B. Edwards. 1987. Characterization of retinal pigment epithelial cells cultured on microporous filters. *Current Eye Research* 6(8):1007–1019.

Higa, K., S. Shimmura, N. Kato et al. 2007. Proliferation and differentiation of transplantable rabbit epithelial sheets engineered with or without an amniotic membrane carrier. *Investigative Ophthalmology and Visual Science* 48(2):597–604.

Higa, K., N. Takeshima, F. Moro et al. 2011. Porous silk fibroin film as a transparent carrier for cultivated corneal epithelial sheets. *Journal of Biomaterials Science. Polymer Edition* 22(17):2261–2276.

Hirami, Y., F. Osakada, K. Takahashi et al. 2009. Generation of retinal cells from mouse and human induced pluripotent stem cells. *Neuroscience Letters* 458(3):126–131.

Honavar, S.G., A.K. Bansal, V.S. Sangwan, and G.N. Rao. 2000. Amniotic membrane transplantation for ocular surface reconstruction in Stevens-Johnson syndrome. *Ophthalmology* 107(5):975–979.

Hsu, W.-C., M.H. Spilker, I.V. Yannas, and P.A.D. Rubin. 2000. Inhibition of conjunctival scarring and contraction by a porous collagen-glycosaminoglycan implant. *Investigative Ophthalmology and Visual Science* 41(9):2404–2411.

Huang, A.J. and S.C. Tseng. 1991. Corneal epithelial wound healing in the absence of limbal epithelium. *Investigative Ophthalmology and Visual Science* 32(1):96–105.

Humayun, M.S., E. De Juan, Jr, M. Del Cerro et al. 2000. Human neural retinal transplantation. *Investigative Ophthalmology and Visual Science* 41(10):3100–3106.

Hynes, S.R. and E.B. Lavik. 2010. A tissue-engineered approach towards retinal repair: Scaffolds for cell transplantation to the subretinal space. *Graefe's Archive for Clinical and Experimental Ophthalmology* 248(6):763–778.

Ide, T., K. Nishida, M. Yamato et al. 2006. Structural characterization of bioengineered human corneal endothelial cell sheets fabricated on temperature-responsive culture dishes. *Biomaterials* 27(4):607–614.

Inoue, T., B.L.K. Coles, K. Dorval et al. 2010. Maximizing functional photoreceptor differentiation from adult human retinal stem cells. *Stem Cells* 28(3):489–500.

Ishino, Y., Y. Sano, T. Nakamura et al. 2004. Amniotic membrane as a carrier for cultivated human corneal endothelial cell transplantation. *Investigative Ophthalmology and Visual Science* 45(3):800–806.

Itaya, H., V. Gullapalli, I.K. Sugino, M. Tamai, and M.A. Zarbin. 2004. Iris pigment epithelium attachment to aged submacular human Bruch's membrane. *Investigative Ophthalmology and Visual Science* 45(12):4520–4528.

Jones, D.L. and A.J. Wagers. 2008. No place like home: Anatomy and function of the stem cell niche. *Nature Reviews Molecular and Cell Biology* 9(1):11–21.

Joyce, N.C. and C.C. Zhu. 2004. Human corneal endothelial cell proliferation: Potential for use in regenerative medicine. *Cornea* 23(8 Suppl.):S8–S19.

Kaplan, H.J., T.H. Tezel, A.S. Berger, M.L. Wolf, and L.V. Del Priore. 1997. Human photoreceptor transplantation in retinitis pigmentosa: A safety study. *Archives of Ophthalmology* 115(9):1168–1172.

Klassen, H. 2006. Transplantation of cultured progenitor cells to the mammalian retina. *Expert Opinion on Biological Therapy* 6(5):443–451.

Klimanskaya, I., J. Hipp, K.A. Rezai et al. 2004. Derivation and comparative assessment of retinal pigment epithelium from human embryonic stem cells using transcriptomics. *Cloning and Stem Cells* 6(3):217–245.

Koizumi, N., N.J. Fullwood, G. Bairaktaris et al. 2000. Cultivation of corneal epithelial cells on intact and denuded human amniotic membrane. *Investigative Ophthalmology and Visual Science* 41(9):2506–2513.

Kokkinaki, M., N. Sahibzada, and N. Golestaneh. 2011. Human induced pluripotent stem-derived retinal pigment epithelium (RPE) cells exhibit ion transport, membrane potential, polarized vascular endothelial growth factor secretion, and gene expression pattern similar to native RPE. *Stem Cells* 29(5):825–835.

Krishna, Y., C.M. Sheridan, D.L. Kent, I. Grierson, and R.L. Williams. 2007. Polydimethylsiloxane as a substrate for retinal pigment epithelial cell growth. *Journal of Biomedical Materials Research Part A* 80(3):669–678.

Krishna, Y., C. Sheridan, D. Kent et al. 2011. Expanded polytetrafluoroethylene as a substrate for retinal pigment epithelial cell growth and transplantation in age-related macular degeneration. *British Journal of Ophthalmology* 95(4):569–573.

Lai, J.-Y., K.-H. Chen, and G.-H. Hsiue. 2007. Tissue-engineered human corneal endothelial cell sheet transplantation in a rabbit model using functional biomaterials. *Transplantation* 84(10):1222–1232.

Lajtha, L.G. 1979. Stem cell concepts. *Differentiation* 14(1–3):23–33.

Lappas, A., A.W.A. Weinberger, A.M.H. Foerster et al. 2000. Iris pigment epithelial cell translocation in exudative age-related macular degeneration. *Graefe's Archive for Clinical and Experimental Ophthalmology* 238(8):631–641.

Lavik, E.B., H. Klassen, K. Warfvinge, R. Langer, and M.J. Young. 2005. Fabrication of degradable polymer scaffolds to direct the integration and differentiation of retinal progenitors. *Biomaterials* 26(16):3187–3196.

Lavker, R.M., S. Miller, C. Wilson et al. 1993. Hair follicle stem cells: Their location, role in hair cycle, and involvement in skin tumor formation. *Journal of Investigative Dermatology* 101(s1):16S–26S.

Lee, S.Y., J.H. Oh, J.C. Kim et al. 2003. In vivo conjunctival reconstruction using modified PLGA grafts for decreased scar formation and contraction. *Biomaterials* 24(27):5049–5059.

Leng, T., P. Wu, N.Z. Mehenti et al. 2004. Directed retinal nerve cell growth for use in a retinal prosthesis interface. *Investigative Ophthalmology and Visual Science* 45(11):4132–4137.

Levis, H.J., R.A. Brown, and J.T. Daniels. 2010. Plastic compressed collagen as a biomimetic substrate for human limbal epithelial cell culture. *Biomaterials* 31(30):7726–7737.

Liao, J.L., J. Yu, K. Huang et al. 2010. Molecular signature of primary retinal pigment epithelium and stem-cell-derived RPE cells. *Human Molecular Genetics* 19(21):4229–4238.

Lim, J.-M., S. Byun, S. Chung et al. 2004. Retinal pigment epithelial cell behavior is modulated by alterations in focal cell-substrate contacts. *Investigative Ophthalmology and Visual Science* 45(11):4210–4216.

Liu, J., H. Sheha, Y. Fu, L. Liang, and S.C.G. Tseng. 2010. Update on amniotic membrane transplantation. *Expert Review of Ophthalmology* 5(5):645–661.

Lloyd, A.W., R.G. Faragher, and S.P. Denyer. 2001. Ocular biomaterials and implants. *Biomaterials* 22(8):769–785.

Lu, L., C.A. Garcia, and A.G. Mikos. 1998. Retinal pigment epithelium cell culture on thin biodegradable poly(DL-lactic-co-glycolic acid) films. *Journal of Biomaterials Science, Polymer Edition* 9(11):1187–1205.

Lu, L., L. Kam, M. Hasenbein et al. 1999. Retinal pigment epithelial cell function on substrates with chemically micropatterned surfaces. *Biomaterials* 20(23–24):2351–2361.

Lu, L., K. Nyalakonda, L. Kam et al. 2001. Retinal pigment epithelial cell adhesion on novel micropatterned surfaces fabricated from synthetic biodegradable polymers. *Biomaterials* 22(3):291–297.

MacLaren, R.E., R.A. Pearson, A. MacNeil et al. 2006. Retinal repair by transplantation of photoreceptor precursors. *Nature* 444(7116):203–207.

MacLaren, R.E., G.S. Uppal, K.S. Balaggan et al. 2007. Autologous transplantation of the retinal pigment epithelium and choroid in the treatment of neovascular age-related macular degeneration. *Ophthalmology* 114(3):561–570.

Majo, F., A. Rochat, M. Nicolas, G.A. Jaoude, and Y. Barrandon. 2008. Oligopotent stem cells are distributed throughout the mammalian ocular surface. *Nature* 456(7219):250–254.

Marzo, A., C. Aruta, and V. Marigo. 2010. PEDF promotes retinal neurosphere formation and expansion in vitro. *Retinal Degenerative Diseases* 664:621–630.

Meller, D., V. Dabul, and S.C.G. Tseng. 2002. Expansion of conjunctival epithelial progenitor cells on amniotic membrane. *Experimental Eye Research* 74(4):537–545.

Mi, S., B. Chen, B. Wright, and C.J. Connon. 2010. Plastic compression of a collagen gel forms a much improved scaffold for ocular surface tissue engineering over conventional collagen gels. *Journal of Biomedical Materials Research Part A* 95 (2):447–453.

Mikkers, H. and J. Frisen. 2005. Deconstructing stemness. *EMBO Journal* 24(15):2715–2719.

Morrison, S.J., N.M. Shah, and D.J. Anderson. 1997. Regulatory mechanisms in stem cell biology. *Cell* 88(3):287–298.

Neeley, W.L., S. Redenti, H. Klassen et al. 2008. A microfabricated scaffold for retinal progenitor cell grafting. *Biomaterials* 29(4):418–426.

Nguyen, P. and S.C. Yiu. 2008. Ocular surface reconstruction: Recent innovations, surgical candidate selection and postoperative management. *Expert Review of Ophthalmology* 3(5):567–584.

Nishida, K., M. Yamato, Y. Hayashida et al. 2004. Functional bioengineered corneal epithelial sheet grafts from corneal stem cells expanded ex vivo on a temperature-responsive cell culture surface. *Transplantation* 77(3):379–385.

Nitschke, M., S. Gramm, T. Götze et al. 2007. Thermo-responsive poly(NiPAAm-co-DEGMA) substrates for gentle harvest of human corneal endothelial cell sheets. *Journal of Biomedical Materials Research Part A* 80A(4):1003–1010.

Oganesian, A., K. Gabrielian, M.S. Verp et al. 1996. A new model of RPE transplantation with microspheres. *Investigative Ophthalmology and Visual Science* 37(3):S115.

Ono, K., S. Yokoo, T. Mimura et al. 2007. Autologous transplantation of conjunctival epithelial cells cultured on amniotic membrane in a rabbit model. *Molecular Vision* 13:1138–1143.

Ooto, S., T. Akagi, R. Kageyama et al. 2004. Potential for neural regeneration after neurotoxic injury in the adult mammalian retina. *Proceedings of the National Academy of Sciences of the United States of America* 101(37):13654–13659.

Patel, B.C.K., N.A. Sapp, and R. Collin. 1998. Standardized range of conformers and symblepharon rings. *Ophthalmic Plastic and Reconstructive Surgery* 14(2):144–145.

Pellegrini, G., O. Golisano, P. Paterna et al. 1999. Location and clonal analysis of stem cells and their differentiated progeny in the human ocular surface. *Journal of Cell Biology* 145(4):769–782.

Pellegrini, G., C.E. Traverso, A.T. Franzi et al. 1997. Long-term restoration of damaged corneal surfaces with autologous cultivated corneal epithelium. *Lancet* 349(9057):990–993.

Pertile, G. and C. Claes. 2002. Macular translocation with 360 degree retinotomy for management of age-related macular degeneration with subfoveal choroidal neovascularization. *American Journal of Ophthalmology* 134(4):560–565.

Phillips, S.J., S.R. Sadda, M.O.M. Tso et al. 2003. Autologous transplantation of retinal pigment epithelium after mechanical debridement of Bruch's membrane. *Current Eye Research* 26(2):81–88.

Piatigorsky, J. 1987. Gene expression and genetic engineering in the lens. Friedenwald lecture. *Investigative Ophthalmology and Visual Science* 28(1):9–28.

Potten, C.S. and M. Loeffler. 1990. Stem cells: Attributes, cycles, spirals, pitfalls and uncertainties. Lessons for and from the crypt. *Development* 110(4):1001–1020.

Princz, M.A., H. Sheardown, and M. Griffith. 2011. Corneal tissue engineering versus synthetic artificial cornea. In *Biomaterials and Regenerative Medicine in Ophthalmology*, T.V. Chirila (Ed.), pp. 134–149. Cambridge, U.K.: Woodhead Publishing Limited.

Prokhorova, T.A., L.M. Harkness, U. Frandsen et al. 2009. Teratoma formation by human embryonic stem cells is site dependent and enhanced by the presence of Matrigel. *Stem Cells and Development* 18(1):47–54.

Proulx, S., J. d'Arc Uwamaliya, P. Carrier et al. 2010. Reconstruction of a human cornea by the self-assembly approach of tissue engineering using the three native cell types. *Molecular Vision* 16:2192–2201.

Radtke, N.D., M.J. Seiler, R.B. Aramant, H.M. Petry, and D.J. Pidwell. 2002. Transplantation of intact sheets of fetal neural retina with its retinal pigment epithelium in retinitis pigmentosa patients. *American Journal of Ophthalmology* 133(4):544–550.

Rafat, M., F. Li, P. Fagerholm et al. 2008. PEG-stabilized carbodiimide crosslinked collagen-chitosan hydrogels for corneal tissue engineering. *Biomaterials* 29(29):3960–3972.

Rajnicek, A.M., L.E. Foubister, and C.D. McCaig. 2008. Alignment of corneal and lens epithelial cells by co-operative effects of substratum topography and DC electric fields. *Biomaterials* 29(13):2082–2095.

Ralph, R.A. 1975. Reconstruction of conjunctival fornices using silicone rubber sheets. *Ophthalmic Surgery* 6(3):55–57.

Rama, P., S. Matuska, G. Paganoni et al. 2010. Limbal stem-cell therapy and long-term corneal regeneration. *New England Journal of Medicine* 363(2):147–155.

Redenti, S., W.L. Neeley, S. Rompani et al. 2009. Engineering retinal progenitor cell and scrollable poly(glycerol-sebacate) composites for expansion and subretinal transplantation. *Biomaterials* 30(20):3405–3414.

Redenti, S., S. Tao, J. Yang et al. 2008. Retinal tissue engineering using mouse retinal progenitor cells and a novel biodegradable, thin-film poly(e-caprolactone) nanowire scaffold. *Journal of Ocular Biology, Diseases, and Informatics* 1(1):19–29.

Reichl, S., M. Borrelli, and G. Geerling. 2011. Keratin films for ocular surface reconstruction. *Biomaterials* 32(13):3375–3386.

Remington, S. and R. Meyer. 2007. Lens stem cells may reside outside the lens capsule: An hypothesis. *Theoretical Biology and Medical Modelling* 4(1):22.

Rimmer, S., C. Johnson, B. Zhao et al. 2007. Epithelialization of hydrogels achieved by amine functionalization and co-culture with stromal cells. *Biomaterials* 28(35):5319–5331.

Sangwan, V.S., G.K. Vemuganti, S. Singh, and D. Balasubramanian. 2003. Successful reconstruction of damaged ocular outer surface in humans using limbal and conjuctival stem cell culture methods. *Bioscience Reports* 23(4):169–174.

Schneider, A.I., K. Maier-Reif, and T. Graeve. 1999. Constructing an in vitro cornea from cultures of the three specific corneal cell types. *In Vitro Cellular and Developmental Biology—Animal* 35(9):515–526.

Schraermeyer, U., V. Enzmann, L. Kohen et al. 1997. Porcine iris pigment epithelial cells can take up retinal outer segments. *Experimental Eye Research* 65(2):277–287.

Schwab, I.R. 1999. Cultured corneal epithelia for ocular surface disease. *Transactions of the American Ophthalmological Society* 97:891–986.

Seigel, G.M. and A. Kummer. 2001. The enigma of lenticular oncology. *Digital Journal of Ophthalmology* 7(4):http://www.djo.harvard.edu/site.php?url=/physicians/oa/360

Sheridan, C., Y. Krishna, R. Williams et al. 2007. Transplantation in the treatment of age-related macular degeneration: Past, present and future directions. *Expert Review of Ophthalmology* 2(3):497–511.

Shortt, A.J. and S.J. Tuft. 2011. Ocular surface reconstruction. *British Journal of Ophthalmology* 95(7):901–902.

Silverman, M.S., S.E. Hughes, T.L. Valentino, and Y. Liu. 1992. Photoreceptor transplantation: Anatomic, electrophysiologic, and behavioral evidence for the functional reconstruction of retinas lacking photoreceptors. *Experimental Neurology* 115(1):87–94.

Singh, S., S. Woerly, and B.J. McLaughlin. 2001. Natural and artificial substrates for retinal pigment epithelial monolayer transplantation. *Biomaterials* 22(24):3337–3343.

Solomon, A., E.M. Espana, and S.C.G. Tseng. 2003. Amniotic membrane transplantation for reconstruction of the conjunctival fornices. *Ophthalmology* 110(1):93–100.

Stanzel, B.V., M.S. Blumenkranz, S. Binder, and M.F. Marmor. 2012. Longterm cultures of the aged human RPE do not maintain epithelial morphology and high transepithelial resistance. *Graefe's Archive for Clinical and Experimental Ophthalmology* 250(2):313–315.

Stanzel, B.V., C. Clemens, S.R. Sanislo et al. 2010. SD-OCT complements histology in evaluation of potential Bruch's membrane prosthetics. *Investigative Ophthalmology and Visual Science* 51(5):5241–5241.

Stanzel, B.V., E.M. Espana, M. Grueterich et al. 2005. Amniotic membrane maintains the phenotype of rabbit retinal pigment epithelial cells in culture. *Experimental Eye Research* 80(1):103–112.

Steedman, M.R., S.L. Tao, H. Klassen, and T.A. Desai. 2010. Enhanced differentiation of retinal progenitor cells using microfabricated topographical cues. *Biomedical Microdevices* 12(3):363–369.

Sun, G., M. Asami, H. Ohta, J. Kosaka, and M. Kosaka. 2006. Retinal stem/progenitor properties of iris pigment epithelial cells. *Developmental Biology* 289(1):243–252.

Sun, K., H. Cai, T.H. Tezel et al. 2007. Bruch's membrane aging decreased phagocytosis of outer segments by retinal pigment epithelium. *Molecular Vision* 13:2310–2319.

Sundback, C.A., J.Y. Shyu, Y. Wang et al. 2005. Biocompatibility analysis of poly(glycerol sebacate) as a nerve guide material. *Biomaterials* 26(27):5454–5464.

Talbot, M., P. Carrier, C.J. Giasson et al. 2006. Autologous transplantation of rabbit limbal epithelia cultured on fibrin gels for ocular surface reconstruction. *Molecular Vision* 12:65–75.

Tan, D.T.H., L.P.K. Ang, and R.W. Beuerman. 2004. Reconstruction of the ocular surface by transplantation of a serum-free derived cultivated conjunctival epithelial equivalent. *Transplantation* 77(11):1729–1734.

Tanioka, H., S. Kawasaki, K. Yamasaki et al. 2006. Establishment of a cultivated human conjunctival epithelium as an alternative tissue source for autologous corneal epithelial transplantation. *Investigative Ophthalmology and Visual Science* 47(9):3820–3827.

Tao, S., C. Young, S. Redenti et al. 2007. Survival, migration and differentiation of retinal progenitor cells transplanted on micro-machined poly(methyl methacrylate) scaffolds to the subretinal space. *Lab on a Chip* 7(6):695–695.

Tezcaner, A., K. Bugra, and V. Hasirci. 2003. Retinal pigment epithelium cell culture on surface modified poly (hydroxybutyrate-co-hydroxyvalerate) thin films. *Biomaterials* 24(25):4573–4583.

Thieltges, F., B.V. Stanzel, Z. Liu, and F.G. Holz. 2011. A nanofibrillar surface promotes superior growth characteristics in cultured human retinal pigment epithelium. *Ophthalmic Research* 46(3):133–140.

Thoft, R.A. 1977. Conjunctival transplantation. *Archives of Ophthalmology* 95(8):1425–1427.

Thumann, G., K.U. Bartz-Schmidt, K. Heimann, and U. Schraermeyer. 1998. Phagocytosis of rod outer segments by human iris pigment epithelial cells in vitro. *Graefe's Archive for Clinical and Experimental Ophthalmology* 236(10):753–757.

Thumann, G., U. Schraermeyer, K.U. Bartz-Schmidt, and K. Heimann. 1997. Descemet's membrane as membranous support in RPE/IPE transplantation. *Current Eye Research* 16(12):1236–1238.

Thumann, G., A. Viethen, A. Gaebler et al. 2009. The in vitro and in vivo behaviour of retinal pigment epithelial cells cultured on ultrathin collagen membranes. *Biomaterials* 30(3):287–294.

Tian, C., T. Zhao, Y. Zeng, and Z. Yin. 2011. Increased Müller cell dedifferentiation after grafting of retinal stem cell in the subretinal space of RCS rats. *Tissue Engineering Part A* 17(19–20):253–259.

Tomita, M., E. Lavik, H. Klassen et al. 2005. Biodegradable polymer composite grafts promote the survival and differentiation of retinal progenitor cells. *Stem Cells* 23(10):1579–1588.

Tropepe, V., B.L.K. Coles, B.J. Chiasson et al. 2000. Retinal stem cells in the adult mammalian eye. *Science* 287(5460):2032–2036.

Tseng, S.C., D. Meller, D.F. Anderson et al. 2002. Ex vivo preservation and expansion of human limbal epithelial stem cells on amniotic membrane for treating corneal diseases with total limbal stem cell deficiency. *Advances in Experimental Medicine and Biology* 506(Pt B):1323–1334.

Tseng, S.C.G., M.A. Di Pascuale, D.T.-S. Liu, Y.Y. Gao, and A. Baradaran-Rafii. 2005. Intraoperative mitomycin C and amniotic membrane transplantation for fornix reconstruction in severe cicatricial ocular surface diseases. *Ophthalmology* 112(5):896–903.e1.

Tsonis, P.A. and K. Del Rio-Tsonis. 2004. Lens and retina regeneration: Transdifferentiation, stem cells and clinical applications. *Experimental Eye Research* 78(2):161–172.

Van Meurs, J.C., E. Ter Averst, L.J. Hofland et al. 2004. Autologous peripheral retinal pigment epithelium translocation in patients with subfoveal neovascular membranes. *British Journal of Ophthalmology* 88(1):110–113.

van Meurs, J.C. and P.R. Van Den Biesen. 2003. Autologous retinal pigment epithelium and choroid translocation in patients with exudative age-related macular degeneration: Short-term follow-up. *American Journal of Ophthalmology* 136(4):688–695.

Vascotto, S.G. and M. Griffith. 2006. Localization of candidate stem and progenitor cell markers within the human cornea, limbus, and bulbar conjunctiva in vivo and in cell culture. *Anatomical Record Part A: Discoveries in Molecular, Cellular, and Evolutionary Biology* 288(8):921–931.

Vastine, D.W., W.B. Stewart, and I.R. Schwab. 1982. Reconstruction of the periocular mucous membrane by autologous conjunctival transplantation. *Ophthalmology* 89(9):1072–1081.

von Recum, H.A., S.W. Kim, A. Kikuchi et al. 1998. Novel thermally reversible hydrogel as detachable cell culture substrate. *Journal of Biomedical Materials Research* 40(4):631–639.

Vossmerbaeumer, U., S. Kuehl, S. Kern et al. 2008. Induction of retinal pigment epithelium properties in ciliary margin progenitor cells. *Clinical and Experimental Ophthalmology* 36(4):358–366.

Vossmerbaeumer, U., S. Ohnesorge, S. Kuehl et al. 2009. Retinal pigment epithelial phenotype induced in human adipose tissue-derived mesenchymal stromal cells. *Cytotherapy* 11(2):177–188.

Vyas, S.M.S. and V.D.O. Rathi. 2009. Combined phototherapeutic keratectomy and amniotic membrane grafts for symptomatic bullous keratopathy. *Cornea* 28(9):1028–1031.

Wagner, M., M.T. Benson, I.G. Rennie, and S. Macneil. 1995. Effects of pharmacological modulation of intracellular signalling systems on retinal pigment epithelial cell attachment to extracellular matrix proteins. *Current Eye Research* 14(5):373–384.

Wang, E., M. Zhao, J.V. Forrester, and C.D. McCaig. 2003. Bi-directional migration of lens epithelial cells in a physiological electrical field. *Experimental Eye Research* 76(1):29–37.

Warfvinge, K., J.F. Kiilgaard, E.B. Lavik et al. 2005. Retinal progenitor cell xenografts to the pig retina: Morphologic integration and cytochemical differentiation. *Archives of Ophthalmology* 123(10):1385–1393.

Wei, Z.G., T.T. Sun, and R.M. Lavker. 1996. Rabbit conjunctival and corneal epithelial cells belong to two separate lineages. *Investigative Ophthalmology and Visual Science* 37(4):523–533.

Wencan, W., Y. Mao, Y. Wentao et al. 2007. Using basement membrane of human amniotic membrane as a cell carrier for cultivated cat corneal endothelial cell transplantation. *Current Eye Research* 32(3):199–215.

Williams, R.L., Y. Krishna, S. Dixon et al. 2005. Polyurethanes as potential substrates for sub-retinal retinal pigment epithelial cell transplantation. *Journal of Materials Science: Materials in Medicine* 16(12):1087–1092.

Wong, D. 2000. Foveal relocation by redistribution of the neurosensory retina. *British Journal of Ophthalmology* 84(4):352–357.

Wong, D. 2004. Case selection in macular relocation surgery for age related macular degeneration. *British Journal of Ophthalmology* 88(2):186–190.

Yokoo, S., S. Yamagami, T. Usui, S. Amano, and M. Araie. 2008. Human corneal epithelial equivalents for ocular surface reconstruction in a complete serum-free culture system without unknown factors. *Investigative Ophthalmology and Visual Science* 49(6):2438–2443.

Zarbin, M.A. 2008. RPE-retina transplantation for retinal degenerative disease. *American Journal of Ophthalmology* 146(2):151–153.

Zhang, X. and D. Bok. 1998. Transplantation of retinal pigment epithelial cells and immune response in the subretinal space. *Investigative Ophthalmology and Visual Science* 39(6):1021–1027.

Zhou, M., J. Leiberman, J. Xu, and R.M. Lavker. 2006. A hierarchy of proliferative cells exists in mouse lens epithelium: Implications for lens maintenance. *Investigative Ophthalmology and Visual Science* 47(7):2997–3003.

Zorlutuna, P., A. Tezcaner, I. Kıyat, A. Aydınlı, and V. Hasırcı. 2006. Cornea engineering on polyester carriers. *Journal of Biomedical Materials Research Part A* 79(1):104–113.

10

Calcium Phosphates as Scaffolds for Mesenchymal Stem Cells

Iain R. Gibson

CONTENTS

10.1 CaPs as Scaffolds for MSCs

10.1.1 CaP Scaffolds Overview

10.1.1.1 Introduction

The use of calcium phosphates (CaPs) as scaffolds for the repair or regeneration of diseased or damaged bone is based largely on their chemical similarity to the mineral component of bone, but also due to the large body of research on the synthesis and processing of scaffolds with the appropriate architecture for the formation of bone. Additionally, many CaPs have a long history of safe clinical use as bone graft materials.

This chapter will summarize the various CaP compounds that have been used as scaffolds in regenerative medicine for bone repair, focusing on the specific chemistries of these compounds and also the types of architectures, specifically their porosities, that have been developed. The effect of scaffold chemistry and porosity on the ability to guide mesenchymal stem cells (MSCs) to form new bone will be described, with a focus on the formation of bone in vivo using human-derived MSCs. This chapter will focus on bone marrow (BM)-derived MSCs (BM-MSCs), but some mention will be given to adipose- and periosteum-derived MSCs, which are an increasingly studied source of adult MSCs for bone regeneration.

Scaffolds or grafts for bone regeneration/repair should ideally exhibit three key properties; they should be osteoconductive, osteoinductive, and osteogenic. A patient's own harvested bone (autograft) is the only true graft material that exhibits all three properties. CaP scaffolds have been shown to be osteoconductive, whereby they support the ingrowth of cells to produce new bone. Of particular interest to the emerging field of bone regenerative medicine is the role of CaP scaffolds in the other two key properties. Certain CaP scaffolds have been shown to be osteoinductive, whereby they play a role in the induction of host progenitor cell differentiation to bone-forming cells, with the resulting formation of new bone; this has been tested extensively in ectopic sites. CaP scaffolds alone are clearly not capable of being osteogenic, which describes the bone-forming ability of cells within a graft. However, CaP scaffolds can be seeded with progenitor cells capable of forming bone, resulting in the formation of an osteogenic scaffold–cell construct that has the potential to form bone. The combination of progenitor cells with CaP scaffolds to form new bone is described in Section 10.1.2, focusing mostly on human BM-MSCs. The main choices that must be addressed in the route to developing a CaP scaffold–MSC construct for regenerating new bone are summarized in Figure 10.1, and this gives a breakdown of the key factors in the research to date in this field.

It is out of the scope of this chapter to discuss the mechanisms of bone formation and regeneration or the cell biology of MSCs in relation to bone formation.

10.1.1.2 CaP Chemistry

A number of different CaP biomaterials have been utilized as scaffolds for regenerative medicine, and these are summarized in Table 10.1. The most commonly used CaP compound to produce scaffolds for bone regeneration is hydroxyapatite (HA), which is described as having a Ca/P molar ratio of 1.67. It has an apatite structure with a crystal structure comparable to that found in the apatite phase that constitutes bone mineral; the differences are that bone mineral apatite is non-stoichiometric, has a variable Ca/P molar ratio and includes other elements in addition to Ca, P, O, and H. The characterization of HA for use as a medical device is guided by a number of ISO and ASTM standards, and there is a large body of data confirming the biocompatibility and osteoconductivity of HA implants. Although this history of clinical use makes HA an attractive candidate for producing scaffolds that can be used to regenerate bone, and the evidence that HA scaffolds support the growth and differentiation of MSCs in culture, it has some perceived disadvantages. These include very slow resorption rates and a lack of evidence

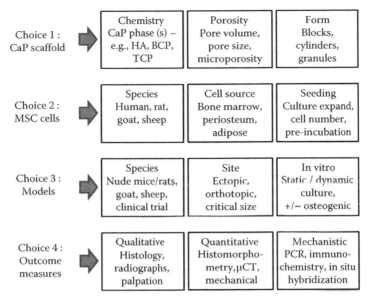

FIGURE 10.1
Pathway to bone regeneration using CaP scaffolds and MSCs—research to date can be summarized as involving four important choices, each with various options.

TABLE 10.1

CaP Compositions Typically Used as Scaffolds for Bone Repair and Regeneration

Name	Abbreviated Name	Chemical Composition	Ca/P Ratio
Hydroxyapatite	HA or Hap	$Ca_{10}(PO_4)_6(OH)_2$	1.67
Beta-tricalcium phosphate	β-TCP	$Ca_3(PO_4)_2$	1.50
Biphasic calcium phosphate	BCP	$Ca_{10}(PO_4)_6(OH)_2$-$Ca_3(PO_4)_2$	>1.50 and <1.67
Octacalcium phosphate	OCP	$Ca_8(PO_4)_4(HPO_4)_2.5H_2O$	1.33
Calcium-deficient (hydroxy)apatite	CDHA	$Ca_{10-x}(PO_4)_{6-x}(HPO_4)_x(OH)_{2-x}$ $(0<x<1)$	>1.50 and <1.67
Substituted hydroxyapatite[a]	Various	Various	Various

[a] Various ions have been substituted for Ca, PO_4, or OH in the HA lattice, such as Sr, SiO_4, CO_3.

of osteoinductivity. Compared to other CaP compositions, HA shows very low resorption rates in vivo (Shimazaki and Mooney 1985; Johnson et al. 1996). While this was historically viewed as an advantageous property when HA was used as a dense implant or coating, the fields of tissue engineering and regenerative medicine have demonstrated a need for scaffolds that initially support new tissue formation but are eventually resorbed and replaced fully by the regenerating tissue. Although a number of resorbable polymers, such as poly(lactic acid) and poly(caprolactone), have been used extensively as scaffolds for MSC culture and bone regeneration for this very reason, other CaP compounds are able to be resorbed in vivo, while still offering a chemistry that is closer to the native bone.

Tricalcium phosphate (TCP) is a compound typically produced by high-temperature synthesis and can exist in a number of polymorphs, all having a Ca/P molar ratio of 1.50. The most widely applied polymorph used to produce scaffolds is the beta-form (β-TCP), which resorbs at a significantly faster rate than HA (Shimazaki and Mooney 1985; Johnson et al. 1996). Another strategy to produce CaP scaffolds with enhanced resorption rates is to produce biphasic mixtures containing both HA and β-TCP phases (typically termed biphasic calcium phosphates [BCPs]). Although these consist of two discrete phases with Ca/P molar ratios of 1.67 and 1.50 for HA and β-TCP, respectively, the overall Ca/P molar ratio of these biphasic compositions can range from 1.67 to 1.50, depending on the relative proportions of the two phases; typical ratios are between 1.57 and 1.62 (Daculsi et al. 2003).

A common theme in HA, β-TCP, and BCPs is their method of synthesis and processing. The typical synthesis method to produce these compositions is the aqueous precipitation of a solution of a calcium salt and a phosphate-containing solution at a controlled pH value, to form an apatite phase whose composition is dependent on the molar ratio of reactants used and the pH of precipitation reaction. The most common method that is suitable for the synthesis of HA, β-TCP, and BCP is the addition of a phosphoric acid solution to a calcium hydroxide solution/slurry at a pH of 8–11 (Akao et al. 1981). A pH of >10, accompanied with a molar ratio of reactants of 1.67, favors the formation of an apatite phase that will, upon high-temperature heat treatment (typically 1100°C–1300°C), produce HA. At a pH of 8 with a molar ratio of reactants of 1.50, a calcium-deficient apatite (CDHA) forms that will produce β-TCP upon high-temperature heat treatment (typically 900°C–1100°C). A combination of an intermediate pH value and reactant molar ratios produces a CDHA that forms a biphasic composition on heating. The high-temperature heat treatment step is not only required to produce the desired final composition from the precipitated product, but it is also required to sinter the CaP material to introduce a degree of mechanical integrity, and also control the porosity/architecture of the scaffold; this will be described in more detail in Section 10.1.1.3.

HA, β-TCP, and BCPs are the CaP compositions that are used almost exclusively in the manufacture of current synthetic CaP bone graft substitutes used clinically (Seebach et al. 2010b; Zannettino et al. 2010; Van der Stok et al. 2011). However, within the more recent field of regenerative medicine, other CaP phases have been used as scaffolds, and these include the CDHA phase (Kasten et al. 2008b; Akiyama et al. 2011) described earlier which can be considered as a precursor to HA, β-TCP, and BCP. A limitation of the use of CDHA to produce scaffolds is that it only exists up to a temperature of approximately 750°C, after which it will convert to HA, β-TCP, or BCP (Gibson et al. 2000). This temperature is too low to sinter the material (see Section 10.1.1.3), resulting in scaffolds with poor mechanical integrity. Another CaP phase that has been reported for use as scaffolds is octacalcium phosphate (OCP), with a Ca/P molar ratio of 1.33, which has been reported to exist as a precursor phase during the mineralization of bone (Brown et al. 1987). OCP is even more temperature sensitive than CDHA, so again to produce a scaffold, high temperatures are avoided, resulting in a scaffold consisting of assemblies of OCP crystals (Liu et al. 2007), rather than fused grains of, for example, HA or β-TCP that are typical in sintered CaP scaffolds. Although this can result in a scaffold with poor mechanical integrity, it has the attraction of producing scaffolds with high surface areas.

As mentioned previously, HA scaffolds are associated with very low resorption rates but have the attraction of having an apatite structure that is comparable to that of bone mineral. To attempt to overcome the low resorption rate of HA, many groups have studied various ion substitutions in the HA structure to produce a chemical composition that more closely resembles that of bone mineral apatite. The most commonly studied so-called substituted HA is carbonate-substituted HA, in which some of the phosphate and/or hydroxyl ions are substituted for carbonate ions (Bonel 1972). These compositions have been shown to have faster resorption rates than HA in vitro and in vivo. Silicon is another element that has been substituted into the HA structure (Gibson et al. 1999), and also the TCP structure (Langstaff et al. 1999), whereby silicon substitutes as a silicate ion for phosphate ions. Such substituted HAs are interesting alternatives to HA and are relatively new in terms of the body of data that has been reported to date, specific to scaffolds for regenerative medicine.

While this section gives an overview of the types of CaP chemistries that have been used in scaffolds, more detailed descriptions of synthesis methods and chemical and physical characterization of these compositions are available in a number of recent reviews (e.g., Hing 2005; Dorozhkin 2007).

10.1.1.3 CaP Scaffold Porosity

Porosity, which is considered the amount (typically quoted as %) of free space within a solid, is an extremely important parameter when designing and discussing scaffolds for the regeneration of tissue, irrespective of the scaffold material or the tissue type being regenerated. Scaffolds with a high degree of porosity allow penetration of new tissue throughout the scaffold, regardless if the tissue is migrating inward from surrounding tissue, or if the new tissue is being formed within the scaffold. For CaP scaffolds for bone regeneration, porosity also enables invasion of the scaffold with blood vessels which are essential for new bone formation. To describe the porosity of a scaffold completely, a number of specific parameters are usually used: total pore volume (total porosity), pore size, interconnectivity of porosity, and size of pore interconnections. The typical "accepted" values of these parameters for porous blocks of CaP scaffolds have evolved from early studies on the formation of new bone found in scaffolds with various porosities. For example, early studies in canines showed that the minimum pore size of calcium aluminate scaffolds that was required to regenerate mineralized bone was 100 µm (Hulbert 1970). However, in addition to pore size, numerous studies have shown in the case of CaP scaffolds that the total pore volume and size of pore interconnections, as well as the CaP chemistry, are all critical parameters that determine bone tissue formation within the scaffolds. For example, implantation of porous cylinders of HA and β-TCP with a fixed total porosity but different pore sizes (50–100 compared to 200–400 µm) in femur and tibia of rabbits showed that more bone was formed in scaffolds with the smaller pore size, and this was attributed to these scaffolds having interconnected pores (20 µm diameter), which were rarely present in the scaffolds with larger pore sizes (Eggli et al. 1988). A similar study in rabbits showed that the size of pore interconnections in HA and β-TCP scaffolds should be greater than 50 µm to favor the formation of new bone in the scaffolds (Lu et al. 1999). Detailed reviews on the role of porosity on osteogenesis in 3D scaffolds (Karageorgiou and Kaplan 2005) and of porosity in CaP scaffolds (Hing 2005) describe these parameters in more depth.

In terms of the production of scaffolds, the complex architecture found in cancellous bone has been a guide to the development of CaP scaffolds over the past 40 years. Indeed, an early and popular method of producing CaP porous scaffolds was to use templates from nature, either animal bones or marine coral. If bovine cancellous bone is heated to remove the organic components, then a sintered macroporous HA scaffold is obtained with an architecture based on the cancellous structure of the original bone (Dard et al. 1994). This should not be confused with hydrothermally treated bovine bone, termed Kiel bone, which was developed in the 1950s, and is considered a xenograft (Maatz and Bauermeister 1957). Alternatively, calcium carbonate–based coral exoskeletons can be hydrothermally converted to a macroporous calcium HA scaffold with an architecture of the coral exoskeleton (Roy and Linnehan 1974). The main disadvantage of these methods is that the ability to produce scaffolds with highly controlled and repeatable porous structures is limited due to batch-to-batch variability in the natural templates.

Porosity can also be introduced in to CaPs by the incorporation of polymer beads in to a compact of the CaP powder. Upon sintering at high temperatures, the polymer beads, such as poly(methyl methacrylate) (PMMA) (Uchida et al. 1984), are burned out, leaving pores throughout the CaP sample. Pore size and total pore volume can be controlled by the diameter and quantity of beads used. A major limitation of this method is a lack of interconnections between the pores.

Another method of introducing porosity is to produce a ceramic slip of the chosen CaP powder in water, and foaming this slip to introduce porosity;

this can be viewed as "bubble generation." The foamed slip can then be cast, dried, and sintered, producing a porous scaffold. A foaming agent is typically employed, and these have included surfactants (Binner and Reichert 1996) and hydrogen peroxide (Driessen et al. 1982).

Based on the "natural template" methods of coral or bovine bone described earlier, an alternative method is to use a porous polymer sponge/foam as a source of porous architecture. This method of making porous ceramics was first described in a patent (Schwartzwalder and Somers 1963) and has been used extensively to make porous CaP scaffolds, for example, porous β-TCP scaffolds using polyurethane foams (Milosevski et al. 1999). The CaP is formed in to a ceramic slip or slurry that can be used to coat and impregnate the polymer sponge/foam. The foam architecture can be designed to the desired pore size and shape. After drying, the ceramic-soaked sponge/foam is sintered, and this burns off the polymer sponge/foam, leaving behind a porous ceramic scaffold. An interesting extension of this method was to form a wax negative mould of cancellous bone and fill the mould with a ceramic slip followed by burning out the wax during sintering (Tancred et al. 1998). This resulted in a positive replica of the architecture of cancellous bone and was carried out using HA, β-TCP, and BCP.

A significant problem encountered using typical polymer foams as negative templates is the very thin struts or walls of the resulting CaP scaffolds, resulting in samples with very poor mechanical integrity. To overcome this, an innovative method of producing polymer negative templates utilizing advanced design/deposition techniques (computer-aided design [CAD] software and computer tomography [CT] images, and stereolithography) resulted in epoxy templates or moulds with channels with a high degree of control of pore size and interconnectivity (Chu et al. 2001). An HA slurry was then cast into the epoxy mould, followed by drying and sintering, which removed the polymer leaving a porous HA scaffolds with highly controlled architectures.

This last example represents an increasing shift in the technology used to fabricate porous scaffolds from introducing pores through removal of matter (beads, sponges) or introduction of "bubbles" (foaming of slips) toward depositing the CaP scaffold layer-by-layer, with the size and architecture being introduced during this process. These techniques can be summarized as solid free-form fabrication (SFF) methods, which can typically be considered as moldless fabrication techniques. A number of these methods have been utilized to produce porous CaP scaffolds, including stereolithography (Levy et al. 1997), selective laser sintering (Tan et al. 2003), 3D printing (Sherwood et al. 2002), fused deposition modeling (Schantz et al. 2005), robocasting (Dellinger et al. 2007), and Bioplotter™ (Rücker et al. 2008) (with the possibility of incorporating cells using this method); it should be noted that some of these techniques require a polymer component for ease of fabrication, resulting in CaP–polymer composite scaffolds. These methods and their applicability to producing scaffolds for regenerative medicine have been reviewed in detail (Hollister 2005).

It should be noted that the vast majority of research utilizing porous CaP scaffolds for bone tissue regeneration have focused on the use of porous blocks, for example, cubes or cylinders, whether this has been for in vitro seeding and expansion of cells or for in vivo determination of bone-forming ability. Typical values of pore volume and pore size of CaP scaffolds discussed in Section 10.1.2 are listed in Table 10.2. Recently, there has been a shift in how porous CaP scaffolds have been utilized in this field, directed largely by the clinical use of synthetic CaP scaffolds as bone graft substitutes, toward using them in a granular form (Seebach et al. 2010a,b; Zannettino et al. 2010; Roberts et al. 2011). Such granules are typically of the size range 0.5–2 mm, which makes cell seeding much more efficient that trying to seed large blocks (e.g., 5–10 mm). These granules are typically produced to have comparable pore sizes, total pore volumes, etc., to blocks and can be packed together to fill a geometrical defect.

Increasing the porosity of a CaP ceramic, while enhancing the bone-forming ability, typically has a negative impact on the mechanical properties of scaffolds. The effect of porosity on the mechanical strength of ceramics is reviewed elsewhere (Montanaro et al. 1998).

10.1.2 CaP Scaffolds with MSCs

10.1.2.1 Introduction

CaP scaffolds implanted in the absence of cells are osteoconductive and, in some cases, can be osteoinductive. Although effective therapies for many bone repair/regeneration indications, it is widely accepted that incorporating cells into the scaffold prior to implantation offers an enhanced strategy for stimulating bone regeneration. This is of particular importance for patients that may present a limited or compromised capacity for bone repair (e.g., elderly, diabetic or chemotherapy-treated patients), or for the repair of defect sites where bone regeneration is problematic (e.g., long bone nonunion fractures). Clinically, the incorporation of cells is regularly achieved by the simple mixing of a synthetic CaP scaffold with the patient's own bone marrow aspirate (BMA). Preclinical data on the role of BMA on the repair of a challenging bone defect (segmental long bone defect in canine radii) showed that HA and TCP scaffolds combined with BMA showed healing that approached that of the control of cancellous autograft, while scaffolds without BMA resulted in nonunion (Johnson et al. 1996). Although this procedure is clinically convenient, BMA provides a nonuniform population of cells, with a very low number (approximately 2000) of connective tissue progenitor cells, or BM-MSCs per ml of aspirate (Muschler et al. 2004), that can vary significantly from patient to patient (Muschler et al. 1997, 2001). For this reason, there has been considerable

TABLE 10.2

Summary of Typical Ranges of Pore Sizes and Total Pore Volumes of Porous CaP Scaffolds Used with MSCs, Implanted Subcutaneously in Nude Mice/Rats (Top) or Cultured In Vitro (Bottom)

Scaffold Type (Composition)	Total Pore Volume (%)	Pore Size (μm)	Specific Surface Area (m²/g)	References
BCP blocks (60% HA to 40% β-TCP)	NR	400	NR	Goshima et al. (1991)
BCP cubes (100%, 76%, 63%, 56%, 20%, 0% HA)	60–70	300–600	NR	Arinzeh et al. (2005)
CDHA	85	200–600	NR	Kasten et al. (2008b)
β-TCP	84	200–600	NR	
HA	69	200–600	NR	
Skelite™ cubes (67% Si–TCP/33% HA)	60	200–500	NR	Mastrogiacomo et al. (2007)
BCP cylinders (NR)	NR	200–450	NR	Bruder et al. (1998)
BCP cubes (60% HA)	NR	200–470	NR	Harris and Cooper (2004)
BCP cubes (20% HA)	NR	170–680	NR	
Coralline HA	NR	200–800	NR	
Bovine-bone-derived HA	NR	100–770	NR	
HA granules	46	NR	0.1	Yuan et al. (2010)
BCP granules (1150)	46	NR	1.0	
BCP granules (1300)	45	NR	0.2	
β-TCP granules	50	NR	1.2	
β-TCP granules	60	100–500	NR	Seebach et al. (2010a)
β-TCP 25 blocks	25	5–200	0.12	Kasten et al. (2008a)
β-TCP 65 blocks	65	5–600	0.18	
β-TCP 75 blocks	75	200–600	0.3	
β-TCP blocks	45	500	NR	Wang et al. (2009)
Coralline HA 200	75	100–200	NR	Mygind et al. (2007)
Coralline HA 500 cylinders	88	300–500	NR	
BCP cylinders (60% HA)	90	NR	NR	Holtorf et al. (2005)

NR, not reported.

interest in combining CaP scaffolds with well-characterized populations of cells in vitro, with the aim of producing a scaffold-cell construct that has the potential to show greater and more reproducible bone-forming potency when implanted in vivo. The following sections will describe strategies to address this.

10.1.2.2 In Vitro Studies

10.1.2.2.1 Introduction

Various in vitro cell culture models have been used to test the role of a range of CaPs on cell proliferation and differentiation; a short review of the response of various MSCs to CaP scaffolds follows. Most studies use BM-MSCs, but recent studies have investigated the response of periosteum-derived and adipose-derived MSCs to CaP scaffolds.

10.1.2.2.2 Role of Scaffold Chemistry on MSCs In Vitro

The response of various CaP chemistries to MSCs has been studied, including coralline-derived HA (Mygind et al. 2007), carbonate-substituted HA (Melville et al. 2006), Si–TCP (Bjerre et al. 2008), OCP (Shelton et al. 2006; Liu et al. 2007), β-TCP (Kasten et al. 2008a; Wang et al. 2009), BCP (Holtorf et al. 2005), and CDHA (Kasten et al. 2008b). While these studies are important steps in evaluating these different scaffolds, the studies tend to be less informative than in vivo studies and lack the ability to test various CaP chemistries in the absence of significant differences in other scaffold physical properties.

10.1.2.2.3 Role of Porosity on MSCs In Vitro

Some more information can be obtained from studies that test the effect of scaffold porosity or surface area on MSC response in vitro. For example, a study demonstrated that the porosity of β-TCP scaffolds affected the osteogenic differentiation of human BM-MSCs in vitro, with osteogenic supplements, and subcutaneous bone formation in vivo in SCID mice (Kasten et al. 2008a). Results showed no significant difference in ALP production by cells cultured on β-TCP scaffolds containing 25%, 65%, or 75% porosity, whereas in vivo ALP activity showed significant differences with different porosities with the trend 65% > 75% > 25%. The authors noted that porosity was clearly not the sole factor in determining in vivo osteogenic differentiation and that differences in surface properties or in pore sizes and distributions between the different samples may also be a factor. Comparing coralline-derived HA scaffolds with different pore sizes showed faster osteogenic differentiation of human MSCs (hMSCs) on scaffolds with pore sizes of 200 μm compared to pore sizes of 500 μm, whereas the larger pore size was associated with higher seeding efficiency and greater cell proliferation (Mygind et al. 2007).

A study compared human BM-MSC response to CDHA and β-TCP scaffolds that had matched total porosities and pore sizes but very different specific surface areas of 48 and <0.5 m²/g, respectively (Kasten et al. 2008b). Cell seeding efficiency was higher on the β-TCP scaffolds with the lower surface area, but no significant difference was observed in osteogenic differentiation, assessed by determining alkaline phosphatase activity.

A recent study compares the response of six commercially available CaP granules, intended for bone grafting applications, to hMSCs

(Seebach et al. 2010b). The chemistries of these granules were different, including β-TCP, BCP, Si–HA, bovine-bone-derived HA, but also the pore sizes and total pore volumes were very different, making correlations with the MSC cell response difficult.

10.1.2.3 In Vivo Studies

10.1.2.3.1 Introduction

Although there are many studies that report the effect of autologous MSCs cultured on CaP scaffolds in various animal models, implantation of hMSCs into immunodeficient mice or rats (e.g., nude mice), in either ectopic or orthotopic defect sites, is now considered the standard preclinical model to assess bone formation using human cells. This strategy tests the efficacy of hMSCs, enabling better translation of the technology to clinical use. The use of culture-expanded MSCs is now also typical, enabling high numbers of cells from a given donor to be obtained.

10.1.2.3.2 MSCs on CaP Scaffolds In Vivo

An early study demonstrated this using rat BM MSCs that were either seeded directly on scaffolds or were expanded in culture using non-osteogenic medium then seeded on scaffolds, which were blocks of porous BCP (60% HA to 40% β-TCP), with mean pore size of 400 μm (Goshima et al. 1991). Scaffolds were implanted subcutaneously, dorsally, and intra-abdominally in syngeneic rats for 1–12 weeks. New bone was observed in the pores of the scaffolds after 2–3 weeks, and early passage cells resulted in bone forming faster than was observed when fresh BM was used.

The same group followed this with a study using culture-expanded human BM-MSCs, using non-osteogenic medium, which were then seeded on to similar porous blocks of BCP and implanted subcutaneously in to nude mice and rats (Haynesworth et al. 1992). As for their previous study, new bone was formed within the pores in both mice and rats from 2 weeks onward, and immunocytochemistry proved that the early bone derived from the (human) donor cells, and not the host (rat) cells.

Porous BCP scaffolds have also been used for the implantation of culture-expanded rat (Kadiyala et al. 1997) and human (Bruder et al. 1998) BM-MSCs, using non-osteogenic medium, into critical size defects in the femora of rats. The BCP was similar to that used in the earlier studies but was used in the form of cylinders. In both studies, significantly more bone was observed in the defect at 8 weeks when the scaffold was combined with expanded MSCs, compared to the scaffold alone.

More recent studies demonstrated the optimum composition of BCP scaffolds for forming new bone when combined with hMSCs and implanted in a subcutaneous site in nude mice. Culture-expanded human BM-MSCs, using non-osteogenic medium, were seeded on 2 mm cubes of BCP containing 20:80 or 60:40 ratios of HA:β-TCP, and also coralline-derived HA and bovine-bone-derived HA and implanted for 5 weeks (Harris and Cooper 2004). Significantly more new bone was formed in defects with BCP scaffolds containing 20% HA than with 60% HA. Although the controls of coralline-derived HA and bovine-bone-derived HA (100% HA) were produced by different methods from the BCP scaffolds, therefore resulting in different physical parameters, it is relevant to note that no new bone was formed in these group at 5 weeks.

Similar results were obtained in a comparable study that utilized porous BCP scaffold 3 mm cubes with varying percentages of HA, and compared with 100% HA and 100% β-TCP controls; these scaffolds were provided by one source, with the pore size (300–600 μm) and total pore volumes (60%–70%) controlled throughout the series (Arinzeh et al. 2005). Using culture-expanded human BM-MSCs seeded on the scaffolds and implanted in nude mice for 6–12 weeks, data from histological scoring showed that no new bone was observed for any scaffolds implanted without cells at both time points, whereas new bone was observed for all scaffolds combined with cells at 6 and 12 weeks. In general, the 100% β-TCP scaffold was associated with the lowest histological scoring at both time points, and the highest scoring was observed for the BCP containing 20% HA.

While a significant number of studies utilize BCP as a scaffold, where the TCP used is in the β-form, incorporation of silicon into the TCP structure results in a BCP where the TCP is present as the α-form; this material was commercialized as Skelite™ and contained 67% α-TCP and 33% HA. BM-MSCs from sheep were culture expanded using non-osteogenic medium then seeded on to Skelite cylinders of the porous scaffold (60% porosity, pore size range 200–500 μm) and implanted in to defects in the sheep tibia (Mastrogiacomo et al. 2007). Results showed new bone formed within the defect when Skelite was implanted with MSCs, but no new bone was formed when Skelite was implanted without MSCs, even by 10 months. The Skelite scaffolds resorbed only when combined with MSCs, so only when new bone was formed, whereas the radiographs showed no scaffold resorption when MSCs were not used.

10.1.2.3.3 MSCs on CaP Granules In Vivo

As mentioned in Section 10.1.1.3, there has been a recent shift in strategy in the field of bone tissue engineering from seeding large blocks or cylinders of porous CaP scaffold to utilizing commercially available synthetic CaP bone graft materials that are used clinically to fill bone defects; these materials are typically presented as porous granules with a typical dimension of 0.5–2 mm. A recent study investigated the bone-forming potency of human BM-MSCs on a number of such granular CaP scaffolds. The study used culture-expanded (passage 4) BM-MSCs seeded on various commercial BCP granules (0.5–1 mm diameter), with HA:TCP ratios ranging from 0% to 60% HA, and implanted these subcutaneously in nude mice for 8 weeks (Zannettino et al. 2010). It should be noted that a detailed analysis of the parameters that describe the porosity and surface area of the different scaffolds was not reported. It was noted that significant ectopic bone formation was only observed when the MSCs were seeded on scaffolds containing at least 15% HA, with negligible new bone observed when the scaffold was 100% β-TCP.

The effect of microstructure and chemistry of HA, BCP, and β-TCP granules seeded with culture-expanded hMSCs on the formation of bone after subcutaneous implantation in nude mice for 6 weeks was studied (Yuan et al. 2010). In addition to the chemistry varying, other parameters such as the amount of microporosity and surface area varied across the samples, but the β-TCP granules and BCP1150 (BCP sintered to have a high level of microporosity) had very comparable physical properties, but the β-TCP granules showed significantly more new bone after 12 weeks. It should be noted that in this study, the granules seeded with MSCs were cultured in vitro in osteogenic medium for 7 days prior to implantation, and the β-TCP granules contained 5%–10% HA; these differences may explain the differences observed in the bone formation observed for BCP and β-TCP in this study compared to other studies. These samples also contain a lower level of total porosity but a much higher level of microporosity than other typical BCP and β-TCP scaffolds described earlier.

Another recent study utilized commercial β-TCP granules (0.7–1.4 mm diameter) with 60% porosity to test the effect of combining endothelial progenitor cells (EPCs) to hMSCs on bone formation in critical size defects in athymic rats for 1, 4, and 8 weeks (Seebach et al. 2010a). Data showed increased vascularization in the defect site after 1 and 4 weeks when EPCs were present and the area of bone formation was significantly greater at 4 weeks in defects containing MSCs + EPCs compared to MSCs alone. Although not a test of the role of the CaP on bone formation, it is clear from this study that the presence of EPCs enhances bone formation, and CaP chemistry and/or porosity may play a role in this.

Although the majority of studies to date have used MSCs derived from BM, other sources of MSCs have been tested for their bone-forming potential. The role of CaP on the bone-forming potential of human periosteum-derived MSCs (hPDCs) in an ectopic site was demonstrated by the subcutaneous implantation of cells that were seeded on a collagen–BCP scaffold (Collagraft) in nude mice; as a control, the CaP component was removed by EDTA treatment, and cells seeded on the remaining collagen scaffold (Eyckmans et al. 2010). The authors found that new bone was only formed when CaP was present in the scaffold, suggesting that the CaP may trigger the osteoinduction process.

Another recent study also used MSCs derived from the periosteum. Culture-expanded hPDCs, using non-osteogenic medium, were seeded at passage 5 on to cylinders (3 mm diameter and 3 mm height) of five commercial porous CaP–collagen scaffolds and implanted subcutaneously in nude mice for 18 days, 4 and 8 weeks (Roberts et al. 2011).

Physical characterization of the scaffolds using μCT showed large differences in surface area and volume fraction of CaP, and ion release studies showed a large variation in Ca^{2+} ion release. Varying levels of new bone were formed in the different scaffolds after 8 weeks, with one of the five materials showing no bone. Based on the data obtained, correlations were reported between bone-forming ability of the hPDCs and the parameters obtained from the physical characterization of the scaffolds. The CaP chemistry of the five scaffolds varies significantly, from 100% bovine-bone-derived HA, BCP, 100% β-TCP, $CaHPO_4$, and this was not considered directly in the modeling.

Another valuable source of adult MSCs is adipose tissue which can easily be harvested by liposuction from a patient to produce autologous cells with the ability to differentiate toward different phenotypes. For bone tissue engineering, human adipose tissue–derived MSCs have been combined with various CaP scaffolds to produce bone. For example, human adipose tissue–derived MSCs were seeded on HA–TCP scaffolds (65%–35%) or Collagraft (a collagen type I and HA:TCP matrix) scaffolds (both from Zimmer), and implanted subcutaneously in nude mice for 6 weeks (Hicok et al. 2004). Histology showed osteoid matrix had formed in HA–TCP scaffolds with cells, whereas a collagen-rich matrix was observed with Collagraft scaffolds with cells, and osteoid was only found in one cell-seeded Collagraft scaffold. Controls in the study consisting of the two scaffolds without cells showed no osteoid formation. The role of the scaffolds, in terms of chemistry and physical architecture, was not reported in this study.

The in vivo bone-forming potential of human adipose tissue–derived MSCs and human BM-MSCs seeded on to β-TCP scaffolds has been compared (Hattori et al. 2006). Both cell types were cultured in osteogenic medium for 2 weeks and were then implanted subcutaneously in nude mice. After 8 weeks of implantation, histological analysis showed comparable levels of new bone formation for both cell types, supporting the suitability of adipose tissue–derived MSCs for bone regeneration.

These studies support the use of adipose tissue as a source of MSCs to combine with CaP scaffolds for bone regeneration. Other tissue sources of adult hMSCs include synovium, dental pulp, periodontal ligament, peripheral blood, and umbilical cord blood.

10.1.2.3.4 Clinical Use of MSCs in CaP Scaffolds

The first published clinical use of culture-expanded hMSCs for repairing large bone defects in three patients involved seeding the expanded BM-MSCs on porous HA scaffolds (60% total porosity and mean pore diameter approximately 610 μm) and implanting in to the bone defects, which were 4 or 7 cm in size (Quarto et al. 2001). All three patients showed newly formed bone bridging the defect on radiographs at 15–18 months after implantation. This study was followed up 6–7 years after surgery and reported no complications and excellent bone integration, but it should be noted that the HA scaffold showed no evidence of resorption or remodeling after this period (Marcacci et al. 2007). A recent article reviews seven clinical studies that all used autologous, culture-expanded MSCs for bone tissue engineering (Chatterjea et al. 2010). Of these seven independent studies, five used CaPs as the scaffold: two HA, one BCP, and two TCP. Due to the small number of patients in each study, ranging from 1 to 6, there is too little information to understand the role of the CaP scaffold used, and there are too many other variables that could determine clinical success. However, these pilot studies, along with the extensive preclinical animal studies, suggest that the combination of MSCs and CaP scaffolds could be an effective therapy for bone regeneration.

10.1.3 Summary and Future Perspectives

The recent developments of CaP scaffolds with engineered chemistries, surface properties, and porosities that can be considered as being "bioinstructive" scaffolds rather than simply osteoconductive scaffolds have opened up new opportunities in the regeneration of bone tissue. Not only are some of these CaP scaffolds osteoinductive in their own right, but evidence also supports the hypothesis that specific properties of CaP scaffolds can be engineered to have a direct effect on the bone forming potential of hMSCs. The increased understanding of MSC biology, particularly from tissues other than BM such as periosteum and adipose tissue, and the development of human adult induced pluripotent stem cells (iPS cells), increases the potential application of CaP scaffolds as not only carriers of these cells but also scaffolds capable of guiding the behavior of these important cell types. Evidence from current and future clinical trials of MSCs that will be published over the forthcoming years will help in the understanding and delivery of MSC-based therapies for bone repair.

References

Akao, M., Aoki, H., and Kato, K. 1981. Mechanical properties of sintered hydroxyapatite for prosthetic applications. *J. Mater. Sci.* 16: 809–812.

Akiyama, N., Takemoto, M., Fujibayashi, S., Neo, M., Hirano, M., and Nakamura, T. 2011. Difference between dogs and rats with regard to osteoclast-like cells in calcium-deficient hydroxyapatite-induced osteoinduction. *J. Biomed. Mater. Res. A* 96A: 402–412.

Arinzeh, T.L., Tran, T., Mcalary, J., and Daculsi, G. 2005. A comparative study of biphasic calcium phosphate ceramics for human mesenchymal stem-cell-induced bone formation. *Biomaterials* 26: 3631–3638.

Binner, J.G.P. and Reichert, J. 1996. Processing of hydroxyapatite ceramic foams. *J. Mater. Sci.* 31: 5717–5723.

Bjerre, L., Bünger, C.E., Kassem, M., and Mygind, T. 2008. Flow perfusion culture of human mesenchymal stem cells on silicate-substituted tricalcium phosphate scaffolds. *Biomaterials* 29: 2616–2627.

Bonel, G. 1972. Contribution a l'étude de la carbonation des apatites. *Ann. Chim.* 7: 127–144.

Brown, W.E., Eidelman, N., and Tomazic, B. 1987. Octacalcium phosphate as a precursor in biomineral formation. *Adv. Dent. Res.* 1: 306–313.

Bruder, S.P., Kurth, A.A., Shea, M., Hayes, W.C., Jaiswal, N., and Kadiyala, S. 1998. Bone regeneration by implantation of purified, culture-expanded human mesenchymal stem cells. *J. Orthop. Res.* 16: 155–162.

Chatterjea, A., Meijer, G., van Blitterswijk, C., and de Boer, J. 2010. Clinical application of human mesenchymal stromal cells for bone tissue engineering. *Stem Cells Int.* doi:10.4061/2010/215625.

Chu, T.M.G., Halloran, J.W., Hollister, S.J., and Feinberg, S.E. 2001. Hydroxyapatite implants with designed internal architecture. *J. Mater. Sci. Mater. Med.* 12: 471–478.

Daculsi, G., Laboux, O., Malard, O., and Weiss, P. 2003. Current state of the art of biphasic calcium phosphate bioceramics. *J. Mater. Sci. Mater. Med.* 14: 195–200.

Dard, M., Bauer, A., Liebendorger, A., Wahlig, H., and Dingeldein, E. 1994. Preparation, physio-chemical and biological evaluations of a hydroxyapatite ceramic from bovine spongiosa. *Acta Odonto. Stom.* 185: 61–69.

Dellinger, J.G., Cesarano III, J., and Jamison, R.D. 2007. Robotic deposition of model hydroxyapatite scaffolds with multiple architectures and multiscale porosity for bone tissue engineering. *J. Biomed. Mater. Res. A.* 82A: 383–394.

Dorozhkin, S.V. 2007. Calcium orthophosphates. *J. Mater. Sci.* 42: 1061–1095.

Driessen, A.A., Klein, C.P.A.T, and de Groot, K. 1982. Preparation and some properties of sintered β-whitlockite. *Biomaterials* 3:113–116.

Eggli, P.S., Muller, W., and Schenk, R.K. 1988. Porous hydroxyapatite and tricalcium phosphate cylinders with two different pore size ranges implanted in the cancellous bone of rabbits. A comparative histomorphometric and histologic study of bony ingrowth and implant substitution. *Clin. Orthop. Relat. Res.* 232: 127–138.

Eyckmans, J., Roberts, S.J., Schrooten, J., and Luyten, F.P. 2010. A clinically relevant model of osteoinduction: A process requiring calcium phosphate and BMP/Wnt signalling. *J. Cell Mol. Med.* 14: 1845–1856.

Gibson, I.R., Best, S.M., and Bonfield, W. 1999. Chemical characterization of silicon-substituted hydroxyapatite. *J. Biomed. Mater. Res.* 44: 422–428.

Gibson, I.R., Rehman, I.U., Best, S.M., and Bonfield, W. 2000. Characterisation of the transformation from Ca-deficient apatite to β-tricalcium phosphate. *J. Mater. Sci. Mater. Med.* 11: 533–539.

Goshima, J., Goldberg, V.M., and Caplan, A.I. 1991. The osteogenic potential of culture-expanded rat marrow mesenchymal cells assayed *in vivo* in calcium phosphate ceramic blocks. *Biomaterials* 12: 253–258.

Harris, C.T. and Cooper, L.F. 2004. Comparison of bone graft matrices for human mesenchymal stem cell-directed osteogenesis. *J. Biomed. Mater. Res.* 68A: 747–755.

Hattori, H., Masuoka, K., Sato, M., Ishihara, M., Asazuma, T., Takase, B., Kikuchi, M., Nemoto, K., and Ishihara, M. 2006. Bone formation using human adipose tissue-derived stromal cells and a biodegradable scaffold. *J. Biomed. Mater. Res.* 76B: 230–239.

Haynesworth, S.E., Goshima, J., Goldberg, V.M., and Caplan, A.I. 1992. Characterization of cells with osteogenic potential from human marrow. *Bone* 13: 81–88.

Hicok, K.C., Du Laney, T.V., Zhou, Y.S., Halvorsen, Y.D., Hitt, D.C., Cooper, L.F., and Gimble, J.M. 2004. Human adipose-derived adult stem cells produce osteoid *in vivo*. *Tissue Eng.* 10: 371–380.

Hing, K.A. 2005. Bioceramic bone graft substitutes: Influence of porosity and chemistry. *Int. J. Appl. Ceram. Technol.* 2: 184–199.

Hollister, S.J. 2005. Porous scaffold design for tissue engineering. *Nat. Mater.* 4: 518–524.

Holtorf, H.L., Sheffield, T.L., Ambrose, C.G., Jansen, J.A., and Mikos, A.G. 2005. Flow perfusion culture of marrow stromal cells seeded on porous biphasic calcium phosphate ceramics. *Ann. Biomed. Eng.* 33: 1238–1248.

Hulbert, S.F., Young, F.A., Mathews, R.S., Klawitter, J.J., Talbert, C.D., and Stelling, F.H. 1970. Potential of ceramic materials as permanently implantable skeletal prostheses. *J. Biomed. Mater. Res.* 4: 433–456.

Johnson, K.D., Frierson, K.E., Keller, T.S., Cook, C., Scheinberg, R., Zerwekh, J., Meyers, L., and Sciadini, M.F. 1996. Porous ceramics as bone graft substitutes in long bone defects: A biomechanical, histological, and radiographic analysis. *J. Orthop. Res.* 14: 351–369.

Kadiyala, S., Jaiswal, N., and Bruder, S.P. 1997. Culture-expanded, bone marrow-derived mesenchymal stem cells can regenerate a critical-sized segmental bone defect. *Tissue Eng.* 3: 173–185.

Karageorgiou, V. and Kaplan, D. 2005. Porosity of 3D biomaterial scaffolds and osteogenesis. *Biomaterials* 26: 5474–5491.

Kasten, P., Beyen, I., Niemeyer, P., Luginbühl, R., Bohner, M., and Richter, W. 2008a. Porosity and pore size of beta-tricalcium phosphate scaffold can influence protein production and osteogenic differentiation of human mesenchymal stem cells: An *in vitro* and *in vivo* study. *Acta Biomater.* 4: 1904–1915.

Kasten, P., Vogel, J., Beyen, I., Weiss, S., Niemeyer, P., Leo, A., and Lüginbuhl, R. 2008b. Effect of platelet-rich plasma on the *in vitro* proliferation and osteogenic differentiation of human mesenchymal stem cells on distinct calcium phosphate scaffolds: The specific surface area makes a difference. *J. Biomater. Appl.* 23: 169–188.

Langstaff, S., Sayer, M., Weaver, L., Smith, T.J.N., Pugh, S.M., Hesp, S.A.M., and Thompson, W.T. 1999. Resorbable bioceramics based on stabilized calcium phosphates. Part I: Rational design, sample preparation, and material characterization. *Biomaterials* 20: 1727–1741.

Levy, R.A., Chu, T.M., Halloran, J.W., Feinberg, S.E., and Hollister, S.J. 1997. CT-generated porous hydroxyapatite orbital floor prosthesis as a prototype bioimplant. *Am. J. Neuroradiol.* 18: 1522–1525.

Liu, Y., Cooper, P.R., Barralet, J.E., and Shelton, R.M. 2007. Influence of calcium phosphate crystal assemblies on the proliferation and osteogenic gene expression of rat bone marrow stromal cells. *Biomaterials* 28: 1393–1403.

Lu, J.X., Flautre, B., Anselme, K., Hardouin, P., Gallur, A., Descamps, M., and Thierry, B. 1999. Role of interconnections in porous bioceramics on bone recolonization *in vitro* and *in vivo*. *J. Mater. Sci. Mater. Med.* 10: 111–120.

Maatz, R. and Bauermeister, A.J. 1957. A method of bone maceration—Results in animal experiments. *J. Bone Joint Surg. Am.* 39: 153–166.

Mastrogiacomo, M., Papadimitropoulos, A., Cedola, A., Peyrin, F., Giannoni, P., Pearce, S.G., Alini, M., Giannini, C., Guagliardi, A., and Cancedda, R. 2007. Engineering of bone using bone marrow stromal cells and silicon-stabilized tricalcium phosphate bioceramic: Evidence for a coupling between bone formation and scaffold resorption. *Biomaterials* 28: 1376–1384.

Marcacci, M., Kon, E., Moukhachev, V., Lavroukov, A., Kutepov, S., Quarto, R., Mastrogiacomo, M., and Cancedda, R. 2007. Stem cells associated with macroporous bioceramics for long bone repair: 6- to 7-year outcome of a pilot clinical study. *Tissue Eng.* 13: 947–955.

Melville, A.J., Harrison, J., Gross, K.A., Forsythe, J.S., Trounson, A.O., and Mollard, R. 2006. Mouse embryonic stem cell colonisation of carbonated apatite surfaces. *Biomaterials* 27: 615–622.

Milosevski, M., Bossert, J., Milosevski, D., and Gruevska, N. 1999. Preparation and properties of dense and porous calcium phosphate. *Ceram. Int.* 25: 693–696.

Montanaro, L., Jorand, Y., Fantozzi, G., and Negro, A. 1998. Ceramic foams by powder processing. *J. Eur. Ceram. Soc.* 18: 1339–1350.

Muschler, G.F., Boehm, C., and Easley, K. 1997. Aspiration to obtain osteoblast progenitor cells from human bone marrow: The influence of aspiration volume. *J. Bone Joint Surg. Am.* 79: 1699–1709.

Muschler, G.F., Nakamoto, C., and Griffith, L.G. 2004. Engineering principles of clinical cell-based tissue engineering. *J. Bone Joint Surg. Am.* 86: 1541–1558.

Muschler, G.F., Nitto, H., Boehm, C.A., and Easley, K.A. 2001. Age- and gender-related changes in the cellularity of human bone marrow and the prevalence of osteoblastic progenitors. *J. Orthop. Res.* 19: 117–125.

Mygind, T., Stiehler, M., Baatrup, A., Li, H., Zou, X., Flyvbjerg, A., Kassem, M., and Bünger, C. 2007. Mesenchymal stem cell ingrowth and differentiation on coralline hydroxyapatite scaffolds. *Biomaterials* 28: 1036–1047.

Quarto, R., Mastrogiacomo, M., Cancedda, R., Kutepov, S.M., Mukhachev, V., Lavroukov, A., Kon, E., and Marcacci, M. 2001. Repair of large bone defects with the use of autologous bone marrow stromal cells. *N. Engl. J. Med.* 344: 385–386.

Roberts, S.J., Geris, L., Kerckhofs, G., Desmet, E., Schrooten, J., and Luyten, F.P. 2011. The combined bone forming capacity of human periosteal derived cells and calcium phosphates. *Biomaterials* 32: 4393–4405.

Roy, D.M. and Linnehan, S.K. 1974. Hydroxyapatite formed from coral skeletal carbonate by hydrothermal exchange. *Nature* 247: 220–222.

Rücker, M., Laschke, M.W., Junker, D., Carvalho, C., Tavassol, F., Mülhaupt, R., Gellrich, N-C., and Menger, M.D. 2008. Vascularization and biocompatibility of scaffolds consisting of different calcium phosphate compounds. *J. Biomed. Mater. Res.* 86A: 1002–1011.

Schantz, J-T., Brandwood, A., Hutmacher, D.W., Khor, H.L., and Bittner, K. 2005. Osteogenic differentiation of mesenchymal progenitor cells in computer designed fibrin-polymer-ceramic scaffolds manufactured by fused deposition modelling. *J. Mater. Sci. Mater. Med.* 16: 807–819.

Schwartzwalder, K. and Somers, A.V. 1963. Method of making porous ceramic articles. U.S. Patent No. 3,090,094.

Seebach, C., Henrich, D., Kähling, C., Wilhelm, K., Tami, A.E., Alini, M., and Marzi, I. 2010a. Endothelial progenitor cells and mesenchymal stem cells seeded onto β-TCP granules enhance early vascularization and bone healing in a critical-sized bone defect in rats. *Tissue Eng. A* 16: 1961–1970.

Seebach, C., Schultheiss, J., Wilhelm, K., Frank, J., and Henrich, D. 2010b. Comparison of six bone-graft substitutes regarding to cell seeding efficiency, metabolism and growth behaviour of human mesenchymal stem cells (MSC) *in vitro. Injury* 41: 731–738.

Shelton, R.M., Liu, Y., Cooper, P.R., Gbureck, U., German, M.J., and Barralet, J.E. 2006. Bone marrow cell gene expression and tissue construct assembly using octacalcium phosphate microscaffolds. *Biomaterials* 27: 2874–2881.

Sherwood, J.K., Riley, S.L., Palazzolo, R., Brown, S.C., Monkhouse, D.C., Coates, M., Griffith, L.G., Landeen, L.K., and Ratcliffe, A. 2002. A three-dimensional osteochondral composite scaffold for articular cartilage repair. *Biomaterials* 23: 4739–4751.

Shimazaki, K. and Mooney, V. 1985. Comparative study of porous hydroxyapatite and tricalcium phosphate as bone substitute. *J. Orthop. Res.* 3: 301–310.

Tan, K.H., Chua, C.K., Leong, K.F., Cheah, C.M., Cheang, P., Abu Bakar, M.S., and Cha, S.W. 2003. Scaffold development using selective laser sintering of polyetheretherketone–hydroxyapatite biocomposite blends. *Biomaterials* 24: 3115–3123.

Tancred, D.C., McCormack, B.A.O., and Carr, A.J. 1998. A synthetic bone implant macroscopically identical to cancellous bone. *Biomaterials* 19:2303–2311.

Uchida, A., Nade, S.M.L., McCartney, E.R., and Ching, W. 1984. The use of ceramics for bone replacement: A comparative study of three different porous ceramics. *J. Bone Joint Surg.* 66-B: 269–275.

Van der Stok, J., Van Lieshout, E.M.M., El-Massoudi, Y., Van Kralingen, G.H., and Patka, P. 2011. Bone substitutes in the Netherlands—A systematic literature review. *Acta Biomater.* 7: 739–750.

Wang, L., Hu, Y-Y., Wang, Z., Li, X., Li, D-C., Lu, B-H., and Xu, S-F. 2009. Flow perfusion culture of human fetal bone cells in large β-tricalcium phosphate scaffold with controlled architecture. *J. Biomed. Mater. Res.* 91A: 102–113.

Yuan, H., Fernandes, H., Habibovic, P., de Boer, J., Barradas, A.M.C., de Ruiter, A., Walsh, W.R., van Blitterswijk, C.A., and de Bruijn, J.D. 2010. Osteoinductive ceramics as a synthetic alternative to autologous bone grafting. *Proc. Nat. Acad. Sci. USA* 107: 13614–13619.

Zannettino, A.C.W., Paton, S., Itescu, S., and Gronthos, S. 2010. Comparative assessment of the osteoconductive properties of different biomaterials in vivo seeded with human or ovine mesenchymal stem/stromal cells. *Tissue Eng. A.* 16: 3579–3587.

11

Bioactive Glasses as Composite Components: Technological Advantages and Bone Tissue Engineering Applications

Elzbieta Pamula, Katarzyna Cholewa-Kowalska,
Mariusz Szula, and Anna M. Osyczka

CONTENTS

11.1 Introduction

Extracellular matrix (ECM) of bone tissue is a typical composite system consisting of a mineral phase (hydroxyapatite [HA]—more precisely bone mineral is a hydroxyl- and calcium-deficient, carbonated apatite) and an organic phase (collagen and noncollagenous proteins such as osteocalcin, osteonectin, bone sialoprotein) (Boskey 2005). This composite ECM structure can be described in terms of hierarchical levels of organization from nanometers, through micrometers, and finally to millimeters scale range (Rho et al. 1998).

According to a widely accepted biomimetic approach, it is beneficial if biomaterial scaffolds for bone tissue engineering mimic properties and functions of the ECM in bone tissue (Owen and Shoichet 2010). It implies that neither one-phase polymeric, ceramic, or metallic material is able to fulfill all the requirements for the ideal bone tissue engineering scaffold. It results from the intrinsic properties of these classic materials. For example, if one considers mechanical properties only, polymeric materials have far too low tensile strength and Young's modulus, ceramic materials are too brittle, while metallic materials are characterized by an excessively high Young's modulus, as compared to bone tissue. The mismatch between the properties of bone tissue and the one-phase implant materials often provoke several unwanted phenomena, for example, stress shielding and aseptic loosening of metallic implants in hip-joint prostheses (Khanuja et al. 2011).

Another parameter, often desired in temporary scaffolds, is susceptibility to degradation and resorption in a biological environment. While degradation kinetics of several polymeric materials (e.g., aliphatic polyesters such as polylactides, polyglycolide, poly-ε-caprolactone (PCL), and their copolymers) can be adjusted to the requirements of bone regeneration rate (Naira and Laurencin 2007), this is rather difficult to achieve for metallic or ceramic materials. The bioactivity and activation of specific cellular responses are probably the most important features of new-generation biomaterials for bone tissue regeneration (Hench and Polak 2002). Except for HA, tricalcium phosphate, bioactive glasses (BGs), and glass ceramics, the majority of the classic one-phase materials have little potential to elicit specific and desirable responses of cells and tissues.

BGs are of particular interest in bone tissue engineering because of their excellent bioactivity, that is, their ability to form a tight bond with hard and soft tissues due to the development of a surface hydroxyl-carbonate apatite (HCA) layer (Cao and Hench 1996, Kokubo and Takadama 2006). The mechanisms of bioactivity of silicate glasses have been extensively studied and the creation of HCA has been proposed to consist of the following stages: (1) an ion exchange with hydrogen ions from the surrounding medium, (2) glass network dissolution resulting in the formation of silanol groups, (3) condensation and repolymerization of silica-rich layer, (4) migration of Ca^{2+} and PO_4^{3-} groups to the surface and formation a $CaO\text{-}P_2O_5$-rich film, and (5) crystallization of the amorphous $CaO\text{-}P_2O_5$ film by incorporation of OH^-, CO_3^{2-} to form HCA layer (Andersson and Karlsson 1991, Hench 1991, Ohtsuki et al. 1992, Hench and West 1996, Ducheyne and Qiu 1999).

Other favorable features of BGs are their osteoconductivity/osteoinductivity (Clark and Hench 1994), for example, the ability to induce osteogenesis (Välimäki and Aro 2006) and promote angiogenesis (Gorustovich et al. 2010) as well as controllable degradability (Hamadouche et al. 2001). The dissolution products of BGs have been shown to contribute to their osteoconductive/inductive properties (Gough et al. 2004, Christodoulou et al. 2005, Varanasi et al. 2009, Hoppe et al. 2011). The disadvantages of BGs and other ceramic materials are their low fracture toughness and strength and high stiffness (Kumta 2006). As a result, unprecedented biological properties of BGs can be fully exploited only in the non-load-bearing medical applications. BGs in the form of granules, powders, and cements are commercially available for periodontology and endodontic surgery applications to enhance healing of bone tissue defects (Pantchev et al. 2009, Pandit et al. 2010). Processing BGs into highly porous scaffolds resulted in a further deterioration of their mechanical properties (El-Ghannam 2005), making them less attractive materials as cell carriers and matrices for filling bone defects.

One way to improve strength and fracture toughness of BGs is to form a crystalline phase in the bioactive bulk glasses in order to develop glass ceramics. However, this approach may impair bioactivity since the newly formed crystal phases have been reported to be scarcely bioactive (Bellucci et al. 2010).

The more promising strategy is to design composites based on polymeric or ceramic materials enriched with BG particles. This approach allows for better exploitation of biological properties of BGs, such as bioactivity. On the other hand, it is possible to control mechanical properties and adapt degradation kinetics of the resulting composite materials to the required applications. Another advantage of composite materials lies in the fact that the composites can be easily processed into three-dimensional (3D) porous scaffolds mimicking ECM of trabecular bone. This is due to the fact that several manufacturing techniques of porous scaffolds from polymers or ceramics matrices have already been established and require only slight modifications to be adapted for the particular composite component systems.

In this chapter, we review the advances in BGs composite research with specific emphasis on the manufacturing methods of BGs, followed by a description of techniques to fabricate porous BG enriched polymer-based composite scaffolds. Finally, we summarize the biological performance of recently developed BG-enriched composites and their perspectives in clinical applications.

11.2 Bioactive Glasses: Fabrication and Properties

There are two main methods of BGs fabrication: a classic one based on melting and a more versatile one based on the sol-gel route.

In the melting method, all silicate glasses are prepared from chemically pure grade powder chemicals (oxides, carbonates), which are melted at 1400°C–1500°C; the latter depends on the chemical composition of the system (Hench et al. 1971, Brink et al. 1997). The first BG from the SiO_2-Na_2O-CaO-P_2O_5 system was developed by Hench et al. (1971), and it is available under the trade name 45S5 Bioglass®. Systematic studies of this four-component system with a constant 6 wt% P_2O_5 allowed the three key compositional features making the glass surface bioactive to be distinguished: (1) less than 58 wt% SiO_2, (2) high Na_2O and CaO contents, (3) high CaO/P_2O_5 ratio (Hench 1991).

Since the development of Bioglass, several BGs differing in composition have been synthesized (Table 11.1). For example, BG S53P4 is commercially available, and it is known under the trade name BonAlive® (Andersson and Kangasniemi 1991). The other type of silicate BG was enriched with MgO and K_2O (Brink et al. 1997, Pirhonen et al. 2006). It was found that, in excess of SiO_2, the addition of B_2O_3 up to 15 wt% or substitution of CaF_2 (up to 12.5 wt%)

TABLE 11.1

Composition of Various BGs (wt%)

	45S5 Bioglass®	S53P4 BonAlive®	13–93	S48P2	2	58S	A2	S2	AgBG
SiO_2	45.0	53	53	48.0	42.23	58.2	38.25	76.6	76
CaO	24.5	20	20	19.0	31.54	32.6	48.20	14.3	19
P_2O_5	6.0	4	4	2.0	2.99	9.2	—	9.1	2
Na_2O	24.5	23	6	28.0	4.36	—	—	—	—
K_2O	—	—	12	—	—	—	—	—	—
MgO	—	—	5	—	8.50	—	—	—	—
B_2O_3	—	—	—	1.5	4.89	—	—	—	—
Al_2O_3	—	—	—	1.5	—	—	—	—	—
CaF_2	—	—	—	—	5.49	—	—	—	—
Ag_2O	—	—	—	—	—	—	—	—	3
Production method	Melting	Melting	Melting	Melting	Melting	Sol-gel	Sol-gel	Sol-gel	Sol-gel
References	Hench et al. (1971)	Andersson and Kangasniem (1991)	Brink et al. (1997), Pirhonen et al. (2006)	Andersson et al. (1990)	Agathopoulos et al. (2006)	Li et al. (1991), Sepulveda et al. (2002)	Laczka and Cholewa (1997)	Laczka et al. (1997)	Bellantone et al. (2000)

for CaO have no significant effect on the creation of HCA layer and bone bonding ability for melted BGs (Andersson et al. 1990, Agathopoulos et al. 2006). In contrast, addition of as little as 2–3 wt% Al_2O_3 inhibits bioactivity, whereas the presence of MgO in the glass composition slows down the formation of HCA (Andersson et al. 1990, Agathopoulos et al. 2006).

In the sol-gel method, synthesis generally requires liquid oxide precursors (i.e., alkoxides) or substances soluble in water/alcohols (i.e., nitrates, chlorides, chelates). This method is based on the hydrolysis and polycondensation reactions which occur in solution in the presence of water and acidic or basic catalysts at room temperature. After gelation and drying, the heat-treatment temperature of BGs usually does not exceed 800°C.

Gel-derived BGs from the ternary system CaO-P_2O_5-SiO_2 (Table 11.1) are most common (Li et al. 1991). In comparison to melted BGs, gel-derived BGs can be composed of as much as 77 wt% SiO_2 (80 mol%) without losing bioactive properties (Pereira et al. 1994, Vallet-Regí et al. 2003). The absence of Na_2O in their composition prevents the rapid pH change of surrounding fluids, which is of key importance in biological applications (Hamadouche et al. 2001). Moreover, the addition of 7 wt% Al_2O_3 does not inhibit bioactive properties but slows the formation rate of the HCA layer (Laczka et al. 2000). Similarly to melted BGs, the addition of MgO to the composition of gel-derived glass results in diminished HCA layer formation (Arcos and Vallet-Regí 2010). Finally, the incorporation of 1–3 wt% Ag_2O introduces antimicrobial properties to the glass without causing a loss its bioactivity (Bellantone et al. 2002).

The advantages of gel-derived BGs are their excellent bioactivity and enhanced susceptibility to degradation due to their unique composition and structure. Gel-derived BGs have much higher surface area and fine porous texture as compared to melted BGs (Li et al. 1991, Pereira et al. 1995, Laczka et al. 2000, Zhong and Greenspan 2000, Hamadouche et al. 2001, Sepulveda et al. 2002). For example, the surface area of gel-derived 58S BG (for chemical composition, see Table 11.1) is 164 m^2/g, while that of melted BG (Bioglass) is 0.17–2.7 m^2/g (Sepulveda et al. 2002). The high specific surface area of gel-derived BGs enriched in silanol groups provides more sites for calcium phosphate nucleation and allows faster release of ions into the solution and their higher local supersaturation. Altogether, this results in faster HCA deposition (Sepulveda et al. 2002).

Recently, the technology of gel-derived BGs manufacturing has been combined with supramolecular chemistry of surfactants to obtain highly ordered mesoporous bioactive glasses (MBGs). Application of nonionic triblock copolymers acting as templates enabled to synthesize MBGs of the CaO-P_2O_5-SiO_2 system containing relatively high P_2O_5 contents (10–30 mol%), extremely high specific surface area ~430 m^2/g, and very small pores of ~4 nm (Zhao et al. 2011). This new approach allows production of a new generation of BGs with enhanced properties, which are not achievable by classical melting or sol-gel methods.

11.3 Bioactive Glasses as Components of Composite Scaffolds

A variety of manufacturing methods have been used to produce porous 3D composite scaffolds for bone tissue engineering that are based on BG-containing polymeric and ceramic matrices. These studies describe the modification of the matrices by the incorporation of BG particles in order to obtain composite materials with enhanced bioactivity, improved mechanical properties, and tailored degradation rates. These features, in turn, should allow the body's own bone to heal, grow, remodel, and eventually replace the graft material. The strategies of matrix modifications with BGs to achieve enhanced properties of the composites are presented in Figure 11.1. While the ceramic-based composites display several clinically attractive properties, polymer-based composites are now becoming prevalent (Rezwan et al. 2006, Boccaccini et al. 2010). The majority of recent studies focus on the synthetic degradable polymer matrices, and only a few studies of bone tissue engineering designated metallic or ceramic matrices with incorporated BG particles have been published (Cholewa-Kowalska et al. 2009, Hatzistavrou et al. 2010, Oksiuta et al. 2009, Jurczyk et al. 2011, Schickle et al. 2011).

The manufacturing techniques for BG/degradable polymer composites include solvent casting/particulate leaching, thermally induced phase

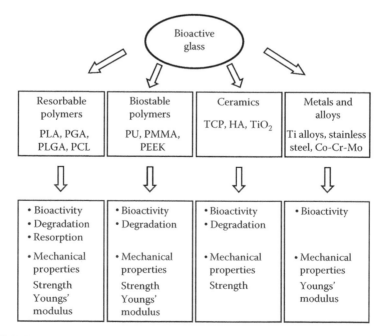

FIGURE 11.1
Strategies of matrix modification with BGs to achieve enhanced properties of the composites.

separation (TIPS), microspheres synthesis, gas foaming, electrospinning, and, most recently, a combination of the polymer templating/salt leaching/ rapid prototyping.

Solvent casting/particulate leaching is a very simple method by which pore size, porosity, and pore interconnectivity in the scaffolds can be controlled by the size of the salt particulates and the ratio of polymer/salt particulates. Porous composite PCL scaffolds containing gel-derived BG (Si/Ca/P 80:15:5, molar ratio) were produced by this technique and were less hydrophobic than scaffolds made of pure PCL. Furthermore, they were capable of inducing a dense and continuous HCA layer after soaking in simulated body fluid (SBF) for 3 weeks, which was not the case for PCL scaffolds (Li et al. 2008). We used solvent casting to produce several composites based on poly(L-lactide-*co*-glycolide) (PLGA) and two bioactive gel-derived BGs: high content silica (S2) and high content lime (A2) (for chemical composition, see Table 11.1); the volume fraction of BG in the composites was 12%, 21%, and 33% (Pamula et al. 2011). The incorporation of either BG increased the tensile strength and Young's modulus, but reduced the elongation at break. Thus, it was possible to control mechanical properties of the composites and adjust them better to the natural bone tissue properties. On the other hand, the rate of HCA layer formation depended on the type of BG. After incubation in SBF, HCA was preferably created on the composites containing A2 BG as soon as after 10 days in SBF, suggesting their higher bioactivity than those containing S2 BG. Incorporation of BG into PLGA matrix also resulted in the modulation of degradation kinetics of the scaffolds and their ability to buffer acidic degradation products of PLGA, which are important features for bone tissue engineering applications. In the same work, the salt particulates were used to produce PLGA composite scaffolds containing 21% volume fraction of either S2 or A2 BG particles. The resulting scaffolds had 85% porosity, pore size of 320–400 μm, and acceptable handling properties (Pamula et al. 2011).

TIPS is a widely used method for the preparation of porous BG/degradable polymer scaffolds. In this method, the homogenous polymer solution containing dispersed BG particles is phase separated by lowering the temperature. The phase separation is based on liquid-liquid demixing, which generates polymer-poor and polymer-rich liquid phases. The subsequent growth and coalescence of the polymer-poor phase result in the creation of pores after a freeze-drying procedure. Boccaccini et al. (2005) developed highly porous PLGA and poly(D,L-lactide) (PDLLA) scaffolds with addition of BG particles (45S5 Bioglass). The scaffolds containing BG were of high porosity >90%; had well defined, oriented, and interconnected 10–100 μm pores; and exhibited enhanced bioactivity as confirmed by SBF tests. Fabbri et al. (2010) found that the mechanical properties of PCL/45S5 Bioglass composites depended on the amount of BG in the composites and the type of solvent that had been used for the preparation of the composites. Specifically, dioxane-prepared composites showed the enhanced stress at deformation with respect to composites prepared with dimethylcarbonate. In a more recent work, the gel-derived

BG (60 wt% SiO_2, 36 wt% CaO, 4 wt% P_2O_5)/poly(L-lactide) (PLLA) composite scaffolds with highly interconnected porous structure were produced by TIPS method (El-Kady et al. 2010). The addition of 50 wt% BG particles into the polymer matrix reduced the size of pores in the scaffolds from ~250 to ~100 μm, enhanced bioactivity, and accelerated degradation in vitro.

Microsphere sintering is a method in which microspheres of a BG/polymer composite are synthesized starting with the emulsion/solvent technique. This is followed by sintering of the microspheres to obtain porous 3D scaffolds (Yao et al. 2005). This method was previously used to produce PLGA/45S5 Bioglass composites with porosity of 40%, pore size of 90 μm, and mechanical properties similar to trabecular bone (Lu et al. 2003). The scaffolds exhibited very good bioactivity, given that a layer of HCA was formed in SBF after a relatively short period of time, that is, 7 days.

Gas foaming method was used to produce PLGA/BG S53P4 (20 wt%) scaffolds. The first step was to prepare the films of polymer/BG composites by solvent casting. Afterward, porous scaffolds were produced by dissolving CO_2 in the polymer/BG films under high pressure in an autoclave (50 bar), followed by a rapid release of the gas from the vessel (Orava et al. 2007). The fabricated scaffolds of pore size 50–500 μm were incubated in TRIS/HCl aqueous solution to evaluate their susceptibility to hydrolytic degradation and in SBF to study bioactivity. The scaffolds' weight loss rate was readily adjusted by the copolymer composition and the addition of BG. This suggests that these are key factors influencing physical and chemical properties of the resulting scaffolds.

Electrospinning is a technique which enables the production of nanofibrous scaffolds from the sol of BGs by a high-voltage electrostatic field. The first publication on the production of nanofibrous BGs (composition 70 mol% SiO_2, 25 mol% CaO, 5 mol% P_2O_5) was released by Kim et al. in 2006. In another work of Kim et al. (2008), BG nanofibers (composition 58 mol% SiO_2, 38 mol% CaO, 4 mol% P_2O_5) of diameter ~320 nm were used to prepare BG nanofibers/PLLA composites. The resulting nanocomposites induced rapid formation of an HCA layer on the composite surface in SBF. As the amount of bioactive nanofiber increased from 5% to 25%, the in vitro bioactivity of the nanocomposites was improved. More recently, gel-derived BG nanofibers of the composition of 60 mol% SiO_2, 36 mol% CaO, and 4 mol% P_2O_5 were manufactured with average diameter of 450 nm (Jo et al. 2009). These fibers were introduced into a PCL matrix, and the resulting composites showed significantly higher elastic modulus and higher bioactivity, as compared with typical composites produced from PCL and BG particles (~4 μm in size) of the same chemical composition.

Rapid prototyping or solid free-form fabrication refers to a group of technologies which can build a physical 3D object using layer-by-layer fusion. This is possible thanks to computer-aided design (CAD) and computer-aided manufacturing (CAM). Yun et al. (2011) manufactured very promising MBG/PCL scaffolds by combination of two pore-forming agents and rapid prototyping. Hence, it was possible to produce hierarchical mesoporous/ macroporous/giant-porous MBG/PCL scaffolds. Nonionic triblock copolymer $EO_{20}PO_{70}EO_{20}$ was chosen as a template to induce the formation of nanometer-range mesopores in BG. The incorporation and leaching of NaCl particles of 25–33 µm size resulted in macroporous BG/PCL composites, whereas rapid prototyping created scaffolds with large uniformly distributed pores of the size of ~500 µm. The scaffolds exhibited good molding capability and handling properties and demonstrated good compressibility. Due to their large specific surface area and 3D interconnected pore structure, the formation of HCA on the scaffolds was observed as soon as after 4 h incubation in SBF; after 24 h, the surface was almost completely covered with the newly formed HCA, confirming a very high bioactivity of scaffolds produced.

11.4 Biological Responses to BG-Enriched Composites

Bone tissue engineering combines cells, growth factors, osteoconductive and osteoinductive scaffolds, and mechanical stimuli. The latter are introduced to tissue-engineered constructs in vitro to create the dynamic environment that accompanies bone tissue formation and remodeling in vivo.

Bone tissue consists of bone-forming osteoblasts and bone-remodeling osteoclasts and osteocytes—the mature osteoblasts embedded in the mineralized extracellular bone matrix, which are believed to sense the mechanical forces that bone tissue is exposed to in vivo (Bonewald 2011). The studies evaluating experimental materials for bone tissue engineering and regeneration purposes often begin with the examinations of their interactions with selected cell populations, mostly osteoblastic cell lines, primary osteoblasts, and osteoprogenitor cells. For tissue engineering purposes, bone marrow mesenchymal and hematopoietic stem cells seem to be the optimal cell source since they can be differentiated into osteoblasts and osteoclasts in vitro, respectively (Augello et al. 2010). Adipose-tissue-derived stem cells are now becoming an alternative to bone marrow and are an easy accessible source of osteoprogenitors (Locke et al. 2011).

Bone exists in two forms—compact or cancellous. The compact form prevails; it builds the long bones of the arm and leg and other bones, where greater strength and rigidity are needed. Compact bone often surrounds highly porous and a sponge-like cancellous bone that is found, for example, in the marrow space. Compact and load-bearing implant materials such as titanium alloys are thus a reasonable choice for the support of diseased or damaged long bones. These implant materials are biocompatible and remain in the body as long as they maintain the proper tissue integration and mechanical function. Bone tissue engineering seeks alternative materials that will serve as scaffolds rather than the mechanical support. They should attract and stimulate osteoprogenitors to populate the scaffolds, form new bone, and shape it to obtain the natural tissue morphology and function. The scaffolds are not supposed to remain in the body permanently but degrade with a rate that matches the rate of new bone formation. They should also allow for bone cell ingrowth and eventually bone tissue vascularization; hence, proper scaffold porosity is a key factor. Other material properties such as macro-, micro-, and nanostructure, surface topography and roughness, material elasticity, and chemistry will contribute to the proper material-tissue integration and overall clinical success of tissue-engineered constructs. Osteoconductive materials stimulate the migration of osteoprogenitors to the site of implantation and permit their differentiation into bone cells. The latter will depend on the osteogenic growth factors and hormones, and correct tissue preconditioning, that are all required for progenitor cells to find their niches, settle, and progress along the osteogenic pathway. Osteoinductive materials are expected to stimulate bone tissue formation on their own and regardless of the tissue environment, for example, they can induce bone formation in heterotopic sites. The classic example of osteoinductive growth factors are bone morphogenetic proteins (BMPs), which were discovered in the early 1960s of the last century in demineralized bone matrix and, when implanted heterotopically in rodents, induced a complete cascade of events starting with the infiltration of bone marrow progenitors and ending with the formation of new bone and its remodeling (Reddi 1981). The osteoinductive material types, their properties, and the mechanisms of their osteoinduction have been recently thoroughly reviewed by Ana et al. (2011). In this paragraph, we summarize the recent work regarding the biological responses to the materials in which a BG was chosen as a modifier of resulting composites to achieve osteoconductive or osteoinductive properties.

Beta-tricalcium phosphate (β-TCP) has been reported for its osteoinductive properties (Kondo et al. 2006), whereas HA has been widely used as bone substitute for its osteoconductivity and similarity to bone mineral (Tanner 2010). BGs are introduced into these matrices for better mechanical properties and improved degradability rather than improved biological performance. Interestingly, the incorporation of gel-derived BG of the composition of 64 mol% SiO_2, 26 mol% CaO, 5 mol% P_2O_5, and 5 mol% MgO into β-TCP improved proliferation of human osteosarcoma G-292 cells compared to pure β-TCP (Hesaraki et al. 2009). We have shown that, compared to pure HA, the incorporation of gel-derived silica-enriched S2 or calcium-enriched A2 BGs into HA base increases viability and alkaline phosphatase (ALP) activity of human bone marrow mesenchymal cells stimulated in culture to osteogenesis by ascorbate and dexamethasone (Cholewa-Kowalska et al. 2009). Thus, incorporation of BG into β-TCP or HA may result in enhanced osteoconductive/inductive properties of such composites.

Several recent investigations have focused on the BG-enriched polymer composites. PLA/Bioglass or PLA/BG composites of different compositions have been investigated in human osteoblast 2D cultures and 3D cultures of human adipose stem cells. Osteoblasts' attachment and spreading were delayed on the PLA/45S5 Bioglass compared to pure polymer surface (Roether et al. 2002). Similarly, PLA/13-93 BG composites decreased proliferation and ALP expression of osteoblasts at early culture stages compared to pure polymer (Ruuttila et al. 2006). In contrast, the studies of Tsigkou et al. (2007) showed that the incorporation of 45S5 Bioglass into a PLA base significantly enhanced ALP activity of human fetal osteoblasts and their expression of several bone-specific markers. 3D scaffolds made of PLA and nanofibrous gel-derived BG (58% SiO_2, 38% CaO, 4% P_2O_5) enhanced proliferation of mouse preosteoblastic MC3T3 cells at early culture stages and increased ALP activity, collagen, and mineral production compared to pure polymer (Kim et al. 2008). In contrast, BG0127 (59.5 wt% SiO_2, 25 wt% CaO, 5 wt% Na_2O, 7.5 wt% K_2O, 3 wt% MgO) enrichment of PLA 3D scaffolds decreased the proliferation rate of human adipose stem cells compared to pure polymer, whereas a specific ALP activity (i.e., ALP per cell) was increased on these composites compared to other materials studied, including PLA-β-TCP (Haimi et al. 2009). The mRNA expression and activity of ALP are not specific to bone tissue, but they increase rapidly at early stages of osteoblast differentiation and both the increased numbers of ALP-positive cells and the increased amounts of ALP per cell are indicative of osteogenic progression. ALP is also important for skeletal mineralization (Orimo 2010). Notably, the previously mentioned studies have been performed in standard culture conditions without the addition of osteogenic growth factors. Thus, the PLA-BG composites display some potential to enhance osteogenesis on their own.

The incorporation of S53P4 BG into PCL base increased rat bone marrow stromal cell proliferation, ALP activity, and bone sialoprotein expression at early stages of osteogenic cultures stimulated with ascorbate, dexamethasone, and β-glycerophosphate (Meretoja et al. 2009). When dynamic culture conditions were applied by a rotating wall vessel, the osteogenic potential of these composite scaffolds was diminished, but penetration of cells into the 3D scaffold interior was achieved instead. Similar results were obtained by Jo et al. (2009) who examined PCL-based composites containing gel-derived BG of the composition 60 mol% SiO_2, 36 mol% CaO, and 4 mol% P_2O_5. Compared to pure polymer films, incorporation of either BG powders or BG nanofibers into a PCL base improved spreading and viability of mouse preosteoblastic MC3T3 cells. Furthermore, cells cultured on composites containing BG nanofibers and stimulated with ascorbate and β-glycerophosphate exhibited significantly higher ALP activity compared to other materials studied. When implanted into rat calvarial bone defects, these composites also showed improved bone regeneration compared to empty and PCL-filled defects.

PLGA-based composites containing 45S5 Bioglass enhanced ALP expression and activity of rat bone marrow mesenchymal stem cells compared to pure polymer, and this was observed in both standard and osteogenic cultures; the latter were stimulated with ascorbate and dexamethasone

(Yao et al. 2005). Lu et al. (2005) showed that PLGA/45S5 Bioglass composites may enhance the proliferation, ALP activity, and matrix mineralization of human osteosarcoma SaOS-2 cells compared to pure polymer, but the effect is dependent on the concentration of BG in the polymer matrix. Wu et al. (2009) investigated the role of MBG and non-MBG additions to PLGA in human osteoblast-like cultures. Both types of BGs enhanced attachment, spreading, and proliferation of cells compared to PLGA, but at later culture stages, a specific ALP activity was significantly increased on PLGA/MBG composites. Our studies have shown that the addition of either silica-enriched S2 BG or calcium-enriched A2 BG to PLGA enhance specific ALP activity and ECM calcification in 2D cultures of human bone marrow mesenchymal stem cells stimulated to osteogenesis by human recombinant BMP-2 (Pamula et al. 2011). The importance of this finding lies in the fact that the BMPs are relatively weak inducers of human mesenchymal stem cell osteogenesis (Osyczka et al. 2004), and thus, PLGA/BG composites may enhance BMP osteogenic response of human mesenchymal stem cells. Interestingly, the effects on the expression of other bone-related genes and osteoclast formation depended on the type of BG that had been incorporated into polymer base in either 2D or 3D cultures (Pamula et al. 2011).

11.5 Clinical Perspectives of BG-Enriched Composites

Tissue engineering may avoid the complications caused by the conventional transplantation. Autografts are nowadays widely applied in reconstructive surgery, but harvesting tissues can lead to complications in donor site healing (infection, necrosis, pathological bone fracture, etc.). Also, the amount of the patient's own tissues is limited (Baino et al. 2009). Allografting or xenografting are the alternative methods used in reconstructive surgery, but they carry the risk of transmitting pathogenic factors and immune-related complications. Recent experimental studies on the application of tissues harvested from transgenic animals (pigs) are a new potentially useful direction in human reconstructive surgery and transplantology (Zapala and Wyszynska-Pawelec 2006).

Composites coated with BG are reliable reconstructive materials for bone defects when biostable scaffolds are needed. These biomaterials combine the ability of BG to bond to surrounding bone and plasticity of composites (Peltola et al. 2012). The dynamic development of biomaterials research (including composites coated with BG) has given a new range of possibilities to many different branches of medical specialties which deal with bone diseases (Tuusa et al. 2007, Baino et al. 2009, Sharma et al. 2009, Kundu et al. 2011, Peltola et al. 2011, Zhu et al. 2011).

The treatment of tumors, fracture fixation, posttraumatic deformations, or congenital malformations in head and neck surgery is often associated with the loss of hard tissues. Implantation of biomaterials seems to be a good and reliable manner of reconstruction which permits to restore even large bone defects without harming potential donor site. In craniomaxillofacial surgery and neurosurgery, the application of BG composites and rapid prototyping technology has brought a satisfying and quick reconstructive method for cranial bone defects with a minimal percentage of postoperative complications (Tuusa et al. 2007).

BG composites can be also used in dentistry or in oral surgery when the treatment is connected with bone defects after, for example, tooth extraction, extirpation of odontogenic cysts, or tumor resection. BG composites are also used in periodontics for guided bone regeneration in patients with bone loss caused by chronic periodontitis. They are yet to be reported as reliable methods of bone augmentation in dental implantology (e.g., sinus lift). Fiber-reinforced composites are also used as dental implants themselves (Tuusa et al. 2007). BG as a component of composites has a range of applications including bone defect filling after trauma treatment and tumor resection in craniomaxillofacial surgery (Peltola et al. 2012).

In cases of orthopedic treatment preservation, the size and shape of an implant is crucial for long-term functional and aesthetic outcomes. The implantation of BG composites provides strong and biostable scaffolds which can replace load-bearing long bones (Tuusa et al. 2007). They have been also recommended as bone graft materials in spinal surgery (Baino et al. 2009).

The other promising clinical application of composites coated with BG is the treatment of arthritis or osteoarthritis. Due to poor regenerative ability of injured adult articular cartilage, the results of surgical treatments of these degenerative diseases are still unsatisfying. Tissue-engineered cartilage can be an alternative treatment option for cartilage lesions. PLGA/45S5 Bioglass composite has been shown to be a biodegradable and osteointegrative material which promotes chondrocyte mineralization and cartilage repair. Furthermore, prefabricated 3D stratified composite scaffolds optimize both cell density and microsphere composition of newly grown bone and cartilaginous tissue (Jiang et al. 2010).

Prefabricated composites with BG can be also used in chronic osteomyelitis treatment when the systemic antibiotic therapy is not efficient due to poor vascular perfusion accompanied by infection of the surrounding tissue. Following surgical debridement of the bone, implantation of composites with BG impregnated with antibiotic allows a high concentration of the antibiotic in the infected area to be maintained for usually 4–6 weeks, long enough to complete the healing process. Different antibiotic-impregnated implants based on various kinds of scaffolds have been explored. Gentamicin-vancomycin impregnated poly-methyl methacrylate has been reported as a drug delivery device in the treatment of intramedullary infections caused by methicillin-resistant *Staphylococcus aureus* (MRSA). BG and ceftriaxone-sulbactam composite drug delivery systems could be safely applied for the treatment of chronic osteomyelitis due to eradication of infection and new bone formation (Cauda et al. 2008, Kundu et al. 2011).

Gerhardt et al. (2011) proved the excellent angiogenic properties of 3D BG-PDLLA composites. These resorbable materials induce the secretion of vascular endothelial growth factor (VEGF) by human fibroblasts, and after about 8 weeks of implantation, scaffolds are infiltrated with young, well-vascularized tissue. The study showed that on composite films containing microsized or nanosized BG, human fibroblasts produced five times more VEGF than on pure PDLLA films. It therefore can be expected that in the near future implantation of PDLLA/BG composites will significantly improve regeneration of hard–soft tissue defects and increase bone formation arising from enhanced vascularization of the bioactive construct even in the case of undernourished and poorly vascularized wounds (Gerhardt et al. 2011).

There are also clinical trials with the application of BG composites as antibiotic delivery systems for osteoarticular tuberculosis (TB) therapy. Strategies for treating tuberculosis consist of multidrug chemotherapy administration for 9 months in adults and 12 months in children. Osteoarticular TB, due to poor blood supply, is particularly resistant to chemotherapy. After surgical bone debridement, the implantation of PLGA/BG composites enriched with antituberculotic drug is used as a controlled drug release system. This method can shorten or even replace the drug therapy following surgical intervention and therefore reduce lesions to hepatic and renal functions. It also decreases the risk of surviving latent TB bacilli in the affected bone sites which would reproduce rapidly in advantageous conditions (Zhu et al. 2011).

11.6 Conclusions

Recent work indicates the great potential of the sol-gel route in manufacture of BGs and BG-enriched composite scaffolds for bone tissue engineering. Particularly, two novel promising approaches are based on (1) supramolecular chemistry of surfactants to obtain a highly ordered MBG structure and (2) electrospinning techniques to create nanofibrous BGs. These two techniques enable production of BG materials with extremely high relative surface area (hundreds m^2/g), which was not possible to achieve by conventional sol-gel route or melting technique. Thanks to this parameter, BGs exhibit excellent bioactivity. Moreover, nanofibrous BGs could be used as a reinforcing phase of several organic or inorganic matrices to enhance mechanical properties of the resulting composites.

In the scaffold technology, there are several advances enabling production of BG-enriched composite scaffolds with hierarchical structure mimicking ECM of bone tissue. These include techniques providing pores in the range of millimeters (rapid prototyping), micrometers (solvent casting/particulate leaching, TIPS, microspheres synthesis, electrospinning, or gas foaming), and nanometers (supramolecular chemistry/polymer templating).

From the biological perspective, the addition of BG into polymer or calcium phosphate matrices proves to be a good strategy to enhance the response of osteoblastic and osteoprogenitor cells, but more studies are required to determine the bone formation and remodeling capacity of such composites. The comparative studies of different matrices with selected types of BGs would also shed more light on the mechanisms by which the BG contributes to the bioactivity and biocompatibility of the resulting composites.

Current preclinical and clinical studies prove the potential of BG composites in various bone-related therapies. Tissue engineering strategies with BG composites will certainly extend their clinical applications.

References

Agathopoulos, S., D. U. Tulyaganova, J. M. G. Ventura, S. Kannana, M. A. Karakassides, and J. M. F. Ferreira. 2006. Formation of hydroxyapatite onto glasses of the CaO–MgO–SiO$_2$ system with B$_2$O$_3$, Na$_2$O, CaF$_2$ and P$_2$O$_5$ additives. *Biomaterials* 27:1832–1840.

Ana, M. C., H. Y. Barradas, C. A. van Blitterswijk, and P. Habibovic. 2011. Osteoinductive biomaterials: Current knowledge of properties, experimental models and biological mechanisms. *Eur Cell Mater* 21:407–429.

Andersson, O. H. and I Kangasniemi. 1991. Calcium phosphate formation at the surface of bioactive glass in vitro. *J Biomed Mater Res* 25:1019–1030.

Andersson, O. H. and K. H. Karlsson. 1991. On the bioactivity of silicate glass. *J Non-Cryst Solids* 129:145–151.

Andersson, O. H., G. Liu, K. H. Karlsson, L. Niemi, I. Miettinen, and I. Juhanoja. 1990. In vivo behaviour of glasses in the SiO$_2$-Na$_2$O-CaO-P$_2$O$_5$-Al$_2$O$_3$-B$_2$O$_3$ system. *J Mater Sci Mater Med* 1:219–227.

Arcos, D. and M. Vallet-Regí. 2010. Sol–gel silica-based biomaterials and bone tissue regeneration. *Acta Biomater* 6:2874–2888.

Augello, A., T. B. Kurth, and C. De Bari. 2010. Mesenchymal stem cells: A perspective from in vitro cultures to in vivo migration and niches. *Eur Cell Mater* 20:121–133.

Baino, F., E. Verne, and Ch. Vitale-Brovarone. 2009. Feasibility, tailoring and properties of polyurethane/bioactive glass composite scaffolds for tissue engineering. *J Mater Sci Mater Med* 20:2189–2195.

Bellantone, M., N. Coleman, and L. L. Hench. 2000. Bacteriostatic action of a novel four-component bioactive glass. *J Biomed Mater Res* 51:484–490.

Bellantone, M., H. D. Williams, and L. L. Hench. 2002. Broad-spectrum bactericidal activity of Ag$_2$O-doped bioactive glass. *Antimicrob Agents Chemother* 46:1940–1945.

Bellucci, D., V. Cannillo, and A. Sola. 2010. An overview of the effects of thermal processing on bioactive glasses. *Sci Sinter* 42:307–320.

Boccaccini, A. R., J. J. Blaker, V. Maquet, R. Jerome, S. Blacher, and J. A. Roether. 2005. Biodegradable and bioactive polymer/Bioglass® composite foams for tissue engineering scaffolds. *Mater Sci Forum* 49:499–506.

Boccaccini, A. R., M. Erol, W. J. Stark, D. Mohn, Z. Hong, and J. F. Mano. 2010. Polymer/bioactive glass nanocomposites for biomedical applications: A review. *Compos Sci Technol* 70:1764–1776.

Bonewald, L. F. 2011. The amazing osteocyte. *J Bone Miner Res* 26:229–238.

Boskey, A. L. 2005. The organic and inorganic matrices. In *Bone Tissue Engineering*, J. O. Hollinger, T. A. Einthorn, B. A. Doll, and C. Sfeir (eds.), pp. 91–123. Boca Raton, FL: CRC Press.

Brink, M., T. Turunen, R-P. Happonen, and A. Yli-Urpo. 1997. Compositional dependence of bioactivity of glasses in the system $Na_2O–K_2O–MgO–CaO–B_2O_3–P_2O_5–SiO_2$. *J Biomed Mater Res* 37:114–121.

Cao, W. and L. L. Hench. 1996. Bioactive materials. *Ceram Int* 22:493–507.

Cauda, V., S. Fiorilli, B. Onida et al. 2008. SBA-15 ordered mesoporous silica inside a bioactive glass–ceramic scaffold for local drug delivery. *J Mater Sci Mater Med* 19:3303–3310.

Cholewa-Kowalska, K., J. Kokoszka, M. Laczka, L. Niedzwiedzki, W. Madej, and A. M. Osyczka, 2009. Gel-derived bioglass as a compound of hydroxyapatite composites. *Biomed Mater* 4:55007.

Christodoulou, I., L. D. Buttery, P. Saravanapavan, G. Tai, L. L. Hench, and J. M. Polak. 2005. Dose- and time-dependent effect of bioactive gel-glass ionic-dissolution products on human fetal osteoblast-specific gene expression. *J Biomed Mater Res B Appl Biomater* 74:529–537.

Clark, A. E. and L. L. Hench. 1994. Calcium phosphate formation on sol-gel derived bioactive glasses. *J Biomed Mater Res* 28:693–698.

Ducheyne, P. and Q. Qiu. 1999. Bioactive ceramics: The effect of surface reactivity on bone formation and bone cell function. *Biomaterials* 20:2287–2303.

El-Ghannam, A. 2005. Bone reconstruction: From bioceramics to tissue engineering. *Expert Rev Med Devices* 2:87–101

El-Kady, A. M., A. F. Ali, and M. M. Farag. 2010. Development, characterization, and in vitro bioactivity studies of sol–gel bioactive glass/poly(L-lactide) nanocomposite scaffolds. *Mater Sci Eng C* 30:120–131.

Fabbri, P., V. Cannillo, A. Sola, A. Dorigato, and F. Chiellini. 2010. Highly porous polycaprolactone-45S5 Bioglass scaffolds for bone tissue engineering. *Compos Sci Technol* 70:1869–1878.

Gerhardt, L. Ch., K. L. Widdows, M. M. Erol et al. 2011. The pro-angiogenic properties of multi-functional bioactive glass composite scaffolds. *Biomaterials* 32:4096–4108.

Gorustovich, A. A., J. A. Roether, and A. R. Boccaccini. 2010. Effect of bioactive glasses on angiogenesis: A review of in vitro and in vivo evidences. *Tissue Eng Part B Rev* 16:199–207.

Gough, J. E., J. R. Jones, and L.L. Hench. 2004. Nodule formation and mineralisation of human primary osteoblasts cultured on a porous bioactive glass scaffold. *Biomaterials* 25:2039–2046.

Haimi, S., N. Suuriniemi, A.-M. Haaparanta et al. 2009. Growth and osteogenic differentiation of adipose stem cells on PLA/Bioactive glass and PLA/β-TCP scaffolds. *Tissue Eng Part A* 15:1473–1480.

Hamadouche, M., A. Meunier, D. C. Greenspan et al. 2001. Long-term in vivo bioactivity and degradability of bulk sol-gel bioactive glasses. *J Biomed Mater Res* 54:560–566.

Hatzistavrou, E., X. Chatzistavrou, L. Papadopoulou et al. 2010. Characterisation of the bioactive behaviour of sol-gel hydroxyapatite-CaO and hydroxyapatite-CaO-bioactive glass composites. *Mater Sci Eng C* 30:497–502.

Hench, L. L. 1991. Bioceramics: From concept to clinic. *J Am Ceram Soc* 74:1487–1501.

Hench, L. L. and J. M. Polak. 2002. Third-generation biomedical materials. *Science* 295:1014–1017.

Hench, L. L., R. J. Splinter, W. C. Allen, and T. K. Greenlee Jr. 1971. Bonding mechanisms at the interface of ceramic prosthetic materials. *J Biomed Mater Res* 5:117–141.

Hench, L. L. and J. K. West. 1996. Biological application of bioactive glasses. *Life Chem Rep* 13:303–343.

Hesaraki, S., M. Safari, and M. A. Shokrgozar. 2009. Development of β-tricalcium phosphate/sol-gel derived bioactive glass composites: Physical, mechanical, and in vitro biological evaluations. *J Biomed Mater Res B Appl Biomater* 91:459–469.

Hoppe A., N. S. Güldal, and A. R. Boccaccini. 2011. A review of the biological response to ionic dissolution products from bioactive glasses and glass-ceramics. *Biomaterials* 32:2757–2774.

Jiang, J., A. Tang, G. A. Ateshian, X. E. Guo, C. T. Hung, and H. H. Lu. 2010. Bioactive stratified polymer ceramic-hydrogel scaffold for integrative osteochondral repair. *Ann Biomed Eng* 38:2183–2196.

Jo, J.-H., E.-J. Lee, D.-S. Shin et al. 2009. In vitro/in vivo biocompatibility and mechanical properties of bioactive glass nanofiber and poly(ε-caprolactone) composite materials. *J Biomed Mater Res B Appl Biomater* 91:213–220.

Jurczyk, K., K. Niespodziana, M. U. Jurczyk, and M. Jurczyk. 2011. Synthesis and characterization of titanium-45S5 Bioglass nanocomposites. *Mater Des* 32:2554–2560.

Khanuja, H. S., J. J. Vakil, M. S. Goddard, and M. A. Mont. 2011. Cementless femoral fixation in total hip arthroplasty. *J Bone Joint Surg Am* 93:500–509.

Kim, H. W., H. E. Kim, and J. C. Knowles. 2006. Production and potential of bioactive glass nanofibers as a next-generation biomaterial. *Adv Funct Mater* 16:1529–1535.

Kim, H. W., H. H. Lee, and G. S. Chun. 2008. Bioactivity and osteoblast responses of novel biomedical nanocomposites of bioactive glass nanofiber filled poly(lactic acid). *J Biomed Mater Res A* 85:651–663.

Kokubo, T. and H. Takadama. 2006. How useful is SBF in predicting in vivo bone bioactivity? *Biomaterials* 27:2907–2915.

Kondo, N., A. Ogose, K. Tokunaga et al. 2006. Osteoinduction with highly purified β-tricalcium phosphate in dog dorsal muscles and the proliferation of osteoclasts before heterotopic bone formation. *Biomaterials* 27:4419–4427.

Kumta, P. N. 2006. Ceramic biomaterials. In *An Introduction in Biomaterials*, S. A. Guelcher and J. O. Hollinger (eds.), pp. 311–339. Boca Raton, FL: CRC Press.

Kundu, B., S. K. Nandi, S. Dasgupta et al. 2011. Macro-to-micro porous special bioactive glass and ceftriaxone–sulbactam composite drug delivery system for treatment of chronic osteomyelitis: An investigation through in vitro and in vivo animal trial. *J Mater Sci Mater Med* 22:705–720.

Laczka, M. and K. Cholewa. 1997. The surface phenomena in gel-derived glasses and glass-ceramics materials of the CaO-P$_2$O$_5$-SiO$_2$ system. *Chem Pap* 51:348–356.

Laczka, M., K. Cholewa, and A. Laczka-Osyczka 1997. Gel-derived powders of CaO-P$_2$O$_5$-SiO$_2$ system as a starting material to production of bioactive ceramics. *J Alloys Comp* 248:42–51.

Laczka, M., K. Cholewa-Kowalska, A. Laczka-Osyczka, B. Turyna, and M. Tworzydlo. 2000. Gel-derived materials of CaO-P$_2$O$_5$-SiO$_2$ system modified by boron, sodium, aluminium, magnesium and fluorine compounds. *J Biomed Mater Res* 52:601–612.

Li, R., A. E. Clark, and L. L. Hench. 1991. An investigation of bioactive glass powders by sol-gel processing. *J Appl Biomater* 2:231–239.

Li, X., J. Shi, X. Dong, L. Zhang, and H. Zeng. 2008. A mesoporous bioactive glass/polycaprolactone composite scaffold and its bioactivity behavior. *J Biomed Mater Res A* 84: 84–91.

Locke, M., V. Feisst, and P. Rod Dunbar. 2011. Concise review: Human adipose-derived stem cells: Separating promise from clinical need. *Stem Cells* 29:404–411.

Lu, H. H., S. F. El-Amin, K. D. Scott, and C. T. Laurencin. 2003. Three-dimensional, bioactive, biodegradable, polymer-bioactive glass composite scaffolds with improved mechanical properties support collagen synthesis and mineralization of human osteoblast-like cells in vitro. *J Biomed Mater Res A* 64:465–474.

Lu, H. H., A. Tang, S. C. Oh, J. P. Spalazzi, and K. Dionisio. 2005. Compositional effects on the formation of a calcium phosphate layer and the response of osteoblast-like cells on polymer-bioactive glass composites. *Biomaterials* 26:6323–6334.

Meretoja, V. V., M. Malin, J. V. Seppala, and T. O. Narhi. 2009. Osteoblast response to continuous phase macroporous scaffolds under static and dynamic culture conditions. *J Biomed Mater Res A* 89:317–325.

Naira, L. S. and C. T. Laurencin. 2007. Biodegradable polymers as biomaterials. *Prog Polym Sci* 32:762–798.

Ohtsuki, C., T. Kokubo, and T. Yamamuro. 1992. Mechanism of apatite formation on CaO-SiO$_2$-P$_2$O$_5$ glasses in simulated body fluid. *J Non-Cryst Solids* 143:84–92.

Oksiuta, Z., J. R. Dabrowski, and A. Olszyna. 2009. Co-Cr-Mo-based composite reinforced with bioactive glass. *J Mater Process Technol* 209:978–985.

Orava, E., J. Korventausta, M. Rosenberg, M. Jokinen, and A. Rosling. 2007. In vitro degradation of porous poly(DL-lactide-co-glycolide) (PLGA)/bioactive glass composite foams with a polar structure. *Polym Degrad Stab* 92:14–23.

Orimo, H. 2010. The mechanisms of mineralization and the role of alkaline phosphatase in health and disease. *J Nippon Med Sch* 77:4–12.

Osyczka, A. M., D. L. Diefenderfer, G. A. Bhargave, and P. S. Leboy. 2004. Different effects of BMP-2 on marrow stromal cells from human and rat bone. *Cells Tissues Organs* 176:109–119.

Owen, S. C. and M. S. Shoichet. 2010. Design of three-dimensional biomimetic scaffolds. *J Biomed Mater Res A* 94:1321–1331.

Pamula, E., J. Kokoszka, K. Cholewa-Kowalska et al. 2011. Degradation, bioactivity, and osteogenic potential of composites made of PLGA and two different sol-gel bioactive glasses. *Ann Biomed Eng* 39(8):2114–2129.

Pandit, N., R. Gupta, and S. Gupta. 2010. A comparative evaluation of biphasic calcium phosphate material and bioglass in the treatment of periodontal osseous defects: A clinical and radiological study. *J Contemp Dent Pract* 11:25–32.

Pantchev, A., E. Nohlert, and A. Tegelberg. 2009. Endodontic surgery with and without inserts of bioactive glass PerioGlas—A clinical and radiographic follow-up. *J Oral Maxillofac Surg* 13:21–26.

Peltola, J., P. K. Vallittu, V. Vuorinen, A. A. J. Aho, A. Puntala, and K. M. J. Aitasalo. 2012. Novel composite implant in craniofacial bone reconstruction. *Eur Arch Otorhinolaryngol* 269(2):623–628.

Pereira, M. M., A. E. Clark, and L. L. Hench. 1994. Calcium-phosphate formation on sol-gel derived bioactive glasses in-vitro. *J Biomed Mater Res* 28:693–698.

Pereira, M. M., A. E. Clark, and L. L. Hench. 1995. Effect of texture on the rate of hydroxyapatite formation on gel-silica surface. *J Am Ceram Soc* 78:2463–2468.

Pirhonen, E., H. Niiranen, T. Niemal, M. Brink, and P. Tormala. 2006. Manufacturing, mechanical characterization, and in vitro performance of bioactive glass 13–93 fibers. *J Biomed Mater Res B Appl Biomater* 77:227–233.

Reddi, A. H. 1981. Cell biology and biochemistry of endochondral bone development. *Coll Relat Res* 1:209–226.

Rezwan, K., Q. Z. Chen, J. J. Blaker, and A. R. Boccaccini. 2006. Biodegradable and bioactive porous polymer/inorganic composite scaffolds for bone tissue engineering. *Biomaterials* 27:3413–3431.

Rho, J. Y., L. Kuhn-Spearing, and P. Zioupos. 1998. Mechanical properties and the hierarchical structure of bone. *Med Eng Phys* 20:92–102.

Roether, J. A., J. E. Gough, A. R. Boccaccini, L. L. Hench, V. Maquet, and L. Jerome. 2002. Novel bioresorbable and bioactive composites based on bioactive glass and polylacatide foams for bone tissue engineering. *J Mater Sci Mater Med* 13.1207–1214.

Ruuttila, P., H. Niiranen, M. Kellomaki, P. Tormala, Y. T. Konttinen, and M. Hukkanen. 2006. Characterization of human primary osteoblast response on bioactive glass (BaG 13–93) coated poly-L,DL-lactide (SR-PLA70) surface in vitro. *J Biomed Mater Res Part B Appl Biomater* 78B:97–104.

Schickle, K., K. Zurlinden, Ch. Bergmann et al. 2011. Synthesis of novel tricalcium phosphate-bioactive glass composite and functionalization with rhBMP-2. *J Mater Sci Mater Med* 22:763–771.

Sepulveda, P., J. R. Jones, and L. L. Hench. 2002. In vitro dissolution of melt-derived 45S5 and sol-gel derived 58S bioactive glasses. *J Biomed Mater Res* 61:301–311.

Sharma, S., V. P. Soni, and J. R. Bellare. 2009. Electrophoretic deposition of nanobiocomposites for orthopedic applications: Influence of current density and coating duration. *J Mater Sci Mater Med* 20:93–100.

Tanner, K. E. 2010. Bioactive ceramic-reinforced composites for bone augmentation. *J R Soc Interface* 7:541–557.

Tsigkou, O., L. L. Hench, A. R. Boccaccini, J. M. Polak, and M. M. Stevens. 2007. Enhanced differentiation and mineralization of human fetal osteoblasts on PDLLA containing Bioglass® composite films in the absence of osteogenic supplements. *J Biomed Mater Res A* 80:837–851.

Tuusa, S. M.-R., M. J. Peltola, T. Tirri, L. V. J. Lassila, and P. K. Vallittu. 2007. Frontal bone defect repair with experimental glass-fiber-reinforced composite with bioactive glass granule coating. *J Biomed Mater Res Part B Appl Biomater* 82:149–155.

Välimäki, V. V. and H. T. Aro. 2006. Molecular basis for action of bioactive glasses as bone graft substitute. *Scand J Surg* 95:95–102.

Vallet-Regí, M., C. V. Ragel, and A. J. Salinas. 2003. Glasses with medical applications. *Eur J Inorg Chem* 1029–1042.

Varanasi, V. G., E. Saiz, P. M. Loomer et al. 2009. Enhanced osteocalcin expression by osteoblast-like cells (MC3T3-E1) exposed to bioactive coating glass (SiO_2-CaO-P_2O_5-MgO-K_2O-NA_2O system) ions. *Acta Biomater* 5:3536–3547.

Wu, C., Y. Ramaswamy, Y. Zhu et al. 2009. The effect of mesoporous bioactive glass on the physiochemical, biological and drug-release properties of poly(DL-lactide-co-glycolide) films. *Biomaterials* 30:2199–2208.

Yao, J., S. Radin, P. S. Leboy, and P. Ducheyne. 2005. The effect of bioactive glass content on synthesis and bioactivity of composite poly(lactic-co-glycolic acid)/bioactive glass substrate for tissue engineering. *Biomaterials* 26:1935–1943.

Yun, H., S. Kim, and E. K. Park. 2011. Bioactive glass–poly (ε-caprolactone) composite scaffolds with 3 dimensionally hierarchical pore networks. *Mat Sci Eng C* 32:198–200.

Zapala, J. and G. Wyszynska-Pawelec. 2006. Present-day and future state of art of reconstruction of the facial skeleton. *Biotechnologia* 72:110–124.

Zhao, S., Y. Li, and D. Li. 2011. Synthesis of CaO-SiO$_2$-P$_2$O$_5$ mesoporous bioactive glasses with high P$_2$O$_5$ content by evaporation induced self assembly process. *J Mater Sci Mater Med* 22:201–208.

Zhong, J. and D. C. Greenspan. 2000. Processing and properties of sol–gel bioactive glasses. *J Biomed Mater Res B Appl Biomater* 53:694–701.

Zhu, M., H. Wang, J. Liu et al. 2011. A mesoporous silica nanoparticulate/β-TCP/BG composite drug delivery system for osteoarticular tuberculosis therapy. *Biomaterials* 32:1986–1995.

12

Processing Metallic Biomaterials for a Better Cell Response

Ioana Demetrescu, Daniela Ionita, and Cristian Pirvu

CONTENTS

12.1 Introduction

Metallic biomaterials are in extensive use in the early twenty-first century throughout medicine, dentistry, and biotechnology [1,2]. The use of such materials is not new: the Romans, Chinese, and Aztecs used biomaterials such as gold for dentistry more than 2000 years ago.

Nowadays, metallic biomaterials, such as stainless steel, cobalt chromium alloys, or titanium and its alloys, are widely used in biomedical applications due to their resistance to corrosion [3], mechanical properties [4], and bio-compatibility [5]. Such properties are the result of structure, composition, and surface topography and can be achieved via various processing proce-dures at micro- and nanolevel.

Taking into account that corrosion resistance is due to electrochemical stability, various methods for obtaining a better passive stratum on metal-lic biomaterials [6], together with methods covering the surface with biomi-metic coatings such as hydroxyapatite (HA) (which is a bone component) have been developed over time [7].

In fact, changes in the materials' behavior from bioinert to bioactive and biointeractive [8] can be achieved via metallic biomaterials processing; such surface modifications induce better cell response as an aspect of in vitro biocompatibility.

Thus, the production of a new generation of smart biomaterials is enabled by a new wave of advances in cell biology, chemistry, and materials science.

The biologically active or bioactive materials such as bioglass and HA-coated materials were developed by Jarcho [9]; numerous studies have continued work on this topic in subsequent years [10,11].

Different surface modification methods have been used to improve surface characteristics, such as mechanical methods (grinding, polishing) [12], chem-ical methods (acidic or alkaline treatment, chemical vapor deposition [CVD]) [13], biochemical methods [14], electrochemical methods (anodic oxidation, electrochemical deposition) [15,16], and physical methods (thermal treatment, physical vapor deposition [PVD], ion implantation, and deposition) [17,18].

Depositions of polymer layers are also recognized to play an important role in biomedical applications [19]. Films of conducting polymers (CPs), deposited on titanium and titanium alloy *substrata*, may be used to graft bio-logically active molecules which accelerate the process of osseointegration and protect the surface of the substratum from corrosion [20].

The present chapter reviews the elaboration and characterization of dif-ferent coatings obtained on metallic biomaterials in order to obtain better bioperformance, as quantified by an increase of stability in bioliquids, less ion release in such environments and better cell response.

12.2 Enhancing Bioperformance, from Natural Passivation to Various Anodic Films as Micro- and Nanostructure

Titanium and its alloys (which are important metallic biomaterials) have native protective oxide layer spontaneously formed on surface [21]. This oxide is responsible for stability in various environments including bioliquids [22]. The protective oxide has a thickness of approximately 2–5 nm and provides corrosion resistance; it is a mixture of TiO, Ti_2O_3, and TiO_2 oxides, with TiO_2 predominating. According to thermodynamic Pourbaix Diagram [23], the passive domain with TiO_2 is a non-corrosion one. This natural TiO_2 passive oxide [21,22] sustains both corrosion resistance and good biocompatibility.

Despite the very good biocompatibility of the native oxide layer of Ti, long-term implant failure may appear due to the lack of osseointegration; in the last decades, many approaches have been made to improve the surface activity of Ti and Ti alloys [14,24]. In order to improve the bioperformance of natural TiO_2, various anodic films at micro- and nanolevel were elaborated and characterized [25,26]. The Ti anodizing induces the growth of oxide film, the reduction of ions release, and the increase in porosity, which may improve cell adhesion and proliferation. The chemical and physical properties of anodic oxides may be changed by monitoring process parameters such as anodic potential, electrolytes composition, temperature, and current density. The growth of oxide thickness modifies surface topography, especially the surface porosity configuration; an anion incorporated in the oxide stratum changes the chemical composition and crystalline structure of TiO_2 which is anatase and rutile [27]. It is known that Ti acts as a valve material, and the growth of oxide film according to Faraday law [28] involves ions migration through oxide. The uniformity of the oxide influences the shape of the defects and their density, as well as the formation of an intermediate suboxide film as TiO_{2-x} between the TiO oxide existing on Ti substrate and superficial TiO_2 film [29]. The final thickness of oxide, d, varies almost linearly with applied voltage according to the relation $d = a \cdot U$, where a is a constant with values between 1.5 and 3 nm/V. Cui et al. [25] have obtained anodic TiO_2 films using as electrolytes H_2SO_4, $H_3PO_4 + Na_2SO_4$, and CH_3COOH, imposing various voltage for 1 min at room temperature. As a result of anodizing, obtained films contain rutile and/or anatase in the case of electrolytes H_2SO_4 and Na_2SO_4. In the case of electrolytes CH_3COOH and H_3PO_4, the titanium oxide is amorphous. Crystalline films as rutile and anatase stimulate formation of an apatite stratum, after immersion in SBF (simulated body fluid) for 7 days, but titanium covered with an amorphous oxide does not induce apatite formation, showing less bioperformance.

The new approach in materials science at nanolevel introduced nano-structures with well-defined porous microchannels as biocompatible supports for culture growth as was described in the literature [30,31]. Usually, surface modifications at nanolevel may induce positive aspects [15,16], such as enhancing corrosion resistance or biocompatibility, but also negative aspects; in particular, toxicology-related problems may appear. For instance, the unintentional toxicological effect as a side effect of nanomaterials [32,33] was intensively discussed after the rapid development of nanotechnology, when nanolevel became interesting for tissue regeneration.

One-dimensional TiO_2 nanoarchitectures structures such as nanofibers, nanowires and, in particular, nanotubes (with a higher surface-to-volume ratio compared to conventional materials) are expected to play important roles in cell adhesion and proliferation. Generally, there are three strategies used in the fabrication of TiO_2 nanotubes as follows: template synthesis [32] method involves the fabrication of a nanoporous template, backfilling with the TiO_2 precursors, then removing the template to yield the resulting nanostructures. Hydrothermal methods [32], prepare individual nanotubes but only randomly aligned tubes. The third method is electrochemical synthesis which was first reported in 1999 as fabrication of porous TiO2 nanostructures in an acid electrolyte containing a small amount of HF. Since then, many research groups have paid considerable attention to this procedure, anodizing being a simple way to easily produce closely packed tube arrays with a self-organized alignment.

The anodizing mechanism in commonly used electrolytes consists in an oxide passive stratum formation according to the equation

$$Ti + 2H_2O \rightarrow TiO_2 + 4H^+$$

In fluoride mixtures, an oxide layer is formed on the surface of titanium as well, but in the presence of F-, the oxide layer partially dissolves and forms pits according to the reaction

$$TiO_2 + 6HF \rightarrow \left[TiF_6\right]^{2-} + 2H_2O + 2H^+$$

These concurrent processes anodic oxidation and dissolution leads to the formation of nanotubes. In fact, the mechanism is a way from passivity breakdown to nanopores and nanotubes.

So far, there have been four generations of anodic titanium oxide nanotubes. In the first generation, TiO_2 nanotubes were fabricated in HF-based aqueous solutions. Some other inorganic acids and fluoride compounds, such as H_3PO_4/NaF, NH_4F/$(NH_4)_2SO_4$, and Na_2SO_4/NaF [16,35], were used as a substitute for HF. Due to the high chemical dissolution rate, the length of the obtained nanotubes was limited to 500 nm. The effect of nanotopographical features of Ti/TiO_2 electrode surface on cell response is the subject of Figure 12.1 where a comparison between cell behavior on TiO_2 native stratum and on first generation nanotubes of 120 nm diameter is presented.

According to Figure 12.1, the cells cultured on nanotube surfaces have produced, released and assembled fibronectin molecules into a fibrillar matrix, thus contributing to stable cell substratum interactions, to a slightly higher extent than the cells grown on native passive TiO_2 surfaces.

Cells on the 120 nm nanotube surface displayed a more elongated polygonal shape, with a higher number of different-sized filopodia, especially at the

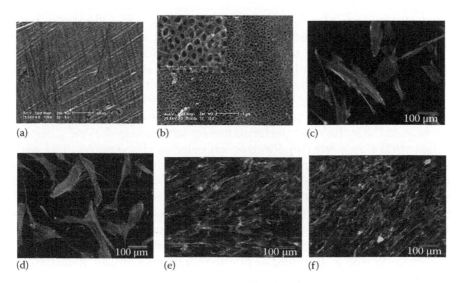

FIGURE 12.1
(I) The top view of (a) native passive TiO₂ and (b) TiO₂ nanotubes and corresponding (II) (c and d) fluorescent micrograph of the actin cytoskeletal organization and morphology of HGF-1cellsgrown for 76h and (III) (e and f) fibronectin immunostaining and nuclei visualization (DAPI staining) in human gingival fibroblasts. (From Demetrescu et al., *Bioelectrochemistry*, 79(1), 122, 2010.)

leading edges. These results are in concordance with studies performed by Brammer et al. [30].

Monitoring the pH of the electrolyte, the second generation of nanotubes was fabricated, and nanotube length was increased to a few micrometers [36]. For the third generation, tens, even hundreds, of micron-length nanotubes were easily prepared using an almost water free polar solution, especially in viscous glycerol [37] or ethylene glycol [38,39]. Studies indicate that, in used hybrid electrolytes (inorganic + organic), the chemical dissolution is lower in comparison to aqueous electrolytes and the formation mechanism of nanotubes is different due especially to the change of viscosity [15]. The last generation of anodic titanium oxides nanotubes is obtained in two or more steps leading to nanotube multilayers [40].

We are able to compare the behavior of nanotubes of the second and third generations by investigating growth on Ti6Al7Nb alloy using anodizing in two kinds of electrolytes: a mixture of fluoride salts for inorganic solution (similar to the mixture mentioned earlier), and a hybrid inorganic + organic components as glycerol +4% H₂O+NH₄F 0.36wt.%. These two kinds of electrolytes are representative for anodic growth of self-organized TiO₂ nanotubes in water-based electrolytes and in viscous electrolytes [35,37–40]. Our experiments [15,16] on the effect of various nanotubes structure indicated that electrochemical stability, ion release and cell behavior are favorably modulated and significantly different as a function of conditions of nanotubes formation, in particular for the nanotubes obtained in hybrid electrolyte with an organic part (which, as expected, may act and react more friendly with protein and other organic components). The differences in the performance of the two kinds of nanoarchitectures are attributed to a favorable surface topography as a result of different formation stages; the differences are also correlated with electrochemical stability. Electrochemical stability, evaluated by open circuit measurements, Tafel plots, cyclic voltammetry and EIS, revealed that the most performant nanostructure is the one fabricated in the hybrid electrolyte [15]. Electrochemically anodized alloys presented a better cell growth than untreated material due to improved electrochemical and surface properties after the specified treatments. Fluorescence micrographs showed that G292 cells displayed normal morphological features on all analyzed surfaces, but differences in cell density and spreading were found. Moreover, MTT assay performed after 24 and 48 h of culture showed a higher number of metabolically active cells on TiO_2 nanotubes obtained in the hybrid electrolyte. It should be pointed out that not all TiO_2 nanoarchitectures have a better biocompatibility: it is their dimensions that are critical for tissue response. Size-selective behavior of mesenchymal stem cells on TiO_2 nanotube arrays was reported as following: cell adhesion and spreading are enhanced for nanotube diameters of approximately 15–30 nm, while a strong decay in cell activity is observed for diameters >50 nm [41].

12.3 Surface Modification of Titanium Alloys for Rapid Osseointegration

The stability of an implant is determined by their osseointegration, which, to a large extent, depends on the chemistry and topography of its surface. Surface modification can alter the surface topography and chemistry, which directly affects the biological reaction to implants, that is, the interaction of implants and biological environments. A number of surface modifications and strategies have been developed to improve the osseointegration of titanium implants and can be divided into physical and chemical treatments as well as a combination of both.

12.3.1 Formation of Bioactive Surfaces by Physiological Method

This method involves the heterogeneous nucleation and growth of bone-like crystals on the surface of implant at physiological temperatures and under specific pH.

To enhance the heterogeneous nucleation of the Ca–P on the titanium implants, high concentration of calcium and phosphate is used in an increasing pH solution to form a thin layer on titanium surface [42].

In electrolytes containing Ca^{2+} ions (in the absence of phosphate ions), small quantities of calcium are absorbed on the surface of the titanium oxide layer after 24 h of immersion. This shows that the presence of phosphate ions is not mandatory for the Ca^{2+} ions to be adsorbed on the oxide surface. The absorption of Ca^{2+} on the surface is most likely due to the electrostatic interaction of Ca^{2+} ions with the negatively charged surface.

In physiological fluids containing phosphate and calcium ions, phosphate is absorbed on the surface of the titanium oxide film, replacing hydroxyl groups OH– attached to the titanium ion as $H_2PO_4^-$ and/or HPO_4^{2-}, thus forming a strong and complex link.

In the vicinity of titanium electrode, the following equilibrium of precipitation reactions has to be taken into consideration according to literature [43]:

$$H_2PO_4^- \leftrightarrow H_2PO_4^{2-} + H^+ \tag{12.1}$$

$$HPO_4^{2-} \leftrightarrow PO_4^{3-} + H^+ \tag{12.2}$$

$$10Ca^{2+} + 6PO_4^{3-} + 2OH^- \leftrightarrow Ca_{10}(PO)_6(OH)_2 \tag{12.3}$$

$$3Ca^{2+} + 2PO_4^{3-} + nH_2O \leftrightarrow Ca_3(PO_4)_2 \cdot nH_2O \tag{12.4}$$

$$8Ca^{2+} + 6HPO_4^{2-} + 5H_2O \leftrightarrow Ca_8H_2(HPO_4)_6 \cdot 5H_2O + 4H^+ \tag{12.5}$$

$$Ca^{2+} + HPO_4^{2-} + 2H_2O \leftrightarrow CaHPO_4 + 2H_2O \tag{12.6}$$

The first and second reaction changes the solution to a weak acid condition which results in a stable solution with high Ca^{2+} concentration. The consumption of hydroxide ions as shown in Equation 12.3 is possibly responsible for the drop of the pH value in the solution.

The implant (TiAlMoFe) is first immersed in five-time concentrated SBF for 24 h at 37°C. A thin (thickness less than 3 μm), dense, and amorphous layer of calcium phosphate uniformly deposited upon the implant surface (Figure 12.2) serves as a seeding substratum for the subsequent growth of a substantial (30–50 μm thick) crystalline latticework.

After this first step, the implants are immersed in Kokubo SBF [44] for various periods of time.

Figure 12.3a shows the initiations of growth of HA on the TiAlVMoFe support which in the initial stage presents a structure of dandelion bud that increases like an inflorescence. The Ca/P ratio obtained by EDS was 1.33, specific for octacalcium phosphate which is a precursor of the HA due to its high similarity with crystals present in bone and teeth [45]. Leng et al. [46] have identified such phase as OCP. According to Liu et al. [47], OCP crystal

FIGURE 12.2
SEM image of TiAlVMoFe immersed in 5SBF for 1 day.

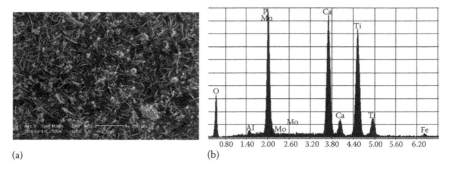

(a)   (b)

FIGURE 12.3
SEM (a) and EDS (b) spectrum of TiAlVMoFe immersed 1 day in 5SBF and other 7 days in SBF.

assemblies seem to enhance osteocalcin expression, which is an important marker of osteoblast phenotype.

12.3.2 Biomimetic Deposition of Apatite on Alkaline- and Heat-Treated Titanium Alloy Surface

Alkaline treatment is conducted to create bioactive layers on titanium surfaces by immersing them in thermal NaOH solution with subsequent heat treatment. The surface layer consists of an irregular and porous sodium titanate and TiO_2 (rutile) [48]. Alkaline treatment generally includes immersion in 5–10 M NaOH at 60°C for 24 h followed by heat treatment at 600°C or 800°C for 1 h. Alkali and subsequent heat treatment probably causes low adhesive strength at the interface of the treated layer (about 1 µm in thickness) and the substratum changes by distinct structures and compositions.

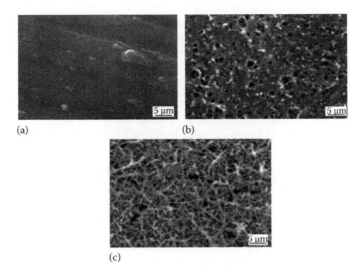

FIGURE 12.4
Surface morphologies (by SEM) for the specimens subjected to (a) control group, (b) 5 M NaOH, and (c) 10 M NaOH treatment at 60°C for 24 h.

Figure 12.4 shows SEM photographs of the surfaces of control specimen and the NaOH-treated Ti6Al4V alloy subjected to 5 and 10 M NaOH treatment at 60°C for 1 day. The SEM photograph revealed that the control specimen had a smooth surface texture with abrasive marks (Figure 12.4a). In contrast, the alkaline-treated specimen had porous surfaces. At the same alkaline-treating temperature, a much porous structure was observed with increasing the concentration of NaOH (Figure 12.4b and c).

Alkaline- and heat-treated titanium alloys exhibit the formation of more porous surfaces. At the same alkaline-treating temperature, a more porous structure was observed with increasing treatment time [49].

The structural change on the titanium surface during alkali and heat treatments and the mechanism of apatite formation on the treated surface in SBFs are described as follows. During the alkali treatment, the TiO$_2$ layer partially dissolves in the alkaline solution because of the attack by hydroxyl groups,

$$TiO_2 + NaOH \rightarrow HTiO_3^- + Na^+$$

This reaction is assumed to proceed simultaneously with hydration of titanium,

$$Ti + 3HO^- \rightarrow Ti(OH)_3^+ + 4e$$

$$Ti(OH)_3^+ + e \rightarrow TiO_2 \cdot H_2O + \tfrac{1}{2}H_2$$

$$Ti(OH)_3^+ + HO^- \leftrightarrow Ti(OH)_4$$

A further hydroxyl attack on the hydrated TiO_2 produces negatively charged hydrates on the surfaces of the substrata,

$$TiO_2 \cdot nH_2O + HO^- \leftrightarrow HTiO_3^- \cdot nH_2O$$

These negatively charged species combine with the alkali ions in the aqueous solution to produce an alkaline titanate hydrogel layer. During heat treatment, the hydrogel layer is dehydrated and densifies to form a stable amorphous or crystalline alkali titanate layer.

The coating of the Ti6Al4V substrata was performed by immersion of the disk-shaped samples in SBF at 370°C for different periods of time. After immersion, apatite was formed on the surface of the specimens treated with 5 M NaOH at 80°C for 3 days when soaked in SBF for 1 day.

The biological activity of the interface is ensured by the ionic exchange, which appears between the ceramic system (phosphate layers or bioglasses) and the physiological fluids. In the physiological fluids, ionic exchange surmises the elimination of sodium ions Na^+ from glass or the layer of sodium titanates and their replacement with H_3O^+ ions produced in the surrounding fluid leads to Ti–OH layer formation [50]. Simultaneously, with increasing pH, the apatite nucleation gets accelerated on increasing the supersaturation of the solution with respect to apatite. Calcium ions are incorporated in the hydrated Ti–OH layer. The positively charged Ca^{2+} may act as nucleation sites for carbonate–HA attaching themselves to the negatively charged $\left(PO_4^{3-}\right)$ and $\left(CO_2^{2-}\right)$ to form Ca–P enriched surface layer which crystallizes to bone-like apatite (carbonate–HA) [51].

The FTIR result of HA powder, alkali- and heat-treated TiAlV, and alkali- and heat-treated TiAlV immersed in SBF is shown in Figure 12.5. The multipleats peaks located around $1000\,cm^{-1}$ originated in phosphate modes. The split bands, mainly at 1028 and $1098\,cm^{-1}$, appear to agree with the formation of a well-crystallized apatite [52,53]. Carbonated bands have been detected

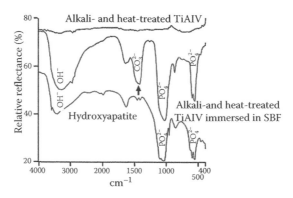

FIGURE 12.5
FTIR spectrum of TiAlV compared with FTIR spectrum of HA powder.

at 875, 1420, and 1453 cm^{-1}. Molecular and adsorbed water bands are also discerned at 1583 and 3400 cm^{-1}. A significant concentration of hydroxyl groups remains in the structure as observed from the intensity of the stretching and vibrational bands at 3568 and 632 cm^{-1}.

12.3.3 Formation of Bioactive Surfaces by Electrochemical HA Deposition

The electrochemical deposition of HA coating is normally performed in an aqueous solution containing calcium and phosphorus species (Ca–P). The conventionally used Ca–P concentrations are in the range of 10^{-1}–10^{-3} M [7,43,54], which are higher than or similar to that of human blood plasma.

The electrolyte used for obtained a HA coating on TiAlVZr has a molar ratio of Ca to P of 1.67. Cathodic polarization was conducted from open circuit potential to −3V (vs. *SCE*) at a rate of −0.6 V/h. Coating was deposited at current densities of 1–20 mA/cm^2 for 50 min at room temperature and pH 4 [7].

Through the electrochemical deposition process of calcium carbonates, uniform and adherent bioactive layers are obtained. The shape of the cathodic deposition curves is presented in Figure 12.6.

Ca^{2+} ions migrate to the cathode where the deposition occurs and can react with ions PO$_4^{3-}$ and OH$^-$ formed on the surface, thus synthesizing the HA deposition. The reactions which underlie the synthesis are as follows:

$$10Ca^{2+} + 6PO_4^{3-} + 2OH^- \leftrightarrow Ca_{10}(PO_4)_6(OH)_2$$

$$3Ca^{2+} + 2PO_4^{3-} + nH_2O \leftrightarrow Ca_3(PO_4)_2 \cdot nH_2O$$

$$8Ca^{2+} + 6HPO_4^{2-} + 5H_2O \leftrightarrow Ca_8H_2(PO_4)_6 \cdot 5H_2O + 4H^+$$

$$Ca^{2+} + 6HPO_4^{2-} + 2H_2O \leftrightarrow CaHPO_4 + 2H_2O$$

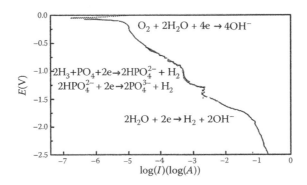

FIGURE 12.6
Cathodic polarization curve for a titanium substrate for obtaining bioactive surfaces.

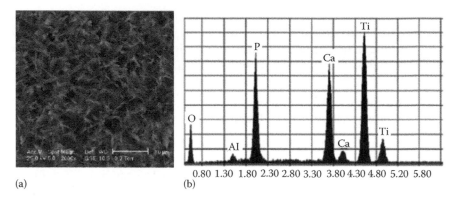

FIGURE 12.7
(a) SEM image and (b) EDS spectrum of HA/Ti–6Al–4V–1Zr.

A SEM image at covered alloy HA/Ti6Al4V1Zr is presented in Figure 12.7 associated with EDS spectrum. SEM micrographs revealed HA coatings to be entirely composed of straight plate-like units with sharp edges. Chemical elemental analysis by EDX of the HA coating detected oxygen, aluminum, phosphor, calcium, and titanium as the main constituent elements, indicating that the coating is a complex deposit on the alloy surface. The Ca/P ratio obtained on the surface was 1.62, almost the same value as the one of the HA (1.67), confirming the formation of the HA.

Shirkanzadeh [55] used cathodic polarization to produce HA nanocrystallites on titanium surfaces at –1.4 V (SCE) in ($Ca(NO_3)_2/NH_4H_2PO_4$, pH = 6.0) at 85°C over 2 h. According to da Silva et al. [56], the electrochemically coated HA surface produced the highest level of cell proliferation compared with uncoated surface after 3 days incubation on human osteoblast-like cells.

12.3.4 HA Coating on Titanium with Nanotubular Anodized TiO$_2$ Intermediate Layer

After the titanium metal was anodically oxidized in 0.5% HF + 5 g/L Na_2HPO_4, for 2 h at 20 V and subjected to heat treatment at 60°C for 1 h, the apatite formation in SBF may be induced due to the increased amount of anatase and/or rutile. In SBFs, the titanium anodically oxidized, inducing apatite formation on its surface, as shown in Figure 12.8 [57]. The induction period of apatite formation decreased with increasing amount of either the anatase or the rutile phase. It is well established that the anatase phase is more efficient in nucleation and growth of apatite than the rutile phase of TiO$_2$ because of better crystal lattice match with HA phase [58].

The overall reaction for anodic oxidation of titanium can be represented as follows.

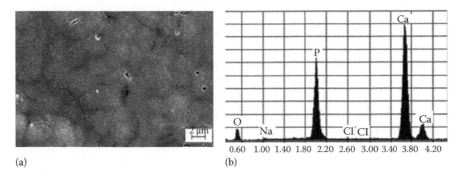

(a) (b)

FIGURE 12.8
(a) SEM image and (b) EDS spectrum of the sample S2 surface after 5 days of immersion in SBF.

At the Ti/Ti oxide interface,

$$Ti \leftrightarrow Ti^{2+} + 2e^-$$

At the Ti oxide/electrolyte interface,

$$2H_2O \leftrightarrow 2O^{2-} + 4H^+$$

$$2H_2O \leftrightarrow O_2 + 4H^+ + 4e^-$$

At both interfaces,

$$Ti^{2+} + 2O_2^{2-} \leftrightarrow TiO_2 + 2e^-$$

Regarding fictionalization of HA coatings in order to stimulate cell adhesion, a well-known strategy is HA modified with peptides as arginine–glycine–aspartate (RDG). Corresponding experiments revealed that mesenchymal stem cell adhesion to HA with adsorbed serum proteins is significantly larger than adhesion to RGD-modified HA; such strategy for regulating cell adhesion to biomaterials as HA is now being debated [59].

12.4 Processing Metallic Biomaterials by Conducting Polymer Electrodeposition

As discussed in the previous sections, modifications of titanium alloy surfaces are often used in order to improve the biological, chemical, and mechanical properties. The coatings based on CPs have important advantages for bio-medical applications, including biocompatibility [60,61], the ability to entrap and controllably release biological molecules [62], and chemical and electro-chemical stability. The biomedical applications include biosensors [63], tissue-engineering scaffolds [64], neural probes [65], or drug-delivery devices [66].

The use of CPs for biomedical applications was triggered in the 1980s by the finding that these materials were compatible with many biological mole-cules such as those used in biosensors. Then, by the mid-1990s, CPs were also shown to modulate the cellular activities, including cell adhesion, migration, DNA synthesis, and protein secretion.

The possibility of polymer film formation directly on a metal biomaterial surface in the electropolymerization process is a great advantage; such films are therefore obtained mostly by the polymerization of appropriates mono-mers. Polymer coatings such as polyaniline (PAni), polypyrrole (PPy), poly-thiophene (PT), etc., have been shown to offer protection for different metals. Among these coatings, PPy presents high potential application owing to the low monomer toxicity, its high stability in oxidized state and ease of synthe-sis in aqueous solutions. Electrodeposited PPy induced various biological applications as HA incorporation, doped systems with species as biotin and hyaluronic acid, or polymer surface modification through a graft of protein sequences able to promote a specific cellular response. It is known that pyr-role monomer is not biocompatible, but cell culture on its polymer reveals good cell viability and adherence.

Pyrrole has a large overpotential for oxidation at the alloy surface. The results from seeming incompatibility of the rather hydrophobic Py monomer and the rather hydrophilic alloy (oxide) surface can be minimized by using surfactants during polymerization. Electrically conductive polypyrrole-coated titanium significantly enhanced osteoblast functions in vitro; this novel surface modification has potential applications to improve perfor-mance of titanium implants in vivo.

12.4.1 Electropolymerization of Conducting Polymer on Titanium Substrata for the Future Development of New Biocompatible Surfaces

CPs can be prepared via chemical or electrochemical polymerization. Electrochemical synthesis is generally preferred because it provides a better control of film thickness and morphology, as well as cleaner polymers when compared to chemical oxidation.

A number of important variables must be considered during electropo-lymerization, including deposition time and temperature, solvent system (water content), electrolyte, electrode system, and deposition charge. Each of these parameters has an effect on film morphology (thickness and topog-raphy), film mechanics, and film conductivity, which are properties that directly influence cell response.

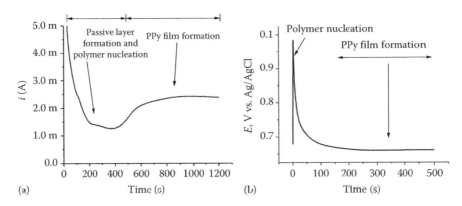

FIGURE 12.9
Potentiostatic (a) and galvanostatic (b) electrodeposition of polypyrrole on titanium electrode in 0.1 mol/L pyrrole in 0.2 M oxalic acid aqueous solution. (From Mindroiu, V.M. et al., *Rev. Chim.* (Bucharest), 61(4), 390, 2010.)

Films of electronically CPs are generally deposited onto a supporting electrode surface (biomaterial surface) by anodic oxidation of the corresponding monomer solution. Different electrochemical techniques can be used, including potentiostatic, galvanostatic, and potentiodynamic methods.

Figure 12.9, containing the potentiostatic and galvanostatic electrodeposition of polypyrrole on titanium electrode, shows an evolution in two steps: polymer nucleation on the metallic surface and polymer formation.

Figure 12.10 presents the first five cycles of the voltammograms obtained on titanium electrode in 0.2 M oxalic acid aqueous solution (Figure 12.10a) and in 0.1 mol/L pyrrole in 0.2 M oxalic acid aqueous solution (Figure 12.10b).

The passive layer formed this way inhibits further oxidation of titanium alloy; the electrode surface is practically blocked after the second cycle (Figure 12.10a).

Figure 12.10b, which corresponds to potentiodynamic polymerization of pyrrole in 0.2 M oxalic acid aqueous solution, presents a gradual increase of the current density in the successive cycles showing a polymerization of pyrrole onto the titanium alloy surface. The obtained films are black, homogeneous, and adherent to the electrode surface.

The ability to control PPy's surface properties such as wettability and charge density creates the potential for modifying neural interactions with the polymer. Two of the most common dopants that are co-deposited with PPy are polystyrenesulfonate (PSS) or sodium dodecylbenzenesulfonate (NaDBS). PPy/PSS and PPy/NaDBS polymers have been used in many applications ranging from actuators to neural electrode coatings and neural substrata.

The wettability of the polymeric film can also be drastically changed by PEG incorporation in PPy–PEG composite films.

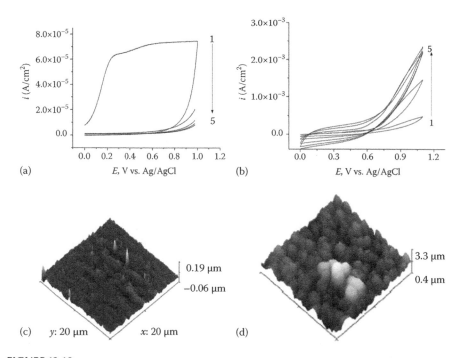

FIGURE 12.10
Cyclic voltammograms obtained on titanium electrode in 0.2 M oxalic acid aqueous solution (a) and in 0.1 mol/L pyrrole in 0.2 M oxalic acid aqueous solution (b), and scan rate of 50 mV/s and corresponding AFM images of the surface after scanning (c, d).

12.4.2 Conducting Polymers Properties Optimization for Biomedical Applications

The two common properties desired for all biomedical applications are biocompatibility and redox stability, but beyond these requirements, CP modification tends to be specific for various applications. Thus, most research has been focusing on biological and physical modification of CPs.

For biosensors, it is important to tune the hydrophilic/hydrophobic character, conductivity, and reactive functionalities for modification of CPs to successfully incorporate biomolecules and to improve detection of different species.

For tissue engineering, it is important to optimize CP properties including biomolecule functionalization, surface roughness, hydrophobicity, three-dimensional geometry, redox stability and degradability.

The wettability of the composite films, estimated by measuring the contact angle, decreased from 61° (in the case of the surface of uncoated Ti) to 32° (for film deposited on Ti that contain polypyrrole coating).

By forming the polymer composite PPy–PEG, it is possible to tune the contact angle value which can decrease with the increase in PEG concentration, through to a quasi total spreading of the drop of water in the last case, with

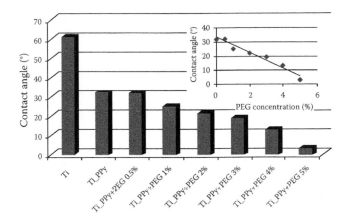

FIGURE 12.11
Contact angle measurement for uncoated titanium, titanium with PPy films and PPy and different concentration of PEG added. Inset—linear evolution of contact angle with PEG concentration on the range 0%–5%. (From Pirvu, C. et al., *Mol. Cryst. Liq. Cryst.*, 522, 425, 2010.)

5% PEG (Figure 12.11). This behavior suggests that PEG concentration plays an important role in determining the surface energy.

When the PPy film is produced in presence of PEG, the dimension of polymeric grains is reduced, resulting in a denser composite layer. In the case of the film with higher concentration of PEG (5%), an agglomeration of the grains and the formation of big clusters were observed. Most likely, this is the reason for which a non-adherent composite film was obtained for PEG concentration higher than 5%.

A quasi linear decrease of contact angle with PEG concentration was observed in the range of 0%–5%,

$$y = 5.46x + 32.95; R^2 = 0.9589$$

where y represents contact angle and x represents PEG concentration, was observed when the PEG concentration increase in the range of 0%–5%.

This fact is a very interesting aspect for the biocompatibility applications, taking into account that wettability is an important contribution to surface biocompatibility.

All these changes have had a positive influence on the biomaterial surface; fluorescence micrographs showed that G292 cells displayed normal morphological features on all analyzed surfaces, but differences in cell density and spreading were found. Moreover, MTT assay performed after 24 and 48 h of culture showed a higher number of metabolically active cells on Ti/PPY–PEG surface compared with Ti/PPy.

It should also be emphasized that, for both PPy- and PPy–PEG-coated titanium surface, cell adhesion and spreading are enhanced compared to uncoated titanium. This strengthens the assessment that the coatings based on CPs present important advantages for biomedical applications.

12.5 Conclusions

In this chapter, a brief overview was given first of the electrochemical elaboration and second of bioperformance characterization of a variety of Ti modified electrodes. It supports the idea that monitoring surface features at macro- and nanolevel, a large variety of architectures can be obtained. A part of them are biomimetic coatings. Biomimetics is derived from the Greek "bios," meaning "life," and "mimesis," meaning "to imitate"; that means application of coatings such as those found in nature on metallic implants.

Research in this domain is aiming to enhance biocompatibility and to ensure a better osseointegration of an implant material. Such changes in the materials' behavior from bioinert to bioactive and biointeractive can be achieved via metallic biomaterials processing and such surface modifications induce better cell response.

Acknowledgment

This work was supported by CNCSIS–UEFISCSU, project number PNII–IDEI PCCE 248/2010.

References

1. Milosev I. 2011. Metallic materials for biomedical applications: Laboratory and clinical studies. *Pure Appl Chem* 83(2):309–324.
2. Kannan S., Balamurugan A., Rajeswari S., Subbaiyan M. 2002. Metallic implants—An approach for long term applications in bone related defects. *Corros Rev* 20(4–5):339–358.
3. Ferreira J. M. F., Balamurugan A., Rajeswari S., Balossier G., Rebelo A. H. S. 2008. Corrosion aspects of metallic implants—An overview. *Mater Corros* 59(11):855–869.
4. Nakai M., Niinomi M., Akahori T. 2008. Mechanically multifunctional properties and microstructure of new beta-type titanium alloy, Ti-29Nb-13Ta-4.6Zr, for biomedical applications. *Adv Mater Res* 10:167–183.

5. Guillemot F. 2005. Recent advances in the design of titanium alloys for orthopedic applications. *Expert Rev Med Dev* 2(6):741–748.
6. de Assis S. L., Wolynec S., Costa I. 2006. Corrosion characterization of titanium alloys by electrochemical techniques. *Electrochim Acta* 51(8–9):1815–1819.
7. Vasilescu E., Vasilescu C., Drob P., Demetrescu I., Ionita D., Prodana M. et al. 2011. Characterisation and corrosion resistance of the electrodeposited hydroxyapatite and bovine serum albumin/hydroxyapatite films on Ti-6Al-4V-1Zr alloy surface. *Corros Sci* 53(3):992–999.
8. Watari F. 2010. Bioreactive nature of nanobiomaterials. *Nanobiomedicine* 1(1):2–8.
9. Cook S. D., Thomas K. A., Kay J. F., Jarcho M. 1988. Hydroxyapatite-coated titanium for orthopedic implant applications. *Clin Orthop Relat Res* (232):225–243.
10. Boccaccini A. R., Stamboulis A. G., Rashid A., Roether J. A. 2003. Composite surgical sutures with bioactive glass coating. *J Biomed Mater Res B* 67B(1):618–626.
11. Boccaccini A. R., Minay E. J., Krause D. 2006. Bioglass® coatings on superelastic NiTi wires by electrophoretic deposition (EPD). *Key Eng Mater* 314:219–224.
12. Huang H. H., Ho C. T., Lee T. H., Lee T. L., Liao K. K., Chen F. L. 2004. Effect of surface roughness of ground titanium on initial cell adhesion. *Biomol Eng* 21(3–5):93–97.
13. Mathur S., Kuhn P. 2006. CVD of titanium oxide coatings: Comparative evaluation of then-nal and plasma assisted processes. *Surf Coat Technol* 201(3–4):807–814.
14. Demetrescu I., Popescu S., Sarantopoulos C., Gleizes A. N., Iordachescu D. 2007. The biocompatibility of titanium in a buffer solution: Compared effects of a thin film of TiO$_2$ deposited by MOCVD and of collagen deposited from a gel. *J Mater Sci Mater Med* 18(10):2075–2083.
15. Demetrescu I., Mindroiu M., Pirvu C., Ion R. 2010. Comparing performance of nanoarchitectures fabricated by Ti6Al7Nb anodizing in two kinds of electrolytes. *Electrochim Acta* 56(1):193–202.
16. Demetrescu I., Pirvu C., Mitran V. 2010. Effect of nano-topographical features of Ti/TiO(2) electrode surface on cell response and electrochemical stability in artificial saliva. *Bioelectrochemistry* 79(1):122–129.
17. Krishnan V., Krishnan A., Remya R., Ravikumar K. K., Nair S. A., Shibli S. M. A. et al. 2011. Development and evaluation of two PVD-coated beta-titanium orthodontic archwires for fluoride-induced corrosion protection. *Acta Biomater* 7(4):1913–1927.
18. Kim K. H., Rautray T. R., Narayanan R., Kwon T. Y. 2010. Surface modification of titanium and titanium alloys by ion implantation. *J Biomed Mater Res B* 93B(2):581–591.
19. Ishihara K., Choi J. Y., Konno T., Matsuno R., Takai M. 2008. Surface immobilization of biocompatible phospholipid polymer multilayered hydrogel on titanium alloy. *Colloid Surf B* 67(2):216–223.
20. Ma C., Qu L. J., Li M. Q., Yang S. Q., Wang J. P. 2007. Electrochemically assisted co-precipitation of electrically conducting polymer with calcium phosphate coatings on Ti alloys. *Key Eng Mater* 336–338:1632–1634.
21. Hanawa T. 1999. In vivo metallic biomaterials and surface modification. *Mater Sci Eng A—Struct* 267(2):260–266.
22. Demetrescu I. 2008. Passive and bioactive films on implant materials and their efficiency in regenerative medicine. *Mol Cryst Liq Cryst* 486:1152–1161.
23. Pourbaix M. 1984. Electrochemical corrosion of metallic biomaterials. *Biomaterials* 5(3):122–134.

24. Wisbey A., Gregson P. J., Peter L. M., Tuke M. 1991. Effect of surface-treatment on the dissolution of titanium-based implant materials. *Biomaterials* 12(5):470–473.
25. Kawashita M., Cui X., Kim H. M., Wang L., Xiong T., Kokubo T. et al. 2009. Preparation of bioactive titania films on titanium metal via anodic oxidation. *Dent Mater* 25(1):80–86.
26. Schmuki P., Macak J. M. 2006. Anodic growth of self-organized anodic TiO_2 nanotubes in viscous electrolytes. *Electrochim Acta* 52(3):1258–1264.
27. Dafonseca C., Traverse A., Tadjeddine A., Belo M. D. 1995. A characterization of titanium anodic oxides by x-ray-absorption spectroscopy and grazing x-ray-diffraction. *J Electroanal Chem* 388(1–2):115–122.
28. Kalra K. C., Singh K. C., Singh M. 1997. Formation and breakdown characteristics of anodic oxide films on valve metal. *Indian J Chem A* 36(3):216–218.
29. Serruys Y., Sakout T., Gorse D. 1993. Anodic-oxidation of titanium in 1m H_2SO_4, studied by Rutherford backscattering. *Surf Sci* 282(3):279–287.
30. Jin S., Brammer K. S., Oh S., Cobb C. J., Bjursten L. M., van der Heyde H. 2009. Improved bone-forming functionality on diameter-controlled TiO(2) nanotube surface. *Acta Biomater* 5(8):3215–3223.
31. Schmuki P., Park J., Bauer S., von der Mark K. 2007. Nanosize and vitality: TiO_2 nanotube diameter directs cell fate. *Nano Lett* 7(6):1686–1691.
32. Grimes C. 2009. *TiO_2 Nanotube Arrays*. New York: Springer.
33. Nel A., Xia T., Madler L., Li N. 2006. Toxic potential of materials at the nanolevel. *Science* 311(5761):622–627.
34. Schmuki P., Roy P. R., P., Berger S. 2011. TiO(2) nanotubes: Synthesis and applications. *Angew Chem Int Ed* 50(13):2904–2939.
35. Taveira L. V., Macak J. M., Tsuchiya H., Dick L. F. P., Schmuki P. 2005. Initiation and growth of self-organized TiO_2 nanotubes anodically formed in NH_4F/ $(NH_4)(2)SO_4$ electrolytes. *J Electrochem Soc* 152(10):B405–B410.
36. Sung Y. E., Kang S. H., Kim J. Y., Kim H. S. 2008. Formation and mechanistic study of self-ordered TiO_2 nanotubes on Ti substrate. *J Ind Eng Chem* 14(1):52–59.
37. Tsuchiya H., Macak J. M., Taveira L., Balaur E., Ghicov A., Sirotna K. et al. 2005. Self-organized TiO_2 nanotubes prepared in ammonium fluoride containing acetic acid electrolytes. *Electrochem Commun* 7(6):576–580.
38. Zhou F., Wang D. A., Yu B., Wang C. W., Liu W. M. 2009. A novel protocol toward perfect alignment of anodized TiO(2) nanotubes. *Adv Mater* 21(19):1964–1967.
39. Xie Z. B., Blackwood D. J. 2010. Effects of anodization parameters on the formation of titania nanotubes in ethylene glycol. *Electrochim Acta* 56(2):905–912.
40. Li H. Y., Wang J. S., Huang K. L., Sun G. S., Zhou M. L. 2011. In-situ preparation of multi-layer TiO(2) nanotube array thin films by anodic oxidation method. *Mater Lett* 65(8):1188–1190.
41. Bauer S., Park J., Faltenbacher J., Berger S., von der Mark K., Schmuki P. 2009. Size selective behavior of mesenchymal stem cells on ZrO(2) and TiO(2) nanotube arrays. *Integr Biol* 1(8–9):525–532.
42. Barrere F., van Blitterswijk C. A., de Groot K., Layrolle P. 2002. Nucleation of biomimetic Ca-P coatings on Ti6Al4V from a SBF × 5 solution: Influence of magnesium. *Biomaterials* 23(10):2211–2220.
43. Peng P., Kumar S., Voelcker N. H., Szili E., Smart R. S., Griesser H. J. 2006. Thin calcium phosphate coatings on titanium by electrochemical deposition in modified simulated body fluid. *J Biomed Mater Res A* 76A(2):347–355.

44. Kokubo T. 1990. Surface-chemistry of bioactive glass-ceramics. *J Non-Cryst Solids* 120(1–3):138–151.
45. Dorozhkin S. V. 2007. Calcium orthophosphates. *J Mater Sci* 42(4):1061–1095.
46. Leng Y., Chen J. Y., Qu S. X. 2003. TEM study of calcium phosphate precipitation on HA/TCP ceramics. *Biomaterials* 24(13):2125–2131.
47. Liu Y., Cooper P. R., Barralet J. E., Shelton R. M. 2007. Influence of calcium phosphate crystal assemblies on the proliferation and osteogenic gene expression of rat bone marrow stromal cells. *Biomaterials* 28(7):1393–1403.
48. Kim H. M., Miyaji F., Kokubo T., Nishiguchi S., Nakamura T. 1999. Graded surface structure of bioactive titanium prepared by chemical treatment. *J Biomed Mater Res* 45(2):100–107.
49. Balamurugan A., Faure J., Benhayoune H., Torres P., Balossier G., Ferreira J. M. F. 2009. Morphological and chemical characterisation of biomimetic bone like apatite formation on alkali treated Ti6Al4V titanium alloy. *Mater Sci Eng C—Mater* 29(4):1252–1257.
50. Kizuki T., Takadama H., Matsushita T., Nakamura T., Kokubo T. 2010. Preparation of bioactive Ti metal surface enriched with calcium ions by chemical treatment. *Acta Biomater* 6(7):2836–2842.
51. Jonasova L., Muller F. A., Helebrant A., Strnad J., Greil P. 2004. Biomimetic apatite formation on chemically treated titanium. *Biomaterials* 25(7–8):1187–1194.
52. Sun L. M., Berndt C. C., Grey C. P. 2003. Phase, structural and microstructural investigations of plasma sprayed hydroxyapatite coatings. *Mater Sci Eng A—Struct* 360(1–2):70–84.
53. Ionita D., Bojin D., Demetrescu I. 2007. The behavior of ceramic coating on titanium using chemical and electrochemical deposition. *Key Eng Mater* 330–332:577–580.
54. Zhang Q. Y., Leng Y., Xin R. L. 2005. A comparative study of electrochemical deposition and biomimetic deposition of calcium phosphate on porous titanium. *Biomaterials* 26(16):2857–2865.
55. Shirkhanzadeh M., Azadegan M. 1993. Hydroxyapatite particles prepared by electrocrystallization from aqueous-electrolytes. *Mater Lett* 15(5–6):392–395.
56. da Silva M. H. P., Soares G. D. A., Elias C. N., Best S. M., Gibson I. R., Disilvio L. et al. 2003. In vitro cellular response to titanium electrochemically coated with hydroxyapatite compared to titanium with three different levels of surface roughness. *J Mater Sci—Mater Med* 14(6):511–519.
57. Ionita D., Mazare A., Portan D., Demetrescu I. 2011. Aspects relating to stability of modified passive stratum on TiO_2 nanostructure. *Met Mater Int* 17(2):321–327.
58. Oh S. H., Finones R. R., Daraio C., Chen L. H., Jin S. H. 2005. Growth of nanoscale hydroxyapatite using chemically treated titanium oxide nanotubes. *Biomaterials* 26(24):4938–4943.
59. Bellis S. L. 2011. Advantages of RGD peptides for directing cell association with biomaterials. *Biomaterials* 32(18):4205–4210.
60. Morozan A., Nastase F., Dumitru A., Nastase C., Vulpe S., Filipescu M. 2009. Plasma processing of polypyrrole-heparin thin films on titanium substrates for biomedical applications. *Phys Status Solidi C* 6(10):2195–2198.
61. De Giglio E., Stefania C., Calvano C. D., Luigia S., Pier G. Z., Silvia C. et al. 2007. A new titanium biofunctionalized interface based on poly(pyrrole-3-acetic acid) coating: proliferation of osteoblast-like cells and future perspectives. *J Mater Sci—Mater Med* 18(9):1781–1789.

62. Lovell A., Kurban Z., Bennington S. M., Jenkins D. W. K., Ryan K. R., Jones M. O. et al. 2010. A solution selection model for coaxial electrospinning and its application to nanostructured hydrogen storage materials. *J Phys Chem C* 114(49):21201–21213.
63. Singh M., Kathuroju P. K., Jampana N. 2009. Polypyrrole based amperometric glucose biosensors. *Sens Actuators B* 143(1):430–443.
64. Kim S. H., Oh A. Y., Jung S. H., Hong H. H., Choi J. H., Hong H. K. et al. 2008. Recent strategies of the regeneration of central nervous system by tissue engineering techniques. *Tissue Eng Regen Med* 5(3):370–387.
65. Richardson-Burns S. M., Hendricks J. L., Martin D. C. 2007. Electrochemical polymerization of conducting polymers in living neural tissue. *J Neural Eng* 4(2):L6–L13.
66. Wallace G., Halldorsson J. A., Little S. J., Diamond D., Spinks G. 2009. Controlled transport of droplets using conducting polymers. *Langmuir* 25(18):11137–11141.
67. Mindroiu V. M., Pirvu C., Popescu S., Demetrescu I. 2010. Polypyrrole electrodeposition on Ti6Al7Nb alloy in aqueous and non-aqueous solutions. *Rev Chim (Bucharest)* 61(4):390–394.
68. Pirvu C., Popescu S., Mindroiu M., Demetrescu I. 2010. Enhancing the stability of PPy film on Ti by PEG incorporation. *Mol Cryst Liq Cryst* 522:425–435.

13

Osteogenic Adult Stem Cells and Titanium Constructs for Repair and Regeneration

Marcus J. Tillotson and Peter M. Brett

CONTENTS

13.1 Introduction

The loss of function of tissues from disease, injury, or aging causes serious health problems as well as a tremendous social and economic cost. There is an important incentive for finding techniques that promote consistent, robust bone formation for repair and regeneration. The substitution of tissues such as bone or cartilage can currently be performed with allograft materials, but this introduces a risk of infection or graft rejection. Autologous bone grafts have, therefore, been considered for augmenting bone regeneration. This can eliminate immunological rejection and unnecessary pathogen transfer. Unfortunately, autologous transplantation provides a limited volume of bone grafts and is affected by potential donor site morbidity. Artificial implants have limitations too due to insufficient bonding to bone and allergic reactions that can be caused through material abrasion. This makes the lifespan for the implant finite and can lead to the need for replacement. Many types of metallic implants have been used including gold, stainless steel and, more recently, chromium alloys.

Several of these have shown significant drawbacks due to corrosion and the subsequent activated immune response. Titanium (Ti) was shown in 1952 to have a biocompatible nature [1,2]. Professor Per-Ingvar Brånemark of Sweden [3] discovered that titanium fuses to bone. This property, osseointegration, involves the direct attachment or connection of osseous tissue to a material without any intervening connective tissue. A direct structural and functional connection between living bone and the load-bearing surface of the artificial implant is thus formed. Ti is often considered the "gold standard" material for repairing damaged bone tissue. For example, Ti screws are widely used in the replacement of hip or knee joints, attachment of metal plates for the mending of badly broken bones; and the insertion of ceramic crowns in the replacement of missing or damaged teeth.

There is now the possibility of engineering tissue substitutes in vitro by seeding multipotent stromal cells (MSCs) or osteoblasts onto Ti constructs. This could provide new strategies for cranial and maxillofacial reconstruction as well as for the restoration of damaged or diseased bone.

13.2 Titanium Properties

There are a number of reasons that Ti is considered to be biocompatible. These can be attributed to its tolerability by the body. It has a tendency not to precipitate phosphates and other minerals from bone, and it has demonstrated strong interactions with adhesion proteins. Under atmospheric conditions, a thin oxide layer spontaneously forms on Ti and Ti alloy surfaces. This has a strong, direct effect on biocompatibility and success of the implant.

Any material that is implanted into the body is considered to be "foreign" and has the potential to cause chemical, physiological or mechanical "insult" to living tissue. Electron exchange occurring at the metal surface, following contact with water and tissue fluid, can lead to denaturation of the tissue surrounding the metallic implant. If metal ions become released during the corrosion process, they are considered to be toxic to living cells. The tendency for a metal to do this can be measured using its polarization resistance (Figure 13.1). The electron exchange processes that lead to the denaturation of macromolecules occur most notably in cobalt–chromium alloys in addition to toxicity effects from copper, nickel, and also stainless steel.

Ti itself is a reactive metal. However, in air and electrolytes, the oxide film at the surface is particularly dense and protects from further corrosion. Oxide ions migrate toward the bulk metal via defects in the crystal structure and react with the counter-ion Ti at the base of the oxide (Figure 13.2).

The dense oxide layer formed provides a positive effect by preventing any metal ions from reaching the outer surface and being released into the

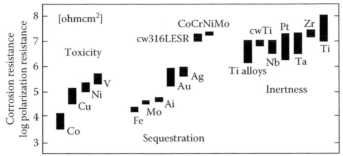

FIGURE 13.1
A diagram showing tissue reaction for various metallic elements and practical alloys, grouped according to "toxicity," "sequestration," and "inertness," against measured polarization resistance. Corrosion resistance is roughly proportional to the measured log of polarization resistance. (Steinemann S.G.: Titanium —The material of choice? *Periodontol. 2000.* 1998. 17. 7–21. Copyright Wiley-VCH Verlag GmbH & Co. KGaA. Reproduced with permission.)

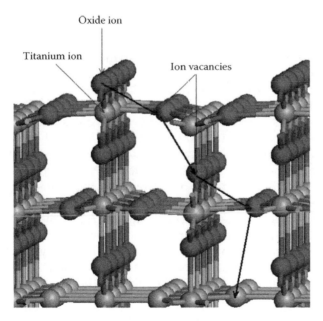

FIGURE 13.2
Migration of oxygen through the thin TiO_2 surface layer toward the bulk metallic structure.

electrolyte. In this way, the unwanted product of corrosion is not the production of free ions. The titanium dioxide TiO_2 in the very thin surface film exists in an amorphous or glassy state and occludes the availability of electrons in the bulk metal toward the electrolyte or living tissue, thus concealing the naked Ti surface from the bone into which it is implanted.

The surface has an amphoteric nature, and for the two structural variants of TiO$_2$, rutile and anatase, the zero point of charge occurs at ≈pH 6. In comparison, the zero point charges of Al$_2$O$_3$ and ZrO$_2$ are ≈pH 9 and ≈pH 2 for SiO$_2$. Hence, at physiological pH values, the surface of TiO$_2$ has no charge, whereas aluminum oxide and zirconium oxide are negative and quartz (SiO$_2$) has a positive charge. The zeta potentials between a dispersed layer of natural phosphate minerals, as found in bone, and the stationary layer of fluid have also been measured at the zero point charge. The measurements indicate a stability of the colloid dispersion of body fluids that is in close contact to the Ti surface. This prevents the precipitation of phosphates and other bone minerals out of solution at a pH value that exists in the body. This acts to prevent further leaching of bone minerals and the weakening of the bone structure [4].

Strong adhesion occurs between bone and connective tissue. The structure of a tissue and its interaction cohesion occurs along different length scales—the space between bone and vessels in soft tissues is in the order of millimeter, the size of cells is in the order of micrometer, and the Debye length, determining the interaction distance of chemical forces, is in the nanometer range. Fibronectin is one of the main adhesion proteins involved in bone cell attachment, and the tripeptide domain of arginine–glycine–aspartic acid (RGD) is the focal point where cells attach to the surface of large macromolecules [5]. TiO$_2$, being amphoteric, has four or five reactive groups of acidic and basic character per nm^2 of surface. The volume of an amino acid molecule in the RGD tripeptide is in the region of 0.1–0.2 nm^3. For amino acids spread out on the surface, about four or five molecules will cover an area of 1 nm^2 which matches well with the number of available bonds on the inorganic substrate. Thus, strong interactions can occur between the RGD tripeptide molecules and the surface. A proton is exchanged and electrostatic energy is gained [6], providing a mechanism for the short-range interaction for adhesion-promoting factors of living tissue to the native oxide of Ti.

The biological factors that determine the contact of the implant to bone and the bonding strength can be influenced through physical modifications (roughening), chemical modifications (surface free energy), and coatings. The roughening of surfaces encourages the entrapment of fibrin protein, adhesion, and mechanical stability of implants in host bone.

13.2.1 Physical Modification: Roughening

Ti is a possible material for use in clinical implants because bone regeneration can occur on the surface with no evidence of host rejection. The initial stage for osteoblastic cells producing bone tissue is cell adhesion followed by proliferation and differentiation. It has been shown that osteoblastic cell adhesion, growth, and differentiation are related directly to surface energy and roughness [7,8]. Osteoblasts have been shown to respond more favorably to roughened surfaces than smooth (Brett et al., 2004) [9].

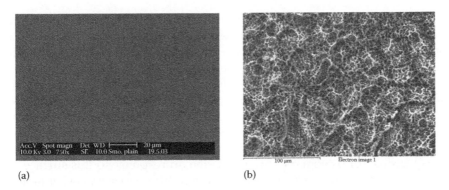

(a) (b)

FIGURE 13.3
SEM micrographs showing the difference in surface roughness of Ti discs. (a) Smooth polished Ti (SMO). (b) SLA. The surface shows wide cavities, 20–40 μm in diameter, produced by the blasting process. Acid etching produces nanometer-sized topography favorable for the adsorption of proteins and the attachment of osteoblastic cells. SEM images taken at the Eastman Dental Institute.

Much attention has been paid to improving the properties of Ti as a biomaterial. Surface roughness has been an important factor for establishing reliable bone-anchored implants. In vitro studies have provided a positive correlation between surface roughness and cellular attachment as well as subsequent osteoblast-like cell activity [10]. This has been supported using in vivo studies by groups such as Buser et al., who measured the mechanical testing strength of the connection between bone and implant [11]. Rønold et al. suggested that an upper limit exists for the correlation between surface roughness and bone fixation [10]. Using an in vivo tensile test procedure, they found that an average surface roughness of 3.62 μm created by grit blasting gave the best functional attachment between bone and implant. It was speculated that a surface with more pits and undercuts, providing a larger surface area, is better for bone cell attachment as seen under scanning electron microscopy (SEM) (Figures 13.3a and b).

Roughened surfaces can be fabricated in various ways to produce defined macro-, micro- and nanotopographies. These include micro-machining, plasma spraying, particle blasting and acid etching. Some surface contamination may exist due to alumina or TiO_2 grit blasting particles. This has been thought by Esposito et al. to hamper the osseointegration process [12].

13.2.2 Chemical Modifications: Surface Free Energy

Ti can be modified to have the same surface roughness but also a higher degree of wettability. SLActive is a surface produced by Straumann AG from previously roughened Ti that is conditioned in nitrogen and immediately preserved in an isotonic saline (NaCl) solution. This maintains a high surface activity, which is otherwise lost on the SLA surface due to reaction with hydrocarbons and carbide in the atmosphere [13]. Wettability and surface energy are key parameters in the adhesion and spreading of osteoblastic cells. A previous study has suggested that faster healing and greater implant stability were achieved with the SLActive implant surface than with conventional, hydrophobic surfaces. In addition, osteoblastic differentiation was enhanced by the most hydrophilic surface. Wall et al. reported a better osteogenic response to SLActive compared with the more hydrophobic SLA surface [7]. This was seen to be related with an increased gene expression of the osteogenic promoter WNT5A among others, which is responsible for osteogenic differentiation, in response to this surface.

Wall et al. have shown that initial stem cell differentiation was enhanced by roughened topographies [7]. This was accompanied by a reduction in cell number early in the culture and increased expression of osteogenic markers. This relates to higher implant survival rates in clinical practice for roughened surfaces compared to polished surfaces.

The surface topographic characters of an implant at the micro- and nanoscales are related to the biological response of a host. The roughness of the micro-roughened surface has an influence on the biocompatibility, mechanical characteristics and interlocking effect between the implant and tissue. A micro-rough surface with a roughness ranging from 10 nm to 10 µm influences increased biocompatibility due to its topography being within the same size range as a cell or a biomacromolecule. A micro-roughness of less than 10 nm at the surface corresponds to defects in the crystal structure, such as vacancy defects, grain boundaries and step and pit. These defects are in the active region and significantly affect the surface energy of biomaterials. This has a further influence on the interface of the implant and the integration with biomolecular material as it influences the adhesion, spreading, and growth of the cell (Figure 13.4).

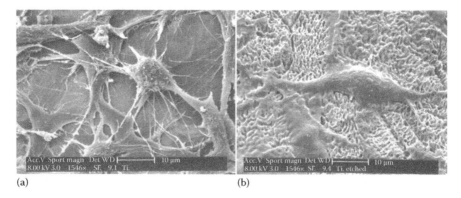

(a)   (b)

FIGURE 13.4
SEM of MSC attachment on Ti surfaces. (a) Smooth polished Ti. (b) Roughened Ti blasted with 180–220 µm particles TiO$_2$ and subsequently acid etched with 0.01 M HCl. (Reproduced with permission from Brett, P.M. et al., *Bone*, 35(1), 124–133, 2004.)

13.2.3 Coatings

The next evolution in metallic implants is likely to be through biochemically inspired specific surface modifications. The trend is to modify surfaces with (bio)chemical moieties in order to

- Trigger cell-selective response, for example, osteogenic cell differentiation
- Provide resistance to bacterial attachment and reducing the need for antibiotic prescription
- Reduce the risk of inflammation
- Improve the reliability and long-term performance of the implant

A summary of how Ti surfaces can be modified to fulfill these objectives is shown in Figure 13.5.

Self-assembled monolayers (SAMs) provide the possibility of controlling the physiochemical properties of the surface (e.g., wettability) to passively influence the interaction between the surface and biological molecules (B). The Langmuir–Blodgett method of deposition can be used to coat the surface with layers of surfactant molecules. Each molecule contains a polar "head" group which interacts with the metal and a nonpolar "tail." For example, *n*-alkyl

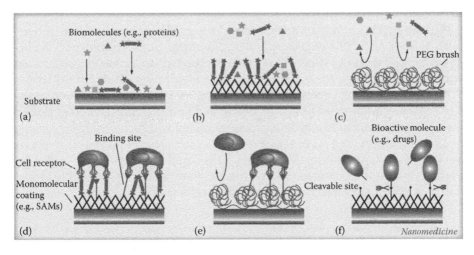

FIGURE 13.5
Interactions between the biological environment and a Ti implant with different surface (bio) chemistries. (a) Bare Ti surface on implantation encounters the non-specific adsorption of proteins. (b) Ti surface with modified physicochemical properties e.g. SAM. (c) Ti surface with non-fouling character reduces the non-specific adsorption of proteins. (d) Ti surface with immobilised biological molecules. (e) Ti surface with immobilised biological molecules of non-fouling character. (f) Ti surface with drug release strategy. (Reproduced from Schuler M. et al., *Nanomedicine*, 1(4), 449–463, 2006. With permission of Future Medicine Ltd.)

phosphates can be used to coat Ti. A lipid film is floated on the surface of a water tank into which a vertical metal slide is immersed. The transfer of the lipid molecules occurs from the horizontal liquid to the vertical solid substrate. This forms a highly ordered surface film of uniform thickness. The LB-film stability can be further improved through cross-linking or internal polymerization.

Monomolecular adlayers (C) provide a surface that is resistant to protein adsorption (i.e., non-fouling) and can reduce the host's nonspecific response to the implant. Polyethylene glycol (PEG) is commonly used to produce a non-fouling surface. Attachment of PEGs can be performed under cloud point conditions or using molecular assembly approaches. The reverse solubility behavior of glycols allows for cloud point deposition and occurs when the temperature of a mixture in water is raised. This causes the glycol molecules to "condense out" and deposit onto a surface [14]. Chemical coupling of the PEGs to the surface is often preferred as this is more robust and provides greater longevity of the anti-fouling response. SAMs have been anchored to metal surfaces via thiol (R-SH) terminal groups [15]. The PEG chains that extend from the surface must be of a sufficiently dense covering in order to be effective. This produces an excluded volume, osmotic and steric repulsion effect to other molecules.

Proteins or peptides can be linked covalently to the surface, introducing specific bioactive groups and binding sites to the biological environment (D). Peptide sequences can be attached covalently via chemical reaction to some of the PEG side chains. For example, a vinyl sulfone–cysteine coupling reaction can be used to attach peptide sequences to PEG side chains in the formation of RGD-peptide-modified PEG coatings.

Further modification with biomolecules can allow selective interactions with tissues and cells (E). Drug-eluting coatings, for example, poly(lactic-*co*-glycolic acid) (PLGA) biodegradable microbeads or lipid vesicles, on the surface can improve their therapeutic benefit through altered biodistribution and pharmacokinetics (F). Drug molecules are covalently linked to the surface and can be released in response to a specific stimulus condition in the environment. For example, a hydrogel loaded with drug contains a cross-linked polymer which possesses a reversible, collapse, and re-swell response to changes in temperature and/or pH. This could be particularly useful for patients with complications in bone formation or wound healing due to systemic conditions such as diabetes or osteoporosis.

13.3 Multipotent Stromal Cells

MSCs, often referred to as adult stem cells, exhibit some stem cell–like behaviors; they exhibit the ability to divide asymmetrically producing two daughter cells: one is a new stem cell with the capability for self-renewal without losing the proliferative capacity with each cell division, and the second is progenitor cell, which has the ability for differentiation and proliferation, but not the capability for self-renewal.

MSCs are functionally associated with tissue regeneration and repair [16]. These are the primary cells recruited to repair damaged bone and are responsible for the tissue regeneration that occurs very early at the bone-implant interface, by differentiating to form the osteoblasts that establish osseointegration through the deposition of the first calcified tissue onto the implant surface [17,18].

For these cells to be used effectively in the tissue engineering process, it is generally accepted that they need to be differentiated into the desired lineage before they are introduced into patients. There are concerns that clinical use of undifferentiated stem cells may lead to uncontrolled proliferation in the formation of cancerous tissue [19]. MSCs can be found in many different tissues, bone marrow (BM), muscle, and adipose.

13.3.1 Bone Marrow–Derived MSCs

The most widely studied source of MSCs is BM. There are two main stem cell populations with distinct progenies that have been identified within BM. These are hematopoietic stem cells (HSCs) and the MSCs. The HSCs include all lymphoid and myeloid lineages that finally produce blood-circulating cells and organ resident cells of the immune response. They are tightly regulated by the BM microenvironment in terms of renewal and differentiation [20]. MSCs are rare in BM representing ~1 in 10,000 nucleated cells. BM-derived MSCs are characterized by their ability to adhere to plastic. MSCs isolated from adult human BM are the obvious choice for bone tissue engineering as these cells are relatively easy to procure and have high potential for osteogenic differentiation. They are also able to be expanded greatly in culture to levels of cell numbers that are therapeutically useful, whilst retaining their growth and multi-lineage potential due to the self-renewing nature of the undifferentiated cells.

The osteogenic differentiation of MSCs can be initiated in vitro by dexamethasone (dex), L-ascorbic acid and inorganic phosphate [21]. During osteogenic differentiation, several genes are upregulated such as alkaline phosphatase (ALP), collagen 1A1 (Col1A1), osteopontin (OP), and bone sialoprotein (BSP) [22]. Runt-related transcription factor 2 (Runx2) has also been identified as a transcription factor involved in osteogenic differentiation, and some genes such as SOX9 are downregulated at the point of osteogenic determination [23]. Runx2 controls various signaling pathways that are used to control the expression of many genes such as the late osteogenic marker, OP [24]. OP is a human gene product that functions as an extracellular, structural glycoprotein in the organic component of bone. Type 1 collagen is the major protein component in bone tissue. It provides a template for mineral deposition. Procollagen is its soluble precursor molecule and comprises two pro-α1 and one pro-α2 chains that are encoded by Col1A1 and collagen 1A2 (Col1A2) genes [25].

The cell source used for tissue engineering should have the capacity to first proliferate and then differentiate in vitro in a manner that can be reproducibly controlled. The type and maturity of the cells substantially influence the robustness and the nature of the regenerative response.

A number of specific surface markers have been proposed as a way of identifying and concentrating MSCs from a population of BMSCs. The identification of these markers may contribute to the clinical application of BMSCs. STRO-1 is a candidate for an MSC-related cell surface marker. For example, Fukiage et al. showed that a fraction of BMSCs with the anti-STRO-1 antibody was shown to be rich in MSC-like cells [26]. Meinel et al. reported that the MSC nature of the cells can also be determined on the basis of the expression of CD105/endoglin (a marker of MSCs, formerly known as SH2) and CD71 (a receptor expressed in the proliferating cells) as well as the lack of expression of CD31 and CD34 (markers of cells that are of endothelial or hematopoietic origin) [27]. Fukiage et al. found CD106 to be a reliable indicator for osteogenic differentiation potential of BMSC [26].

Osteoblasts derived from osteoprogenitor cells take on the task of extracellular matrix formation. The osteoblasts secrete collagen and other proteins which are gradually mineralized. With ECM formation, there is an increase in the resistance of mass transfer to the secreting cells. The osteoblasts buried within the mineralized matrix differentiate into osteocytes. These osteocytes now obtain mass transport via compression-derived fluid convection through much smaller channels, the lacuno-canaliculi (lacunae: approximately 5 μm radius, canaliculi: approximately 200 nm radius). Despite the reduced porosity, the fluid flow in the lacuno-canaliculi remains significant to carry out mass transport of signaling molecules, nutrients to, and wastes from, the osteocytes due to stronger induced fluid compression effects. In addition, this generates an important mechanical stimulus in the mineralized matrix. The osteocytes act as mechanosensory cells by producing signaling molecules in the bone tissue in response to load-induced strains.

ALP activity is an important marker of early osteogenic differentiation, bone formation, and matrix mineralization. It indicates the commitment of mesenchymal stem cells toward the osteogenic phenotype [28]. The synthesis of ECM components such as osteocalcin and type-1 collagen can also provide evidence of the ability of cells to exhibit an osteoblastic phenotype as does any observed deposits of calcified matrix in vitro. An increase in the expression of osteogenic markers SPP1, RUNX2, and BSP as well as the osteogenic promoter WNT5A may be used to indicate an osteogenic response [7].

13.4 Bioreactors

Stem cells can be used to engineer a variety of tissues in bioreactors using natural or synthetic scaffolds to produce precisely shaped constructs. It is necessary to increase the limited number of stem cells harvested from a single patient to be used in an autologous transplantation. Human MSCs are the cells of choice for a number of reasons. Firstly, MSCs are self-renewable, multipotent progenitor cells with the capacity to differentiate into several distinct mesenchymal lineages, including bone, cartilage, adipose, tendon, and muscle tissue [20]. Osteogenic differentiation of MSCs results in the formation of osteoblast cells required for bone growth. They are the primary cells to be recruited to the site of an injury and initiate the healing process through the production of calcified tissue [29]. In this way, MSCs are thought to be the reason why bone can heal itself without scarring [30]. Using adult stem cells in tissue engineering applications also has an advantage since MSC populations expand much more readily than mature cell types [31]. For example, osteoblasts can only be obtained in small numbers and posses low proliferation capacity ex vivo, whereas theoretically unlimited numbers of MSCs can be obtained.

MSCs can be pre-differentiated on applicable scaffolds by addition of osteogenic differentiation factors, such as BMPs or combinations of dexamethasone, β-glycerol phosphate, and ascorbic acid, and then implanted in vivo. However, Ti surfaces have a degree of osteogenic stimulations and might be used to select a population of cells from the MSC population that are predetermined to differentiate along the osteogenic lineage. This could reduce the need for the use of chemical osteogenic factors.

Three-dimensional (3D) tissue generation is one of the main challenges of tissue engineering. Static cell culture has limitations because the cells tend to proliferate in monolayers. This leads to much reduced supplies of oxygen and nutrients, resulting in lower cell densities compared to known tissue structures. The supply of oxygen and nutrients is dependent on diffusion to the cells as expressed by the Fick equation:

$$J_i = -D_i \frac{dc}{dx}$$

where
J_i is the diffusive flux of component i (mol m^{-2} s^{-1})
D_i is the is the diffusion coefficient or diffusivity of i (m^2 s^{-1})
c (for ideal mixtures) is the concentration (mol m^{-3})
x is the diffusion distance (m)

The diffusion coefficient D is proportional to the squared velocity of the diffusing particles, which depends on the viscosity of the fluid and the size of the particles according to the Stokes–Einstein relation. For biological molecules, the diffusion coefficients normally range from 10^{-11} to 10^{-10} m^2 s^{-1}.

The use of bioreactors has been found to be necessary in order to provide the physiochemical requirements for the large mass of cells that are required to generate 3D tissues for graft purposes.

The most commonly used bioreactors for 3D cell growth are perfusion culture techniques, spinner flasks, and rotating wall vessel systems. Diederichs et al. used a ZRP® system for cultivating MSCs on a ceramic scaffold coated in zirconium oxide [32]. This system provides a rotating bed with a perfusion mode where the scaffolds are moved through a cell culture medium and the overlay atmosphere alternatively in order to optimize oxygen and nutrient supply to the cells. The ceramic scaffold is microporous which ensures a laminar flow of media to the cells. ZrO_2-based materials are already proven to be biocompatible and have been in use as a clinical biomaterial for some time. The reactor is placed inside a GMP breeder which provides a sterile and tempered atmosphere which is controlled using integrated sensors for pO_2, pH, and temperature. Cultivation under this dynamic system provided greater proliferation rates compared to static cultivation measured by glucose consumption and lactate production.

DNA assays, taken from in vitro cell constructs, can be used indirectly to reflect the cell proliferation and viability. The quantity of DNA present directly reflects the cell number. However, when the cells are embedded in a mineralized matrix, the DNA cannot be completely released into a solution. Glucose consumption and lactate production, therefore, are better methods of estimating cell proliferation in tissue-engineered bone. This method was used by Diederichs et al. in the determination of cell viability using a rotating bed bioreactor system to cultivate MSCs. This study was able to enhance cellular proliferation and bone lineage-specific growth within a ceramic macroporous scaffold [32].

Each type of tissue construct, for example, bone, requires a unique bioreactor design. This means that a tissue-specific bioreactor needs to be designed on the basis of a comprehensive understanding of biological and engineering aspects. These engineering aspects include the properties of reliability, reproducibility, scalability, and safety. In regard to bone implants, it is necessary to generate 3D tissue constructs from isolated and proliferated cells. The process involves seeding of cells on to microporous scaffolds, sufficient nutrient supply within the resulting tissues, and mechanical stimulation of the developing tissues. Scaling up the procedure to clinically relevant sizes has the additional challenge of maintaining cell viability in larger constructs, during the culture and upon grafting. The bioreactor must enable effective control of environmental factors such as oxygen tension, pH, temperature and shear stress and allow aseptic operation during sampling. The bioreactor system should be an automated process in order to provide reproducible, statistically relevant data from basic studies and for the subsequent routine manufacture of tissues for clinical applications.

The first step is the proliferation of cells in order to establish a sufficient number of primary undifferentiated cells to take forward to 3D tissue culture. Expansion from a small number of cells resulting from a biopsy, by many orders of magnitude, is required. The cell seeding process establishes a 3D culture on a microporous scaffold or similar construct for implantation.

The seeding needs to be at high density and with a homogeneous distribution of cells; however, typically static seeding in petri dishes yields engineered constructs comprising a thin tissue-like layer at the base of the scaffold due to gravitational settling of the cells. Convective mixing in a spinner flask bioreactor or convective flow in perfusion bioreactors improves cell seeding density and homogeneity, thereby improving the tissue architecture of the construct. High initial cell densities have also been associated with enhanced tissue formation features, including bone mineralization [33].

Of critical importance to the tissue construct is the mass transfer of oxygen and nutrients to and the removal of toxic metabolites from the cells. This is typically achieved by diffusion in 2D constructs at low cell density without an internal blood/nutrient supply. However, as the tissue constructs become larger than 100–200 μm, oxygen cannot be supplied by diffusion alone. In fact, the difficulty of supplying cells at high densities (around 1×10^6 cells cm^{-3}) in 3D constructs with sufficient nutrients is the major limitation for engineering large tissues. Thus, bioreactors that perfuse cell-seeded constructs have beneficial effects on cell survival, proliferation, and tissue formation in the scaffold. Alternatively, a pre-vascularization system consisting of a capillary network (and a lymphatic network) can be engineered as part of the creation of a larger structure. The main vascularization strategies include the following:

1. Delivery of vascular endothelial growth factors (VEGFs) by controlled release delivery systems such as microspheres or cell-containing microcapsules.
2. Construction of fabricated micro-channel networks can be used to generate arborised networks with the capacity for perfusion.
3. Use of materials which induce blood vessel to grow without exogenous growth factors.
4. Endothelial cells can be seeded within the tissue construct to create a vascular network.

The most studied approach is through the use of growth factors to promote angiogenesis, the body's natural process for generating new capillaries from the endothelial cells of preexisting blood vessels. This creates a localized, functional increase in vascularization. Synthetic polymer delivery vehicles can be employed to provide targeted delivery of the relevant angiogenic growth factor. Incorporation of the proteins into polymer capsules has the advantage of protecting the proteins from destruction and provides localized, sustained release of the growth factors. Microspheres can be made of copolymers of lactic and glycolic acid (PLGA) or hydrogels made from alginate using the double-emulsion technique. These systems allow for the localized delivery of angiogenic molecules such as VEGF. VEGF is a central molecule in the stimulation of the proliferation, migration, and tube formation of endothelial cells in vitro, as well as being capable of producing a pronounced angiogenic response in vivo.

Bone cells at various stages of differentiation are highly responsive to mechanical forces and fluid shear. Fluid flow enhances the cells potential to form bone, and this can be seen through the expression of alkaline phosphate and an increase in density of the tissue matrix. The cells adapt to their perceived environment by altering their matrix attachment or orientation.

Diffusional limitations of mass transport can severely curtail efforts to engineer bone. To minimize diffusional gradients within engineered constructs, two systems involving flow and mixing of culture medium can be used. These are typically spinner flasks and perfusion cartridges.

Spinner flasks promote mass transport at the construct surfaces, while molecular diffusion remains the dominant mechanism within the construct. Meinel et al. observed that bone formation occurred mainly at the construct periphery, within a 0.5–1 mm thick zone [27]. Medium perfusion, however, has a distinct advantage because it promotes mass transfer throughout the construct volume. This method utilizes interstitial flow that is a physiological requirement for bone formation as it provides the appropriate nutritional support. Diffusional distances are minimized, in comparison to the spinner flask, thereby maintaining the concentrations of nutrients and metabolites at the desired levels. In addition, the perfusion bioreactor exposes the cultured cells to controllable hydrodynamic shear. Consequently, bone formation occurs throughout the construct volume. A study by Bancroft et al. supports the use of a nondegradable material for strong mineralization in perfused cartridges [34]. The orientation of the formed bone occurs in the direction of the fluid flow. Meinel et al. also reported that bone rods and plates extended in the direction of axial flow from one construct surface to another [27].

A final approach that might be of use is the preparation of constructs using the previous methods to achieve a good distribution of osteoblastic cells and mineral matrix deposition followed by implantation into the host at a site distant from the final site of implant. This would allow the construct to develop a capillary system under the natural environment in vivo before final implantation into the site of need. This is a process that has been successfully used by Warnke et al. [35] in the reconstruction of a mandible in a patient after surgery to remove a large tumor.

13.5 Summary

Human MSCs have demonstrated the potential to differentiate into multiple mesenchymal tissues including bone. These cells can be isolated from BM and enriched for osteogenic characteristics following contact with modified Ti surfaces in vivo.

Bioreactors provide a mechanically active environment that can be used to control and improve engineered tissue structures. The first step in the process involves the proliferation of cells to establish a tissue culture.

The cells could then be seeded on either a 3D microporous scaffold or to the 2D surface of a Ti implant. Adequate perfusion in the bioreactor has a beneficial effect on cell survival, differentiation, and proliferation within the scaffold. Mechanical stimulation is also a necessary requirement, thus incorporating both biological and biomechanical factors, for bone tissue formation. Potential microporous scaffolds include collagen or phosphate and phosphosilicate bioglasses. These are fully degradable and are replaced by the engineered tissue. These scaffolds have low mechanical strength so have limited use in load-bearing areas such as the maxillofacial region. In this regard, solid Ti provides a suitable material for use in load-bearing structures of the body, and it integrates completely with surrounding bone. The TiO_2 surface can allow cell adhesion and growth without causing chemical harm to surrounding living tissue. Surface modifications such as roughness and wettability enhance the osteogenic integration and may improve fixation times. In addition, further modifications to the surface could be used to incorporate drug release mechanisms that can help to prevent infection and encourage the healing process following implantation.

References

1. Steinemann S.G., Titanium—The material of choice? *Periodontol. 2000*, 1998. **17**: 7–21.
2. Schuler M. et al., Biomedical interfaces: Titanium surface technology for implants and cell carriers. *Nanomedicine*, 2006. **1**(4): 449–463.
3. Sullivan R.M., Implant dentistry and the concept of osseointegration: A historical perspective. *J. Calif. Dent. Assoc.*, 2001. **29**(11): 737–745.
4. Somasundaran P., *Surface Chemical Characteristics and Adsorption Properties of Apatite*. New York: Plenum Press, 1984: pp. 129–149.
5. Pierschbacher M.D. and Ruoslahti E., Cell attachment activity of fibronectin can be duplicated by small synthetic fragments of the molecule. *Nature*, 1984. **309**(5963): 30–33.
6. Boehm H.P., Acidic and basic properties of hydroxylated metal oxide surfaces. *Discuss. Faraday Soc.*, 1971. **52**: 264–275.
7. Wall I. et al., Modified titanium surfaces promote accelerated osteogenic differentiation of mesenchymal stromal cells in vitro. *Bone*, 2009. **45**(1): 17–26.
8. Le Guehennec L. et al., Osteoblastic cell behaviour on different titanium implant surfaces. *Acta Biomater.*, 2008. **4**(3): 535–543.
9. Brett P.M. et al., Roughness response genes in osteoblasts. *Bone*, 2004. **35**(1): 124–133.

10. Rønold H.J., Lyngstadaas S.P., and Ellingsen J.E., Analysing the optimal value for titanium implant roughness in bone attachment using a tensile test. *Biomaterials*, 2003. **24**(25): 4559–4564.

11. Buser D. et al., Influence of surface characteristics on bone integration of titanium implants—A histomorphometric study in miniature pigs. *J. Biomed. Mater. Res.*, 1991. **25**(7): 889–902.

12. Esposito M. et al., Biological factors contributing to failures of osseointegrated oral implants. (II). Etiopathogenesis. *Eur. J. Oral. Sci.*, 1998. **106**(3): 721–764.

13. Mill M., Straumann launches SLActive implant technology, setting new standards in implant dentistry. http://www.medicalnewstoday.com/articles/31120.php (accessed September 25, 2005).

14. Saeki S., Kuwahara N., Nakata M., and Kaneko M., Upper and lower critical solution temperatures in poly(ethylene glycol) solutions. *Polymer*, 1976. **17**: 685–689.

15. Dalsin J.L., Lin L.J., Tosatti S., Voros J., Textor M., and Messersmith P.B., Protein resistance of titanium oxide surface modified by biologically inspired mPEG-DOPA. *Langmuir*, 2005. **21**: 640–646.

16. Caplan A. All MSCs are pericytes? *Cell Stem Cell*, 2008. **3**(3): 229–230.

17. Cuomo A., Virk M., Petrigliano F., Morgan E., and Lierberman J., Mesenchymal stem cell concentration and bone repair: Potential pitfalls from bench to bedside. *J. Bone Joint Surg. Am.*, 2009. **91**: 1073–1083.

18. Davies J., Understanding peri-implant endosseous healing. *J. Dent. Educ.*, 2003. **67**(8): 932–949.

19. Davies J., Mechanisms of endosseous integration. *Int. J. Prosthodont.*, 1998. **11**(5): 391–401.

20. Park P.C., Selvarajah S., Bayani J., Zielenska M., and Squire J.A., Stem cell enrichment approaches. *Semin. Cancer Biol.*, 2007. **17**: 257–264.

21. Muraglia A., Cancedda R., and Quarto R., Clonal mesenchymal progenitors from human bone marrow differentiate in vitro according to a hierarchical model. *J. Cell Sci.*, 2000. **113**(7): 1161–1166.

22. Marie P.J. and Fromigue O., Osteogenic differentiation of human marrow-derived mesenchymal stem cells. *Regen. Med.*, 2006. **1**(4): 539–548.

23. Jonason J.H. et al., Post-translational Regulation of Runx2 in bone and cartilage. *J. Dent. Res.*, 2009. **88**(8): 693–703.

24. Weiss M.J., Lafferty M.A., Slaughter C., Raducha M., and Harris H., Isolation and characterization of a cDNA encoding a human liver/bone/kidney-type alkaline phosphatase. *Proc. Natl. Acad. Sci. USA*, 1986. **87**: 2220–2224.

25. Van der Dolder J. and Jansen J.A., Enrichment of osteogenic cell populations from rat bone marrow stroma. *Biomaterials*, 2007. **28**(2): 249–255.

26. Fukiage K. et al., Expression of vascular cell adhesion molecule-1 indicates the differentiation potential of human bone marrow stromal cells. *Biochem. Biophys. Res. Commun.*, 2008. **365**(3): 406–412.

27. Meinel L. et al., Bone tissue engineering using human mesenchymal stem cells: Effects of scaffold material and medium flow. *Ann. Biomed. Eng.*, 2004. **32**(1): 112–122.

28. Lian J.B. and Stein G.S., The developmental stages of osteoblast growth and differentiation exhibit selective responses of genes to growth factors (TGF beta 1) and hormones (vitamin D and glucocorticoids). *J. Oral. Implantol.*, 1993. **19**(2): 95–105; discussion 136–137.

29. Thibodeau G.A. and Patton K.T., *Anatomy and Physiology*, 6th edn. St Louis, MO: Mosby Elsevier.
30. Sommerfeldt D.W. and Rubin C.T., Biology of bone and how it orchestrates the form and function of the skeleton. *Eur. Spine J.*, 2001. **10**: 86–95.
31. Alhadlaq A. and Mao J.J., Tissue-engineered osteochondral constructs in the shape of an articular condyle. *J. Bone Joint Surg. Am.*, 2005. **87**(5): 936–944.
32. Diederichs S. et al., Dynamic cultivation of human mesenchymal stem cells in a rotating bed bioreactor system based on the ZRP platform. *Biotechnol. Prog.*, 2009, **25**(6): 1762–1771.
33. Martin I., Wendt D., and Heberer M., The role of bioreactors in tissue engineering. *Trends Biotechnol.*, 2004. **22**(2): 80–86.
34. Bancroft G.N. et al., Fluid flow increases mineralized matrix deposition in 3D perfusion culture of marrow stromal osteoblasts in a dose-dependent manner. *Proc. Natl. Acad. Sci. USA*, 2002. **99**(20): 12600–12605.
35. Warnke P.H. et al., Growth and transplantation of a custom vascularised bone graft in a man. *Lancet*, 2004. **364**(9436):766–770.

14

Stem Cell Response to Biomaterial Topography

Luong T.H. Nguyen, Susan Liao, Casey K. Chan,
and Seeram Ramakrishna

CONTENTS

Stem cells possess their abilities to self-renew and differentiate into specific lineages in response to appropriate signals. In the body, the presentation of chemical, topographical, mechanical, and electrical stimuli cues in the surrounding fluid and extracellular matrix (ECM) provides the guidance for stem cell response. Recently, much attention has been paid to topographical cues as a useful tool for controlling stem cell fate and guiding stem cell differentiation. The use of topographically guided scaffolds for supporting stem cells has a great advantage over the use of chemical reagents due to allowing cells to grow and differentiate in the absence of potentially harmful inducing reagents. Different sizes of topographies, ranging from macro-, micro-, to nanoscale features, have their own potentials to affect cellular behaviors. Besides, surface geometries such as smooth/flat, groove/ridge, pit/pore, and disordered/ordered structure significantly influence cellular morphology, proliferation, and differentiation ability. Additionally, dimensionality of the substrate topographies (two dimension [2D] vs. three dimension [3D]) has a remarkable impact on cell fate and signaling cascade. Understanding influences of topographical cues on stem cell behaviors plays an important role in designing suitable scaffolds for tissue regeneration to expedite expansion and differentiation of stem cells without changing the plasticity nature of stem cells.

14.1 Introduction

Stem cells have the capacity to self-renew and differentiate into various lineages which would have tremendous potentials in research and clinical treatment. There are two main types of stem cells commonly used: embryonic stem cells (ESCs) and mesenchymal stem cells (MSCs). ESCs possess high proliferative capability (Evans and Kaufman 1981; Martin 1981) and are able to form three embryonic germ layers (endoderm, mesoderm, and ectoderm) (Thomson et al. 1998). However, using ESCs encounters some problems related to safety issues such as ESC rejection and the risk of tumorigenicity as well as ethical and religious issues such as the harvesting of donor oocytes and destruction of the blastocyst. Meanwhile, MSCs can be readily obtained with less controversy from many sources such as bone marrow (Pittenger et al. 1999), umbilical cord blood (Wang et al. 2004), and adipose tissue (Jeon et al. 2008). Besides, autologous MSCs surmount immune rejection and carcinogenesis (Chen et al. 2008).

In the body, almost all tissue cells reside in a 3D environment where various stimuli cues in the surrounding fluid and ECM provide the guidance for cellular responses. There are different cues which can regulate stem cell fate: (1) chemical cues (through soluble supplements added into the environment), (2) topographical cues (surface geometry, size effect, and dimensionality),

(3) mechanical cues (various stress stimuli applied to substrate and/or cell construct), and (4) electrical or electromagnetic cues (through application of electrical or electromagnetic currents/fields to stimulate substrate and/or cell construct) (Ngiam et al. 2011) (Figure 14.1A).

Currently, the most common method used to control the differentiation ability of stem cells is the chemical treatment such as the supply of cytokines and growth factors. However, it would be better to induce the stem cell differentiation without those soluble factors because of their noticeable disadvantages: (1) generally derived from animal sources, (2) expensive, (3) difficult to control the optimal concentration for an efficient differentiation without side effects, and (4) not fully understood mechanism (Lee et al. 2010).

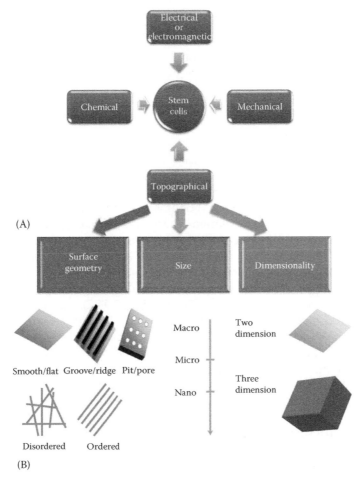

FIGURE 14.1
(A) Various cues regulating stem cell fates: chemical, topographical, and electrical or electromagnetic cues; and (B) three domains of topographical cues: surface geometry, size, and dimensionality.

In this chapter, we focus on analyzing the influence of topographical cues on controlling stem cell fate and guiding stem cell differentiation. As shown in Figure 14.1B, topographical cues include surface geometry (smooth/flat, groove/ridge, pit/pore, and disordered/ordered structure), size (from macro-, micro-, to nanoscale), and dimensionality (2D vs. 3D). Understanding these issues plays an important role in designing suitable scaffolds for tissue regeneration.

14.2 Surface Geometry

With numerous numbers of biomaterials applied in research and medical treatment up to now, many types of surface geometry has been recognized. The surface geometry has significantly affected stem cell behaviors such as morphology, proliferation, and differentiation.

14.2.1 Cell Morphology and Proliferation

14.2.1.1 Morphology

The substrate geometry has significantly affected the morphology and orientation of stem cells. Compared to substrates patterned with hexagonal posts, the ones with square posts were shown to be more influential in directing the orientation of MSCs (Poellmann et al. 2010). On substrates with aligned grooves, MSCs were elongated and aligned along the nanogrooved patterns as indicated by focal adhesions and actin cytoskeleton network (Zhu et al. 2005; Fujita et al. 2009; Yim et al. 2010). The microtubule alignment and elongation were also observed on nanogrooved patterns, whereas a square-localized distribution of α-tubulin/actin fibers was induced by nanogrid patterns (Martino et al. 2009) (Figure 14.2). Similarly, MSCs on microgrooved patterns were oriented parallel to the grooves, while the cells on unpatterned and 5-μm-deep micropitted films were oriented randomly and formed clumps (Kenar et al. 2008). STRO-1+ skeletal stem cells also stretched along the microgrooved surface, while the cells on a flat surface spread and had a tendency toward a fibroblast-like morphology (Kantawong et al. 2010). Using biocompatible shape-memory surfaces, Le et al. demonstrated a switch of MSC morphology from a highly aligned to a stellate shape in response to a surface transformation between a channel array and a planar surface (Le et al. 2011). Not only cell body shapes, but the nuclei of MSCs were also dramatically changed in interaction with the nanogrooves (Chalut et al. 2010). The nuclei were aligned with the cytoskeleton and elongated parallel to the nanografting.

Similar to the regular patterns, MSCs cultured on aligned nanofibers fabricated by electrospinning technique showed elongated cell morphology

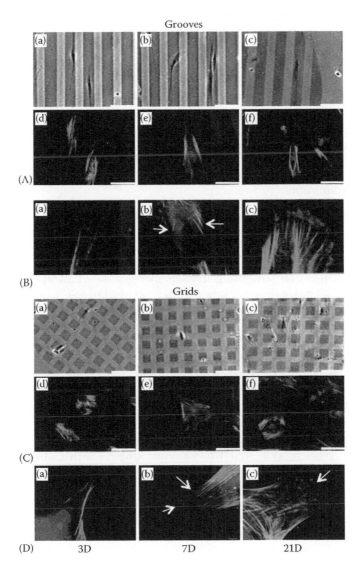

FIGURE 14.2
Representative images of MSCs seeded on a C-H groove and grid nanopatterns (20× and 60× immersion oil objective; scale bar: 100 μm) for 3, 7, and 21 days. (A) Grooves, (a–c) bright field images, (d–f) α-tubulin; (B) Grooves, (a–c) vinculin; (C) Grids, (a–c) bright field images, (d–f) α-tubulin; (D) Grids, (a–c) vinculin. This figure indicated the influence of surface geometries (groove and grid) on the MSC morphology. (Reprinted from Martino, S. et al., *Tissue Eng. Part A*, 15(10), 3139, 2009, Copyright 2009, with permission from Mary Ann Liebert.)

and aligned orientation along the main fibrous axis (Meinel et al. 2009; Ma et al. 2010; Nathan et al. 2011). The nuclear geometry was also considerably affected by the fibrous orientation (Nathan et al. 2011). Changes in nuclear shape were quantified by nuclear aspect ratio (NAR) and orientation angle (θ). The NAR was the ratio of lengths of the long to short axes of the nucleus, and θ was the angle between the direction of fiber alignment and the nuclear long axis. When MSCs were seeded on aligned scaffolds, the values of NAR and θ were approximately 1.7° and 21°, respectively. Meanwhile, with the cells on random scaffolds, those values were 1.5° and 45°, respectively. As such, the nuclei were more elongated and parallel to the aligned fibers.

14.2.1.2 Proliferation

The proliferation of stem cells has been regulated by the substrate geometry. Cell adhesion and proliferation of MSCs were shown to be superior on the island-patterned poly-L-lactic acid (PLLA) membranes than on sunken-patterned membranes (Lee et al. 2009). MSCs on 3D texture PDMS scaffolds with $10\,\mu m$ diameter/height posts increased cell numbers compared to those on 3D smooth scaffolds (Mata et al. 2009). ESCs on the nanohybrids of fibrous poly(ε-caprolactone) (PCL) and Ca-deficient hydroxyapatite nanocrystals (d-HAp) proliferated at the same rate as the cells on standard tissue culture polystyrene (TCPS) plates without signs of cytotoxicity as well as maintained pluripotency markers (Bianco et al. 2009). As such, the nanohybrid scaffolds provided adequate supports for the ESC proliferation. In another study, the proliferation of ESC-derived progeny was reduced by 15-μm-, but not 5-μm-high, microprojections (Biehl et al. 2009). Besides, the porous silicon with $10{-}100\,nm$ nanostructures was shown to inhibit the stem cell proliferation (Osminkina et al. 2011). Meanwhile, nanoporous structures on microfibers significantly increased the cell spreading and proliferation of MSCs in comparison with smooth microfibers (Moroni et al. 2006).

Taken together, surface geometry of biomaterials is able to guide the morphology of cell bodies as well as cell nuclei. Besides, it can affect the proliferation of stem cells in such different ways.

14.2.2 Cell Differentiation

14.2.2.1 With Soluble Differentiation Inducers

Choosing suitable surface geometry has been reported to regulate the differentiation ability of stem cells in the presence of soluble differentiation inducers. To induce the osteogenic differentiation of stem cells, some chemical cues have been widely used such as growth factors (bone morphogenetic protein [BMP] and fibroblast growth factor [FGF]) and osteogenic supplements (dexamethasone [dex], β-glycerophosphate, ascorbic acid, and vitamin D). Over a time course of the osteogenic differentiation of MSCs, there are expressions of different osteoblastic markers which reflect the maturation and the mineralization of the ECM including early marker (alkaline phosphatase [ALP]), middle markers (osteopontin [OPN], bone sialoprotein [BSP], etc.), and later marker (osteocalcin [OCN], etc.). MSCs cultured on hydrogenated amorphous carbon (C-H) film designed with nanogrooved/grid patterns preserved their multidifferentiation properties during culture (Martino et al. 2009). On micropatterned pNIPAM films (channels with 10 μm groove width, 2 μm ridge width, and 20 μm depth), MSCs showed higher specific ALP activity than on TCPS plates (Ozturk et al. 2009). Similarly, MSCs on 3D texture PDMS scaffolds with 10 μm diameter/ height posts had higher expression of ALP and similar expression of OCN in comparison with those on 3D smooth scaffolds (Mata et al. 2009). MSCs cultured on PCL nanofibers also increased ALP activity compared to the cells on smooth PCL surfaces (Ruckh et al. 2010). Besides, the results of Von Kossa and calcium staining as well as SEM-EDX indicated acceleration in the calcium phosphate mineralization. The intra- and extracellular levels of OPN and OCN were also increased on the nanofibrous scaffolds after 3 weeks of culture. Highly porous nanofibrous PLLA was shown to be an excellent candidate scaffold for osteochondral defect repair (cartilage/bone composite) (Hu et al. 2009). The calcium deposition and the gene expression of early chondrogenic commitment marker Sox-9 were higher on these scaffolds than on smooth films. Compared to the random nanofibers, the aligned nanofibers enhanced the extent of mineralization (Ma et al. 2010). However, there was no significant difference between random and aligned nanofibers in ALP activity as well as the expression levels of OPN and OCN. Interestingly, the PLLA nanofibers were shown to recover osteogenic abilities of MSCs which had been severely affected by serial passage (from passage 2 to passage 8) (Nguyen et al. 2011).

In the neuronal differentiation of MSCs or neuron stem cells in the presence of nerve growth factor (NGF) or retinoic acid, aligned nanofibrous meshes obviously increased the elongation along the major fibrous axis as well as the expression of neuronal markers (such as microtubule-associated protein 2 [MAP2]) in comparison with random nanofibrous meshes (Cho et al. 2010; Lim et al. 2010). Similar observations were found on nanografting of 350 nm width with unpatterned surfaces as controls (Yim et al. 2007).

As such, designing proper surface geometry would help to increase the differentiation of MSCs toward certain cell lineages which is first induced by soluble differentiation inducers.

14.2.2.2 Without Soluble Differentiation Inducers

Surface geometry has not only regulated the differentiation in the presence of soluble differentiation supplements but also has been able to direct the differentiation of stem cells without those factors. Instead of using soluble supplements, 2 wt.% puerariae powder—a natural herb with the ability to support the osteogenesis—was combined with PCL microgrooved surface (Kantawong et al. 2010). This scaffold influenced the osteogenic differentiation of STRO-1+ skeletal stem cells, enriched in osteoprogenitors, and represented in an earlier bone cell population, as indicated by positive OCN staining. Meanwhile, control flat PCL sheets as well as grooved PCL and 1 wt.% puerariae powder-impregnated grooved PCL did not express this marker. The importance of topography in directing the osteogenic differentiation of MSCs was obviously shown in the study of Grazino et al. (2007). MSCs cultured on substrates enriched with pits or microcavities (80–120 μm in diameter and 40–100 μm in depth) expressed higher ALP activity as well as released larger amounts of BMP-2 and VEGF into the culture medium than on the smooth scaffolds. When transplanted into rats, this type of scaffold elicited superior bone formation. In another study, Jang et al. showed that microgrooved surfaces preferred the MSC differentiation into smooth muscle–like cells, whereas flat substrates provided a suitable environment for the differentiation into osteo-related cells (Jang et al. 2011). The neuronal differentiation was able to be induced by nanostructured scaffolds (Lee et al. 2010). The 350 nm ridge/groove patterned arrays alone could effectively and rapidly induce the neuronal differentiation of ESCs without any soluble inducers. The cells were positive for neuronal markers such as Tuj1, HuC/D, and MAP2 on the fifth day.

To clarify the role of surface geometry in directing osteogenic differentiation, MSCs cultured on nanoscale topographies (pitted surfaces and raised islands) were compared with the cells treated with dex (Dalby et al. 2008). Interestingly, the efficiency of bone formation on the nanotopographies was similar to that with dex. Additionally, unlike treated with dex, the cells on the nanoscale substrates did not suppress the angiogenesis which is important to supplying nutrients to new tissues. As such, by altering the matrix shape, a similar cellular response to chemical stimulation can be achieved, maybe by a different mechanism. MSCs treated with dex performed wide-ranging and unspecific actions including a large number of canonical and functional pathways stimulated. Meanwhile, the nanotopography was more selective in the canonical pathways affected. Dalby et al. also showed that the nanoscale disorder of PMMA embossed with 120-nm-diameter, 100-nm-deep nanopits over 1 cm^2 stimulated MSCs to produce bone minerals in vitro in the absence of osteogenic inducers (Dalby et al. 2007b). This approach had a similar efficiency to that of the cells cultured in osteogenic media, but in a distinct differentiation profile.

In the neuronal differentiation, interestingly, MSCs cultured on aligned nanofibers/nanograftings without NGF/retinoic acid also indicated much higher elongation levels than random nanofibers/unpatterned surfaces with those chemical cues (Yim et al. 2007). Besides, without the soluble cues, the aligned nanofibers were shown to be sufficient to drive the activation of canonical Wnt signaling which is crucial for the neurogenesis of stem cells (Cho et al. 2010). These evidences demonstrated that the topographical cues were important in allowing stem cells to differentiate into neuron cells.

14.2.3 Mechanism

Tissue-specific cell differentiation and its mechanism is a key question in cell biology (Lim et al. 2010). The cells possess the same genome and transcription factor pool, but in different parts of the body, they express a vast array of different and distinct phenotypes (Dalby et al. 2008). As such, the topography may play an important role in tissue-specific development through mechanotransductive pathways. Some studies showed that the interaction of topography with stem cells was enhanced in comparison with terminally differentiated cells (Getzenberg 1994; Hart et al. 2007). The exquisite sensitivity of the stem cells to their microenvironment provided further evidence with the importance of topographical environment to tissue-specific differentiation.

By observing dynamic behaviors of living MSCs, Fujita et al. indicated that the retracting phase of cell protrusions played an important role in cell alignment (Fujita et al. 2009). Cell protrusions perpendicular to the nanogrooves demonstrated a trend toward retraction more rapid than those parallel to the pattern. Meanwhile, the filopodia probing phase did not cause significant influence on cell alignment.

The ability of surface geometry to dictate selective differentiation may be derived from its influence on cell morphology. MSCs cultured on hydrogels with lamellar or hexagonal surface wrinkles in the same media (1:1 adipogenic induction/osteogenic media) exhibited different morphologies (Guvendiren and Burdick 2010). The cells on the lamellar wrinkles spread following the shape of the pattern, whereas the cells on the hexagonal wrinkles remained rounded with low spreading. As a result, the osteogenic differentiation was observed on the lamellar pattern, while the adipogenic differentiation was obtained on the hexagonal pattern.

The changes in the gene expression of stem cells on a specific topography leading to targeted differentiation might be caused by nuclear deformation and focal contacts (Thomas et al. 2002; Guvendiren and Burdick 2010). It has been hypothesized that stem cells can recognize several focal adhesion points on the surface. The changes in focal adhesions can lead to changes in cytoskeletal organization (Chew et al. 2008). In recent years, there have been increasing evidence which indicate filopodia (or microspikes) as a crucial factor to contact guidance (Dalby et al. 2004b; Yim et al. 2010). Filopodia are driven by actin cytoskeleton and forming integrin-containing adhesions. The cells use filopodia in response to nanotopography for the specific activation of adhesion and cytoskeleton-related pathways (Hart et al. 2007). The guidance of filopodia by nanotopography may alter mechanical forces within the cells which can affect the interphase nucleus organization and genomic regulation (Dalby et al. 2004a, 2007a).

Further studies should be done to clarify the intermediate steps connecting intracellular structural changes and signaling pathways in response to external topographical cues.

14.3 Size

In addition to topographical properties which play an important role in cellular functions, the topographical size also affects stem cell fate. Its influence is normally nonlinear manner.

14.3.1 Cell Morphology and Proliferation

14.3.1.1 Morphology

The morphology of stem cells has been shown to be controlled by the size of the topography. The gap size of hydrogel micropatterned array considerably affected the morphology of MSCs. Narrow gaps (5 μm) between posts directed cell extension and tended to elongate the cell bodies in the direction of gaps (Poellmann et al. 2010). In comparison with MSCs on conventional coarse grained surface, the cells on refine grained surface enhanced the attachment and spreading in the initial stages (up to 24 h) (Estrin et al. 2011). MSCs cultured on nanofibrous scaffolds also attached and spread more rapidly than on microfibrous PCL (Binulal et al. 2010). On microfibers, MSCs were more globular with less cell spreading, whereas the cells on nanofibers were significantly flattened and spread even after 4 h. In contrast, the spreading of MSCs was confined on nanopillar surfaces (~20 nm) compared to micropillar surfaces (Brammer et al. 2011). Within nanoscale, neuron stem cells on random nanofibrous matrix with 283 nm diameter stretched multidirectionally, whereas the cells on larger fibers (749 and 1452 nm) extended along a single fibrous axis (Christopherson et al. 2009). When the diameters of vertically oriented TiO_2 nanotubes increased from 15 to 100 nm, the morphology of MSCs cultured in the medium with or without osteogenic supplements was dramatically changed (Park et al. 2007; Oh et al. 2009). With the diameters smaller than 30 nm, the cells were somewhat flat and rounded, and they became progressively elongated as the diameters were increased to 50 nm or beyond. Extraordinary elongation was found on the nanotubes with 70 and 100 nm diameters.

14.3.1.2 Proliferation

Different fibrous diameters (1, 4, 10, 21, and 270 μm) were fabricated by elec-trospinning to investigate the influence of diameter on the attachment and proliferation of MSCs (Moroni et al. 2006). The results showed that 10 μm was an optimum diameter to support these cellular properties. In another study, MSCs increased the adhesion and proliferation on the island-patterned PLLA membranes with diameter of 100 μm in comparison with those of 60 μm (Lee et al. 2009).

With different grain sizes, the proliferation of MSCs on the 24 and 50 nm substrates was significantly lower than on the substrates of 200 and 1500 nm after 3 or more consecutive days of culture (Dulgar-Tulloch et al. 2009). After 7 or more days of culture, the cell proliferation was enhanced on 200 nm substrates in comparison with 1500 nm substrates.

The diameters of random nanofibrous type I collagen also affected the migration of MSCs (Shih et al. 2006). The distances of cell migration were 56.7%, 37.3%, and 46.3% for 50–200, 200–500, and 500–1000 nm, respectively, in comparison with those on the TCPS plates. The adhesion, migration, and proliferation of neuron stem cells on random nanofibrous matrix with 283 nm diameter were reduced compared to those on larger fibers (749 and 1452 nm) (Christopherson et al. 2009).

With sub-100 nm scale, nanopillar surfaces (~20 nm) significantly increased the adhesion and growth of MSCs compared to micropillar surfaces (Brammer et al. 2011). In another study, vertically oriented ZrO_2 and TiO_2 nanotubes with sub-100 nm range in diameter (15–100 nm) were used to demonstrate the size-specific reaction to these nanoscale patterned surfaces of MSCs cultured in the medium with or without osteogenic inducers (Park et al. 2007; Bauer et al. 2009; Oh et al. 2009). Cell adhesion and proliferation were shown to be dependent on the nanotube size with a maximum cellular activity at the diameter of 15 nm. A strong decay in the adhesion and prolif-eration was observed on the nanotubes with larger than 50 nm in diameter.

In addition to the diameter of grains/fibers/tubes/pillars, the poros-ity of scaffolds has been known to affect cell behaviors. With different porosities of β-tricalcium phosphate (β-TCP) ceramics investigated (25%, 65%, and 75%), the higher porosities increased the protein content of MSCs (Kasten et al. 2008). MSCs seeded on coralline hydroxyapatite (HA) scaf-folds with 500 μm pores had higher final cell number than with 200 μm (Bjerre et al. 2011).

14.3.2 Cell Differentiation

14.3.2.1 With Soluble Differentiation Inducers

The topographical size has shown its role in modulating the differentiation of stem cells in the presence of soluble differentiation agents. HA granules calcified from red algae, varying in grain sizes (10–100, 200–500, and 600–1000 μm), were found to affect the osteogenic potential of MSCs (Weißenböck et al. 2006). The cells grown on the smallest grain size had significantly higher ALP activity than on the others. Titania surfaces with different grain sizes (50, 200, and 1500 nm) were used to culture MSCs in the osteogenic medium (containing dex, ascorbic acid, and glycerol phosphate) (Dulgar-Tulloch et al. 2011). Almost no osteogenic differentiation was detected on 50 nm substrates. Meanwhile, on the 200 nm grain size titania, the cells differentiated earlier with greater extent than on either 50 or 1500 nm substrates, as evidenced by ALP activity and calcium content. In contrast, without those osteogenic supplements, the differentiation was not observed on any grain size surface. In another study with a different type of materials, the osteogenic differentiation of MSCs which was evaluated by ALP activity and the expression of OPN and OCN was not affected by the diameters of random nanofibrous type I collagen (50–200, 200–500, and 500–1000 nm) (Shih et al. 2006). Sub-100 nm diameters of vertically oriented TiO_2 nanotubes significantly affected the osteogenic differentiation of MSCs in the medium with osteogenic inducers (Park et al. 2007). After 2 weeks of culture, OCN was obviously observed on 15 nm nanotubes, but rarely detectable on 100 nm nanotubes.

In the presence of retinoic acid, MSCs cultured on nanografting of 350 nm width showed higher expression of neuronal markers than on micropatterns (Yim et al. 2007). Within nanoscale, it was also shown that nanofibers with 480 nm diameter yielded the highest fraction of Tuj1+ cells for the neuronal differentiation of neuron stem cells compared to nanofibers with the diameters of 260 and 930 nm (Lim et al. 2010). Under the differentiation condition (in 1 μM retinoic acid and 1% FBS), neuron stem cells on random nanofibrous matrix with larger diameters (749 and 1452 nm) preferentially differentiate into neuronal lineage, whereas the cells on the smaller diameter, 283 nm, preferentially differentiate into oligodendrocytes (Christopherson et al. 2009) (Figure 14.3). It might be due to more elongated neuron stem cells on the larger diameters.

Many studies have shown the influence of porosity on the differentiation of stem cells. The cells cultured on HA composites in the presence of dex after 2 weeks showed higher ALP activity and OCN protein than those without dex, and osteogenic ability increased with increasing porosity (from 30% to 70%) (Okamoto et al. 2006). The results of in vivo implantation of the resultant bone tissue grafts after 1 week indicated an expanded bone formation for only the scaffold with 70% porosity. The bone formation was only observed morphologically in the constructs with 50% and 30% porosity after 2 and 4 weeks, respectively. In another study, the osteogenic differentiation of MSCs occurred primarily and earlier in cultured coralline HA with 200 μm pores than in 500 μm pores as indicated by ALP activity and other osteogenic markers (Bjerre et al. 2011). For the chondrogenic differentiation, PCL scaffolds with a pore size range of 370–400 μm provided an

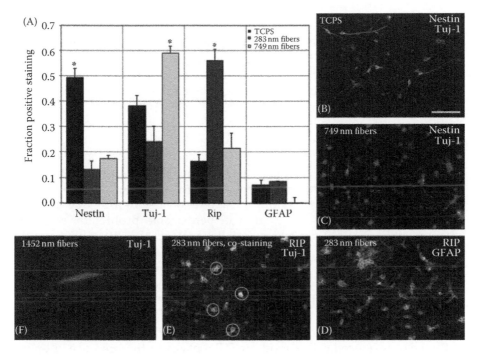

FIGURE 14.3
Immunostaining of neuron stem cells cultured on various substrates: TCPS (B), 749 nm fibers (C), 283 nm fibers (D) and 1452 nm fibers (F). Cell nuclei were stained using DAPI. Quantification of staining results is shown in (A). Different markers were used for staining: Tuj-1 (neuronal marker), RIP (oligodendrocyte marker), GFAP (astrocyte marker), and nestin (neural stem/progenitor cell marker). All images were captured 200×, with scale bar = 100 µm. Circled cells on 283 nm fibrous mesh were double-stained for RIP and Tuj-1 (E). This figure demonstrated the importance of nanofibrous diameters (283, 749, and 1452 nm) to the differentiation of neuron stem cells. (Reprinted from *Biomaterials*, 30(4), Christopherson, G.T., Song, H., and Mao, H.Q., The influence of fiber diameter of electrospun substrates on neural stem cell differentiation and proliferation, 556–564, Copyright 2009, with permission from Elsevier.)

optimum environment for the differentiation of adipose-derived stem cells (ASCs) in comparison with other smaller pore size groups (90–105, 190–220, and 300–320 µm) as evidenced by the expression of positive markers (type I collagen and Sox-9) and safari-O staining (Oh et al. 2010). Similarly, fibrous meshes with pores in microscale considerably enhanced the chondrogenic differentiation of MSCs compared to the meshes with nanoscale pores as shown by specific chondrogenic markers (Shanmugasundaram et al. 2011). In contrast, different porosities (25%, 65%, and 75%) of β-TCP ceramics did not influence the osteogenic differentiation of MSCs (Kasten et al. 2008). Besides, different pore sizes of poly(D,L-lactic-co-glycolic acid) foams in the range of 150–710 µm did not affect the osteogenic differentiation of MSCs (Ishaug et al. 1997).

14.3.2.2 Without Soluble Differentiation Inducers

In the absence of soluble differentiation inducers, topographical size has played an important role in directing stem cells to a certain targeted lineage. The grooved topographies of identical groove depths with different widths (10, 25, and 100 µm) were used to investigate the influence of groove width on osteogenic differentiation of MSCs (Biggs et al. 2008). Grooved/ridge arrays of 100 µm were found to induce the upregulation of genes related to skeletal development and increase osteo-specific function, which were not indicated by 10 µm grooved/ridge arrays. Similarly, on 100-µm-wide grooved substrates, STRO-1+ skeletal stem cells were stimulated to differentiation into osteoblasts through the upregulation of osteo-specific genes, Ets and Stat1 (Biggs et al. 2009). On the contrary, the cells on the 10-µm-wide grooved substrates induced the upregulation of peroxisome proliferator-activated receptor γ (PPARγ) which acts sequentially to trigger the differentiation into adipocytes.

C-H grooved patterns with different width/spacing ridges including 80/40, 40/30 and 30/20 µm were used to demonstrate the role of topographical size on the neuronal differentiation of MSCs in the absence of differentiating agents (D'Angelo et al. 2010). Interestingly, 2.0% ± 0.5% Tuj1+ cells were observed on the 40/30 µm pattern which accounted for approximately 40% of the neural-induced cells found on TCPS plates in the presence of brain-derived neurotrophic factor (BDNF). The cells on this pattern also showed long-neurite-like protrusions. As such, the 40/30 µm pattern was sufficient to induce cellular elongation and trigger cell signaling associated with neural phenotypes.

The ability of disorder nanoscale to induce the osteogenic differentiation of MSCs was shown in many studies. Gadegaard et al. performed an optimization experiment, and the results showed that the arrays of pillars 35 nm tall with a diameter of 193 nm and a disorder of ±30 nm provided optimal conditions for the stimulation of MSCs to differentiate into osteoblastic lineage (Gadegaard et al. 2008). In another study, MSCs on nanopillar surfaces (~20 nm) were induced to form large 3D cell aggregates which had upregulated osteogenic specific matrix components compared to micropillar surfaces (Brammer et al. 2011). Without any inducing reagents, the cells on nanopillar structures significantly enhanced alizarin red quantification and the OPN expression. In another study, the ECM aggregates of MSCs were also found on vertically oriented TiO_2 nanotubes (Oh et al. 2009). The protein aggregates were abundant on 30-nm-diameter nanotubes, but much less on the nanotubes with larger diameters (70 and 100 nm). However, the aggregates did not increase the osteogenic function as shown in the study of Brammer et al. (2011). The larger diameters significantly induced the selective differentiation into osteoblastic-like cells instead. Meanwhile, the cells on 30-nm-diameter nanotubes had no noticeable differentiation as indicated by ALP, OPN, and OCN expression. The response trend of MSCs in the osteogenic differentiation in increasing nanotube diameters in this study (without osteogenic inducers) was different from those in the study of Park et al. (with osteogenic inducers) (Park et al. 2007) where the cells on the 100-nm-diameter nanotube had lower osteogenic potential than on the 15 nm diameter.

14.3.3 Mechanism

It was hypothesized that for fibers with diameters closer to the cell dimensions, the fibrous diameter would become less predominant in limiting cellular properties which could lead to a significantly higher amount of cells on the scaffolds (Moroni et al. 2006). However, when the fibrous diameters come too close to the cell dimension (~20 μm), the cell motility was enhanced (Smilenov et al. 1999; Dalby et al. 2003), resulting in a lower proliferation rate.

The higher protein content of MSCs, which indicates higher proliferation, on the higher porosities might be due to the facilitated transport of oxygen and nutrients (Kasten et al. 2008).

The ability of topographical size to control the differentiation of stem cells may be explained through a phenomenon called contact guidance. Reducing cellular spreading and focal adhesion formation on 10-μm-wide grooved substrates may induce the adipocyte differentiation, while increasing cell spreading on 100-μm-wide grooved substrates may induce the osteogenic function (Biggs et al. 2009). The role of nanotopography in controlling stem cell fate has been discussed in many previous literatures (Curtis and Wilkinson 1999; Ravichandran et al. 2009; McNamara et al. 2010). The nanofibers were more effective in promoting cell spreading might be due to the fact that integrins on the cell surface engaged with proper adhesion proteins on the matrix and transformed to a spread configuration (Takagi 2004; Salsmann et al. 2006; Streuli 2009). Another possibility was that the protein conformation on the nanofibers might be more favorable for cell–matrix interaction (Binulal et al. 2010). Additionally, the large surface-to-volume ratio of nanostructures could promote the adsorption of serum proteins, provide more binding sites for receptors, and thus enhance cellular functions (Stevens and George 2005).

On the other hand, although MSCs on nanopillar arrays (~20 nm) were less spreading than on micropillar arrays, the osteogenic differentiation was induced on the nanostructure in the absence of osteogenic inducers (Brammer et al. 2011). One possible explanation was the nanoscale disorder of the nanopillar array. It was claimed that the disordered nanotopography of random circular nanostructures induced changes in the adhesion formations and cell morphology affecting cytoskeleton tension and mechanotransductive pathways which could induce and direct the osteogenic differentiation of osteoprogenitor cells (Dalby et al. 2007b).

The changes in cell morphology might be a reason resulting in the significant induction of osteogenic differentiation MSCs on vertically oriented

TiO$_2$ nanotubes with 100 nm diameter (Oh et al. 2009). The cells on these tubes were extraordinarily more elongated than on 30-nm-diameter nanotubes. This elongated morphology might cause cellular cytoskeletal tension and stress on MSCs. Another explanation was that on the larger diameters, because of the reduced cell adhesion and proliferation, much less ECM aggregates were induced. As a result, the cells were forced to elongate and stretch to search for protein aggregates and thus guided/forced to specifically differentiate into osteoblast cells. It was shown that when the stem cells were stressed, the osteogenic differentiation was enhanced (Dalby et al. 2006, 2007b; Engler et al. 2006). This suggested mechanism agreed to the general notion that when the stem cells were busy with the adhesion and proliferation, their functions were expected to be reduced (Pittenger et al. 1999; McBeath et al. 2004), and when the stem cells were stressed, they had a tendency to differentiate to a specific lineage to accommodate the stress.

The opposite results and a different proposed mechanism were shown in the study of Park et al. (2007). It was hypothesized that the maximum cellular activity on 15-nm-diameter nanotubes might be due to the fact that this scale is comparable to the integrin size (Park et al. 2007; Bauer et al. 2009). Using electron micrographic imaging of individual integrin molecules, the predicted size of surface occupancy by the head of an integrin heterodimer was about 10 nm in diameter (Takagi et al. 2002). Cell–matrix interactions are mostly mediated by integrin. The adhesion of cells to ECM causes clustering of integrins into focal adhesion complexes and consequently activates intracellular signaling cascades into the nucleus and cytoskeleton (Schlaepfer et al. 1999; DeMali et al. 2003). As such, the 15 nm spacing would allow or force the integrin clustering into the possible closet packing which could lead to optimum integrin activation. The dramatically reduced cellular activity on 100 nm nanotubes might be caused by the prevention of integrin clustering and focal adhesion complex formation (Park et al. 2007). These were different from the findings of Oh et al. (2009). However, it is noted that the medium used for culturing MSCs in these two studies was different. Park et al. supplemented osteogenic inducers in the medium, whereas a normal medium without any osteogenic inducer was used by Oh et al. As such, the presence of osteogenic inducers might cause different responses of MSCs to nanotube diameters. Further studies should be done to investigate this phenomenon.

The higher osteogenic abilities were observed on the higher porosity constructs might be due to their high pore interconnection (Okamoto et al. 2006). The significantly higher chondrogenic differentiation was also found on scaffolds with larger pore sizes (370–400 μm compared to 90–320 μm, microscale compared to nanoscale) (Oh et al. 2010; Shanmugasundaram et al. 2011). One possible reason was that the higher permeability of oxygen into the scaffolds resulted in higher oxygen tension which can increase the chondrogenesis in vitro. Additionally, the larger pore sizes could promote cell aggregation and cell–cell contact which are necessary for chondrogenesis. Although some other studies showed that the porosity (25%–75%) and pore size (150–710 µm) did not affect the osteogenic differentiation of MSCs, suitable values of those factors are significantly important because of their roles in facilitating blood vessel development for oxygen/nutrient supplies and tissue ingrowth in vivo (Ishaug et al. 1997; Kasten et al. 2008).

14.4 Dimensionality

Up to now, most studies have been performed on 2D scaffolds such as TCPS plates or other 2D substrates because of their convenient handling and analyzing (Martins and Kolega 2006; Lee et al. 2008). However, almost all tissue cells reside in a 3D environment in the body. The lack of structural cues in 2D cultures has forced cells to adapt to flat surfaces which can cause alterations and reduce cellular functions (Zhang et al. 2005; Albrecht et al. 2006; Lee et al. 2007, 2008). So, developing 3D scaffold and studying cellular responses to the 3D culture play key roles in tissue engineering.

14.4.1 Cell Proliferation

ESCs and MSCs could be expanded and maintained in their undifferentiated state in 3D scaffolds for at least 20 days without subculturing as indicated by surface markers (Li et al. 2003; Cao et al. 2010, Storm et al. 2010). While MSCs grown on 2D surfaces (TCPS plates) had a higher initial rate and stopped their proliferation as the confluence reached (~15th day), the cells in porous 3D polyethylene terephthalate (PET) matrix grew at a stable and lower rate, but lasted for a longer time and reached a higher final cell number (Xie et al. 2001; Cao et al. 2010). Other 3D scaffolds also demonstrated greater cell numbers than TCPS plates (Ishaug et al. 1997; Hosseinkhani et al. 2006a,b; Liu et al. 2006). During cell culture, MSCs formed connections with surrounding porous matrix leading to 3D aggregates (Maeng et al. 2010). A bioreactor perfusion system has been often used to culture 3D scaffolds for promoting the culture of stem cells. The attachment and proliferation of cardiac stem cells in 3D collagen-PGA (nanoscale in fibrous diameter and microscale in pore size) were significantly enhanced by the bioreactor system compared to 3D static and 2D systems (Hosseinkhani et al. 2010).

14.4.2 Cell Differentiation

14.4.2.1 With Soluble Differentiation Inducers

After 3 weeks of culture in the osteogenic medium, ESCs in 3D porous PLGA had significantly higher expressions of ALP and OCN than those on TCPS plates coated with gelatin (2D control) (Tian et al. 2008). The constructs were subsequently implanted in rabbits, and the new bone formation was detected at the implantation site after 4 weeks. In the 3D self-assembling peptide scaffold, ESCs also significantly promoted the differentiation into osteoblast-like cells in comparison with 2D scaffolds (TCPS plates) as evidenced by higher levels of ALP, collagen I marker, OPN, and calcium phosphate deposits (Garreta et al. 2006). Similarly, culturing MSCs in a 3D fibrous network formed by self-assembly of peptide-amphiphile molecules led to higher expressions of ALP and OCN than on TCPS plates (Hosseinkhani et al. 2006a,b). In the presence of transforming growth factor (TGF)-β1, compared to ESCs grown on TCPS plates coated with gelatin (2D substrates), the cells in poly(ethylene glycol) (PEG)-based hydrogels upregulated the expression of cartilage-relevant markers (Hwang et al. 2006). In these 3D scaffolds, decreasing levels of ectodermal and endodermal cell lineage markers suggested that the ability of ESCs to differentiate into different lineages might be limited after the chondrogenic differentiation. 3D collagen-coated PLGA scaffolds were used for the hepatocyte differentiation of MSCs (Li et al. 2010). Most hepatocyte-specific markers (albumin [ALB], α-fetoprotein, cytokeratin 18, hepatocyte nuclear factor 4α, and cytochrome P450) of MSCs cultured in the 3D scaffolds were expressed 1 week earlier than those on TCPS plates. As such, the 3D collagen-coated PLGA scaffolds provided a suitable environment for the hepatocyte differentiation of MSCs. Many other studies have shown the importance of 3D scaffolds to improving the differentiation of ESCs into hepatocytes (Levenberg et al. 2003; Imamura et al. 2004; Baharvand et al. 2006). 3D fibrous matrices were also used for the hematopoietic differentiation of ESCs (Li et al. 2003). The cells in the 3D matrix had higher expression of c-kit, a cell surface marker of hematopoietic progenitor, suggesting a better-directed hematopoietic differentiation than those on TCPS plates.

14.4.2.2 *Without Soluble Differentiation Inducers*

Three-dimensional MSCs/biphasic calcium phosphate (BCP) constructs were formed after few days of culturing MSCs with BCP particles (140–200 μm) (Cordonnier et al. 2010). In proliferative medium, the 3D constructs expressed much more BSP and BMP-2 than the cells on TCPS plates. As such, BCP particles alone could induce the osteogenic differentiation of MSCs which is useful for in vivo conditions. However, the osteogenic ability of the 3D constructs was slightly lower than 2D cultures in the osteogenic medium. Using porous β-TCP matrix helped to significantly increase the osteogenic potential of ASCs (Marino et al. 2010). The cells cultured in this 3D scaffold in proliferative medium had considerably higher ALP, OPN, and OCN expression than on TCPS plates in differentiation medium. As such, β-TCP matrices showed their efficiency in triggering the differentiation of ASCs toward osteoblastic lineage without the need for differentiative media.

14.4.3 Mechanism

Due to larger surface provided by the interconnected pores and the 3D nature of the constructs, 3D scaffolds had a more significant growth potential than 2D surfaces, which is important to long-term culture of stem cells (Ishaug et al. 1997; Cao et al. 2010).

A 3D culture system had many advantages over conventional 2D system. This architecture closely mimics tissue architecture in vivo. Besides, in comparison with 2D environments, 3D environments provide another dimension for cell adhesion and external mechanical inputs. The enhancement in cell adhesion remarkably affects cell contraction, integrin ligation, and related intracellular signals (Roskelley et al. 1994; Knight et al. 2000). 3D matrices also provide an optimal template for cell–cell communication and cell migration within a 3D space (Tian et al. 2008). Cell–cell and cell–matrix interactions are important in regulating many cell signaling pathways to efficiently promote tissue development (Geiger et al. 2001). As such, engineered 3D structures might give appropriate structured patterns to cells leading to the correct ECM organization and the increase of their proliferation and differentiation abilities (Kale et al. 2000; Ferrera et al. 2002).

To guide tissue growth in well-defined configurations, it is necessary to fabricate 3D scaffolds with hierarchical geometrical resolutions. With well-designed, this structure can promote particular biological responses solely through selective physical stimuli without destabilizing surrounding delicate biochemical environment, and consequently can enhance and direct specific cell behaviors and tissue growth (Mata et al. 2009). 3D scaffolds with bimodal pore size distributions might also be ideal scaffolds for tissue engineering with enhanced cellular responses and vascularization (Salerno et al. 2010; Sundararaghavan et al. 2010).

14.5 Conclusion

Different aspects of biomaterial topography including surface geometry, size, and dimensionality showed their significant importance to controlling stem cell behaviors such as morphology, proliferation, and differentiation. The biomaterial topography not only can modulate the cell differentiation in the presence of soluble differentiation agents but also can induce the differentiation in the absence of those factors. As such, understanding the responses of stem cells to biomaterial topography can provide a valuable guidance to design "smart" biomaterials which are able to direct the tissue-specific differentiation for stem cells. Based on the earlier findings, the development of 3D aligned nanofibrous scaffolds might bring novel impacts on directing stem cell fate due to the integration of the advantages of nanostructures, aligned patterns, and 3D environment.

Acknowledgment

The authors thank National Research Foundation (NRF), Technion grant (R-398-001-065-592), Singapore, for the financial support.

References

Albrecht, D. R., G. H. Underhill, T. B. Wassermann, R. L. Sah, and S. N. Bhatia. 2006. Probing the role of multicellular organization in three-dimensional microenvironments. *Nat Methods* 3 (5): 369–375.

Baharvand, H., S. M. Hashemi, S. Kazemi Ashtiani, and A. Farrokhi. 2006. Differentiation of human embryonic stem cells into hepatocytes in 2D and 3D culture systems in vitro. *Int J Dev Biol* 50 (7): 645–652.

Bauer, S., J. Park, J. Faltenbacher, S. Berger, K. von der Mark, and P. Schmuki. 2009. Size selective behavior of mesenchymal stem cells on ZrO(2) and TiO(2) nanotube arrays. *Integr Biol* 1 (8–9): 525–532.

Bianco, A., E. Di Federico, I. Moscatelli, A. Camaioni, I. Armentano, L. Campagnolo, M. Dottori, J. M. Kenny, G. Siracusa, and G. Gusmano. 2009. Electrospun poly(ε-caprolactone)/Ca-deficient hydroxyapatite nanohybrids: Microstructure, mechanical properties and cell response by murine embryonic stem cells. *Mater Sci Eng C* 29 (6): 2063–2071.

Biehl, J. K., S. Yamanaka, T. A. Desai, K. R. Boheler, and B. Russell. 2009. Proliferation of mouse embryonic stem cell progeny and the spontaneous contractile activity of cardiomyocytes are affected by microtopography. *Dev Dyn* 238 (8): 1964–1973.

Biggs, M. J., R. G. Richards, N. Gadegaard, C. D. Wilkinson, R. O. Oreffo, and M. J. Dalby. 2009. The use of nanoscale topography to modulate the dynamics of adhesion formation in primary osteoblasts and ERK/MAPK signalling in STRO-1+ enriched skeletal stem cells. *Biomaterials* 30 (28): 5094–5103.

Biggs, M., R. Richards, S. McFarlane, C. Wilkinson, R. Oreffo, and M. Dalby. 2008. Adhesion formation of primary human osteoblasts and the functional response of mesenchymal stem cells to 330 nm deep microgrooves. *J R Soc Interface* 5 (27): 1231–1242.

Binulal, N. S., M. Deepthy, N. Selvamurugan, K. T. Shalumon, S. Suja, U. Mony, R. Jayakumar, and S. V. Nair. 2010. Role of nanofibrous poly(caprolactone) scaffolds in human mesenchymal stem cell attachment and spreading for in vitro bone tissue engineering—Response to osteogenic regulators. *Tissue Eng Part A* 16 (2): 393–404.

Bjerre, L., C. Bünger, A. Baatrup, M. Kassem, and T. Mygind. 2011. Flow perfusion culture of human mesenchymal stem cells on coralline hydroxyapatite scaffolds with various pore sizes. *J Biomed Mater Res A* 97 (3): 251–263.

Brammer, K. S., C. Choi, C. J. Frandsen, S. Oh, and S. Jin. 2011. Hydrophobic nanopillars initiate mesenchymal stem cell aggregation and osteo-differentiation. *Acta Biomater* 7 (2): 683–690.

Cao, Y., D. Li, C. Shang, S. T. Yang, J. Wang, and X. Wang. 2010. Three-dimensional culture of human mesenchymal stem cells in a polyethylene terephthalate matrix. *Biomed Mater* (Bristol, England) 5 (6): 065013.

Chalut, K. J., K. Kulangara, M. G. Giacomelli, A. Wax, and K. W. Leong. 2010. Deformation of stem cell nuclei by nanotopographical cues. *Soft Matter* 6 (8): 1675–1681.

Chen, Y., J. Z. Shao, L. X. Xiang, X. J. Dong, and G. R. Zhang. 2008. Mesenchymal stem cells: A promising candidate in regenerative medicine. *Int J Biochem Cell Biol* 40 (5): 815–820.

Chew, S. Y., R. Mi, A. Hoke, and K. W. Leong. 2008. The effect of the alignment of electrospun fibrous scaffolds on Schwann cell maturation. *Biomaterials* 29 (6): 653–661.

Cho, Y. I., J. S. Choi, S. Y. Jeong, and H. S. Yoo. 2010. Nerve growth factor (NGF)-conjugated electrospun nanostructures with topographical cues for neuronal differentiation of mesenchymal stem cells. *Acta Biomater* 6 (12): 4725–4733.

Christopherson, G. T., H. Song, and H. Q. Mao. 2009. The influence of fiber diameter of electrospun substrates on neural stem cell differentiation and proliferation. *Biomaterials* 30 (4): 556–564.

Cordonnier, T., P. Layrolle, J. Gaillard, A. Langonné, L. Sensebé, P. Rosset, and J. Sohier. 2010. 3D environment on human mesenchymal stem cells differentiation for bone tissue engineering. *J Mater Sci Mater Med* 21 (3): 981–987.

Curtis, A. and C. Wilkinson. 1999. New depths in cell behaviour: Reactions of cells to nanotopography. *Biochem Soc Symp* 65: 15–26.

D'Angelo, F., I. Armentano, S. Mattioli, L. Crispoltoni, R. Tiribuzi, G. Cerulli, C. Palmerini, J. Kenny, S. Martino, and A. Orlacchio. 2010. Micropatterned hydrogenated amorphous carbon guides mesenchymal stem cells towards neuronal differentiation. *Eur Cell Mater* 20: 231–244.

Dalby, M. J., A. Andar, A. Nag, S. Affrossman, R. Tare, S. McFarlane, and R. O. C. Oreffo. 2008. Genomic expression of mesenchymal stem cells to altered nanoscale topographies. *J R Soc Interface* 5 (26): 1055–1065.

Dalby, M. J., S. Childs, M. O. Riehle, H. J. H. Johnstone, S. Affrossman, and A. S. G. Curtis. 2003. Fibroblast reaction to island topography: Changes in cytoskeleton and morphology with time. *Biomaterials* 24 (6): 927–935.

Dalby, M. J., N. Gadegaard, P. Herzyk, D. Sutherland, H. Agheli, C. D. W. Wilkinson, and A. S. G. Curtis. 2007a. Nanomechanotransduction and interphase nuclear organization influence on genomic control. *J Cell Biochem* 102 (5): 1234–1244.

Dalby, M. J., N. Gadegaard, R. Tare, A. Andar, M. O. Riehle, P. Herzyk, C. D. W. Wilkinson, and R. O. C. Oreffo. 2007b. The control of human mesenchymal cell differentiation using nanoscale symmetry and disorder. *Nat Mater* 6 (12): 997–1003.

Dalby, M. J., D. McCloy, M. Robertson, H. Agheli, D. Sutherland, S. Affrossman, and R. O. C. Oreffo. 2006. Osteoprogenitor response to semi-ordered and random nanotopographies. *Biomaterials* 27 (15): 2980–2987.

Dalby, M. J., M. O. Riehle, H. Johnstone, S. Affrossman, and A. S. G. Curtis. 2004a. Investigating the limits of filopodial sensing: A brief report using SEM to image the interaction between 10 nm high nano-topography and fibroblast filopodia. *Cell Biol Int* 28 (3): 229–236.

Dalby, M. J., M. O. Riehle, D. S. Sutherland, H. Agheli, and A. S. G. Curtis. 2004b. Changes in fibroblast morphology in response to nano-columns produced by colloidal lithography. *Biomaterials* 25 (23): 5415–5422.

DeMali, K. A., K. Wennerberg, and K. Burridge. 2003. Integrin signaling to the actin cytoskeleton. *Curr Opin Cell Biol* 15 (5): 572–582.

Dulgar-Tulloch, A. J., R. Bizios, and R. W. Siegel. 2009. Human mesenchymal stem cell adhesion and proliferation in response to ceramic chemistry and nanoscale topography. *J Biomed Mater Res A* 90 (2): 586–594.

Dulgar-Tulloch, A., R. Bizios, and R. Siegel. 2011. Differentiation of human mesenchymal stem cells on nano- and micro-grain size titania. *Mater Sci Eng C* 31 (2): 357–362.

Engler, A. J., S. Sen, H. L. Sweeney, and D. E. Discher. 2006. Matrix elasticity directs stem cell lineage specification. *Cell* 126 (4): 677–689.

Estrin, Y., E. P. Ivanova, A. Michalska, V. K. Truong, R. Lapovok, and R. Boyd. 2011. Accelerated stem cell attachment to ultrafine grained titanium. *Acta Biomater* 7 (2): 900–906.

Evans, M. and M. Kaufman. 1981. Establishment in culture of pluripotential cells from mouse embryos. *Nature* 292: 154–156.

Ferrera, D., S. Poggi, C. Biassoni, G. R. Dickson, S. Astigiano, O. Barbieri, A. Favre, A. T. Franzi, A. Strangio, A. Federici, and P. Manduca. 2002. Three-dimensional cultures of normal human osteoblasts: Proliferation and differentiation potential in vitro and upon ectopic implantation in nude mice. *Bone* 30 (5): 718–725.

Fujita, S., M. Ohshima, and H. Iwata. 2009. Time-lapse observation of cell alignment on nanogrooved patterns. *J R Soc Interface* 6 (Suppl 3): S269–S277.

Gadegaard, N., M. J. Dalby, M. O. Riehle, and C. D. W. Wilkinson. 2008. Optimizing substrate disorder for bone tissue engineering of mesenchymal stem cells. *J Vac Sci Technol B: Microelectron Nanometer Struct* 26 (6): 2554.

Garreta, E., E. Genové, S. Borrós, and C. E. Semino. 2006. Osteogenic differentiation of mouse embryonic stem cells and mouse embryonic fibroblasts in a three-dimensional self-assembling peptide scaffold. *Tissue Eng* 12 (8): 2215–2227.

Geiger, B., A. Bershadsky, R. Pankov, and K. M. Yamada. 2001. Transmembrane cross-talk between the extracellular matrix—Cytoskeleton crosstalk. *Nat Rev Mol Cell Biol* 2 (11): 793–805.

Getzenberg, R. H. 1994. Nuclear matrix and the regulation of gene expression: Tissue specificity. *J Cell Biochem* 55 (1): 22–31.

Graziano, A., R. d'Aquino, M. G. Cusella-De Angelis, G. Laino, A. Piattelli, M. Pacifici, A. De Rosa, G. Papaccio, and J. Najbauer. 2007. Concave pit-containing scaffold surfaces improve stem cell-derived osteoblast performance and lead to significant bone tissue formation. *PloS One* 2 (6): e496.

Guvendiren, M. and J. A. Burdick. 2010. The control of stem cell morphology and differentiation by hydrogel surface wrinkles. *Biomaterials* 31 (25): 6511–6518.

Hart, A., N. Gadegaard, C. D. W. Wilkinson, R. O. C. Oreffo, and M. J. Dalby. 2007. Osteoprogenitor response to low-adhesion nanotopographies originally fabricated by electron beam lithography. *J Mater Sci Mater Med* 18 (6): 1211–1218.

Hosseinkhani, H., M. Hosseinkhani, S. Hattori, R. Matsuoka, and N. Kawaguchi. 2010. Micro and nano-scale in vitro 3D culture system for cardiac stem cells. *J Biomed Mater Res A* 94 (1): 1–8.

Hosseinkhani, H., M. Hosseinkhani, and H. Kobayashi. 2006a. Proliferation and differentiation of mesenchymal stem cells using self-assembled peptide amphiphile nanofibers. *Biomed Mater* (Bristol, England) 1 (1): 8–15.

Hosseinkhani, H., M. Hosseinkhani, F. Tian, H. Kobayashi, and Y. Tabata. 2006b. Osteogenic differentiation of mesenchymal stem cells in self-assembled peptide-amphiphile nanofibers. *Biomaterials* 27 (22): 4079–4086.

Hu, J., K. Feng, X. Liu, and P. X. Ma. 2009. Chondrogenic and osteogenic differentiations of human bone marrow-derived mesenchymal stem cells on a nanofibrous scaffold with designed pore network. *Biomaterials* 30 (28): 5061–5067.

Hwang, N. S., M. S. Kim, S. Sampattavanich, J. H. Baek, Z. Zhang, and J. Elisseeff. 2006. Effects of three-dimensional culture and growth factors on the chondrogenic differentiation of murine embryonic stem cells. *Stem Cells* 24 (2): 284–291.

Imamura, T., L. Cui, R. Teng, K. Johkura, Y. Okouchi, K. Asanuma, N. Ogiwara, and K. Sasaki. 2004. Embryonic stem cell-derived embryoid bodies in three-dimensional culture system form hepatocyte-like cells in vitro and in vivo. *Tissue Eng* 10 (11–12): 1716–1724.

Ishaug, S. L., G. M. Crane, M. J. Miller, A. W. Yasko, M. J. Yaszemski, and A. G. Mikos. 1997. Bone formation by three-dimensional stromal osteoblast culture in biodegradable polymer scaffolds. *J Biomed Mater Res* 36 (1): 17–28.

Jang, J. Y., S. W. Lee, S. H. Park, J. W. Shin, C. Mun, S. H. Kim, D. H. Kim, and J. W. Shin. 2011. Combined effects of surface morphology and mechanical straining magnitudes on the differentiation of mesenchymal stem cells without using biochemical reagents. *J Biomed Biotechnol* 2011: 860652.

Jeon, O., J. W. Rhie, I. K. Kwon, J. H. Kim, B. S. Kim, and S. H. Lee. 2008. In vivo bone formation following transplantation of human adipose-derived stromal cells that are not differentiated osteogenically. *Tissue Eng Part A* 14 (8): 1285–1294.

Kale, S., S. Biermann, C. Edwards, C. Tarnowski, M. Morris, and M. W. Long. 2000. Three-dimensional cellular development is essential for ex vivo formation of human bone. *Nat Biotechnol* 18 (9): 954–958.

Kantawong, F., K. E. Burgess, K. Jayawardena, A. Hart, M. O. Riehle, R. O. Oreffo, M. J. Dalby, and R. Burchmore. 2010. Effects of a surface topography composite with puerariae radix on human STRO-1-positive stem cells. *Acta Biomater* 6 (9): 3694–3703.

Kasten, P., I. Beyen, P. Niemeyer, R. Luginbühl, M. Bohner, and W. Richter. 2008. Porosity and pore size of beta-tricalcium phosphate scaffold can influence protein production and osteogenic differentiation of human mesenchymal stem cells: An in vitro and in vivo study. *Acta Biomater* 4 (6): 1904–1915.

Kenar, H., A. Kocabas, A. Aydinli, and V. Hasirci. 2008. Chemical and topographical modification of PHBV surface to promote osteoblast alignment and confinement. *J Biomed Mater Res A* 85 (4): 1001–1010.

Knight, B., C. Laukaitis, N. Akhtar, N. A. Hotchin, M. Edlund, and A. R. Horwitz. 2000. Visualizing muscle cell migration in situ. *Curr Biol* 10 (10): 576–585.

Le, D. M., K. Kulangara, A. F. Adler, K. W. Leong, and V. S. Ashby. 2011. Dynamic topographical control of mesenchymal stem cells by culture on responsive poly(ε-caprolactone) surfaces. *Adv Mater* 23 (29): 3278–3283.

Lee, J., M. J. Cuddihy, and N. A. Kotov. 2008. Three-dimensional cell culture matrices: State of the art. *Tissue Eng Part B, Rev* 14 (1): 61–86.

Lee, G. Y., P. A. Kenny, E. H. Lee, and M. J. Bissell. 2007. Three-dimensional culture models of normal and malignant breast epithelial cells. *Nat Methods* 4 (4): 359–365.

Lee, M. R., K. W. Kwon, H. Jung, H. N. Kim, K. Y. Suh, K. Kim, and K. S. Kim. 2010. Direct differentiation of human embryonic stem cells into selective neurons on nanoscale ridge/groove pattern arrays. *Biomaterials* 31 (15): 4360–4366.

Lee, I. C., Y. T. Lee, B. Y. Yu, J. Y. Lai, and T. H. Young. 2009. The behavior of mesen-chymal stem cells on micropatterned PLLA membranes. *J Biomed Mater Res A* 91 (3): 929–938.

Levenberg, S., N. F. Huang, E. Lavik, A. B. Rogers, J. Itskovitz-Eldor, and R. Langer. 2003. Differentiation of human embryonic stem cells on three-dimensional poly-mer scaffolds. *Proc Natl Acad Sci USA* 100 (22): 12741–12746.

Li, Y., D. A. Kniss, L. C. Lasky, and S. T. Yang. 2003. Culturing and differentia-tion of murine embryonic stem cells in a three-dimensional fibrous matrix. *Cytotechnology* 41 (1): 23–35.

Li, J., R. Tao, W. Wu, H. Cao, J. Xin, J. Guo, L. Jiang, C. Gao, A. A. Demetriou, D. L. Farkas, and L. Li. 2010. 3D PLGA scaffolds improve differentiation and function of bone marrow mesenchymal stem cell-derived hepatocytes. *Stem Cells Dev* 19 (9): 1427–1436.

Lim, S. H., X. Y. Liu, H. Song, K. J. Yarema, and H. Q. Mao. 2010. The effect of nanofiber-guided cell alignment on the preferential differentiation of neural stem cells. *Biomaterials* 31 (34): 9031–9039.

Liu, H., S. F. Collins, and L. J. Suggs. 2006. Three-dimensional culture for expansion and differentiation of mouse embryonic stem cells. *Biomaterials* 27 (36): 6004–6014.

Ma, J., X. He, and E. Jabbari. 2010. Osteogenic differentiation of marrow stromal cells on random and aligned electrospun poly(L-lactide) nanofibers. *Ann Biomed Eng* 39(1): 14–25.

Maeng, Y. J., S. W. Choi, H. O. Kim, and J. H. Kim. 2010. Culture of human mesenchy-mal stem cells using electrosprayed porous chitosan microbeads. *J Biomed Mater Res A* 92 (3): 869–876.

Marino, G., F. Rosso, G. Cafiero, C. Tortora, M. Moraci, M. Barbarisi, and A. Barbarisi. 2010. Beta-tricalcium phosphate 3D scaffold promote alone osteogenic differen-tiation of human adipose stem cells: In vitro study. *J Mater Sci Mater Med* 21 (1): 353–363.

Martin, G. R. 1981. Isolation of a pluripotent cell line from early mouse embryos cul-tured in medium conditioned by teratocarcinoma stem cells. *Proc Natl Acad Sci USA* 78 (12): 7634–7638.

Martino, S., F. D'Angelo, I. Armentano, R. Tiribuzi, M. Pennacchi, M. Dottori, S. Mattioli, A. Caraffa, G. Cerulli, and J. Kenny. 2009. Hydrogenated amorphous carbon nanopatterned film designs drive human bone marrow mesenchymal stem cell cytoskeleton architecture. *Tissue Eng Part A* 15 (10): 3139–3149.

Martins, G. G. and J. Kolega. 2006. Endothelial cell protrusion and migration in three-dimensional collagen matrices. *Cell Motil Cytoskeleton* 63 (2): 101–115.

Mata, A., E. J. Kim, C. A. Boehm, A. J. Fleischman, G. F. Muschler, and S. Roy. 2009. A three-dimensional scaffold with precise micro-architecture and surface micro-textures. *Biomaterials* 30 (27): 4610–4617.

McBeath, R., D. M. Pirone, C. M. Nelson, K. Bhadriraju, and C. S. Chen. 2004. Cell shape, cytoskeletal tension, and RhoA regulate stem cell lineage commitment. *Dev Cell* 6 (4): 483–495.

McNamara, L. E., R. J. McMurray, M. J. P. Biggs, F. Kantawong, R. O. C. Oreffo, and M. J. Dalby. 2010. Nanotopographical control of stem cell differentiation. *J Tissue Eng* 2010: 120623.

Meinel, A. J., K. E. Kubow, E. Klotzsch, M. Garcia-Fuentes, M. L. Smith, V. Vogel, H. P. Merkle, and L. Meinel. 2009. Optimization strategies for electrospun silk fibroin tissue engineering scaffolds. *Biomaterials* 30 (17): 3058–3067.

Moroni, L., R. Licht, J. de Boer, J. R. de Wijn, and C. A. van Blitterswijk. 2006. Fiber diameter and texture of electrospun PEOT/PBT scaffolds influence human mesenchymal stem cell proliferation and morphology, and the release of incorporated compounds. *Biomaterials* 27 (28): 4911–4922.

Nathan, A. S., B. M. Baker, N. L. Nerurkar, and R. L. Mauck. 2011. Mechano-topographic modulation of stem cell nuclear shape on nanofibrous scaffolds. *Acta Biomater* 7 (1): 57–66.

Ngiam, M., L. T. Nguyen, S. Liao, C. K. Chan, and S. Ramakrishna. 2011. Biomimetic nanostructured materials—Potential regulators for osteogenesis? *Ann Acad Med Singapore* 40 (5): 213–222.

Nguyen, L. T., S. Liao, S. Ramakrishna, and C. K. Chan. 2011. The role of nanofibrous structure in osteogenic differentiation of human mesenchymal stem cells with serial passage. *Nanomedicine* 6 (6): 961–974.

Oh, S., K. S. Brammer, Y. S. J. Li, D. Teng, A. J. Engler, S. Chien, and S. Jin. 2009. Stem cell fate dictated solely by altered nanotube dimension. *Proc Natl Acad Sci USA* 106 (7): 2130–2135.

Oh, S. H., T. H. Kim, G. I. Im, and J. H. Lee. 2010. Investigation of pore size effect on chondrogenic differentiation of adipose stem cells using a pore size gradient scaffold. *Biomacromolecules* 11 (8): 1948–1955.

Okamoto, M., Y. Dohi, H. Ohgushi, H. Shimaoka, M. Ikeuchi, A. Matsushima, K. Yonemasu, and H. Hosoi. 2006. Influence of the porosity of hydroxyapatite ceramics on in vitro and in vivo bone formation by cultured rat bone marrow stromal cells. *J Mater Sci Mater Med* 17 (4): 327–336.

Osminkina, L., E. Luckyanova, M. Gongalsky, A. Kudryavtsev, A. Gaydarova, R. Poltavtseva, P. Kashkarov, V. Timoshenko, and G. Sukhikh. 2011. Effects of nanostructurized silicon on proliferation of stem and cancer cell. *Bull Exp Biol Med* 151(1): 79–83.

Ozturk, N., A. Girotti, G. T. Kose, J. C. Rodríguez-Cabello, and V. Hasirci. 2009. Dynamic cell culturing and its application to micropatterned, elastin-like protein-modified poly(N-isopropylacrylamide) scaffolds. *Biomaterials* 30 (29): 5417–5426.

Park, J., S. Bauer, K. von der Mark, and P. Schmuki. 2007. Nanosize and vitality: TiO₂ nanotube diameter directs cell fate. *Nano Lett* 7 (6): 1686–1691.

Pittenger, M. F., A. M. Mackay, S. C. Beck, R. K. Jaiswal, R. Douglas, J. D. Mosca, M. A. Moorman, D. W. Simonetti, S. Craig, and D. R. Marshak. 1999. Multilineage potential of adult human mesenchymal stem cells. *Science* 284 (5411): 143–147.

Poellmann, M. J., P. A. Harrell, W. P. King, and A. J. Wagoner Johnson. 2010. Geometric microenvironment directs cell morphology on topographically patterned hydrogel substrates. *Acta Biomater* 6 (9): 3514–3523.

Ravichandran, R., S. Liao, C. C. Ng, C. K. Chan, M. Raghunath, and S. Ramakrishna. 2009. Effects of nanotopography on stem cell phenotypes. *World J Stem Cells* 1 (1): 55–66.

Roskelley, C. D., P. Y. Desprez, and M. J. Bissell. 1994. Extracellular matrix-dependent tissue-specific gene expression in mammary epithelial cells requires both physical and biochemical signal transduction. *Proc Natl Acad Sci USA* 91 (26): 12378–12382.

Ruckh, T. T., K. Kumar, M. J. Kipper, and K. C. Popat. 2010. Osteogenic differentiation of bone marrow stromal cells on poly(epsilon-caprolactone) nanofiber scaffolds. *Acta Biomater* 6 (8): 2949–2959.

Salerno, A., D. Guarnieri, M. Iannone, S. Zeppetelli, and P. A. Netti. 2010. Effect of micro- and macroporosity of bone tissue three-dimensional-poly(epsilon-caprolactone) scaffold on human mesenchymal stem cells invasion, proliferation, and differentiation in vitro. *Tissue Eng Part A* 16 (8): 2661–2673.

Salsmann, A., E. Schaffner-Reckinger, and N. Kieffer. 2006. RGD, the Rho'd to cell spreading. *Eur J Cell Biol* 85 (3–4): 249–254.

Schlaepfer, D. D., C. R. Hauck, and D. J. Sieg. 1999. Signaling through focal adhesion kinase. *Prog Biophys Mol Biol* 71 (3–4): 435–478.

Shanmugasundaram, S., H. Chaudhry, and T. L. Arinzeh. 2011. Microscale versus nanoscale scaffold architecture for mesenchymal stem cell chondrogenesis. *Tissue Eng Part A* 17 (5–6): 831–840.

Shih, Y. R. V., C. N. Chen, S. W. Tsai, Y. J. Wang, and O. K. Lee. 2006. Growth of mesenchymal stem cells on electrospun type I collagen nanofibers. *Stem Cells* 24 (11): 2391–2397.

Smilenov, L. B., A. Mikhailov, R. J. Pelham, E. E. Marcantonio, and G. G. Gundersen. 1999. Focal adhesion motility revealed in stationary fibroblasts. *Science* 286 (5442): 1172–1174.

Stevens, M. M. and J. H. George. 2005. Exploring and engineering the cell surface interface. *Science* 310 (5751): 1135–1138.

Storm, M. P., C. B. Orchard, H. K. Bone, J. B. Chaudhuri, and M. J. Welham. 2010. Three-dimensional culture systems for the expansion of pluripotent embryonic stem cells. *Biotechnol Bioeng* 107 (4): 683–695.

Streuli, C. H. 2009. Integrins and cell-fate determination. *J Cell Sci* 122 (Pt 2): 171–177.

Sundararaghavan, H. G., R. B. Metter, and J. A. Burdick. 2010. Electrospun fibrous scaffolds with multiscale and photopatterned porosity. *Macromol Biosci* 10 (3): 265–270.

Takagi, J. 2004. Structural basis for ligand recognition by RGD (Arg-Gly-Asp)-dependent integrins. *Biochem Soc Trans* 32 (Pt 3): 403–406.

Takagi, J., B. M. Petre, T. Walz, and T. A. Springer. 2002. Global conformational rearrangements in integrin extracellular domains in outside-in and inside-out signaling. *Cell* 110 (5): 599–611.

Thomas, C. H., J. H. Collier, C. S. Sfeir, and K. E. Healy. 2002. Engineering gene expression and protein synthesis by modulation of nuclear shape. *Proc Natl Acad Sci USA* 99 (4): 1972–1977.

Thomson, J. A., J. Itskovitz-Eldor, S. S. Shapiro, M. A. Waknitz, J. J. Swiergiel, V. S. Marshall, and J. M. Jones. 1998. Embryonic stem cell lines derived from human blastocysts. *Science* 282 (5391): 1145–1147.

Tian, X. F., B. C. Heng, Z. Ge, K. Lu, A. J. Rufaihah, V. T. W. Fan, J. F. Yeo, and T. Cao. 2008. Comparison of osteogenesis of human embryonic stem cells within 2D and 3D culture systems. *Scand J Clin Lab Invest* 68 (1): 58–67.

Wang, H. S., S. C. Hung, S. T. Peng, C. C. Huang, H. M. Wei, Y. J. Guo, Y. S. Fu, M. C. Lai, and C. C. Chen. 2004. Mesenchymal stem cells in the Wharton's jelly of the human umbilical cord. *Stem Cells* 22 (7): 1330–1337.

Weißenböck, M., E. Stein, G. Undt, R. Ewers, G. Lauer, and D. Turhani. 2006. Particle size of hydroxyapatite granules calcified from red algae affects the osteogenic potential of human mesenchymal stem cells in vitro. *Cells Tissues Organs* 182 (2): 79–88.

Xie, Y., S. T. Yang, and D. A. Kniss. 2001. Three-dimensional cell-scaffold constructs promote efficient gene transfection: Implications for cell-based gene therapy. *Tissue Eng* 7 (5): 585–598.

Yim, E. K. F., E. M. Darling, K. Kulangara, F. Guilak, and K. W. Leong. 2010. Nanotopography-induced changes in focal adhesions, cytoskeletal organization, and mechanical properties of human mesenchymal stem cells. *Biomaterials* 31 (6): 1299–1306.

Yim, E. K. F., S. W. Pang, and K. W. Leong. 2007. Synthetic nanostructures inducing differentiation of human mesenchymal stem cells into neuronal lineage. *Exp Cell Res* 313 (9): 1820–1829.

Zhang, S., F. Gelain, and X. Zhao. 2005. Designer self-assembling peptide nanofiber scaffolds for 3D tissue cell cultures. *Semin Cancer Biol* 15 (5): 413–420.

Zhu, B., Q. Lu, J. Yin, J. Hu, and Z. Wang. 2005. Alignment of osteoblast-like cells and cell-produced collagen matrix induced by nanogrooves. *Tissue Eng* 11 (5–6): 825–834.

15

Growth Factors, Stem Cells, Scaffolds and Biomaterials for Tendon Regeneration

James Zhenggui Tang, Guo-Qiang Chen, and Nicholas R. Forsyth

CONTENTS

15.1 Introduction

Tendon injuries can result from almost any physical activity. The global burden of tendon injuries is largely due to the cost of care of the disabled young male population in the low- and middle-income countries (LMICs). Tendon injuries account for considerable morbidity and often prove disabling for several months, despite what is considered appropriate management (Almekinders and Almekinders, 1994). Tendon injuries can be acute or chronic and are caused by intrinsic or extrinsic factors, either alone or in combination (Sharma and Maffulli, 2006). Tendon rupture is an acute injury in which extrinsic factors are predominant. Tendinopathy refers to a disease of a tendon, largely associated with the overuse of tendon during work- or sports-related activities. It is a chronic disorder associated with both intrinsic and extrinsic stimuli. Tendon healing occurs in three overlapping phases: the initial tissue inflammatory phase (approximately 1 week) (Koob, 2002; Sharma and Maffulli, 2005), cell proliferative stage (in the first 6 weeks) (Koob, 2002; Sharma and Maffulli, 2005; Cheung et al., 2010), and tissue remodeling phases (consolidation stage from 6 to ~10 weeks and maturation stage from 10 weeks to ~1 year) (Sharma and Maffulli, 2005; James et al., 2008). Tissue inflammation stimulates the recruitment of fibroblasts and inflammatory cells to the injury site, and regulates cell migration and the expression and responses to growth factors, which can promote angiogenesis. Cell proliferation results in increased collagen matrix production and cell–matrix interaction. Tissue remodeling encourages extracellular matrix (ECM) maturation, terminates cell proliferation, and stimulates collagen I synthesis. The whole tendon healing process requires molecular stimuli, either intrinsic or extrinsic; recruits specific cell types; forms and grows ECM, especially collagen type I in a proper stratification; and requires mechanical stimuli for tissue remodeling.

The key factors in the three stages of the tendon healing process are summarized figuratively in Figure 15.1. The duration of the inflammation stage is approximately 1 week. It is overlapped by the cell proliferation stage. Responding to the trauma of the tissue, the exudative, the oozing of fluid and other materials from cell and tissues, brings two major components, fibroblasts and inflammatory cells, into the site of injury, removing the diseased

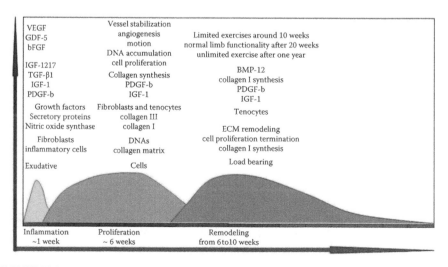

FIGURE 15.1
Schematic diagram of the three stages and key factors in the healing process of tendon injury.

tissue and activating the repair function of the wound. At this stage, molecular stimuli are recruited for cell proliferation. These include molecular stimuli such as growth factors, cytokines, and other secretory proteins. The second stage, cell proliferation, involves two major cell types, fibroblasts and tenocytes. Regulated by growth factors such as insulin growth factors (IGFs) and platelet-derived growth factors (PDGFs), collagen III and collagen I are produced, acting as components of the ECM to support further cell growth. The duration of cell proliferation stage is approximately 6 weeks. The final stage, tissue remodeling, depends largely on the formation of an appropriate collagen I matrix. The duration of this stage is approximately 4 weeks and is highly dependent on the patient with outcomes ranging from limited exercise capability to unlimited exercise capability of the patient in about 1 year. Therefore, this stage manipulates the load-bearing function of the regenerated tissue.

This chapter focuses on three components: growth factors as molecular stimuli, stem cells for generating tenocytes, and scaffolds introducing load bearing for tissue remodeling. Although it is not a comprehensive review of all factors in tendon healing process, the three components including approaches in academic research and clinical practices will give a light touch of this promising area for tissue regeneration consideration.

15.2 Growth Factors

Growth factors as molecular stimuli are featured throughout the tendon healing process (Molloy et al., 2003). The first batch of growth factors and inflammatory molecules are produced by cells within the blood clot and the second batch of growth factors stimulates fibroblast proliferation. Vascular endothelial growth factor (VEGF) becomes active almost immediately following tissue injury. IGF-I is located in all the three stages of the healing process. Transforming growth factor β (TGF-β) is also active during inflammation. PDGF is produced shortly after tendon damage and helps to stimulate the production of other growth factors. Basic fibroblast growth factor (bFGF) is both a powerful stimulator of angiogenesis and a regulator of cellular migration and proliferation. These five growth factors, namely, VEGF, IGF-I, TGF-β, PDGF, and bFGF, well characterized during tendon healing, and are selected to highlight the molecular stimuli in tendon healing process.

15.2.1 Vascular Endothelial Growth Factor

VEGF is a dimerized, 45 kDa peptide that normally attracts endothelial cells in wound healing (Pufe et al., 2005). Five important VEGF isoforms with 121, 145, 165, 189, and 205 amino acids can be generated as a result of alternative splicing from the single VEGF gene (Petersen et al., 2003). Expression of the VEGF gene can be upregulated in response to both biological and biomechanical stimuli, including hypoxia (Deroanne et al., 1997) and other growth factors (Gelberman et al., 1991). It was found that the VEGF mRNA levels in the canine intrasynovial flexor tendon elevate after day 4 following injury, peak at day 7, and steadily decline back to baseline by day 21 (Boyer et al., 2001). VEGF is also upregulated in degenerative tendon tissue (Pufe et al., 2001; Petersen et al., 2002). These suggest that VEGF plays an important role in the cell proliferation stage as VEGF mediates angiogenesis and enhances revascularization in the degenerative tissue or the site of injury. VEGF promotes angiogenesis in vivo and renders the microvasculature hyperpermeable to circulating macromolecules (Senger et al., 1983; Ferrara, 1999; Neufeld et al., 1999).VEGF is produced at its highest levels only after the inflammatory phase, at which time it is a powerful stimulator of angiogenesis.

15.2.2 Insulin Growth Factor-I

IGF-I has three isoforms, IGF-IEa, IGF-IEb, and IGF- IEc, all of which contain a signal peptide, a mature IGF-I peptide and a different E domain in at the COOH terminus of the IGF-I proprotein. IGF-IEa is the circulating form of IGF-I that is released from the liver, and IGF-IEc, known as mechano growth factor (MGF), is the tissue isoform of IGF-I released from skeletal muscle cells (Philippou et al., 2007). The different isoforms have slightly different biological actions. IGF-IEa stimulates terminal differentiation of muscle cells into myotubes and promotes stem-cell-mediated muscle regeneration. MGF is smaller than IGF-IEa, not glycosylated and less stable than IGF-IEa in the interstitial fluid. MGF is damage sensitive, controls local tissue repair, and is more potent than IGF-IEa at causing hypertrophy (Philippou et al., 2007). The mature IGF-I contains 70 highly conserved amino acid residues and is often associated with their binding proteins (IGF-BPs) in circulating plasma (Rosenzweig, 2004). IGF-I binding to specific receptors and presenting on target cells is a natural stimulator of cell growth and differentiation and also a potent inhibitor of cell apoptosis. IGF-I has been shown to be highly expressed during the early inflammatory phase in a number of animal tendon healing models, acting as an aid in the proliferation and migration of fibroblasts with a noticeable increase in the subsequent collagen production (Molloy et al., 2003).

15.2.3 Transforming Growth Factor β

TGF-β is a cytokine that plays a key role in acute inflammation and wound healing (Uysal and Mizuno, 2010). During rotator cuff repair, the inflammatory stage begins within the first week and is characterized by the infiltration of neutrophils and macrophages recruitment. The macrophages secrete TGF-β1, which increases the proteinase activity and collagen formation

(Hays et al., 2008). The early recruitment of macrophages contributes to the formation of fibrovascular scar tissue; it is therefore likely that TGF-β1 secretion is involved in scar tissue formation. There are three 25 kDa homodimeric mammalian isoforms of TGF-βs, namely, TGF-β1, TGF-β2, and TGF-β3 (Bottinger et al., 1997). TGF-β1 and TGF-β2 are present in adult healing which involves scar tissue formation but are rarely found in embryonic scar-free healing in which TGF-β3 is dominant (Ferguson and O'Kane, 2004). TGF-β is also active during inflammation, and has a variety of effects including the regulation of cellular migration and proliferation and fibronectin binding interactions. Growth factors lack biological activity unless they bind to their specific receptors, so it follows that TGF-β receptors are also upregulated during tendon healing. Levels of TGF-β receptors peaked 2 weeks after surgical repair and had decreased by day 56 (Ngo et al., 2001).

15.2.4 Platelet-Derived Growth Factor

PDGFs are a group of dimeric polypeptide isoforms made up from three types of structurally similar subunits. Activities are mediated through interactions with two related tyrosine kinase receptors, one of which binds all three PDGF chains, and one of which binds only one (Ronnstrand et al., 2001). PDGF-b, which stimulates increased levels of type I collagen, is present in low levels throughout the healing process (Spindler et al., 2003; Kobayashi et al., 2006). It is thought to play a significant role in the early stages of healing where it induces the synthesis of other growth factors, such as IGF-I (Lynch et al., 1989). PDGF plays an important role in vitro during tissue remodeling (Yoshikawa and Abrahamsson, 2001). It was observed to stimulate both collagen and non-collagen protein productions, as well as DNA synthesis, in a dose-dependent manner. PDGF exerts its effects almost immediately after injury occurs, triggering the healing cascades seen during inflammation that mark the beginning of healing process, whereas IGF-I and bFGF are important during the intermediate and later phases, particularly during cell proliferation and angiogenesis (Molloy et al., 2003). PDGF, a chemotactic and mitotic factor for fibroblasts, also induces the synthesis of collagen type I.

15.2.5 Basic Fibroblast Growth Factor

bFGF is a single-chain polypeptide composed of 146 amino acids and is a member of the heparin-binding growth factor family (Rubini et al., 1994). It has been shown to promote cellular proliferation and collagen synthesis in vivo (Molloy et al., 2003). Collagen type III expression and cellular proliferation increase after 7 days, increasing dosage of bFGF injected to a defect at the midpart of the patellar tendon (Chan et al., 2000). In vitro and in vivo studies have shown that bFGF is both a powerful stimulator of angiogenesis and a regulator of cellular migration and proliferation. A rabbit flexor tendon model was investigated to localize and quantify bFGF mRNA during tendon healing (Chang et al., 1998). In situ hybridization showed that bFGF expression was increased in both the tendon parenchyma and the tendon sheath from the first postoperative day and remained elevated up to the last time point, day 56. IGF-I, PDGF, and bFGF have vital functions during the early and intermediate stage of healing, during which they aid in the migration and proliferation of fibroblasts and stimulate ECM synthesis.

15.3 Stem Cells

Several main types of stem cells are currently described: embryonic stem cells (ESCs), birth-tissue-derived stem cells (BTSCs), adult stem cells (adipose-derived stem cells (ADSCs)), human bone marrow aspirate-adherent stem cells (hMSCs), tendon stem cells (TSCs), etc. (Thomson et al., 1998; Pittenger et al., 1999; Goodwin et al., 2001; Gimble and Guilak, 2003; Bi et al., 2007). These stem cell types vary in their capacity to form different cell types. Broadly, the hierarchy runs from the pluripotent ESC to the potentially pluripotent BTSC to the multipotent hMSC and TSC. Stem cell differentiation into tendon cells in our opinion can only be concluded when positive accompanying expression of tenomodulin can be demonstrated (Jelinsky et al., 2010). Tenomodulin was identified as a unique marker of tendon during extensive microarray profiling of mouse, rat, and human tendon tissue.

The lack of overall success with either conservative or surgical approaches to tendon repair and the clear successes in the equine model suggest that targeting cell therapeutics toward tendon repair in humans deserves further investigation (Smith and Webbon, 2005). In principle, the primary sources of human cells for use in therapeutic repair of tendons are autologous (tenocytes, adult stem cells), allogenic (tenocytes, adult stem cells, BTSCs), or from other sources (ESC).

15.3.1 Embryonic Stem Cell

It is clear that ESCs participate developmentally in all aspects of chimeric animals (Bradley et al., 1984). In spite of this, there are an extremely limited number of reports describing the differentiation of hESC in tendon cells. Those examples which do exist are thus far reliant on first differentiating the hESC into an hMSC-like cell or a connective tissue progenitor (CTP). Protocols describe spontaneous differentiation during either monolayer or embryoid body culture before continuous culture in high serum concentrations (15% or 20% fetal bovine serum [FBS]) and addition of ascorbic acid and dexamethasone in one instance (Chen et al., 2009;Cohen et al., 2010). During continuous culture as confluent sheets, the hESC-hMSC and CTP can both be induced to form either cell tubes by rolling up or through spontaneous formation. In both in vitro and in vivo models, tendon-linked gene expression is noted for hESC-hMSC (in vitro; SCX, SIX1, EPHA4, EYA2, in vivo; COLIII, COLIV, TENC, SDF1, GDF5) and for CTP (SCX, COLIII, COLXII, DCN, BLCN, ELTN, TENC in addition to transcripts representative of bone, fat, and cartilage). In vivo experimentation with hESC-MSC and CTP in tendon repair models both provided good data with histological evidence of neo-tendon formation. It is important to note however that expression of TNMD is not indicated in either model.

15.3.2 Birth-Tissue-Derived Stem Cell

To our knowledge, there are no reports in the literature which describe the tenogenic differentiation of BTSC.

15.3.3 Adipose-Derived Stem Cell

Similar to BTSC, there is a paucity of data regarding the tenogenic differentiation of hADSC. The tenogenic differentiation of rat ADSC following supplementation with GDF-5 on poly(D,L-lactide-*co*-glycolide) (PLAGA) fiber scaffolds provides an exception, albeit an exception derived from an alternate species (James et al., 2011). In this instance, rADSC displayed significant upregulation of SCX and COL1 and nonsignificant modulation of TNMD expression levels when differentiated on 3D scaffolds supplemented with 100 ng/mL GDF-5.

15.3.4 Human Bone Marrow Aspirate-Adherent Stem Cell

hMSC have been described as being tendon-capable. Until recently, these descriptions originate almost entirely from animal model studies involving mouse, rabbit, rhesus monkey, and horse (Awad et al.,1999; Pittenger et al., 1999; Smith and Webbon, 2005; Wang et al., 2005; Hoffmann et al., 2006). The horse provides the proof-of-principle model that hMSC can produce tendon tissue and that cell therapeutics can provide repair back to a pre-injury level (Smith and Webbon, 2005).

The scarcity of data regarding human stem cell capacity to form tendon-capable cells or tendon itself makes conclusions relatively difficult to draw. At present, there are three reports which describe the differentiation of hMSC into tenocytes or tendon tissue. In the first of these, upregulation of scleraxis and histological staining consistent with tendon was described when hMSC were placed into dynamically loaded collagen constructs (Kuo and Tuan, 2008). Similar to this, hMSC placed into fixed-length fibrin gels, supplemented with TGF-β3, and with the application of uniaxial force formed tendon-like tissue which displayed fibril formation and upregulation of COL1, TGF-β3, and Smad2 (Kapacee et al., 2010). The final observation relied on in vivo transplantation of hMSC into intramuscular pouches (or subcutaneously) of immunologically compromised mice. The hMSCs were preloaded through lentiviral infection with biologically active Smad8 with and without BMP2 (Shahab-Osterloh et al., 2010). In this elegant approach, it was apparent that Smad8 drove tendon formation only while Smad8/BMP2 stimulated the formation of bone, tendon, and osteotendinous junctions. These observations strongly reinforce the pivotal role that Smad8 plays in tendon differentiation (Bullough et al., 2008).

15.3.5 Tendon Stem Cell

An alternative autologous option to the tenocyte could be the tendon stem/progenitor cell (TSC) (Bi et al., 2007). This unique population of cells has been identified in patients up to the age of 12 years of age and a number of model species (Rui et al., 2010; Zhang and Wang, 2010). These cells are multipotent and capable of regenerating tendon tissue, in animal models, after transplantation. Evidence from the rat suggests that TSC also exhibit an age-related decline in their in vivo frequency and functionality where tenogenic capacity is lost while adipogenic potency is increased (Zhou et al., 2010). It remains to be determined if TSC persists beyond early adolescence in bipedal species such as primates and humans.

15.3.6 Tenocytes

Tenocytes represent an obvious cell choice. Tenocytes, derived from tendon progenitors, secrete large amounts of collagens which aggregate into basic tendon fibrils (Shukunami et al., 2006). Two primary problems with the tenocyte as a cell therapeutic for tendon repair are (1) tendons are relatively acellular, containing few tenocytes, and (2) tenocytes do not proliferate for long in in vitro culture, and during this period, they dedifferentiate and lose the characteristic tenocyte morphology (Yao et al., 2006). It remains unclear if a redifferentiation capacity remains after this point.

15.4 Scaffolds and Materials

Tendons are brilliant white in color. The ECM network of tendon is made of collagen, glycosaminoglycans, cells such as tenocytes and small molecules of particular functions (Sharma and Maffulli, 2005). The load-bearing element of tendon is the collagen fiber. Collagen fibers are uniaxially oriented with a slight wavy structure. Collagen is arranged in hierarchical levels of increasing complexity, beginning with tropocollagen, a triple-helix polypeptide chain, which unites into fibrils; fibers (primary bundles); fascicles (secondary bundles); tertiary bundles; and the tendon itself (Astrom, 1997; Jozsa and Kannus, 1997; Movin et al., 1998). Soluble tropocollagen molecules form cross-links to create insoluble collagen molecules, which aggregate to form collagen fibrils.

Tendon transfers force from muscle to bone. As a load-bearing element, collagen fibers in tendon have dynamic balanced structures responding to the states of loading and unloading. Tendon strain up to 2% initiates a flattening of the crimp pattern (Viidik, 1973; Butler et al., 1978; Hess et al., 1989). Intramolecular sliding of collagen triple helices occurs when strained up beyond 2% and the fibers become more parallel (Mosler et al., 1985; Zernicke and Loitz, 2002). If the strain remains <4%, the tendon behaves in an elastic fashion and returns to its original length when unloaded (Curwin and Stanish, 1984). As the strain levels rise to between 4% and 8%, the collagen fibers start to slide past one another as the intermolecular cross-links fail, and at strain levels greater than 8%, macroscopic ruptures occur because of tensile failure of the fibers and interfibrillar shear failure (O'Brien, 1992). The tensile strength of tendons is related to their thickness and collagen content. A tendon with an area of $1\,cm^2$ is capable of bearing 500–1000 kg (Elliott, 1965; Oakes et al., 1998), approximately 5–10 MPa. The rate of collagen metabolism is relatively slow and the turnover time for tendon collagen being from 50 to 100 days (approximately 7–14 weeks) (Curwin and Stanish, 1984). There is normally a balance between synthesis and breakdown, but synthesis will exceed degradation during growth and following injury.

As stated previously, tendon injuries can be chronic and acute. Chronic injuries start as a micro-damage of collagen fibers, falling into the strain failure of anywhere that the strain is beyond 4%. When the damage of collagen fibers exceeds the limited healing capacity of tendon tissue, the accumulation of micro-damage results in the rupture of the tendon. A form of acute injuries can be a tendon rupture resulting from any sudden impact. The appropriate treatment of tendon injuries varies. In this chapter, scaffolds and materials will be described exclusively for use with the acute tendon rupture which may or may not need augmentation.

Scaffolds for treating acute tendon rupture include suture and mesh. These scaffolds are used to join the ruptured tendon and bear the minimal force incurred during the healing process. Sutures are used to join the ruptured tendon (Strickland, 2005). Sutures are, in general, categorized according to the type of material (natural or synthetic), the lifetime of the material in the body (absorbable or nonabsorbable), and the form in which they were made (braided, twisted, and monofilament) (Moy et al., 1992; Chu et al., 1996). Two variables in material properties are important when considering the use of one suture type rather than another: (1) the ultimate strength or load to failure of the suture and (2) the ability of the suture to withstand abrasion against the anchor (Edward et al., 2005). Suture materials for tendon regeneration purposes must have their mechanical compatibility with the tendon considered, for example, suture strength within 5–10 MPa would meet the requirements of the load-bearing healing tendon. Coupled to this, the suture should also display a degradation profile consistent with the increasing strength of the newly formed load-bearing collagen fibrils. As the tendon has a healing process which lasts for approximately 14 weeks, any suture degrading in less than 14 weeks (<4.5 months) could be problematic.

Natural sutures are made of catgut or reconstituted collagen (RC) or from cotton, silk, or linen (Pillai and Sharma, 2010). Catgut and RC are the two absorbable natural sutures available in the current market. Catgut sutures cannot be used for tendon regeneration because their tensile strength is only retained in the first 4–5 days and completely lost after 2 weeks (Bennett, 1988). Collagen sutures in a recent review (Pillai and Sharma, 2010) were not mentioned in relation to tendon repair and healing.

Synthetic absorbable sutures include (1) polyglycolide (PGA) and its derivatives, (2) poly(lactide-*co*-glycolide) (PLGA, monocryl), (3) polyglactin 910 (Vicryl), (4) polydioxanone (PDO), (5) polytrimethylene carbonate (Maxon), and (6) a copolymer of glycolide, dioxanone, and trimethylene carbonate. PGA and its derivatives maintain most of their strength for 2 weeks and disintegrate only after 5–8 weeks. PLGA retains over 90% of its initial strength at the end of 1 week while chromic collagen had lost nearly 50% of its strength (Dunlap et al., 1976). The rapid loss of the strength of PLGA, the copolymer, in 10 days and complete loss in 21 days indicate that the degradation profile of the suture along with its strength is critical for the selection of proper sutures for tendon repair and healing. Kangas et al. (2001) compared the degradation profiles of two sutures, polytrimethylene carbonate (Maxon) and poly-L/D-lactide (PLDLA), in vitro and in vivo. Maxon suture in vivo retained its initial strength in 2 weeks, followed by high loss of the initial strength at week 4 with a complete loss of its strength at week 6. PLLDA sutures retained almost 70% of its initial strength in 6 weeks and 42% in 13 weeks in vivo. PLLDA suture therefore was concluded as a suitable suture for wounds that require healing time of up to 28 weeks. O'Broin et al. (1993) compared the performance of the absorbable suture, PDS (Ethicon Ltd, Edinburgh, United Kingdom), with nonabsorbable Prolene sutures. Both have a tensile strength around and below 10 MPa that is mechanically compatible with the tendon tissue mentioned earlier. PDS sutures in the left flexor digitorum longus tendons of rabbits lost half of their initial strength after 3–4 weeks. The integrity of the repair at this point was approximately eight times stronger than the original suture repair from not only the contribution of the suture material but also the net gain of the collagen fibril content introduced in the healing repair phase of the cell proliferation. There is no significant difference seen in PDS or Prolene repair strength throughout all groups.

In summary, sutures for tendon repair and healing can be made from natural, nonabsorbable, and absorbable materials. Nonabsorbable sutures phased out sutures from natural materials such as catgut with the advantages being biocompatibility of the material and the repeatability in production.

Absorbable sutures are promising in tendon healing compared to nonabsorbable sutures. However, surgeons still prefer nonabsorbable sutures for both safety and convenience, not considering the leftover of foreign materials and further consequences. Because of the complexity of the tendon healing process, tendon regeneration can only achieve half of the original strength.

15.5 Future Perspectives

Tendon healing approaches depend on the size and scale of the injury in the diseased or damaged tissue. Three typical scenarios of injuries and their subsequent healing approaches are described here to foresee the potential future approaches. In the first scenario, micro-damage in the tendon has occurred as a result of chronic overuse. The second scenario describes a rupture of the tendon tissue as a result of a sudden impact. The third and final scenario describes a large diseased area which arises as a consequence of micro-damage accumulation, which requires a further surgical removal, leaving a gap between the two dissected tendon tissue ends.

The chronic overuse of the tendon tissue is described under the term of "tendinopathy," although other terms like "tendinosis" and "tendinitis" are used for the same purpose (Sharma and Maffulli, 2005). Tendinopathy can be inflammatory or degenerative or both and is the closest fit to the first scenario. One of the tendon healing approaches for tendinopathy is using growth factors to stimulate the cell proliferation in the first few weeks. These growth factors include TGF-βs, IGF-I, PDGF- b, and BMP-12 (Cheung et al., 2010). Foreseeing the involvement of delivering growth factors in the first week of the injury as described in Figure 15.1, products such as platelet-rich plasma (PRP), which help by stimulating repair by the delivery of biological factors and promoting healing, are very promising although not conclusive (Foster et al., 2009; Borchers, 2011).

PRP enhances the recruitment, proliferation, and differentiation of cells involved in tissue regeneration. Growth factors identified within PRP include platelet-derived epidermal growth factor (PD-EGF), PDGF A +B, TGF-β1, VEGF, endothelial cell growth factor (ECGF), and bFGF (Foster et al., 2009). Compared to growth factors which are involved in the inflammation stage, PRP has no detectable IGF-I. More investigations of PRP-related products are in demand and are necessary for clinical practices to give a conclusive answer to an optimized tendon healing approach.

Paoloni et al. (2011) reviewed PRP injections based on basic science and animal studies and small case series reports for ligament or tendon injuries. Despite the interest and apparent widespread use of PRP injections, there is a lack of high-level evidence regarding randomized clinical trials assessing the efficacy of PRP in treating ligament and tendon injuries. Scientific studies should be performed to assess clinical indications, efficacy, and safety of PRP, and this will require appropriately powered randomized controlled trials with adequate and validated clinical and functional outcome measures and sound statistical analysis. Other aspects of PRP use that need to be determined are (1) volume of injection/application, (2) most effective preparation, (3) buffering/activation, (4) injection technique (1 depot vs. multiple depots), (5) timing of injection to injury, (6) single application versus series of injections, and (7) the most effective rehabilitation protocol to use after PRP injection. With all proposed treatments, the doctor and the patient should weigh up potential benefits of treatment, potential risks, and costs. Based on the limited publications to date and theoretical considerations, the potential risks involved with PRP are fortunately very low. However, benefits remain unproven to date, particularly when comparing PRP with other injections for ligament and tendon injuries.

Tendon rupture either caused by a sudden impact or by the removal of diseased tissue as described in the scenario 2 and 3 will use sutures and/or meshes to close gap. Sutures used in clinical practices will gradually move from nonabsorbable to absorbable materials (Pillai and Sharma, 2010). Along with academic investigations, surgical practices will likely become adapted to use absorbable sutures. Filling the gap with composites, which mimic tendon, growth factors, stem cells, and fibers, could be another approach for tissue augmentation. These approaches could be compromised by high costs, and it may take a long time to achieve the objectives.

The concept of using fiber composites for tendon tissue engineering is not new but still requires substantial research before clinical adoption. The general concept is to use fibers to bear the force required for tendon tissue function and the bioactivation of the fiber to attract cell proliferation and/or differentiation. The fibers are often combined with natural polymers such as collagen to form bioactive fibers. Ladd et al. (2011) co-electrospun poly(ε-caprolactone)/collagen and poly(L-lactide)/collagen fibers with fiber diameters from 452 to 549 nm. Scaffolds exhibited regional variations in mechanical properties with moduli from 4.490 to 27.62 MPa and generally withstood cyclic testing, although with some evidence of hysteresis. Video analysis demonstrated scaffold strain profiles exhibited similar trends to native muscle–tendon junctions (MTJ). The scaffolds were cytocompatible and accommodated cell attachment and myotube formation. The properties engineered into these scaffolds make them attractive candidates for tissue engineering of MTJs. Sawaguchi et al. (2010) investigated the effect of cyclic 3D strain on cell proliferation and collagen synthesis of fibroblast-seeded chitosan-hyaluronan hybrid polymer fiber. The results indicated that multidimensional cyclic mechanical strain to mimic the in vivo physiological condition has the potential to improve or accelerate tissue regeneration in ligament and tendon tissue engineering using in vitro 3D scaffolds. An extension of the previously mentioned reports saw Sahoo et al. (2010) demonstrate that bFGF could be successfully incorporated within blend-electrospun nanofibers and released in a bioactive form over a 1 week period. The released bioactive bFGF activated tyrosine phosphorylation signaling within seeded BMSCs, resulting in BMSC proliferation, upregulated gene expression of tendon/ligament-specific ECM proteins, increased production and deposition of collagen and tenascin-C, reduced multipotency of the BMSCs, and induced tendon-/ligament-like fibroblastic differentiation, indicating their potential in tendon/ligament tissue engineering applications.

In addition to growth factors, stem cells, and scaffolds, physical therapy can be a promising approach to achieving greater collagen fibril strength in the regenerated tendon. In a recent review, physical modalities for tendon healing are summarized (Sharma and Maffulli, 2005) as (1) extracorporeal shock wave therapy for generating an increase in growth factor levels of TGF-β1 and persistently elevated levels of IGF-1; (2) pulsed magnetic fields for the improved collagen fiber alignment in a rat Achilles tendinopathy model as a consequence of an increased force to breakage in the anode stimulated group compared with controls and a cathode-stimulated group; (3) direct current applied to rabbit tendons in vitro increased type-I-collagen production and decreased adhesion formation; and (4) laser phototherapy increased collagen production in rabbits subjected to tenotomy and surgical repair. Therefore, physical therapy in combination with the correct balance of growth factors, stem cells, and scaffolds could provide a viable, future alternative to current practices.

A potential sensing and signal transduction mechanism for physical therapy is mechanotransduction, the physiological process where cells sense and respond to mechanical loads (Khan and Scott, 2009). Mechanotransduction is generally broken down into three steps: (1) mechanocoupling, (2) cell–cell communication, and (3) the effector response. Tendon is a dynamic, mechanoresponsive tissue. One of the major load-induced responses shown both in vitro (Banes et al., 1995) and in vivo (Olesen et al., 2006; Heinemeier et al., 2007; Scott et al., 2007) in tendon is an upregulation of IGF-I. This upregulation of IGF-I is associated with cellular proliferation and matrix remodeling within the tendon.

Taken together, the prospects for improved tendon regeneration in the future appear strong. As greater understanding of tendon biology combines with growth factor awareness, stem cell biology utility, scaffold compatibility and functionality, and the benefits offered by physical therapy, the patient's prospects and options are likely to improve and increase, respectively.

Acknowledgments

The authors of this chapter thank Dr. Yang Wang for her timely help in literature search.

Abbreviations

ADSCs	adipose-derived stem cells
bFGF	basic fibroblast growth factor
BTSCs	birth-tissue-derived stem cells
CTP	connective tissue progenitor
ECGF	endothelial cell growth factor
ECM	extracellular matrix
ESCs	embryonic stem cells
FBS	fetal bovine serum
hMSCs	human bone marrow aspirate-adherent fraction stem cells
IGFs	insulin growth factors
LMICs	low-and middle-income countries
PD-EGF	platelet-derived epidermal growth factor
PDGFs	platelet-derived growth factors
PDO	polydioxanone
PGA	polyglycolide
PLDLA	poly-L/D-lactide
PLGA	poly(lactide-*co*-glycolide)
PRP	platelet-rich plasma
RC	reconstituted collagen
TGF-β	transforming growth factor β
TSCs	tendon stem cells
VEGF	vascular endothelial growth factor

References

Almekinders LC, Almekinders SV. 1994. Outcome in the treatment of chronic overuse sports injuries: A retrospective study. *J Orthop Sports Phys Ther* 19: 157–161.

Astrom M. 1997. On the nature and etiology of chronic achilles tendinopathy (thesis). Lund, Sweden: University of Lund.

Awad HA et al. 1999. Autologous mesenchymal stem cell-mediated repair of tendon. *Tissue Eng* 5: 267–277.

Banes AJ et al. 1995. PDGF-BB, IGF-I and mechanical load stimulate DNA synthesis in avian tendon fibroblasts in vitro. *J Biomech* 28: 1505–1513.

Bennett R.G. 1988. Selection of wound closure materials. *J Am Acad Dermatol* 18: 619–637.

Bi Y et al. 2007. Identification of tendon stem/progenitor cells and the role of the extracellular matrix in their niche. *Nat Med* 13: 1219–1227.

Borchers JR. 2011. Emerging issues in sport medicine. *Clin J Sport Med* 21(1): 1–2.

Bottinger EP, Letterio J, Roberts AB. 1997. Biology of TGF-Beta in knockout and transgenic mouse models. *Kidney Int* 51: 1355–1360.

Boyer MI et al. 2001. Quantitative variation in vascular endothelial growth factor mRNA expression during early flexor tendon healing: An investigation in a canine model. *J Orthop Res* 19 (5): 869–872.

Bradley A et al. 1984. Formation of germ-line chimaeras from embryo-derived teratocarcinoma cell lines. *Nature* 309: 255–256.

Bullough R et al. 2008. Tendon repair through stem cell intervention: Cellular and molecular approaches. *Disabil Rehabil* 2008: 1–6.

Butler DL et al. 1978. Biomechanics of ligaments and tendons. *Exerc Sport Sci Rev* 6: 125–181.

Chan BP et al. 2000. Effects of basic fibroblast growth factor (bFGF) on early stages of tendon healing: A rat patellar tendon model. *Acta Orthop Scand* 71(5): 513–518.

Chang J et al. 1998. Molecular studies in flexor tendon wound healing: The role of basic fibroblast growth factor gene expression. *J Hand Surg [Am]* 23A (6): 1052–1059.

Chen X et al. 2009. Stepwise differentiation of human embryonic stem cells promotes tendon regeneration by secreting fetal tendon matrix and differentiation factors. *Stem Cells* 27: 1276–1287.

Cheung EV et al. 2010. Strategies in biologic augmentation of rotator cuff repair. *Clin Orthop Relat Res* 468: 1476–1484.

Chu C-C, von Fraunhofer JA, Greisler HP. 1996. *Wound Closure Biomaterials and Devices*. Boca Raton, FL: CRC Press, Inc.

Cohen S et al. 2010. Repair of full-thickness tendon injury using connective tissue progenitors efficiently derived from human embryonic stem cells and fetal tissues. *Tissue Eng Part A* 16(10): 3119–3137.

Creaney L, Hamilton B. 2008. Growth factor delivery methods in the management of sports injuries: The state of play. *Br J Sports Med* 42(5): 314–320.

Curwin S, Stanish WD. 1984. *Tendinitis, Its Etiology and Treatment*. Lexington, MA: Collamore Press.

Deroanne CF et al. 1997. Angiogenesis by fibroblast growth factor 4 is mediated through an autocrine up-regulation of vascular endothelial growth factor expression. *Cancer Res* 57(24): 5590–5597.

Dunlap WA et al. 1976. Laboratory and clinical evaluation of a new synthetic absorbable suture for ophthalmic surgery. *Adv Ophthalmol* 33: 49–61.

Edward G et al. 2005. Suture anchors and tacks for shoulder surgery, part 1: Biology and biomechanics. *Am J Sports Med* 33(12): 1918–1923.

Elliott DH. 1965. Structure and function of mammalian tendon. *Biol Rev Camb Philos Soc* 40: 392–421.

Ferguson MW, O'Kane S. 2004. Scar-free healing: From embryonic mechanisms to adult therapeutic intervention. *Philos Trans R Soc Lond B Biol Sci* 359(1445): 839–850.

Ferrara N. 1999. Molecular and biological properties of vascular endothelial growth factor. *J Mol Med* 77: 527–543.

Foster TE et al. 2009. Platelet-rich plasma: From basic science to clinical applications. *Am J Sports Med* 37: 2259–2272.

Gelberman RH et al. 1991. The revascularization of healing flexor tendons in the digital sheath: A vascular injection study in dogs. *J Bone Joint Surg Am* 73(6): 868–881.

Gimble J, Guilak, F. 2003. Adipose-derived adult stem cells: Isolation, characterization, and differentiation potential. *Cytotherapy* 5: 362–369.

Goodwin HS et al. 2001. Multilineage differentiation activity by cells isolated from umbilical cord blood: Expression of bone, fat, and neural markers. *Biol Blood Marrow Transplant* 7: 581–588.

Hays PL et al. 2008. The role of macrophages in early healing of a tendon graft in a bone tunnel. *J Bone Joint Surg Am* 90: 565–579.

Heinemeier KM et al. 2007. Short-term strength training and the expression of myostatin and IGF-I isoforms in rat muscle and tendon: Differential effects of specific contraction types. *J Appl Physiol* 102: 573–581.

Hess GP et al. 1989. Prevention and treatment of overuse tendon injuries. *Sports Med* 8: 371–384.

Hoffmann A et al. 2006. Neotendon formation induced by manipulation of the Smad8 signalling pathway in mesenchymal stem cells. *J Clin Invest* 116: 940–952.

James R et al. 2008. Tendon: Biology, biomechanics, repair, growth factors, and evolving treatment options. *J Hand Surg* 33A: 102–112.

James R et al. 2011. Tendon tissue engineering: Adipose-derived stem cell and GDF-5 mediated regeneration using electrospun matrix systems. *Biomed Mater* 6: 025011.

Jelinsky SA et al. 2010. Tendon-selective genes identified from rat and human musculoskeletal tissues. *J Orthop Res* 28: 289–297.

Jozsa LG, Kannus P. 1997. *Human Tendons: Anatomy, Physiology, and Pathology.* Champaign, IL: Human Kinetics Publishers.

Kangas J et al. 2001. Comparison of strength properties of poly-L/D-lactide (PLDLA) 96/4 and polygluconate (Maxon®) sutures: In vitro, in the subcutis, and in the achilles tendon of rabbits. *J Biomed Mater Res* 58: 121–126.

Kapacee Z et al. 2010. Synthesis of embryonic tendon-like tissue by human marrow stromal/mesenchymal stem cells requires a three-dimensional environment and transforming growth factor beta3. *Matrix Biol* 29: 668–677.

Khan KM, Scott A. 2009. Mechanotherapy: How physical therapists' prescription of exercise promotes tissue repair. *Br J Sports Med* 43: 247–251.

Kobayashi M et al. 2006. Expression of growth factors in the early phase of supraspinatus tendon healing in rabbits. *J Shoulder Elbow Surg* 15: 371–377.

Koob TJ. 2002. Biomimetic approaches to tendon repair. *Comparative Biochem Physiol Part A* 133: 1171–1192.

Kuo CK, Tuan, RS. 2008. Mechanoactive tenogenic differentiation of human mesenchymal stem cells. *Tissue Eng Part A* 14: 1615–1627.

Ladd MR et al. 2011. Co-electrospun dual scaffolding system with potential for muscle-tendon junction tissue engineering. *Biomaterials* 32(6): 1549–1559.

Lynch SE et al. 1989. Growth factors in wound healing: Single and synergistic effects on partial thickness porcine skin wounds. *J Clin Invest* 84(2): 640–646.

Molloy T. et al. 2003. The roles of growth factors in tendon and ligament healing. *Sports Med* 33(5): 381–394.

Mosler E et al. 1985. Stress-induced molecular rearrangement in tendon collagen. *J Mol Biol* 182: 589–596.

Movin T et al. 1998. Intratendinous alterations as imaged by ultrasound and contrast medium-enhanced magnetic resonance in chronic achillodynia. *Foot Ankle Int* 19: 311–317.

Moy RL et al. 1992. A review of sutures and suturing techniques. *J Dermatol Surg Oncol* 18: 785–795.

Ngo M et al. 2001. Differential expression of transforming growth factor-beta receptors in a rabbit zone II flexor tendon wound healing model. *Plast Reconstr Surg* 108: 1260–1267.

Neufeld G et al. 1999. Vascular endothelial growth factor and its receptors. *FASEB J* 13: 9–22.

Oakes BW et al. 1998. Correlation of collagen fibril morphology and tensile modulus in the repairing and normal rabbit patella tendon. *Trans Orthop Res Soc* 23: 24.

O'Brien M. 1992. Functional anatomy and physiology of tendons. *Clin Sports Med* 11: 505–520.

O'Broin ES et al. 1993. Absorbable sutures in tendon repair. *J Hand Surg (Brit Eur Vol)* 20B(4): 505–508.

Olesen JL et al. 2006. Expression of insulin-like growth factor I, insulin-like growth factor binding proteins, and collagen mRNA in mechanically loaded plantaris tendon. *J Appl Physiol* 101: 183–188.

Paoloni J et al. 2011. Platelet-rich plasma treatment for ligament and tendon injuries. *Clin J Sport Med* 21(1): 37–45.

Petersen W et al. 2002. Angiogenesis in fetal tendon development: Spatial and temporal expression of the angiogenic peptide vascular endothelial cell growth factor. *Anat Embryol (Berlin)* 205: 263–270.

Petersen W et al. 2003. The splice variants 120 and 164 of the angiogenic peptide vascular endothelial cell growth factor (VEGF) are expressed during Achilles tendon healing. *Arch Orthop Trauma Surg* 123: 475–480.

Philippou A et al. 2007. The role of insulin-like growth factor 1 (IGF-1) in skeletal muscle physiology. *In Vivo* 21: 45–54.

Pillai CKS, Sharma CP. 2010. Absorbable polymeric surgical sutures: Chemistry, production, properties, biodegradability, and performance. *J Biomater Appl* 25: 291–366.

Pittenger MF et al. 1999. Multilineage potential of adult human mesenchymal stem cells. *Science* 284: 143–147.

Pufe T et al. 2001. The angiogenic peptide vascular endothelial growth factor is expressed in foetal and ruptured tendons. *Virchows Arch* 439: 579–585.

Pufe T et al. 2005. The influence of biomechanical parameters on the expression of VEGF and endostatin in the bone and joint system. *Ann Anat* 187(5–6): 461–472.

Ronnstrand L et al. 2001. Mechanisms of platelet-derived growth factor-induced chemotaxis. *Int J Cancer* 91 (6): 757–762.

Rosenzweig SA. 2004. What's new in the IGF-binding proteins? *Growth Horm IGF Res* 14: 329–336.

Rubini M et al. 1994. Platelet-derived growth factor increases the activity of the promoter of the insulin-like growth factor-I (IGFI) receptor gene. *Exp Cell Res* 211: 374–379.

Rui YF et al. 2010. Isolation and characterization of multipotent rat tendon-derived stem cells. *Tissue Eng Part A* 16: 1549–1558.

Sahoo S et al. 2010. Bioactive nanofibers for fibroblastic differentiation of mesenchymal precursor cells for ligament/tendon tissue engineering applications. *Differentiation* 79(2): 102–110.

Sawaguchi N et al. 2010. Effect of cyclic three-dimensional strain on cell proliferation and collagen synthesis of fibroblast-seeded chitosan-hyaluronan hybrid polymer fiber. *J Orthop Sci* 15(4): 569–577.

Scott A et al. 2007. Tenocyte responses to mechanical loading in vivo: A role for local insulin-like growth factor 1 signaling in early tendinosis in rats. *Arthritis Rheum* 56: 871–881.

Senger DR et al. 1983. Tumor cells secrete a vascular permeability factor that promotes accumulation of ascites fluid. *Science* 25: 983–985.

Shahab-Osterloh S et al. 2010. Mesenchymal stem cell-dependent formation of heterotopic tendon-bone insertions (osteotendinous junctions). *Stem Cells* 28: 1590–1601.

Sharma P, Maffulli N. 2005. Tendon injury and tendinopathy: Healing and repair. *J Bone Joint Surg Am* 87: 187–202.

Sharma P, Maffulli N. 2006. Biology of tendon injury: Healing, modelling and remodelling. *J Musculoskelet Neuronal Interact* 6(2): 181–190.

Shukunami C et al. 2006. Scleraxis positively regulates the expression of tenomodulin, a differentiation marker of tenocytes. *Dev Biol* 298: 234–247.

Smith RK, Webbon, PM. 2005. Harnessing the stem cell for the treatment of tendon injuries: Heralding a new dawn? *Br J Sports Med* 39: 582–584.

Spindler KP et al. 2003. The biomechanical response to doses of TGF-beta 2 in the healing rabbit medial collateral ligament. *J Orthop Res* 21: 245–249.

Strickland JW. 2005. The scientific basis for advances in flexor tendon surgery. *J Hand Ther* 18: 94–110.

Thomson JA et al. 1998. Embryonic stem cell lines derived from human blastocysts. *Science* 282: 1145–1147.

Uysal AC, Mizuno H. 2010. Tendon regeneration and repair with adipose derived stem cells. *Current Stem Cell Res Ther* 5: 161–167.

Viidik A. 1973. Functional properties of collagenous tissues. *Int Rev Connect Tissue Res* 6: 127–215.

Wang QW et al. 2005. Mesenchymal stem cells differentiate into tenocytes by bone morphogenetic protein (BMP) 12 gene transfer. *J Biosci Bioeng* 100: 418–422.

Yao L, Bestwick, CS, Bestwick, LA, Maffulli, N, Aspden, RM. 2006. Phenotypic drift in human tenocyte culture. *Tissue Eng* 12: 1843–1849.

Yoshikawa Y, Abrahamsson S. 2001. Dose-related cellular effects of platelet-derived growth factor-BB differ in various types of rabbit tendons in vitro. *Acta Orthop Scand* 72(3): 287–292.

Zernicke RF, Loitz BJ. 2002. Exercise-related adaptations in connective tissue. In: *The Encyclopaedia of Sports Medicine. Strength and Power in Sport*. Vol. 3. ed. Komi PV, pp. 93–113. Boston, MA: Blackwell Scientific Publications.

Zhang J, Wang, JH. 2010. Characterization of differential properties of rabbit tendon stem cells and tenocytes. *BMC Musculoskelet Disord* 11: 10.

Zhou Z et al. 2010. Tendon-derived stem/progenitor cell aging: Defective self-renewal and altered fate. *Aging Cell* 9: 911–915.

16

Biomaterials and Stem Cells for Myocardial Repair

Jiashing Yu, Chao-Min Cheng, and Randall J. Lee

CONTENTS

16.1 Introduction

Cardiovascular diseases account for 12 million deaths annually worldwide. Myocardial infarction (MI) continues to be a significant problem in industrialized countries and is becoming an increasingly significant problem in developing countries. MI is one of the leading causes of morbidity and mortality in the United States. Approximately 1.3 million clinical cases of nonfatal MI—an annual incidence rate of approximately 600 cases per 100,000 people—have been reported per year (Fenton 2007). The rapid development of myocardial necrosis, due to a critical imbalance between oxygen supply and the demand of the myocardium, is one of the pathological factors leading to MI. An MI potentially results from plaque rupture with thrombus formation in a coronary artery, resulting in an acute reduction of blood supply to a portion of the myocardium (Alpert et al. 2000). Approximately

500,000–700,000 deaths are caused by ischemic heart disease annually in the United States.

16.1.1 Physiology and Pathology

MI is defined as death or necrosis of myocardial cells. Myocardial infarction occurs when myocardial ischemia exceeds a critical threshold and overwhelms myocardial cellular repair mechanisms which are designed to maintain normal operating functions and hemostasis. Ischemia at this critical threshold for an extended time period results in irreversible myocardial apoptosis (Cotran et al. 1994, Rubin 1995, Bajzer 2002). Time-dependent degradation of extracellular matrix (ECM) following the myocardial cell death is mainly regulated by matrix metalloproteinases (MMPs), which are well known to function in the extracellular environment of single cells and degrade both matrix and non-matrix proteins as well. MMPs are modulated by their tissue inhibitors (TIMPs); the activities of MMPs and TIMPs in the extracellular domain play central roles in morphogenesis, wound healing, and tissue repair as well as remodeling in response to the inflammatory response after myocardial infarction (Nagase et al. 2006, Spinale 2007). This weakening of the collagen scaffold results in wall thinning and ventricular dilatation. The law of Laplace can explain the biophysical relation between ventricular wall tension and the pressure in the ventricle. In the case of myocardial remodeling following an MI, the rule specifies that a dilated ventricle leads to wall thinning resulting in increased wall tension to maintain the same blood pressure (Tonnessen and Knudsen 2005). The increased wall stress leads to increased myocardial oxygen demand, causing increased myocardial apoptosis and a vicious downward cycle of myocardial remodeling resulting in heart failure.

16.1.2 Clinical Treatments

Currently available therapeutic approaches to treat heart failure include pharmacological therapies, mechanical devices, and surgeries (Braunwald and Bristow 2000, Swedberg et al. 2005). Antithrombotic agents (e.g., aspirin and heparin) are commonly used to prevent the formation of thrombus associated with myocardial infarction and inhibit platelet functions through blocking cyclooxygenase, preventing subsequent platelet aggregation. Platelet aggregation blockers have been shown to reduce mortality (Jones 2008). Other pharmacological agents such as vasodilators (e.g., ACE inhibitors and ARBs) reduce cardiac work through decreasing preload and afterload. In addition, beta-adrenergic blocking agents, inhibitor of chronotropic and inotropic responses, can be used to stimulate vasodilatory as well and reduce blood pressure, which decreases myocardial oxygen demand.

Mechanical devices and surgical approaches (e.g., artificial hearts or constraint devices) provide the nonbiological or chemical opportunity to address cardiovascular diseases, with limited success as alternatives to heart transplantation for the treatment of final-stage heart diseases. Recently, the FDA has approved the first fully implantable heart device, AbioCor™ (Abiomed Inc., Danvers, MA). However, this device is a permanent implant which would take over the pumping function of the heart and is designed only for patients who have no alternatives (Dowling et al. 2001, Lederman et al. 2002). Devices such as CorCap™ Cardiac Support Device (Acorn Cardiovascular™) and the Paracor HeartNET™ provide a mesh wrap which can be implanted around the patient's heart. CorCap and Paracor HeartNET are constraint devices which can only provide external support to the heart but does not repair the injured tissues. The devices did decrease left ventricular (LV) size, but the effects were seen months after the surgical procedure (Mann et al. 2007, Klodell et al. 2008). Moreover, concomitant mitral valve replacement confound the beneficial effects of the constraint devices (Franco-Cereceda et al. 2004, Olsson et al. 2005). Therefore, the further development of other strategies to treat severe cardiomyopathy is needed.

16.2 Cardiac Tissue Engineering

Cardiac tissue engineering aims to repair damaged myocardium by combing cell biology, material science, and engineering principles (Zammaretti and Jaconi 2004, Christman and Lee 2006). One of the major issues in cardiac tissue engineering is to understand the influence of extracellular matrix on cell function and then mimic the function of extracellular matrix in vitro for developing physiologically relevant artificial tissues or biomaterials. ECM not only provides the scaffolding for single cells but also regulates cell functions (e.g., motility, division, or apoptosis) through receptor-modulated pathways (Huang et al. 2007). Engineered cardiac constructs composed of scaffold materials alone or combined with cells or growth factors have been widely investigated and developed (Christman and Lee 2006).

16.2.1 Cell Therapy

Various types of stem cells have been investigated for the purpose of treating myocardial infarction. Different stem cell sources have been considered such as animals (e.g., mouse, rat, or human), different developmental stages of the species (e.g., embryonic, fetal, or adult), different tissues or organs (e.g., hematopoietic, mesenchymal, skeletal, or neural), and potential to differentiate (e.g., totipotent, pluripotent, or multipotent) (Penn and Mal 2006). Cell transplantation has been exhibited to beneficially influence irreversible myocyte death. Somatic progenitor cells have been viewed as "tissue stem cells" due to their capabilities of differentiating into mature cell types, such as cardiomyocytes, fibroblasts, as well as myoblasts (Siminiak and Kurpisz 2003). Adult bone marrow stem cell (BMC) is also one of the typical cell sources which can be isolated from bone marrow or mobilized through specific cytokines.

Skeletal myoblast was the first cell type to be transplanted into ischemic myocardium, and significant improvement of heart function was observed (Menasche 2008, Formigli et al. 2009). Christman et al. injected skeletal myoblasts with fibrin scaffolds; the preservation of cardiac function, the reduction of infarct expansion, and angiogenesis were found (Christman et al. 2004). However, the lack of electrical coupling of transplanted myoblasts with host tissues is the major limitation with skeletal myoblast therapy (Menasche et al. 2008). Furthermore, fibroblast is another common cell type, which can be obtained easily from autologous tissues. Fibroblast transplantation has been demonstrated in both acute and chronic MI models. Schuh et al. injected fibroblasts into myocardium after 4 weeks postinfarction; the improvement of fractional shortening was maintained after 8 weeks of transplantation (Schuh et al. 2009). More recently, the modified fibroblast has been investigated; for instance, it has been shown that cardiac fibroblasts expressing vascular endothelial growth factor (VEGF) mitigated cardiac dysfunction and also increased neovascularization (Goncalves et al. 2009). Fibroblast cell sheets cocultured with endothelial progenitor cells (EPCs) have also been demonstrated to enhance cell grafting efficiency, compared with fibroblasts themselves (Kobayashi et al. 2008a,b). A long-standing hypothesis described that the fibroblast transplantation into damaged myocardium may strengthen the scar in order to prevent LV remodeling (Leor et al. 1997, El Oakley et al. 2001, Lee et al. 2004). Although the tremendous effects have been proved in both myoblast and fibroblast implantations, the barrier that limits the further investigation is still what has been accentuated—the lack of gap junctions with host myocytes. Numerous experimental studies have shown that the transplantation of skeletal myoblasts improved systolic and diastolic performances; however, the drawbacks in this approach are arrhythmogenic and the lack of gap junctions.

BMCs can be subcategorized into hematopoietic stem cells (HSCs) and mesenchymal stem cells (MSCs). In general, BMCs with CD34+ and CD133 markers and endothelial phenotype cells are considered as HSCs. EPCs which can be isolated from blood and bone marrow have been shown to promote neovascularization, likely through the secretion of growth factors, and possibly myogenesis (Rehman et al. 2003). However, the ability of BMC to transdifferentiate into cardiac myoblasts has been challenged (Balsam et al. 2004, Murry et al. 2004), resulting in the notion that the beneficial effects of BMC therapy is due to the release of chemokines and cytokines. Both HSCs and EPCs have been also used to accelerate neovascularization in ischemic tissues and re-endothelialization in injured blood vessels (Urbich and

Dimmeler 2004, Wu et al. 2004). In the rat acute MI model, the injection of ex vivo expanded EPCs significantly improved blood flow and cardiac functions and reduced LV scarring (Kawamoto et al. 2001).

Clinical studies have shown exciting promise of autologous stem cells for myocardial repair and regeneration without major untoward effects. Studies have investigated the transplantation of MSCs in myocardial repair; these experiments indicated that MSCs are able to differentiate into cardiomyocytes and improve LV function and remodeling (Makino et al. 1999, Toma et al. 2002, Schuleri et al. 2008). Forrester et al. also reported that allogeneic MSCs engrafted to peri-infarct area can transform into cardiomyocytes in vivo based on the developmental patterns, structural characteristics, and expressions of cardiac-specific proteins such as actin cytoskeleton and troponin (Forrester et al. 2003). Lunde et al. clinically treated 100 patients having acute anterior wall ST-elevation MI (STEMI) with autologous mononuclear BMCs (mBMCs). One year later, the LV function and functional capacity of 49 patients improved. This promising outcome suggests the feasibility of clinical applications through using adult BMC transplantation (Lunde et al. 2005). Despite these promising advances, only modest improvements in cardiac function have been reported (Strauer et al. 2005, Schachinger et al. 2006a, b) that may be due to lack of specific targeting and/or retention and survival of the BMCs at the myocardial injury site.

Human embryonic stem cells (hESCs) have been shown to have the ability to differentiate into structural and functional cardiomyocytes (Gepstein 2002, Leor et al. 2007). A number of reports have shown that the pluripotent capacity of hESCs not only stabilized myocardial infarcted hearts in rodents but also preserved the cardiac functions (Laflamme et al. 2007, Cao et al. 2008, Graichen et al. 2008). Two individual research groups have also demonstrated the repeatable protocol of directed guided differentiation of ESCs into cardiomyocytes and brought it into clinical trials (Behfar et al. 2008, Murry and Keller 2008). However, the relevant ethical issue in embryonic stem cell (ESC) research is still debated. Additionally, other potential stem cell candidates without the ethical issue are resident cardiac stem cells and induced pluripotent stem cells (iPS). Cardiosphere is the typical example of cardiac stem cells, which is a spherical cluster of cells that can be obtained from a cardiac biopsy. Cardiosphere-derived cells cultured in vitro can form other populations of resident cardiac progenitors such as c-kit+ and SCA-1+ cells (Smith et al. 2007). Several studies have indicated the effects of cardiac stem cells on improving LV function, remodeling, and reducing infarct size (Beltrami et al. 2003, Li et al. 2003, Oh et al. 2003). Furthermore, the most exciting breakthrough in stem cell research is the induction of pluripotent stem cells from fibroblasts explored by Takahashi and Yamanaka (2006). Zhang et al. demonstrated a successful study of differentiating iPS into functional myocytes (Zhang et al. 2009). Nelson et al. also highlighted the potential of iPS to treat acute MI (Nelson et al. 2009). Although the strategy of using iPS is promising, more studies are needed to confirm the implication before

TABLE 16.1

Various Cell Sources

Sources	Cell Types	References
	Fibroblast	Oh et al. (2003), Takahashi and Yamanaka (2006), Zhang et al. (2009)
	Skeletal myoblast	Christman et al. (2004), Beltrami et al. (2003), Li et al. (2009)
	Cardiomyocytes	Kobayashi et al. (2008a,b), Shimizu et al. (2002)
	ESC	Zimmermann et al. (2006), Kofidis et al. (2005a,b), Kellar et al. (2005), Jin et al. (2009), Piao et al. (2007), Yu et al. (2009)
	HSC	Lee et al. (2004), Andrieu-Soler et al. (2005), Nillesen et al. (2007)
	MSC	Rehman et al. (2003), Urbich and Dimmeler (2004), Kobayashi et al. (2008a,b), Shimizu et al. (2006), Huang et al. (2009)
	Cardiac stem cell	Masters et al. (2005), Lionetti et al. (2010), Dobner et al. (2009), Zhang (2003)
	iPS	Okano et al. (1993, 1995)

carrying out clinical trials. Various cell sources that have been evaluated are listed in Table 16.1.

Despite the enthusiasm and advances of stem cell therapy for myocardial repair and regeneration, technological hurdles need to be overcome before stem cell therapy can become a viable option in the treatment of myocardial injury. The type of stem cell for myocardial repair as well as the optimal time for stem cell transplantation still requires investigation. Technological hurdles as reliability of cell isolation to obtain a pure cell population and expansion of cells to produce sufficient numbers to effectively treat heart failure need to be overcome. Likewise, practical issues as stem cell delivery, engraftment, electrical integration, and safety need to be assessed.

16.2.2 Gene and Protein Therapy

Therapeutic proteins have potential to be used as medical treatments—we call it "protein therapy"—and for instance, growth hormone or insulin can be applied to patients for specific protein malfunctions or deficiency. Protein therapy can also be used to inhibit or activate signaling pathways such as blocking blood supply to tumors or increasing new blood vessel formation. Over the past decade, there have been great developments in gene therapy, which can be considered as a form of protein therapy. In contrast to direct administration of proteins, gene therapy has been proposed to deliver therapeutic genes to target cells by carriers, called vectors.

Gene therapy works by placing into a cell a defined gene to either replace a defective gene or to increase the amount of a specific gene in a targeted cell/tissue in order to produce a higher amount of the desired protein. Although transfecting a specific gene into the host cell may provide more sustained and effective therapeutic effects, there are still a lot of uncertainties behind this approach (i.e., safety issues such as host immune responses and oncogenesis, and design and efficiency of appropriate vectors) (Hernandez and Evers 1999). Compared with these unsolved issues, protein delivery provides a less risky and relative easy way to be administrated to patients; however, protein denaturation is the major obstacle in protein therapy. Recently, the approach of using biopolymers as drug delivery systems has been researched. Cai et al. and Lee et al. individually used chitosan polymers loaded with TGF-β1 for controlled release of growth factor. They both found that these polymers enhanced chondrocyte proliferation and ECM synthesis compared with nonencapsulated TGF-β1 (Lee et al. 2004, Cai et al. 2007). Andrieu-Soler et al. encapsulated recombinant human glial cell line–derived neurotrophic factor (rhGDNF) in poly(D,L-lactide-co-glycolide) (PLGA) spheres and injected them into mice with retinal disease. These spheres delayed retinal degeneration when compared to control spheres (Andrieu-Soler et al. 2005). Nillesen et al. implanted collagen–heparin scaffolds containing both FGF2 and VEGF subcutaneously into rats. These scaffolds increased blood vessel formation and maturation (Nillesen et al. 2007). Lee et al. encapsulated the heat shock protein (HSP27) in PLGA microspheres and mixed them with alginate. The protein maintained its bioactivity and recovered the proliferation of cardio myoblasts cultured under hypoxic conditions (Davis et al. 2005a,b). These microspheres have not yet been investigated in vivo.

Other researchers have focused on delivery of growth factors into the heart. Arras et al. injected fibroblast growth factor (FGF) microspheres into porcine hearts via catheter. The FGF successfully translocated into the endothelial cell nuclei, suggesting that angiogenesis can be promoted without immune reactions (Arras et al. 1998). Kobayashi et al. investigated the effects of an erythropoietin (EPO)-gelatin hydrogel on a rabbit myocardial infarction model. The results showed that the EPO-hydrogel delivery system improved LV remodeling and function and induced angiogenesis without causing any observable side effects (Kobayashi et al. 2008a,b).

16.2.3 Biomaterials

A wide range of materials from synthetic organic materials (e.g., polyglycolic acid [PGA], polylactic acid, or polyethylene glycol [PEG]) to naturally derived materials (e.g., gelatin, collagen, or matrigel) have been used for myocardial repair (Eschenhagen et al. 1997, Dar et al. 2002, Nugent and Edelman 2003, Christman and Lee 2006). The similar structural characteristics to biological tissues and their ability to degrade over time allow biopolymers to provide the matrix for tissue regeneration. For example, compared with a bioengineered cardiac graft, the injectable scaffolds, which maintain the liquid form until after injection, provide a flexible approach. After solidification in vivo, the engineered biopolymer matrices have the potential to provide mechanical support and promote cell migration and proliferation and angiogenesis (Christman et al. 2004, Kofidis et al. 2004, Dai et al. 2005, Huang et al. 2005).

It is crucial to find the appropriate polymeric material for the myocardium repair, because the construct should be mechanically robust but also pliable and meet the physiologically relevant environment for cardiomyocytes (McDevitt et al. 2003). Furthermore, injecting a hydrogel is also a common method to introduce the scaffold directly into the injured myocardium. Using an injectable hydrogel to repair both acute and chronic MI models has been demonstrated. Hydrogels not only produce neovascularization but also sustained increase of wall thickness and improvement of heart function (Christman et al. 2004, Huang et al. 2005, Kofidis et al. 2005a,b, Landa et al. 2008, Yu et al. 2009a,b). Additionally, biomaterial-based tissues have been constructed and implanted successfully in vivo to regenerate tissues (e.g., bone or cartilage) and organs (e.g., bladder). To date, myocardium is still one of the most challenging tissues that we can construct and regenerate through tissue engineering approaches because this tissue has a low oxygen level and is composed of thin and complex multilayers (Jeon et al. 2007, Pattison et al. 2007, Khan et al. 2008). The high oxygen demand and heterogeneity of cardiac muscles limit the application of conventional tissue engineering approaches for myocardial regeneration. However, three-dimensional scaffold-based patches have been implanted in the epicardium, resulting in successful angiogenesis (Kellar et al. 2001).

ECM not only provides the scaffolding and serves as a delivery matrix for cells but can also regulate cell function. Signaling transduction induced by cell adhesion or cell–substrate interaction through the transmembrane proteins such as integrin or syndecan-4 (Bellin et al. 2009) can modulate gene expressions in a living cell and also influence cellular behaviors such as migration, differentiation, and proliferation, as well as apoptosis (Hynes 1992, Vogel and Baneyx 2003). Surface modifications of biomaterials with molecule-recognition sites can enhance the interactions between cells and their adherent substrates (Evangelista et al. 2007). Specific ligands for cell adhesion such as arginine–glycine–aspartic acid (RGD) sequences can be immobilized on biomaterial surface to promote cell anchorage (Rosso et al. 2005). Peptide-modified biomaterials not only coexist with tissues but can also influence the cardiac microenvironment at the molecular level, which is critical for tissue reconstruction (Davis et al. 2005a,b, Hennessy et al. 2008). The biomaterials that have been studied in myocardial tissue engineering are summarized in the Table 16.2.

16.3 Functional Biomaterial Approaches

Materials as the scaffolding are the basis of myocardial tissue engineering strategies. These materials can be classified into several categories based on the methods to introduce the engineered tissues, either alone or cocktail

TABLE 16.2

Current Biomaterials Used in Myocardial Tissue Engineering

Sources	Materials	References
Naturally derived materials	Collagen	Huang et al. (2005), Hennessy et al. (2008), Penn and Mal (2006), Siminiak and Kurpisz (2003)
	Chitosan	Formigli et al. (2009), Menasche et al. (2008)
	Fibrin	Huang et al. (2005), Christman et al. (2004), Menasche et al. (2008), Huang et al. (2009)
	Gelatin	Schuh et al. (2009), Goncalves et al. (2009)
	Alginate	Pattison et al. (2007), Kellar et al. (2001), Kobayashi et al. (2008a,b), Lee et al. (2004), Yu et al. (2009)
	Matrigel	Huang et al. (2005), Kofidis et al. (2004), Hennessy et al. (2008), Siminiak and Kurpisz (2003)
	HA	Leor et al. (1997), El Oakley et al. (2001)
	MC	Rehman et al. (2003), Urbich and Dimmeler (2004)
	Self-assembling peptides	Wu et al. (2004)
Synthetic materials	PEG	Kawamoto et al. (2001)
	PGA	Makino et al. (1999)
	PLGA	Toma et al. (2002), Schuleri et al. (2008)
	Vicryl	Forrester et al. (2003)
	Poly(caprolactone) (PCL)	Lunde et al. (2005), Gepstein (2002)
	PIPAAm	Leor et al. (2007), Laflamme et al. (2007), Graichen et al. (2008), Cao et al. (2008)

therapy with cells, genes, or cytokines. The different approaches for applying tissue engineering for myocardial repair consists of in vitro approach, ex vivo approach, and in situ approach.

16.3.1 In Vitro Approach

Biomaterial-based scaffolds seeded with cells in vitro and the subsequent implantation of these scaffolds in vivo was the first generation of myocardial tissue engineering. Li et al. isolated fetal cardiomyocytes, stomach smooth muscle cells, and skin fibroblasts and then cultured them in gelatin meshes. The results showed different cell types can grow and proliferate in gelatin scaffolds up to 1 month while biomaterials degraded overtime (Li et al. 2000). An in vivo study on cryoinjured rats was also conducted; 5 weeks later, the cardiomyocyte–gelatin meshes formed a beating graft after the implantation (Li et al. 1999). Leor et al. constructed a three-dimensional patch composed of alginate scaffolds and fetal cardiomyocytes and then grafted into the infarcted myocardium. The biografts increased neovascularization and attenuated LV dilatation and failure (Leor et al. 2000). Except for seeding the cells into porous scaffolds, Zimmermann et al. mixed the cardiac myocytes with collagen type-I, matrigel and culture medium to form the engineered heart tissue (EHT) maintaining contractility after 28 days posttransplantation; moreover, this graft showed electric coupling with the host myocardium (Zimmermann et al. 2002, 2006). Kofidis et al. also used the collagen type-I as the scaffold but mixed it with ESCs, and similar beneficial effects were obtained (Kofidis et al. 2005a,b). Synthetic materials have also been utilized to form a cardiac patch. Kellar et al. created Dermagraft, which was a human tissue composed of human dermal fibroblasts cultured on a Vicryl mesh (poly(glycolide)/lactide). The Dermagraft implantation onto SCID (severe combined immune deficiency) mice infarcted myocardium attenuated the deterioration of ejection function compared with the control group as well as revascularization (Kellar et al. 2005). Jin et al. investigated the effects of MSCs with elastic biodegradable poly(lactide-co-epsilon-caprolactone) (PLCL) scaffolds on a rat MI model. The MSCs in PLCL groups had significant higher ejection fraction and decreased of infarct area compared with the control (Jin et al. 2009). Piao et al. used another PCL-based copolymer, poly-glycolide-co-caprolactone (PGCL); they seeded the scaffolds with bone-marrow-derived mononuclear cells (BMMNC). The PGCL-based cardiac patch influenced cell migration and differentiation to cardiomyocytes. The lessening of LV remodeling and progressive LV systolic dysfunction was also observed (Piao et al. 2007).

16.3.2 Ex Vivo Approach

The ex vivo approach for myocardial tissue engineering requires the biopolymer scaffold during in vitro culture to form a cell-sheet tissue. Cells secret ECM molecules (e.g., fibronection or vitronectin) ex vivo and then can be easily detached through a specific material before implantation. The cell-sheet engineering technique was proposed by Okano. One of the temperature-responsive polymers, poly(N-isopropyl acrylamide) (PIPAAm), is hydrophilic below 32°C but becomes hydrophobic above 32°C. With this unique property, single cell sheets detached from the cell culture dish and remained intact (Okano et al. 1993, 1995). Shimizu et al. created neonatal rat cardiomyocyte sheets and also transplanted them into the epicardium of rats. Long-term survival of pulsatile cardiac grafts was confirmed up to 12 weeks (Shimizu et al. 2002a,b, 2006). They also found that the layered embryonic cardiomyocyte sheets pulsed spontaneously and synchronously, altering their characteristic pulsing frequency with applied electric stimulation transmitted across the sheets (Shimizu et al. 2002a,b). Coculture of endothelial cells and cardiomyocytes cell sheets improved the previous therapeutic effect by increasing revascularization in ischemic hearts (Sekine et al. 2008). More recently, Chen et al. developed another cell-sheet harvest system by using biologically derived methylcellulose (MC) hydrogel. The MC/PBS (phosphate buffered saline) hydrogel is aqueous at 4°C but solidifies at the temperature of ~25°C (Chen et al. 2006). They successfully created an MSC sheet and transplanted the fragments of this cell sheet into skeletal muscles of a syngeneic rat model by local injection. Significantly, more MSCs retained in the local skeletal muscle for the group injected with fragmented MSC sheets than that injected with dissociated MSCs (Chen et al. 2007). Wei et al. examined the MSCs cell sheet on a rat MI model, and the MSCs cardiac patch restored the dilated LV and preserved cardiac functions after infarction. In addition, expressions of angiogenic cytokines (e.g., bFGF, vWF, and PDGF-B) and cardioprotective factors (e.g., IGF-1 and HGF) were significantly increased in the patched group (Wei et al. 2008). Chen et al. further constructed a sandwich cardiac patch by layering porous acellular bovine pericardia and MSCs sheets. Restoration of cardiac functions and neovascularization was shown in a chronic MI syngeneic Lewis rat model (Chen et al. 2008).

16.3.3 In Situ Approach

Another tissue engineering approach is the utilization of an injectable biomaterial which can be delivered to the injured myocardium in situ. These biomaterials can also act as a matrix for cells and genes, or as a drug delivery vehicle. In addition, in situ engineering offers a less invasive tissue engineering approach for myocardial reconstruction, avoiding surgery for the implantation of in vitro engineered grafts. Mechanically, injectable biopolymers may deform due to the dynamically loaded myocardial environment and align their matrices with the host tissue, providing the structural support to the weak LV walls (Christman and Lee 2006). Christman et al. initially demonstrated the in situ myocardial engineering approach which showed that fibrin glue with or without skeletal myoblasts can prevent negative remodeling and stimulate angiogenesis after an acute MI in rat model (Christman et al. 2004). Huang et al. also confirmed that the acellular injection of matrigel, collagen type-I, and fibrin enhanced the neovascularization (Huang et al. 2005). Except for making the biograft in vitro as described earlier, Kofidis et al. examined the in situ approach of mixing matrigel and ESCs to obtain the similar beneficial effects on myocardial function (Kofidis et al. 2004). Hyaluronic acid (HA)—a polysaccharide structure and a major component of the cardiac jelly during heart morphogenesis—is also an employable biological material for cardiac tissue repair (Masters et al. 2005). Injection of a hyaluronan mixed with ester of butyric and retinoic acid into infarcted rat hearts was also found to be capable of providing substantial cardiovascular repair and the recovery of myocardial performance (Lionetti et al. 2010). Some of synthetic materials such as nondegradable PEG are also considered as injectable biomaterials. For example, Dobner et al. injected PEG gel into permanent left anterior descending artery ligation rat model; the results indicated the retardation of the acute MI stage, while no prevention of the dilation at the later stage was observed (Dobner et al. 2009). Another novelty of in situ tissue engineering strategy is the use of self-assembling peptides. Self-assembling peptides can be designed with alternating hydrophobic and hydrophilic domains of various amino acids. These peptides maintain aqueous under low pH or low osmolarity condition. When they are injected in situ, due to the changes of the physiological ionic strength and pH, single peptides transform into fibril structures and then form hydrated gels (Zhang 2003, Davis et al. 2005a,b). Davis et al. demonstrated that nanofiber microenvironments created through self-assembling peptides were detectable within the myocardium; additionally, these microenvironments recruited progenitor cells that expressed endothelial markers (Davis et al. 2005a,b).

Compared with in vitro approach, the main advantage of in situ approach is that it allows the engineered tissues to directly cross talk with the injured myocardium. The barrier of biograft and scar area limits the regeneration of the host tissue itself. Studies have shown that the injectable scaffolds have the capability of influencing the local tissues based on their chemical and physical properties and had the potential to create new myocardium. Especially, the scar is no longer considered as dead but an active tissue. The cytokine effects and metabolic activities of myofibroblasts demonstrated the neovascularization and tissue regeneration prospects (Sun et al. 2002). Therefore, the in situ approach carries greater flexibility of repairing the heart from a global organ construct perspective.

The majority of tissue engineering approaches for myocardial repair and regeneration have been aimed at treating the acute MI and prevention of the negative remodeling that result in progressive LV dilatation, LV wall thinning, and a decrease in LV function. It was suggested that biopolymers alter the material properties of the LV, leading to the development of in situ tissue engineering for chronic heart failure (Christman et al. 2004). In a chronic MI model, Yu et al. showed that alginate injected into the LV aneurysm was able to reshape and restore LV geometry, resulting in sustainably improved cardiac function (Yu et al. 2009a,b). Moreover, the RGD-peptide-functionalized alginate further increased the capability of neovascularization (Yu et al. 2009a,b). Consistent with the law of Laplace, it has been hypothesized that increasing LV wall thickness leads to decreased wall stress, resulting in improved myocardial function (Dai et al. 2005, Wall et al. 2006, Wenk et al. 2009, Yu et al. 2009a,b).

16.4 Summary and Future Directions

Numerous studies have demonstrated that various tissue engineering approaches can be used to repair damaged myocardium with various cell types, scaffolds, genes, and drug therapies. Although several beneficial effects (e.g., the improvement of LV function, neovascularization, prevention of infarct expansion, and stem cell recruitments) have been exhibited, there are still questions that remain to be investigated; for example, the underlining mechanism of cardiac repair needs to be clarified. It seems that most cell types could work. However, only a few reports have shown that the transplanted cell differentiated into cardiomyocytes. Detailed mechanisms behind single cells that modulate cardiac repair are the major guidance for further study.

Biomaterials have been proven in all studies to play an essential role in cardiac tissue regeneration (with/without combining the scaffolds with cells, gene, and drugs). Although investigators have demonstrated the promising potential of cell therapy, the major obstacle is cell survival after transplanted in vivo—most cells die when injected alone (Itescu et al. 2003). Even if injected cells survive, cells do not integrate with the native tissue naturally. The transplanted cells tended to form an isolated area instead of physiologically coupling with the myocardium (Dowell et al. 2003, Rubart et al. 2003, Murry et al. 2004). It has also been demonstrated that acellular approach may be sufficient to recruit autologous stem cells, thicken LV wall, increase ejection fraction, and stimulate blood vessels formation (Christman et al. 2004, Huang et al. 2005, Gaballa et al. 2006, Landa et al. 2008, Dobner et al. 2009, Ryan et al. 2009, Yu et al. 2009a,b). Biomaterial scaffolds do not simply provide structural support to the weak LV wall; they also create a microenvironment for cardiac tissue repair. Cell–substrate interaction is a series of signaling cascade. An optimal matrix material is important for cell attachment, proliferation, and differentiation. The ideal substrate may differ between the various types of cells (Huang et al. 2007). Instead of simple materials, the innovation of the custom-designed microenvironment is the main focus in the future research directions.

For instance, as described previously, specific peptide sequences such as RGD can be conjugated to alginate with basic chemistry methods (Rowley et al. 1999). RGD interacts with integrin $\alpha_v\beta_3$ and influences endothelial cell migration (Nam et al. 2003, Nisato et al. 2003). Kong et al. demonstrated that different ligand densities and spaces altered cell growth and gene expression (Kong et al. 2007). Comisar et al. showed that different nanopatterns of peptides on the substrate influenced osteoblast behaviors (Comisar et al. 2007). Boontheekul et al. investigated the effects of molecular weight of hydrogel on matrix degradation and cellular responses (Kong et al. 2004, Boontheekul et al. 2007). Except for the surface modification, the interactions between cells and biomaterials also alter dramatically depending on the mechanical properties of biomaterials (Engler et al. 2006). Nevertheless, these studies have not been tailored made for myocardial tissue. More detailed investigations of suitable degradation rate, specific peptide sequences, porosity, and other chemical and mechanical factors that influence the microenvironment between cells and their adherent substrates are still waiting to be understood.

The innovation of so-called smart biomaterials has attracted research-ers' interests. MMPs recognize specific amino acids sequences, and MMPs are potential triggers for smart biomaterial behaviors (Anderson et al. 2004, Behfar et al. 2008). For example, hydrogels containing both MMP-degradable sites and tethered adhesive ligands have been synthesized. The MMP sites enable local cells to control matrix remodeling such that these cells replace the transplanted materials with tissue in bone repair (Lutolf et al. 2003). Since it has been proved that various MMPs contribute to adverse myocardial dis-ease after the MI (Spinale 2007), a custom-designed biomatrix sensitive to MMPs concentration can be a potential time-release smart material. Lee et al. demonstrated a growth factor VEGF-binding alginate hydrogel sensitive to mechanical stimuli (Lee et al. 2000). Similarly, the concept can be applied on the dynamic mechanically loaded myocardium.

Acknowledgments

This work was partially funded by NIH R21 HL084121, NIH R44 HL079720, the California Institute of Regenerative Medicine (RC1-00124-1) (R.J.L.), and LoneStar Heart Inc. (R. J. Lee is a consultant to LoneStar Heart Inc). The authors would also like to thank the start-up funds from the Department of Chemical Engineering, National Taiwan University, Taiwan (to J. Yu), and the Institute of Nanoengineering and Microsystems, National Tsing Hua University, Taiwan (to C.-M. Cheng), respectively.

References

Alpert JS, Thygesen K, Antman E, Bassand JP. Myocardial infarction redefined—A consensus document of The Joint European Society of Cardiology/American College of Cardiology Committee for the redefinition of myocardial infarction. *J Am Coll Cardiol* 2000;36(3):959–969.

Anderson DG, Burdick JA, Langer R. Materials science. Smart biomaterials. *Science* 2004;305(5692):1923–1924.

Andrieu-Soler C, Aubert-Pouessel A, Doat M, Picaud S, Halhal M, Simonutti M et al. Intravitreous injection of PLGA microspheres encapsulating GDNF promotes the survival of photoreceptors in the rd1/rd1 mouse. *Mol Vis* 2005;11:1002–1011.

Arras M, Mollnau H, Strasser R, Wenz R, Ito WD, Schaper J et al. The delivery of angiogenic factors to the heart by microsphere therapy. *Nat Biotechnol* 1998;16(2):159–162.

Bajzer C. Acute myocardial infarction. In *Medicine Index*. Cleveland, OH: Cleveland Clinic Foundation, 2002, pp. 222–226.

Balsam, LB, Wagers, AJ, Christensen, JL, Kofidis, T, Weissman, IL, Robbins, RC. Haematopoietic stem cells adopt mature haematopoietic fates in ischaemic myocardium. *Nature* 2004;428:668–673.

Behfar A, Faustino RS, Arrell DK, Dzeja PP, Perez-Terzic C, Terzic A. Guided stem cell cardiopoiesis: Discovery and translation. *J Mol Cell Cardiol* 2008;45(4):523–529.

Bellin RM, Kubicek JD, Frigault MJ, Kamien AJ, Steward RL, Jr., Barnes HM et al. Defining the role of syndecan-4 in mechanotransduction using surface-modification approaches. *Proc Natl Acad Sci USA* 2009;106(52):22102–22107.

Beltrami AP, Barlucchi L, Torella D, Baker M, Limana F, Chimenti S et al. Adult cardiac stem cells are multipotent and support myocardial regeneration. *Cell* 2003;114(6):763–776.

Boontheekul T, Hill EE, Kong HJ, Mooney DJ. Regulating myoblast phenotype through controlled gel stiffness and degradation. *Tissue Eng* 2007;13(7):1431–1442.

Braunwald E, Bristow MR. Congestive heart failure: Fifty years of progress. *Circulation* 2000;102(20 Suppl 4):IV14–IV23.

Cai DZ, Zeng C, Quan DP, Bu LS, Wang K, Lu HD et al. Biodegradable chitosan scaffolds containing microspheres as carriers for controlled transforming growth factor-beta1 delivery for cartilage tissue engineering. *Chin Med J (Engl)* 2007;120(3):197–203.

Cao F, Wagner RA, Wilson KD, Xie X, Fu JD, Drukker M et al. Transcriptional and functional profiling of human embryonic stem cell-derived cardiomyocytes. *PLoS One* 2008;3(10):e3474.

Chen CH, Chang Y, Wang CC, Huang CH, Huang CC, Yeh YC et al. Construction and characterization of fragmented mesenchymal-stem-cell sheets for intramuscular injection. *Biomaterials* 2007;28(31):4643–4651.

Chen CH, Tsai CC, Chen W, Mi FL, Liang HF, Chen SC et al. Novel living cell sheet harvest system composed of thermoreversible methylcellulose hydrogels. *Biomacromolecules* 2006;7(3):736–743.

Chen CH, Wei HJ, Lin WW, Chiu I, Hwang SM, Wang CC et al. Porous tissue grafts sandwiched with multilayered mesenchymal stromal cell sheets induce tissue regeneration for cardiac repair. *Cardiovasc Res* 2008;80(1):88–95.

Christman KL, Fok HH, Sievers RE, Fang Q, Lee RJ. Fibrin glue alone and skeletal myoblasts in a fibrin scaffold preserve cardiac function after myocardial infarction. *Tissue Eng* 2004;10(3–4):403–409.

Christman KL, Lee RJ. Biomaterials for the treatment of myocardial infarction. *J Am Coll Cardiol* 2006;48(5):907–913.

Comisar WA, Kazmers NH, Mooney DJ, Linderman JJ. Engineering RGD nanopatterned hydrogels to control preosteoblast behavior: A combined computational and experimental approach. *Biomaterials* 2007;28(30):4409–4417.

Cotran RS, Kumar V, Robbins SL. *Robbins Pathologic Basis of Disease.* 5th edn. Philadelphia, PA: WB Saunders Co., 1994.

Dai W, Wold LE, Dow JS, Kloner RA. Thickening of the infarcted wall by collagen injection improves left ventricular function in rats: A novel approach to preserve cardiac function after myocardial infarction. *J Am Coll Cardiol* 2005;46(4):714–719.

Dar A, Shachar M, Leor J, Cohen S. Optimization of cardiac cell seeding and distribution in 3D porous alginate scaffolds. *Biotechnol Bioeng* 2002;80(3):305–312.

Davis ME, Hsieh PC, Grodzinsky AJ, Lee RT. Custom design of the cardiac microenvironment with biomaterials. *Circ Res* 2005a;97(1):8–15.

Davis ME, Motion JP, Narmoneva DA, Takahashi T, Hakuno D, Kamm RD et al. Injectable self-assembling peptide nanofibers create intramyocardial microenvironments for endothelial cells. *Circulation* 2005b;111(4):442–450.

Dobner S, Bezuidenhout D, Govender P, Zilla P, Davies N. A synthetic non-degradable polyethylene glycol hydrogel retards adverse post-infarct left ventricular remodeling. *J Card Fail* 2009;15(7):629–636.

Dowell JD, Rubart M, Pasumarthi KB, Soonpaa MH, Field LJ. Myocyte and myogenic stem cell transplantation in the heart. *Cardiovasc Res* 2003;58(2):336–350.

Dowling RD, Etoch SW, Stevens KA, Johnson AC, Gray LA, Jr. Current status of the AbioCor implantable replacement heart. *Ann Thorac Surg* 2001;71(3 Suppl):S147–S149; discussion S183–S144.

El Oakley RM, Ooi OC, Bongso A, Yacoub MH. Myocyte transplantation for myocardial repair: A few good cells can mend a broken heart. *Ann Thorac Surg* 2001;71(5):1724–1733.

Engler AJ, Sen S, Sweeney HL, Discher DE. Matrix elasticity directs stem cell lineage specification. *Cell* 2006;126(4):677–689.

Eschenhagen T, Fink C, Remmers U, Scholz H, Wattchow J, Weil J et al. Three-dimensional reconstitution of embryonic cardiomyocytes in a collagen matrix: A new heart muscle model system. *Faseb J* 1997;11(8):683–694.

Evangelista MB, Hsiong SX, Fernandes R, Sampaio P, Kong HJ, Barrias CC et al. Upregulation of bone cell differentiation through immobilization within a synthetic extracellular matrix. *Biomaterials* 2007;28(25):3644–3655.

Fenton DE. Myocardial infarction. eMedicine Clinical Reference, 2007 Sep.

Formigli L, Zecchi-Orlandini S, Meacci E, Bani D. Skeletal myoblasts for heart regeneration and repair: State of the art and perspectives on the mechanisms for functional cardiac benefits. *Curr Pharm Des* 2009;24:24.

Forrester JS, Price MJ, Makkar RR. Stem cell repair of infarcted myocardium: An overview for clinicians. *Circulation* 2003;108(9):1139–1145.

Franco-Cereceda A, Lockowandt U, Olsson A, Bredin F, Forssell G, Owall A et al. Early results with cardiac support device implant in patients with ischemic and non-ischemic cardiomyopathy. *Scand Cardiovasc J* 2004;38(3):159–163.

Gaballa MA, Sunkomat JN, Thai H, Morkin E, Ewy G, Goldman S. Grafting an acellular 3-dimensional collagen scaffold onto a non-transmural infarcted myocardium induces neo-angiogenesis and reduces cardiac remodeling. *J Heart Lung Transplant* 2006;25(8):946–954.

Gepstein L. Derivation and potential applications of human embryonic stem cells. *Circ Res* 2002;91(10):866–876.

Goncalves GA, Vassallo PF, Dos Santos L, Schettert IT, Nakamuta JS, Becker C et al. Intramyocardial transplantation of fibroblasts expressing vascular endothelial growth factor attenuates cardiac dysfunction. *Gene Ther* 2009;10:10.

Graichen R, Xu X, Braam SR, Balakrishnan T, Norfiza S, Sieh S et al. Enhanced cardiomyogenesis of human embryonic stem cells by a small molecular inhibitor of p38 MAPK. *Differentiation* 2008;76(4):357–370.

Hennessy KM, Clem WC, Phipps MC, Sawyer AA, Shaikh FM, Bellis SL. The effect of RGD peptides on osseointegration of hydroxyapatite biomaterials. *Biomaterials* 2008;24:24.

Hernandez A, Evers BM. Functional genomics: Clinical effect and the evolving role of the surgeon. *Arch Surg* 1999;134(11):1209–1215.

Huang NF, Lam A, Fang Q, Sievers RE, Li S, Lee RJ. Bone marrow-derived mesenchymal stem cells in fibrin augment angiogenesis in the chronically infarcted myocardium. *Regen Med* 2009;4(4):527–538.

Huang NF, Lee RJ, Li S. Chemical and physical regulation of stem cells and progenitor cells: Potential for cardiovascular tissue engineering. *Tissue Eng* 2007;13(8):1809–1823.

Huang NF, Yu J, Sievers R, Li S, Lee RJ. Injectable biopolymers enhance angiogenesis after myocardial infarction. *Tissue Eng* 2005;11(11–12):1860–1866.

Hynes RO. Integrins: Versatility, modulation, and signaling in cell adhesion. *Cell* 1992;69(1):11–25.

Itescu S, Schuster MD, Kocher AA. New directions in strategies using cell therapy for heart disease. *J Mol Med* 2003;81(5):288–296.

Jeon YH, Choi JH, Sung JK, Kim TK, Cho BC, Chung HY. Different effects of PLGA and chitosan scaffolds on human cartilage tissue engineering. *J Craniofac Surg* 2007;18(6):1249–1258.

Jin J, Jeong SI, Shin YM, Lim KS, Shin H, Lee YM et al. Transplantation of mesenchymal stem cells within a poly(lactide-co-epsilon-caprolactone) scaffold improves cardiac function in a rat myocardial infarction model. *Eur J Heart Fail* 2009;11(2):147–153.

Jones RH. The year in cardiovascular surgery. *J Am Coll Cardiol* 2008;51(17):1707–1718.

Kawamoto A, Gwon HC, Iwaguro H, Yamaguchi JI, Uchida S, Masuda H et al. Therapeutic potential of ex vivo expanded endothelial progenitor cells for myocardial ischemia. *Circulation* 2001;103(5):634–637.

Kellar RS, Landeen LK, Shepherd BR, Naughton GK, Ratcliffe A, Williams SK. Scaffold-based three-dimensional human fibroblast culture provides a structural matrix that supports angiogenesis in infarcted heart tissue. *Circulation* 2001;104(17):2063–2068.

Kellar RS, Shepherd BR, Larson DF, Naughton GK, Williams SK. Cardiac patch constructed from human fibroblasts attenuates reduction in cardiac function after acute infarct. *Tissue Eng* 2005;11(11–12):1678–1687.

Khan Y, Yaszemski MJ, Mikos AG, Laurencin CT. Tissue engineering of bone: Material and matrix considerations. *J Bone Joint Surg Am* 2008;90(Suppl 1):36–42.

Klodell CT, Jr., Aranda JM, Jr., McGiffin DC, Rayburn BK, Sun B, Abraham WT, Pae WE, Jr., Boehmer JP, Klein H, Huth C. Worldwide surgical experience with the Paracor HeartNet cardiac restraint device. *J Thorac Cardiovasc Surg* 2008;135:188–195.

Kobayashi H, Minatoguchi S, Yasuda S, Bao N, Kawamura I, Iwasa M et al. Postinfarct treatment with an erythropoietin-gelatin hydrogel drug delivery system for cardiac repair. *Cardiovasc Res* 2008a;9:9.

Kobayashi H, Shimizu T, Yamato M, Tono K, Masuda H, Asahara T et al. Fibroblast sheets co-cultured with endothelial progenitor cells improve cardiac function of infarcted hearts. *J Artif Organs* 2008b;11(3):141–147.

Kofidis T, de Bruin JL, Hoyt G, Ho Y, Tanaka M, Yamane T et al. Myocardial restoration with embryonic stem cell bioartificial tissue transplantation. *J Heart Lung Transplant* 2005a;24(6):737–744.

Kofidis T, de Bruin JL, Hoyt G, Lebl DR, Tanaka M, Yamane T et al. Injectable bioartificial myocardial tissue for large-scale intramural cell transfer and functional recovery of injured heart muscle. *J Thorac Cardiovasc Surg* 2004;128(4):571–578.

Kofidis T, Lebl DR, Martinez EC, Hoyt G, Tanaka M, Robbins RC. Novel injectable bioartificial tissue facilitates targeted, less invasive, large-scale tissue restoration on the beating heart after myocardial injury. *Circulation* 2005b;112(9 Suppl):I-173–I-177.

Kong HJ, Hsiong S, Mooney DJ. Nanoscale cell adhesion ligand presentation regulates nonviral gene delivery and expression. *Nano Lett* 2007;7(1):161–166.

Kong HJ, Kaigler D, Kim K, Mooney DJ. Controlling rigidity and degradation of alginate hydrogels via molecular weight distribution. *Biomacromolecules* 2004;5(5):1720–1727.

Laflamme MA, Chen KY, Naumova AV, Muskheli V, Fugate JA, Dupras SK et al. Cardiomyocytes derived from human embryonic stem cells in pro-survival factors enhance function of infarcted rat hearts. *Nat Biotechnol* 2007;25(9):1015–1024.

Landa N, Miller L, Feinberg MS, Holbova R, Shachar M, Freeman I et al. Effect of injectable alginate implant on cardiac remodeling and function after recent and old infarcts in rat. *Circulation* 2008;117(11):1388–1396.

Lederman DM, Kung RT, McNair DS. Therapeutic potential of implantable replacement hearts. *Am J Cardiovasc Drugs* 2002;2(5):297–301.

Lee JE, Kim SE, Kwon IC, Ahn HJ, Cho H, Lee SH et al. Effects of a chitosan scaffold containing TGF-beta1 encapsulated chitosan microspheres on in vitro chondrocyte culture. *Artif Organs* 2004;28(9):829–839.

Lee KY, Peters MC, Anderson KW, Mooney DJ. Controlled growth factor release from synthetic extracellular matrices. *Nature* 2000;408(6815):998–1000.

Leor J, Aboulafia-Etzion S, Dar A, Shapiro L, Barbash IM, Battler A et al. Bioengineered cardiac grafts: A new approach to repair the infarcted myocardium? *Circulation* 2000;102(19 Suppl 3):III56–III61.

Leor J, Gerecht S, Cohen S, Miller L, Holbova R, Ziskind A et al. Human embryonic stem cell transplantation to repair the infarcted myocardium. *Heart* 2007;93(10):1278–1284.

Leor J, Prentice H, Sartorelli V, Quinones MJ, Patterson M, Kedes LK et al. Gene transfer and cell transplant: An experimental approach to repair a 'broken heart'. *Cardiovasc Res* 1997;35(3):431–441.

Li RK, Jia ZQ, Weisel RD, Mickle DA, Choi A, Yau TM. Survival and function of bioengineered cardiac grafts. *Circulation* 1999;100(19 Suppl):II63–II69.

Li Z, Lee A, Huang M, Chun H, Chung J, Chu P et al. Imaging survival and function of transplanted cardiac resident stem cells. *J Am Coll Cardiol* 2009; 53(14):1229–1240.

Li RK, Yau TM, Weisel RD, Mickle DA, Sakai T, Choi A et al. Construction of a bioengineered cardiac graft. *J Thorac Cardiovasc Surg* 2000;119(2):368–375.

Lionetti V, Cantoni S, Cavallini C, Bianchi F, Valente S, Frascari I et al. Hyaluronan mixed esters of butyric and retinoic acid affording myocardial survival and repair without stem cell transplantation. *J Biol Chem* 2010;285(13):9949–9961.

Lunde K, Solheim S, Aakhus S, Arnesen H, Abdelnoor M, Forfang K. Autologous stem cell transplantation in acute myocardial infarction: The ASTAMI randomized controlled trial. Intracoronary transplantation of autologous mononuclear bone marrow cells, study design and safety aspects. *Scand Cardiovasc J* 2005;39(3):150–158.

Lutolf MP, Weber FE, Schmoekel HG, Schense JC, Kohler T, Muller R et al. Repair of bone defects using synthetic mimetics of collagenous extracellular matrices. *Nat Biotechnol* 2003;21(5):513–518.

Makino S, Fukuda K, Miyoshi S, Konishi F, Kodama H, Pan J et al. Cardiomyocytes can be generated from marrow stromal cells in vitro. *J Clin Invest* 1999; 103(5):697–705.

Mann DL, Acker MA, Jessup M, Sabbah HN, Starling RC, Kubo, SH. Clinical evaluation of the CorCap Cardiac Support Device in patients with dilated cardiomyopathy. *Ann Thorac Surg* 2007;84:1226–1235.

Masters KS, Shah DN, Leinwand LA, Anseth KS. Crosslinked hyaluronan scaffolds as a biologically active carrier for valvular interstitial cells. *Biomaterials* 2005;26(15):2517–2525.

McDevitt TC, Woodhouse KA, Hauschka SD, Murry CE, Stayton PS. Spatially organized layers of cardiomyocytes on biodegradable polyurethane films for myocardial repair. *J Biomed Mater Res A* 2003;66(3):586–595.

Menasche P. Skeletal myoblasts for cardiac repair: Act II? *J Am Coll Cardiol* 2008;52(23):1881–1883.

Menasche P, Alfieri O, Janssens S, McKenna W, Reichenspurner H, Trinquart L et al. The Myoblast Autologous Grafting in Ischemic Cardiomyopathy (MAGIC) trial: First randomized placebo-controlled study of myoblast transplantation. *Circulation* 2008;117(9):1189–1200.

Murry CE, Keller G. Differentiation of embryonic stem cells to clinically relevant populations: Lessons from embryonic development. *Cell* 2008;132(4):661–680.

Murry CE, Soonpaa MI, Reinecke H, Nakajima H, Nakajima HO, Rubart M et al. Haematopoietic stem cells do not transdifferentiate into cardiac myocytes in myocardial infarcts. *Nature* 2004;428(6983):664–668.

Nagase H, Visse R, Murphy G. Structure and function of matrix metalloproteinases and TIMPs. *Cardiovasc Res* 2006;69(3):562–573.

Nam JO, Kim JE, Jeong HW, Lee SJ, Lee BH, Choi JY et al. Identification of the alphav-beta3 integrin-interacting motif of betaig-h3 and its anti-angiogenic effect. *J Biol Chem* 2003;278(28):25902–25909.

Nelson TJ, Martinez-Fernandez A, Yamada S, Perez-Terzic C, Ikeda Y, Terzic A. Repair of acute myocardial infarction by human stemness factors induced pluripotent stem cells. *Circulation* 2009;120(5):408–416.

Nillesen ST, Geutjes PJ, Wismans R, Schalkwijk J, Daamen WF, van Kuppevelt TH. Increased angiogenesis and blood vessel maturation in acellular collagen-heparin scaffolds containing both FGF2 and VEGF. *Biomaterials* 2007; 28(6):1123–1131.

Nisato RE, Tille JC, Jonczyk A, Goodman SL, Pepper MS. alphav beta 3 and alphav beta 5 integrin antagonists inhibit angiogenesis in vitro. *Angiogenesis* 2003;6(2):105–119.

Nugent HM, Edelman ER. Tissue engineering therapy for cardiovascular disease. *Circ Res* 2003;92(10):1068–1078.

Oh H, Bradfute SB, Gallardo TD, Nakamura T, Gaussin V, Mishina Y et al. Cardiac progenitor cells from adult myocardium: Homing, differentiation, and fusion after infarction. *Proc Natl Acad Sci USA* 2003;100(21):12313–12318.

Okano T, Yamada N, Okuhara M, Sakai H, Sakurai Y. Mechanism of cell detachment from temperature-modulated, hydrophilic-hydrophobic polymer surfaces. *Biomaterials* 1995;16(4):297–303.

Okano T, Yamada N, Sakai H, Sakurai Y. A novel recovery system for cultured cells using plasma-treated polystyrene dishes grafted with poly(N-isopropylacrylamide). *J Biomed Mater Res* 1993;27(10):1243–1251.

Olsson A, Bredin F, Franco-Cereceda A. Echocardiographic findings using tissue velocity imaging following passive containment surgery with the Acorn CorCap cardiac support device. *Eur J Cardiothorac Surg* 2005;28(3):448–453.

Pattison M, Webster TJ, Leslie J, Kaefer M, Haberstroh KM. Evaluating the in vitro and in vivo efficacy of nano-structured polymers for bladder tissue replacement applications. *Macromol Biosci* 2007;7(5):690–700.

Penn MS, Mal N. Stem cells in cardiovascular disease: Methods and protocols. *Methods Mol Med* 2006;129:329–351.

Piao H, Kwon JS, Piao S, Sohn JH, Lee YS, Bae JW et al. Effects of cardiac patches engineered with bone marrow-derived mononuclear cells and PGCL scaffolds in a rat myocardial infarction model. *Biomaterials* 2007;28(4):641–649.

Rehman J, Li J, Orschell CM, March KL. Peripheral blood "endothelial progenitor cells" are derived from monocyte/macrophages and secrete angiogenic growth factors. *Circulation* 2003;107(8):1164–1169.

Rosso F, Marino G, Giordano A, Barbarisi M, Parmeggiani D, Barbarisi A. Smart materials as scaffolds for tissue engineering. *J Cell Physiol* 2005;203(3):465–470.

Rowley JA, Madlambayan G, Mooney DJ. Alginate hydrogels as synthetic extracellular matrix materials. *Biomaterials* 1999;20(1):45–53.

Rubart M, Pasumarthi KB, Nakajima H, Soonpaa MH, Nakajima HO, Field LJ. Physiological coupling of donor and host cardiomyocytes after cellular transplantation. *Circ Res* 2003;92(11):1217–1224.

Rubin E, Farber JL. *Essential Pathology*. 2nd edn. Philadelphia, PA: JB Lippincott Co., 1995.

Ryan LP, Matsuzaki K, Noma M, Jackson BM, Eperjesi TJ, Plappert TJ et al. Dermal filler injection: A novel approach for limiting infarct expansion. *Ann Thorac Surg* 2009;87(1):148–155.

Schachinger V, Erbs S, Elsasser A, Haberbosch W, Hambrecht R, Holschermann H et al. 2006a. Intracoronary bone marrow-derived progenitor cells in acute myocardial infarction. *N Engl J Med* 355:1210–1221.

Schachinger V, Erbs S, Elsasser A, Haberbosch W, Hambrecht R, Holschermann H et al. 2006b. Improved clinical outcome after intracoronary administration of bone-marrow-derived progenitor cells in acute myocardial infarction: Final 1-year results of the REPAIR-AMI trial. *Eur Heart J* 27:2775–2783.

Schuh A, Liehn EA, Sasse A, Schneider R, Neuss S, Weber C et al. Improved left ventricular function after transplantation of microspheres and fibroblasts in a rat model of myocardial infarction. *Basic Res Cardiol* 2009;104(4):403–411.

Schuleri KH, Amado LC, Boyle AJ, Centola M, Saliaris AP, Gutman MR et al. Early improvement in cardiac tissue perfusion due to mesenchymal stem cells. *Am J Physiol Heart Circ Physiol* 2008;294(5):H2002–H2011.

Sekine H, Shimizu T, Hobo K, Sekiya S, Yang J, Yamato M et al. Endothelial cell coculture within tissue-engineered cardiomyocyte sheets enhances neovascularization and improves cardiac function of ischemic hearts. *Circulation* 2008;118(14 Suppl):S145–S152.

Shimizu T, Sekine H, Isoi Y, Yamato M, Kikuchi A, Okano T. Long-term survival and growth of pulsatile myocardial tissue grafts engineered by the layering of cardiomyocyte sheets. *Tissue Eng* 2006;12(3):499–507.

Shimizu T, Yamato M, Akutsu T, Shibata T, Isoi Y, Kikuchi A et al. Electrically communicating three-dimensional cardiac tissue mimic fabricated by layered cultured cardiomyocyte sheets. *J Biomed Mater Res* 2002a;60(1):110–117.

Shimizu T, Yamato M, Isoi Y, Akutsu T, Setomaru T, Abe K et al. Fabrication of pulsatile cardiac tissue grafts using a novel 3-dimensional cell sheet manipulation technique and temperature-responsive cell culture surfaces. *Circ Res* 2002b;90(3):e40.

Siminiak T, Kurpisz M. Myocardial replacement therapy. *Circulation* 2003;108(10): 1167–1171.

Smith RR, Barile L, Cho HC, Leppo MK, Hare JM, Messina E et al. Regenerative potential of cardiosphere-derived cells expanded from percutaneous endomyocardial biopsy specimens. *Circulation* 2007;115(7):896–908.

Spinale FG. Myocardial matrix remodeling and the matrix metalloproteinases: Influence on cardiac form and function. *Physiol Rev* 2007;87(4):1285–1342.

Strauer BE, Brehm M, Zeus T, Bartsch T, Schannwell C, Antke C et al. 2005. Regeneration of human infarcted heart muscle by intracoronary autologous bone marrow cell transplantation in chronic coronary artery disease: The IACT Study. *J Am Coll Cardiol* 46:1651–1658.

Sun Y, Kiani MF, Postlethwaite AE, Weber KT. Infarct scar as living tissue. *Basic Res Cardiol* 2002;97(5):343–347.

Swedberg K, Cleland J, Dargie H, Drexler H, Follath F, Komajda M et al. [Guidelines for the diagnosis and treatment of chronic heart failure: Executive summary (update 2005)]. *Rev Esp Cardiol* 2005;58(9):1062–1092.

Takahashi K, Yamanaka S. Induction of pluripotent stem cells from mouse embryonic and adult fibroblast cultures by defined factors. *Cell* 2006;126(4):663–676.

Toma C, Pittenger MF, Cahill KS, Byrne BJ, Kessler PD. Human mesenchymal stem cells differentiate to a cardiomyocyte phenotype in the adult murine heart. *Circulation* 2002;105(1):93–98.

Tonnessen T, Knudsen CW. Surgical left ventricular remodeling in heart failure. *Eur J Heart Fail* 2005;7(5):704–709.

Urbich C, Dimmeler S. Endothelial progenitor cells: Characterization and role in vascular biology. *Circ Res* 2004;95(4):343–353.

Vogel V, Baneyx G. The tissue engineering puzzle: A molecular perspective. *Annu Rev Biomed Eng* 2003;5:441–463.

Wall ST, Walker JC, Healy KE, Ratcliffe MB, Guccione JM. Theoretical impact of the injection of material into the myocardium: A finite element model simulation. *Circulation* 2006;114(24):2627–2635.

Wei HJ, Chen CH, Lee WY, Chiu I, Hwang SM, Lin WW et al. Bioengineered cardiac patch constructed from multilayered mesenchymal stem cells for myocardial repair. *Biomaterials* 2008;29(26):3547–3556.

Wenk JF, Wall ST, Peterson RC, Helgerson SL, Sabbah HN, Burger M, Stander N, Ratcliffe MB, Guccione JM. A method for automatically optimizing medical devices for treating heart failure: Designing polymeric injection patterns. *J Biomech Eng* 2009;131(12):121011.

Wu X, Rabkin-Aikawa E, Guleserian KJ, Perry TE, Masuda Y, Sutherland FW et al. Tissue-engineered microvessels on three-dimensional biodegradable scaffolds using human endothelial progenitor cells. *Am J Physiol Heart Circ Physiol* 2004;287(2):H480–H487.

Yu J, Christman KL, Chin E, Sievers RE, Saeed M, Lee RJ. Restoration of left ventricular geometry and improvement of left ventricular function in a rodent model of chronic ischemic cardiomyopathy. *J Thorac Cardiovasc Surg* 2009a;137(1):180–187.

Yu J, Gu Y, Du KT, Mihardja S, Sievers RE, Lee RJ. The effect of injected RGD modified alginate on angiogenesis and left ventricular function in a chronic rat infarct model. *Biomaterials* 2009b;30(5):751–756.

Zammaretti P, Jaconi M. Cardiac tissue engineering: Regeneration of the wounded heart. *Curr Opin Biotechnol* 2004;15(5):430–434.

Zimmermann WH, Didie M, Wasmeier GH, Nixdorff U, Hess A, Melnychenko I et al. Cardiac grafting of engineered heart tissue in syngeneic rats. *Circulation* 2002;106(12 Suppl 1):I151–I157.

Zimmermann WH, Melnychenko I, Wasmeier G, Didie M, Naito H, Nixdorff U et al. Engineered heart tissue grafts improve systolic and diastolic function in infarcted rat hearts. *Nat Med* 2006;12(4):452–458.

Zhang S. Fabrication of novel biomaterials through molecular self-assembly. *Nat Biotechnol* 2003;21(10):1171–1178.

Zhang J, Wilson GF, Soerens AG, Koonce CH, Yu J, Palecek SP et al. Functional cardiomyocytes derived from human induced pluripotent stem cells. *Circ Res* 2009;104(4):e30–e41.

17

Perinatal Stem Cells in Regenerative Medicine

Bridget M. Deasy, Jordan E. Anderson, Kelley J. Colopietro, and Yong Li

CONTENTS

17.1 Introduction

Perinatal sources of stem cells offer primitive cells with robust potential while eliminating the controversy of embryonic or induced pluripotent stem cells. The most extensively used perinatal stem cells—cord-blood (CB)-derived hematopoietic stem cells (HSCs)—have been used successfully in clinic for numerous years. Umbilical cord (UC) blood is a reliable source of stem cells for successful hematopoietic stem cell transplantations (HSCT). Since the first CB transplantation years ago in France, thousands of cell transplantations have been reported.[1] Allogeneic CB-derived HSCT has been used to treat both adults and children with hematologic diseases.[1-5] Success of these perinatal stem cells can be attributed to several advantages including (1) accessibility and the relative ease of procurement, (2) the absence of risks for donors, (3) the reduced likelihood of transmitting infections, and (4) the ability to store fully tested and HLA-typed transplants.[1] While the most intensely investigated perinatal stem cells are blood-derived HSCs used

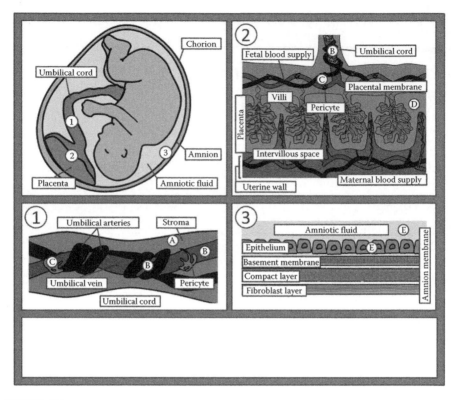

FIGURE 17.1
Sources of perinatal stem cells from the human UC, vascularization, placenta, amnion, and amniotic fluid. (A) Source of umbilical cord mesenchymal stem cells. (B) Source of umbilical cord hematopoietic stem cells. (C) Source of pericyte stem cells. (D) Source of placenta stem cells. (E) Source of amnionic stem cells.

clinically for blood disorders, mesenchymal stem cells (MSCs) with differentiation potential toward other lineages, including myogenic, osteogenic, and neurogenic, have also been identified in CB.[6–11] Further, new investigations are revealing that additional stem cell niches are present in the postnatal tissues.[12–15]

More recently, several other sources have been established including amniotic fluid and tissue, placenta, and UC stroma (Figure 17.1). These sources may provide benefits similar to that of the UC blood, namely, reduced immunorejection and ability to bank cells prospectively. These sources also expand on the range of diseases and disorders that may be treated with perinatal stem cells. While UC blood stem cells are successful for blood and cancer disorders, these new sources show potential to treat additional disorders including neurologic, immunogenetic, and musculoskeletal disorders, among others. In this chapter, we highlight the current use of UC blood stem cells, and we describe recent reports related to the newly described perinatal stem cells.

17.2 Umbilical Cord Hematopoietic Stem Cells

HSCs are characterized as cells that are capable of long-term self-renewal and differentiation into multiple lineages including all hematopoietic lineages.[16,17] HSCs are found in bone marrow (BM), peripheral blood (PB), UC blood, placenta blood, and the fetal hematopoietic system.[18] The nature of HSCs constitutes their use as therapeutic agents for the repair of hematopoiesis after infection or therapeutic ablations such as chemotherapy.[17] At present, UC-HSCs are the only perinatal stem cells that are commonly used in the clinical setting.[19]

HSCs are identified by the specific presence and absence of cell surface markers and proteins. Cell surface markers of human HSCs include CD34+, CD59+, Thy1 (CD90)+, CD38$^{low/-}$,c-kit$^{-/low}$, and lin$^-$ (*Lin$^-$ cells lack 13–14 different mature blood-lineage markers*).

UC blood containing HSCs is collected upon delivery of the infant. After transection of UC from the umbilicus, blood is collected from the placental end of the UC. CB can be collected with the placenta in situ or postdelivery of the placenta.[19,20] Cells can then be left unseparated, separated by gravity or by methylcellulose or using Ficoll-Hypaque technique.[20] Studies have shown that the CB should not be separated or washed before or after freezing in order to maintain the maximum number of HSCs and suggest that CB can be safely infused into donors immediately after thawing.[20–24] HSCs also may be isolated from CB using fluorescence-activated cell sorting (FACS).[16,18]

Several characteristics make HSCs therapeutically beneficial. HSC chemotaxis is crucial to their role as therapeutic agents.[17] HSCs have a high capacity for migration in response to specific chemokines. HSCs have high proliferation rates in the presence of serum and cytokines such as SF, GM-CSF, G-CSF, IL-3, and Epo.[16] In general, UC-HSCs have increased proliferation and replating potential in vitro as compared to BM-HSCs.[16,25] UC-HSCs are not influenced by the BM niche, and their young age characterizes increased resilience.[26] A single UC-HSC is capable of giving rise to a colony of 9.2×10^4 cells, and HSCs from CB can be sustained in vitro for several weeks. UC blood HSCs have an increased ability to sustain long-term hematopoiesis in vitro as compared to BM- and PB-HSCs.[26] Also, transplantation with UC blood mononuclear cells (MNCs) has demonstrated a greater capacity for chimerism than transplantation with BM MNCs.[25]

UC-HSCs are capable of giving rise to all hematopoietic cell types (Table 17.1). In vivo murine studies using FlkSwitch mice (created by crossing Flk2-Cre BAC transgenic mice to mT/mG mice) showed that HSCs can differentiate into all hematopoietic lineages including megakaryocyte and erythroid cell types through Flk2+ nonself-renewing progenitor cells.[16,20,27] CB HSCs cultured in vitro in serum-free medium containing stem cell factor, Flt3 ligand, megakaryocyte growth and development factor, and granulocyte colony-stimulating factor can undergo differentiation into the

TABLE 17.1

Differentiation Potential of Human Perinatal Stem Cells

Cell Type	Differentiation Potential
Umbilical cord hematopoietic stem cells UC-HSCs	Hematopoietic (monocytes, macrophages, neutrophils, basophils, eosinophils, erythrocytes, megakaryocytes/platelets, dendritic cells, T-cells, B-cells, NK-cells)
Umbilical cord mesenchymal stem cells UC-MSCs	Osteogenic, chondrogenic, myogenic (skeletal and cardio-), adipogenic, endothelial, neurogenic
Amniotic mesenchymal stem cells AMSC	Osteogenic, chondrogenic, myogenic (skeletal and cardio-), adipogenic, neurogenic, angiogenic, pancreatic
Chorionic mesenchymal stem cells CSMC	Osteogenic, chondrogenic, myogenic (skeletal), adipogenic, neurogenic
Amniotic fluid stem cells AECs	Osteogenic, adipogenic, myogenic, neurogenic, endothelial, and hepatic
AEC	Osteogenic, adipogenic, myogenic, neurogenic, endothelial, and hepatic phenotypes
HSCs from chorionoamniotic membrane	Hematopoietic

Several perinatal cell types can be isolated form full-term healthy newborn deliveries. UC tissue and placenta have different regions in these tissues; stem cells may derive from the different regions of the tissue. Listed here are the numerous lineages that perinatal cells may differentiate into or express markers characteristic of differentiated cell types.

myeloid and NK, B-, and T-lymphoid pathways.[26,28] Mutilineage differentiation to myeloid and NK, B-, and T-lymphoid pathways was confirmed in vivo by evaluating the long-term engraftment of the BM of NOD-SCID mice.[25,28] While murine studies have shown that BM HSCs have the potential to differentiate into muscle, bone, and blood vessels, controversy surrounds the lineages outside of hematopoietic types for human HSCs.[16,18,27] In sex-crossed and whole liver transplant studies, BM-HSCs differentiated into mature hepatocytes in mice[29–31]; similar findings were reported for BM-HSCs in humans.[30,32,33] BM-HSCs give rise to muscle cells in immunodeficient and mdx mice[34–36] and also epithelial cells of the liver, lung, GI tract, and skin in murine studies.[30,36]

For years, UC blood has been regarded as an excellent source of hematopoietic stem and progenitor cells.[1,37] In fact, the donation and banking of UC blood has increased in popularity, providing patients with a new source of allogenic donors[38,39] in addition to accelerating the identification time of appropriate but nonrelated donors. Further, studies have shown that UC blood containing HSCs from a single donor is sufficient for hematopoietic reconstitution in children.[20] HSCs are routinely used as treatments for cancers and other disorders of the blood and immune systems.[16,18] Fanconi's anemia, leukemia, thalassemia, X-linked adrenoleukodystrophy, myelodys-plastic syndrome, Hurler's disease, and non-Hodgkin's lymphoma have all been treated with UC-HSCs.[19] Because of their multilineage differentiation capabilities, HSCs have the potential to be used as cell therapy for a variety of diseases. Particularly, HSCs' differentiation capacity into epithelial cells may make them a viable option for tissue repair especially of the liver.[30,36]

17.3 Umbilical Cord Stromal Mesenchymal Stem Cells

The human UC tissue is an abundant, readily accessible, and noncontro-versial source of postnatal tissue, which presumably contains stem and progenitor cells involved in development. The cord is comprised of two arteries and one vein surrounded by a proteoglycan-rich Wharton's jelly (WJ). During gestation, the human cord develops to 50–60 cm in length or ~40–50 g at birth and provides the fetus with nutrient rich, oxygenated blood.[40] As described earlier, UC-HSCs are long regarded as an excellent source of HSCs.[1,37] More recently, the UC itself, which is routinely discarded following birth or after UC blood collection, was investigated as a potential source of cells with multilineage differentiation potential beyond the hema-topoietic lineage.

Various groups have reported that UC stromal-derived cells express a variety of MSC markers including CD90, CD105, CD73, CD44, CD29, and CD49.[13,41–43] However, methods for obtaining UC stromal-derived cells are variable among groups, and it remains to be determined if there are different populations being isolated, as a consistent set of markers and characteriza-tion panel has yet to be established for UC stromal-derived cells. Further, there are different regions within the cord that express MSC markers[43,44]; therefore, the origin of UC-derived cells may vary with isolation method. Nevertheless, the first reports that cultured human UC tissue-derived cells are able to express markers similar to MSCs occurred just recently.[13,45,46] Adherent cells isolated from the WJ matrix were examined for their expres-sion of MSC markers at approximately 7 days postisolation. Covas et al. iso-lated cells from the UC veins, and MSC marker expression was examined after 3 weeks of cell culture.[46] However, it is not clear if these markers were upregulated after isolation.[45,47,48] A third region, the perivascular region, which immediately surrounds the vessels and is part of the WJ matrix, also has been the focus of UC cell isolation.[42,44,49] Recent reports describe the dif-fuse expression of the pericyte marker, CD146, in the UC vessels and sur-rounding region.[43,44] MSC markers and CD146 were detected on UC cells approximately 3 weeks after isolation[44] and at the time of isolation.[43] The in situ presence of cells expressing MSC markers, CD90, CD105, CD73, and CD44 in the cord was also demonstrated.[43]

The UC-MSC-like cells have been examined in a number of differentiation assays. They have been shown to be capable of expressing markers of adipocyte, chondrocyte, osteocyte, cardiomyocyte, skeletal myocyte, neuronal cell, and endothelial cell.[13]

Finally, high proliferation rates have also been reported for UC-stromal cells,[43,47,50–52] and enhanced performance in comparison to BM or adipose-derived cells.[50,51,53,54] Together, these features suggest that this perinatal mesenchymal cell source is amendable to cell technologies and regenerative medicine, particularly in light of the challenges associated with primitive cell sourcing.[55]

17.4 Placental Stem Cells

Stem cells have also been isolated from full-term placentas and demonstrate both mesenchymal and HSC characteristics. Generally, human placental cells with mesenchymal characteristics can be grouped as amniotic mesenchymal stromal cells (AMSC), chorionic mesenchymal stromal cells (CMSC), and amniotic epithelial cells (AECs). Amniotic fluid stem cells may also be isolated. Most recently, studies show that the human placenta serves as potent hematopoietic niche throughout development,[14] and the full-term placenta contains hematopoietic progenitors and HSCs.[14,15]

17.4.1 Mesenchymal Cells from Amnion and Chorion: AMSC and CMSC

Since 2004, a small number of groups have been isolating human AMSC and CMSC with MSC characteristics from full-term placentas.[56–63] In both cases, the placental membranes are mechanically and enzymatically digested to access the stem cells. The surface marker profiles of cultured AMSC and CMSC are comparable to BM mesenchymal stromal cells. Yen et al. showed that placenta-derived multipotent cells express several markers of MSCs—including CD105/endoglin/SH-2, SH-3, and SH-4, and several ESC markers (SSEA-4, TRA-1-61, and TRA-1-80)—while lacking expression of hematopoietic, endothelial, and trophoblastic markers.[64] Similarly, Fukuchi reported adherent, placental-derived cells that were CD45^{low}CD31$^-$AC133$^-$CD54$^+$CD29$^+$CD44+.[65] It has recently been proposed that at least some of the chorionic cells may derive from perivascular sources of the chorionic villi.[66] Like MSCs from other sources, AMSC and CMSC differentiate toward musculoskeletal or mesodermal lineages—osteogenic, chondrogenic, and adipogenic.[60–62,66,67] AMSCs were also shown to differentiate to all three germ layers—ectoderm (neural[60,68]), mesoderm (skeletal muscle, cardiomyocytic, and endothelial[67,69–71]), and endoderm (pancreatic[72]). Placental mesenchymal cells show adipogenic, osteogenic, neurogenic, hepatogenic, and pancreatogenic potential.[65,73,74] Bailo et al. also showed that human placental amniotic and chorionic cells could successfully engraft in multiple organs and tissues in vivo in neonatal swine and rats. Human chimerism was detected in the brain, lung, BM, thymus, spleen, kidney, and liver after either intraperitoneal or intravenous transplantation of human amnion and chorion cells.[56]

MSC-like CD34⁻ cells were also obtained from the amniotic epithelium[75] (described in more detail later), and pluripotent stem cell markers, such as Oct-4, nanog, SSEA-3, -4, and TRA 1–60, 1–81, were detected in the amnion epithelial-derived cells.[74]

17.4.2 Amniotic Fluid Stem Cells

Amniotic stem cells may also be obtained from the amniotic fluid that surrounds the fetus. Amniotic fluid stem cells usually are obtained during amniocentesis performed in the second trimester of a pregnancy. Surface marker characterization of the progenitor cells from amniotic fluid demonstrates expression of typical stem cell markers—human embryonic stage-specific marker SSEA4, and Oct—yet do not express SSEA1, SSEA3, CD4, CD8, CD34, CD133, C-MET, ABCG2, NCAM, BMP4, TRA1-60, and TRA1-81.[76] This suggested that amniotic fluid not only may be used as a diagnostic tool but also may be a source of therapeutic cells for a multitude of congenital and adult disorders.

Amniotic fluid stem cells are obtained by amniocentesis. These cells have been shown to express Oct4, hTERT, SSEA-1, SSEA-4, CD117, and telomerase,[77] but not SSEA-3. The stem cells also are positive for MSC markers, including CD29, CD44, CD73, CD90, and CD105, but negative for hematopoietic lineage markers, including CD34, CD45, and CD133.[78,79] These characteristics suggest that human amniotic stem cells represent an intermediate stage between pluripotent embryonic stem cells and lineage-restricted adult stem cells.[76]

Progenitor cells derived from amniotic fluid are pluripotent and have been shown to differentiate into osteogenic, adipogenic, myogenic, neurogenic, endothelial, and hepatic phenotypes in vitro.[76] Atala's group used genetically modified amniotic cells[76] to show for the first time, in clonal assays, that differentiation along six distinct lineages (adipogenic, osteogenic, myogenic, endothelial, neurogenic, and hepatic) can be induced within the stem cells.[76,80]

17.4.3 Amniotic Epithelial Cells

AECs are generally obtained by mechanically stripping the amniotic membrane from the chorion and digesting with enzymes.[68,75,81,82] The cells have been shown to have stem cell characteristics as noted by cell surface markers and pluripotent differentiation; however, long-term proliferation has not been shown.

Human full-term-derived AECs express key MSC markers including CD105, CD90, CD73, CD44, and CD29 and are also positive for HLA-A,B,C, CD13, CD10, CD166, CD49d, and CD49e.[12,69,75] The cells are negative for hematopoietic markers CD34, CD45, and HLA-DR. Interestingly, CD90 was reported at low levels immediately after isolation, but it increases as cells are in culture.[75] The cells also showed pluripotent marker expression of OCT-4, Sox-2, and nanog.[75] In a study by Akle et al., the absence of HLA-A, B, C, and DR antigens or β2-microglobulin suggested a mechanism to explain that acute immune rejection does not occur after the transplantation of human AECs to human volunteers.[81]

Important for their clinical use in regenerative medicine, several investigators have shown that AECs are pluripotent. Tamagawa et al. created an in vitro xenogeneic chimera of human amnion cells with mouse embryonic stem cells,[83] and human donor cell contribution was detected in all three germ layers. Sakuragawa et al. reported the expression of differentiation markers for both neural stem cells and neuron and glial cells.[84] Subsequently, they and others found that human AECs could synthesize and release acetylcholine[84,85] and dopamine.[86,87]

AECs also appear to undergo hepatic differentiation as they synthesize both albumin and alpha-fetoprotein, express a number of liver transcription factors, and perform hepatic functions. Tamagawa et al. first showed that human cells could express albumin and alpha-fetoprotein and could be detected in vivo in hepatic parenchyma after transplantation into the livers of severe combined immunodeficiency (SCID) mice.[88] Others later confirmed these findings and showed differentiated AEC could store glucagon and express several critical liver-enriched transcription factors, such as hepatocyte nuclear factor (HNF) 3 and HNF4alpha.[75,89,90]

A report by Wei et al. suggests that AECs may also differentiate to pancreatic-like cells. AEC stimulated with nicotinamide increased insulin secretion and could normalize blood glucose levels in hyperglycemic-induced diabetic mice.[72]

More recently, AEC differentiation to musculoskeletal lineages has been examined. Several groups[60,62,69] demonstrated the cell capacity to give rise to mesodermal lineages (adipogenic, myogenic, chondrogenic, and osteogenic). Together, the differentiation studies suggest the pluripotency of these cells and their capacity to be used in regenerative therapies.

17.4.4 Placental Stem Cells: Hematopoietic Potential

Recently, reports show that stem cells with hematopoietic potential may be obtained from the placenta.[14,15] The group of Barcena et al. reported that the placenta acts as a hematopoietic organ during embryonic and fetal development. Additionally, cells isolated from the chorionic villi and the chorioamniotic membrane of the full-term placenta contain CD34+ CD45low that show potential for colony-forming units culture (CFU-C) with myeloid and erythroid potential in clonal assays, and they generated CD56(+) natural killer cells and CD19(+)CD20(+)sIgM(+) B cells in polyclonal liquid cultures.[15] The study by Dzierzak's group supported the hematopoietic role of placental MSCs and further showed that placenta-derived stromal cell lines are pericyte-like cells and support human hematopoiesis. In situ localization via immunostaining of placenta revealed hematopoietic cells in close contact with pericytes/perivascular cells and suggest that these cells derive from the vasculature.[14] In light of the clinical success of UC blood cell transplantations, the additional source of HSCs from placental tissue for cell banking is valuable as cord HSCs are often limited in numbers for an individual donor.

17.5 Use of Perinatal Stem Cells in Cell Therapeutics

The current clinical success of UC blood-derived HSCs for hematologic disorders and cancers suggests that other perinatal stem cells may likewise be successful in a broad range of therapeutic applications. In particular, MSCs have potential to be used not only to regenerate specific tissues as described earlier, but their ability to act as paracrine signaling cells may provide another advantage for the perinatal MSCs.

With the explosion of stem cell studies in the 1990s, the field of cell therapeutics has centered on identification of stem cells and their sources using the defining characteristics of self-renewal and multilineage differentiation. Exciting new insights in recent years suggest that stem cells may do much more than differentiate. Indeed, even when differentiation does not occur, transplantation of stem cells still leads to tissue repair or other therapeutic effect.[91–93] The therapeutic benefits may be due to the less well-investigated yet unique characteristics of robust stem cells—cell survival and subsequent paracrine signaling with host cells and the environment. The stem cells' intrinsic ability to survive cell transplantation may be directly related to the bioactive factors that they secrete to promote healing of host tissue. Trophic activity of MSCs in particular have been shown to inhibit tissue damage (including cell death or fibrosis formation) at the site of injury, stimulate proliferation of host cells that contribute to tissue repair, modulate the inflammation and immune response, and stimulate angiogenesis in some cases.[92]

Several animal studies show that therapies involving cell transplantation are associated with large amounts of cell death; therefore, cell survival after transplantation is a critical step in the in vivo pathway to tissue repair. Cell survival may entail overcoming immune rejection, inflammatory stress, or physical stresses of cell manipulation. Cells may survive by being immunoprotected, or by expressing higher levels of antioxidant or antiapoptotic genes, or by modulating the host cells and environment, all characteristics that may be present in perinatal stem cells. Whatever cellular mechanism may be in place or tissue engineering approach may be used, the ability of stem cells to survive the transplant is a key factor in stem cell-mediated tissue repair. After successful cell delivery, stem cells may contribute to tissue repair via self-renewal, (multi)lineage differentiation, or the unique activity of paracrine signaling.

An emerging finding among transplantation studies is that significant tissue repair is often observed despite the low donor cell differentiation and integration within the regenerated tissue. Recently, studies suggest that donor stem cells may participate in repair by signaling with host cells. Paracrine signaling by transplanted donor cells (either systemic or local) appears to represent another pathway to cell-mediated repair. The trophic signaling or release of cytokines or other signaling molecules may provide an impetus for host cells to participate in the repair, perhaps by having an effect in the local microenvironment and/or by inducing a systemic effect or by mediating an inflammatory response. Recent studies have shown that pairing HSCs with MSCs has increased BM homing of hematopoietic progenitors as compared to HSCs alone[94] and that perinatal MSCs obtained from the UC could augment engraftment of human UC-HSCs in mice undergoing cell transplantation with limited numbers of human UC blood cells.[95] These studies suggest that HSCs' regenerative capacity can be enhanced by pairing them with MSCs. Further, MSCs that elicit a minimal immune reactivity may have anti-inflammatory and immunomodulatory effects. A comparative analysis of the immunomodulatory properties of MSCs derived from adult human tissues including BM, adipose tissues (AT), umbilical CB, and cord WJ showed that AT-MSCs, CB-MSCs, and WJ-MSCs effectively suppressed mitogen-induced T-cell proliferation as effectively as did BM-MSCs.[96] These findings suggest that MSCs derived from perinatal sources could be substituted for BM-MSCs for treatment of allogeneic conflicts.

17.6 Conclusions

Regenerative medicine restores the structure and function of damaged tissues and organs by cell therapy or tissue engineering approaches. This approach may treat a large number of diseases and disorders that currently have limited options. According to the U.S. Department of Health and Human Services, the worldwide market for regenerative medicine was estimated to be $500 billion.[97] To develop regenerative medicine technologies, a number of new stem cell sources have emerged. The most promising appears to be primitive stem cells that are harvested from newborn tissue that would otherwise be discarded after healthy full-term delivery, namely, cells from the amniotic fluid and tissue, placenta, and UC stroma. While largely untested at this early point in their development, these sources may provide benefits similar to that of the UC-HSCs, particularly lack of immunorejection and ability to bank cells prospectively. In addition to their differentiation capacity, stem cells are therapeutically effective due to their immunomodulatory capacity.

While UC-HSCs have helped thousands of patients worldwide, these new perinatal sources expand on the range of diseases and disorders that may be treated. UC blood stem cells are successful for blood and cancer disorders; these new sources show potential to treat additional disorders, such as neuronal, hepatic, and musculoskeletal conditions.

References

1. Gluckman, E. and Rocha, V. History of the clinical use of umbilical cord blood hematopoietic cells. *Cytotherapy* 7, 219–227 (2005).
2. Kogler, G. et al. Hematopoietic transplant potential of unrelated cord blood: Critical issues. *J Hematother* 5, 105–116 (1996).
3. Yu, L. C. et al. Unrelated cord blood transplant experience by the pediatric blood and marrow transplant consortium. *Pediatr Hematol Oncol* 18, 235–245 (2001).
4. Laughlin, M. J. et al. Hematopoietic engraftment and survival in adult recipients of umbilical-cord blood from unrelated donors. *N Engl J Med* 344, 1815–1822 (2001).
5. Cohen, Y., Kreiser, D., Mayorov, M., and Nagler, A. Unrelated and related cord blood banking and hematopoietic graft engineering. *Cell Tissue Bank* 4, 29–35 (2003).
6. Erices, A., Conget, P., and Minguell, J. J. Mesenchymal progenitor cells in human umbilical cord blood. *Br J Haematol* 109, 235–242 (2000).
7. Gang, E. J. et al. Skeletal myogenic differentiation of mesenchymal stem cells isolated from human umbilical cord blood. *Stem Cells* 22, 617–624 (2004).
8. Lee, O. K. et al. Isolation of multipotent mesenchymal stem cells from umbilical cord blood. *Blood* 103, 1669–1675 (2004).
9. Goodwin, H. S. et al. Multilineage differentiation activity by cells isolated from umbilical cord blood: Expression of bone, fat, and neural markers. *Biol Blood Marrow Transplant* 7, 581–588 (2001).
10. Pesce, M. et al. Myoendothelial differentiation of human umbilical cord blood-derived stem cells in ischemic limb tissues. *Circ Res* 93, e51–e62 (2003).
11. Kakinuma, S. et al. Human umbilical cord blood as a source of transplantable hepatic progenitor cells. *Stem Cells* 21, 217–227 (2003).
12. Parolini, O. et al. Concise review: Isolation and characterization of cells from human term placenta: Outcome of the first international Workshop on Placenta Derived Stem Cells. *Stem Cells* 26, 300–311 (2008).
13. Can, A. and Karahuseyinoglu, S. Concise review: Human umbilical cord stroma with regard to the source of fetus-derived stem cells. *Stem Cells* 25, 2886–2895 (2007).

14. Robin, C. et al. Human placenta is a potent hematopoietic niche containing hematopoietic stem and progenitor cells throughout development. *Cell Stem Cell* **5**, 385–395 (2009).

15. Barcena, A. et al. The human placenta is a hematopoietic organ during the embryonic and fetal periods of development. *Dev Biol* **327**, 24–33 (2009).

16. Lu, L., Xiao, M., Shen, R. N., Grisby, S., and Broxmeyer, H. E. Enrichment, characterization, and responsiveness of single primitive CD34 human umbilical cord blood hematopoietic progenitors with high proliferative and replating potential. *Blood* **81**, 41–48 (1993).

17. Rossi, L., Challen G. A., Sirin, O., Lin, K. K., and Goodell, M. A. Hematopoietic stem cell characterization and isolation. *Methods Mol Biol* **750**, 47–59 (2011).

18. In Stem Cell Information (National Institutes of Health, U.S. Department of Health and Human Services, Bethesda, MD, 2011). http://stemcells.nih.gov/index.asp

19. Centrulo, C. L., Centrulo, K. J., Centrulo, C. L., Taghizadeh, R. R., and Sherley, J. L. Expanding the therapeutic potential of umbilical cord blood hematopoietic stem cells. In *Perinatal Stem Cells,* Centrulo, C. L. and K. Centrulo (eds.), Hoboken, NJ: Wiley-Blackwell (2009).

20. Broxmeyer, H. E., Douglas, G. W., Hangoc, G., Cooper, S., Bard, J., English, D., Arny, M., Thomas, L., and Boyse, E. A. Human umbilical cord blood as a potential source of transplantable hematopoietic stem/progenitor cells. *Proc Natl Acad Sci USA* **86(10)**, 3828–3832 (1989).

21. Chow, R. et al. Analysis of hematopoietic cell transplants using plasma-depleted cord blood products that are not red blood cell reduced. *Biol Blood Marrow Transplant* **13**, 1346–1357 (2007).

22. Laroche, V., McKenna, D., Moroff, G., Schierman, T., Kadidlo, D., and McCullough, J. Cell loss and recovery in umbilical cord blood processing: A comparison of postthaw and postwash samples. *Transfusion* **45**, 1909–1916 (2005).

23. Hahn, T. et al. Use of nonvolume-reduced (unmanipulated after thawing) umbilical cord blood stem cells for allogeneic transplantation results in safe engraftment. *Bone Marrow Transplant* **32**, 145–150 (2003).

24. Nagamura-Inoue, T. et al. Wash-out of DMSO does not improve the speed of engraftment of cord blood transplantation: Follow-up of 46 adult patients with units shipped from a single cord blood bank. *Transfusion* **43**, 1285–1295 (2003).

25. Kim, D. K., Fujiki, Y., Fukushima, T., Ema, H., Shibuya, A., and Nakauchi, H. Comparison of hematopoietic activities of human bone marrow and umbilical cord blood CD34 positive and negative cells. *Stem Cells* **17**, 286–294 (1999).

26. De Smedt, M., Leclercg, G., Vandekerckhove, B., Kerre, T., Taghon, T., and Plum, J. T-lymphoid differentiation potential measured in vitro is higher in CD34+CD38−/lo hematopoietic stem cells from umbilical cord blood than from bone marrow and is an intrinsic property of the cells. *Haematologica* **96**, 646–654 (2011).

27. Boyer, S. W., Schroeder, A. V., Smith-Berdan, S., and Forsberg, E. C. All hematopoietic cells develop from hematopoietic stem cells through Flk2/Flt3-positive progenitor cells. *Cell Stem Cell* **9**, 64–73 (2011).

28. Kobari, L., Pflumio, F., Giarratana, M., Li, X., Titeux, M., Izac, B., Leteurtre, F., Coulombel, L., and Douay, L. In vitro and in vivo evidence for the long-term multilineage (myeloid, B, NK, and T) reconstitution capacity of ex vivo expanded human CD34+ cord blood cells. *Exp Hematol* **28**, 1470–1480 (2000).

29. Petersen, B. E., Bowen, W. C., Patrene, K. D., Mars, W. M., Sullivan, A. K., Murase, N., Boggs, S. S., Greenberger, J. S., and Goff, J. P. Bone marrow as a potential source of hepatic oval cells. *Science* **284**, 1168–1170 (1999).
30. Krause, D. S., Theise, N. D., Collector, M. I., Henegariu, O., Hwang, S., Gardner, R., Neutzel, S., and Sharkis, S. J. Multi-organ, multi-lineage engraftment by a single bone marrow-derived stem cell. *Cell* **105**, 369–377 (2001).
31. Theise, N. D., Nimmakayalu, M., Gardner, R., Illei, P. B., Morgan, G., Teperman, L., Henegariu, O., and Krause, D. S. Liver from bone marrow in humans. *Hepatology* **32**, 11–16 (2000).
32. Alison, M. R., Poulsom, R., Jeffery, R., Dhillon, A. P., Quaglia, A., Jacob, J., Novelli, M., Prentice, G., Williamson, J., and Wright, N. A. Hepatocytes from non-hepatic adult stem cells. *Nature* **406**, 257 (2000).
33. Theise, N. D., Badve, S., Saxena, R., Henegariu, O., Sell, S., Crawford, J. M., and Krause, D. S. Derivation of hepatocytes from bone marrow cells in mice after radiation-induced myeoablation. *Hepatology* **31**, 235–240 (2000).
34. Ferrari, G. et al. Muscle regeneration by bone marrow-derived myogenic progenitors. *Science* **279**, 1528–1530 (1998).
35. Gussoni, E. et al. Dystrophin expression in the mdx mouse restored by stem cell transplantation. *Nature* **401**, 390–394 (1999).
36. Lagasse, E., Connors, H., Al Dhalimy, M., Reitsma, M., Dohse, M., Osborne, L., Wang, X., Finegold, M., Weissman, I. L., and Grompe, M. Purified hematopoietic stem cells can differentiate into hepatocytes in vivo. *Nat Med* **6**, 1229–1234 (2000).
37. Broxmeyer, H. E. et al. Human umbilical cord blood as a potential source of transplantable hematopoietic stem/progenitor cells. *Proc Natl Acad Sci USA* **86**, 3828–3832 (1989).
38. Wagner, J. E. et al. Successful transplantation of HLA-matched and HLA-mismatched umbilical cord blood from unrelated donors: Analysis of engraftment and acute graft-versus-host disease. *Blood* **88**, 795–802 (1996).
39. Nauta, A. J. et al. Enhanced engraftment of umbilical cord blood-derived stem cells in NOD/SCID mice by cotransplantation of a second unrelated cord blood unit. *Exp Hematol* **33**, 1249–1256 (2005).
40. Benirschke, K., Kaufmann, P., and Baergen, R. N. *Pathology of the Human Placenta* (Springer, New York, 2006).
41. Weiss, M. L. et al. Human umbilical cord matrix stem cells: Preliminary characterization and effect of transplantation in a rodent model of Parkinson's disease. *Stem Cells* **24**(3), 781–792 (2005).
42. Sarugaser, R., Lickorish, D., Baksh, D., Hosseini, M. M., and Davies, J. E. Human umbilical cord perivascular (HUCPV) cells: A source of mesenchymal progenitors. *Stem Cells* **23**, 220–229 (2005).
43. Schugar, R. C. et al. High harvest yield, high expansion, and phenotype stability of CD146 mesenchymal stromal cells from whole primitive human umbilical cord tissue. *J Biomed Biotechnol* **2009**, 789526 (2009).
44. Baksh, D., Yao, R., and Tuan, R. S. Comparison of proliferative and multilineage differentiation potential of human mesenchymal stem cells derived from umbilical cord and bone marrow. *Stem Cells* **25**, 1384–1392 (2007).
45. Wang, H. S. et al. Mesenchymal stem cells in the Wharton's jelly of the human umbilical cord. *Stem Cells* **22**, 1330–1337 (2004).

46. Covas, D. T., Siufi, J. L., Silva, A. R., and Orellana, M. D. Isolation and culture of umbilical vein mesenchymal stem cells. *Braz J Med Biol Res* **36**, 1179–1183 (2003).
47. Karahuseyinoglu, S. et al. Biology of stem cells in human umbilical cord stroma: In situ and in vitro surveys. *Stem Cells* **25**, 319–331 (2007).
48. Conconi, M. T. et al. CD105(+) cells from Wharton's jelly show in vitro and in vivo myogenic differentiative potential. *Int J Mol Med* **18**, 1089–1096 (2006).
49. Zebardast, N., Lickorish, D., and Davies, J. E. Human umbilical cord perivascular cells (HUCPVC): A mesenchymal cell source for dermal wound healing. *Organogenesis* **6**, 197–203 (2010).
50. Capelli, C. et al. Minimally manipulated whole human umbilical cord is a very rich source of clinical-grade human mesenchymal stromal cells expanded in human platelet lysate. *Cytotherapy* **13**(7), 786–801 (2011).
51. Hatlapatka, T. et al. Optimization of culture conditions for the expansion of umbilical cord-derived mesenchymal stem or stromal cell-like cells using xeno-free culture conditions. *Tissue Eng Part C Methods* **17**, 485–493 (2011).
52. Can, A. and Balci, D. Isolation, culture, and characterization of human umbilical cord stroma-derived mesenchymal stem cells. *Methods Mol Biol* **698**, 51–62 (2011).
53. Carvalho, M. M., Teixeira, F. G., Reis, R. L., Sousa, N., and Salgado, A. J. Mesenchymal stem cells in the umbilical cord: Phenotypic characterization, secretome and applications in central nervous system regenerative medicine. *Curr Stem Cell Res Ther* **6**(3), 221–228 (2011).
54. Hass, R., Kasper, C., Bohm, S., and Jacobs, R. Different populations and sources of human mesenchymal stem cells (MSC): A comparison of adult and neonatal tissue-derived MSC. *Cell Commun Signal* **9**, 12 (2011).
55. Daley, G. Q. et al. Ethics. The ISSCR guidelines for human embryonic stem cell research. *Science* **315**, 603–604 (2007).
56. Bailo, M. et al. Engraftment potential of human amnion and chorion cells derived from term placenta. *Transplantation* **78**, 1439–1448 (2004).
57. Bilic, G., Ochsenbein-Kolble, N., Hall, H., Huch, R., and Zimmermann, R. In vitro lesion repair by human amnion epithelial and mesenchymal cells. *Am J Obstet Gynecol* **190**, 87–92 (2004).
58. Evangelista, M., Soncini, M., and Parolini, O. Placenta-derived stem cells: New hope for cell therapy? *Cytotechnology* **58**, 33–42 (2008).
59. Parolini, O. et al. Toward cell therapy using placenta-derived cells: Disease mechanisms, cell biology, preclinical studies, and regulatory aspects at the round table. *Stem Cells Dev* **19**, 143–154 (2010).
60. Portmann-Lanz, C. B. et al. Placental mesenchymal stem cells as potential autologous graft for pre- and perinatal neuroregeneration. *Am J Obstet Gynecol* **194**, 664–673 (2006).
61. Soncini, M. et al. Isolation and characterization of mesenchymal cells from human fetal membranes. *J Tissue Eng Regen Med* **1**, 296–305 (2007).
62. Wolbank, S. et al. Dose-dependent immunomodulatory effect of human stem cells from amniotic membrane: A comparison with human mesenchymal stem cells from adipose tissue. *Tissue Eng* **13**, 1173–1183 (2007).
63. Zhang, X. et al. Mesenchymal progenitor cells derived from chorionic villi of human placenta for cartilage tissue engineering. *Biochem Biophys Res Commun* **340**, 944–952 (2006).
64. Yen, B. L. et al. Isolation of multipotent cells from human term placenta. *Stem Cells* **23**, 3–9 (2005).

65. Fukuchi, Y. et al. Human placenta-derived cells have mesenchymal stem/progenitor cell potential. *Stem Cells* **22**, 649–658 (2004).
66. Park, T. S. et al. Placental perivascular cells for human muscle regeneration. *Stem Cells Dev* **20**, 451–463 (2011).
67. Alviano, F. et al. Term amniotic membrane is a high throughput source for multipotent mesenchymal stem cells with the ability to differentiate into endothelial cells in vitro. *BMC Dev Biol* **7**, 11 (2007).
68. Sakuragawa, N. et al. Evidence for active acetylcholine metabolism in human amniotic epithelial cells: Applicable to intracerebral allografting for neurologic disease. *Neurosci Lett* **232**, 53–56 (1997).
69. Ilancheran, S. et al. Stem cells derived from human fetal membranes display multilineage differentiation potential. *Biol Reprod* **77**, 577–588 (2007).
70. Ventura, C. et al. Hyaluronan mixed esters of butyric and retinoic acid drive cardiac and endothelial fate in term placenta human mesenchymal stem cells and enhance cardiac repair in infarcted rat hearts. *J Biol Chem* **282**, 14243–14252 (2007).
71. Zhao, P. et al. Human amniotic mesenchymal cells have some characteristics of cardiomyocytes. *Transplantation* **79**, 528–535 (2005).
72. Wei, J. P. et al. Human amnion-isolated cells normalize blood glucose in streptozotocin-induced diabetic mice. *Cell Transplant* **12**, 545–552 (2003).
73. Chang, C. M. et al. Placenta-derived multipotent stem cells induced to differentiate into insulin-positive cells. *Biochem Biophys Res Commun* **357**, 414–420 (2007).
74. Chien, C. C. et al. In vitro differentiation of human placenta-derived multipotent cells into hepatocyte-like cells. *Stem Cells* **24**, 1759–1768 (2006).
75. Miki, T., Lehmann, T., Cai, H., Stolz, D. B., and Strom, S. C. Stem cell characteristics of amniotic epithelial cells. *Stem Cells* **23**, 1549–1559 (2005).
76. Delo, D. M., De Coppi, P., Bartsch, G., Jr., and Atala, A. Amniotic fluid and placental stem cells. *Methods Enzymol* **419**, 426–438 (2006).
77. Mosquera, A. et al. Simultaneous decrease of telomere length and telomerase activity with ageing of human amniotic fluid cells. *J Med Genet* **36**, 494–496 (1999).
78. Han, W., Zhong-Ying, D., and Hua-Yan, W. Growth and identification of human amniotic fluid stem cells and analysis of their influencing factors. *J Agric Biotechnol* **16**, 804–809 (2008).
79. Prusa, A. R., Marton, E., Rosner, M., Bernaschek, G., and Hengstschlager, M. Oct-4-expressing cells in human amniotic fluid: A new source for stem cell research? *Hum Reprod* **18**, 1489–1493 (2003).
80. Siegel, N., Rosner, M., Hanneder, M., Freilinger, A., and Hengstschlager, M. Human amniotic fluid stem cells: A new perspective. *Amino Acids* **35**, 291–293 (2008).
81. Akle, C. A., Adinolfi, M., Welsh, K. I., Leibowitz, S., and McColl, I. Immunogenicity of human amniotic epithelial cells after transplantation into volunteers. *Lancet* **2**, 1003–1005 (1981).
82. Terada, S. et al. Inducing proliferation of human amniotic epithelial (HAE) cells for cell therapy. *Cell Transplant* **9**, 701–704 (2000).
83. Tamagawa, T., Ishiwata, I., and Saito, S. Establishment and characterization of a pluripotent stem cell line derived from human amniotic membranes and initiation of germ layers in vitro. *Human Cell* **17**, 125–130 (2004).

84. Sakuragawa, N., Thangavel, R., Mizuguchi, M., Hirasawa, M., and Kamo, I. Expression of markers for both neuronal and glial cells in human amniotic epithelial cells. *Neurosci Lett* **209**, 9–12 (1996).
85. Elwan, M. A. and Sakuragawa, N. Evidence for synthesis and release of catecholamines by human amniotic epithelial cells. *Neuroreport* **8**, 3435–3438 (1997).
86. Kakishita, K., Elwan, M. A., Nakao, N., Itakura, T., and Sakuragawa, N. Human amniotic epithelial cells produce dopamine and survive after implantation into the striatum of a rat model of Parkinson's disease: A potential source of donor for transplantation therapy. *Exp Neurol* **165**, 27–34 (2000).
87. Kakishita, K., Nakao, N., Sakuragawa, N., and Itakura, T. Implantation of human amniotic epithelial cells prevents the degeneration of nigral dopamine neurons in rats with 6-hydroxydopamine lesions. *Brain Res* **980**, 48–56 (2003).
88. Sakuragawa, N. et al. Human amniotic epithelial cells are promising transgene carriers for allogeneic cell transplantation into liver. *J Hum Genet* **45**, 171–176 (2000).
89. Davila, J. C. et al. Use and application of stem cells in toxicology. *Toxicol Sci* **79**, 214–223 (2004).
90. Takashima, S., Ise, H., Zhao, P., Akaike, T., and Nikaido, T. Human amniotic epithelial cells possess hepatocyte-like characteristics and functions. *Cell Struct Funct* **29**, 73–84 (2004).
91. Compte, M. et al. Tumor immunotherapy using gene-modified human mesenchymal stem cells loaded into synthetic extracellular matrix scaffolds. *Stem Cells* **27**, 753–760 (2009).
92. Caplan, A. I. and Dennis, J. E. Mesenchymal stem cells as trophic mediators. *J Cell Biochem* **98**, 1076–1084 (2006).
93. Chidgey, A. P., Layton, D., Trounson, A., and Boyd, R. L. Tolerance strategies for stem-cell-based therapies. *Nature* **453**, 330–337 (2008).
94. Méndez-Ferrer, S., Michurina, T. V., Ferraro, F., Mazloom, A. R., Macarthur, B. D., Lira, S. A., Scadden, D. T., Ma'ayan, A., Enikolopov, G. N., and Frenette, P. S. Mesenchymal and haematopoietic stem cells form a unique bone marrow niche. *Nature* **466**, 829–834 (2010).
95. Friedman, R. et al. Umbilical cord mesenchymal stem cells: Adjuvants for human cell transplantation. *Biol Blood Marrow Transplant* **13**, 1477–1486 (2007).
96. Yoo, K. H. et al. Comparison of immunomodulatory properties of mesenchymal stem cells derived from adult human tissues. *Cell Immunol* **259**, 150–156 (2009).
97. U. S. Department of Health and Human Services. 2020: A new vision-a future for regenerative medicine. http://www.hhs.gov/reference/newfuture.shtml#head2 (2006).

18

Adult Stem Cell Survival Strategies

Melanie Rodrigues, Linda G. Griffith, and Alan Wells

CONTENTS

18.1 Introduction

18.1.1 Adult Stem Cells: Regenerative Potential

Stem cells are found in adult tissue at low frequencies and are capable of differentiating into specific tissues based on their differentiation potential, location, stimuli, and relocation (Huh et al. 2006). Multipotent stromal cells or mesenchymal stem cells (MSCs) present in the adult bone marrow are self-renewing clonal precursors of non-hematopoietic stromal tissue and are capable of differentiating into mesenchymal lineages of bone, cartilage, fat, tendon, muscle, endothelial cells, and marrow stroma (Pittenger et al. 1999). These can be obtained by marrow aspiration. There are also stem and progenitor cells found in adipose tissue, called adipose-derived stem cells (ASCs) obtained at a higher frequency compared to bone-marrow-derived stem cells, 0.5% compared to 0.01%. These cells which differ in expression of certain cell surface markers (Fraser et al. 2006) possess similar multilineage differentiation potential though there is a predilection toward fat tissue (Brayfield et al. 2010). ASCs are called by several other names in literature including processed lipoaspirate cells (LPA) and adipose-tissue-derived stromal cells. Likewise, the muscle also holds postnatal stem cells called muscle-derived stem cells (MDSCs); these cells can differentiate spontaneously into cardiac and skeletal muscle and by modification ex vivo with growth factors can differentiate into the osteogenic and chondrogenic lineages (Usas and Huard 2007). There also exist stem cells with more restricted differentiation ability like neural stem cells (NSCs) found in the adult hippocampus and lateral ventricle which can differentiate into neurons and glia (astrocytes and oligodendrocytes) (Massirer et al. 2011). There are other less well clinically characterized epithelial stem cells in tissues with high homeostasis, turnover rates, and regeneration such as the skin which form hair follicles (Zhang et al. 2009), sebaceous glands (Jensen et al. 2009), and epidermal cells (reviewed [Barker et al. 2010]). Other stem cells are found in the mammary and prostate glands (Shackleton et al. 2006; Stingl et al. 2006).

Gastrointestinal stem cells, which help renew cells of the stomach, small intestine, and colon, are readily identifiable in the crypts of the GI tract (reviewed [Barker et al. 2010; Mills and Shivdasani 2011]). Colonic stem cells hold special status as they are necessary for normal organ homeostasis during adult life (Shaker and Rubin 2010). This evidence of ongoing physiological functioning is shared only with hematopoietic stem cells.

The need for organ stem cells does not appear to be universal. For instance, skin can be regenerated from basal keratinocytes and dermal fibroblasts (Werner et al. 2007). Still in these situations, the adnexal structures of the skin (hair follicles, sebaceous glands) are not formed in the absence of the epidermal stem cells from the hair follicle bulge (Yang and Peng 2010). A second organ in which the parenchymal cells may reconstitute the organ

TABLE 18.1

Adult Stem Cell Types and Differentiation Potential

Stem Cell Type	Tissue of Residence	Differentiation Potential
MSCs/multipotent stromal cells	Bone marrow	Bone, cartilage, muscle, marrow, adipose tissue, tendon, connective tissue, vasculature (Pittenger et al. 1999)
ASCs/LPA	Adipose tissue	Adipose tissue (Zuk et al. 2002; Choi et al. 2006) bone (Haimi et al. 2009; McCullen et al. 2009), muscle (Kim et al. 2006; Marra et al. 2011), cartilage (Xu et al. 2007), cardiovascular cells (Planat-Benard et al. 2004)
MDSCs	Muscle tissue	Muscle, cartilage, bone
NSCs	Subgranular zone of the hippocampus and subventricular zone of the lateral ventricle (Alvarez-Buylla and Lim 2004; Suh et al. 2007)	Neurons and glia (Duan et al. 2008)
Colonic stem cells	Base of crypts of Lieberkuhn (Bjerknes and Cheng 1999, 2002)	Intestinal epithelial cells and Paneth cells
Stomach stem cells	Pyloric and corpus region (Nomura et al. 1998)	Epithelial cells of the stomach (McDonald et al. 2008)
Mammary stem cells	Terminal end buds (Kenney et al. 2001)	Mammary epithelial cells
Prostrate stem cells	Basal layer of prostatic epithelium (reviewed [Miki 2010])	Prostatic epithelial cells

is the liver. Hepatocytes have been shown to have the capacity to repopulate injured livers via serial transplantation (Fausto 1997). In fact, the existence of liver stem cells in adults has not been conclusively demonstrated. A comprehensive list of adult stem cells, with their origin and differentiation potential, is listed in Table 18.1.

Thus, while the role of adult stem cells is not certain in most instances, their presence allows for novel interventions.

18.1.2 Adult Stem Cells: Clinical Need

There exists a strong clinical need for replacement to conventional transplantation procedures due to problems of donor availability, cost, and rejection. Adult stem cells are great candidates for such replacement therapy due to their potential for expansion and differentiation into the multiple cell types that constitute an organ. A single cell type will not be sufficient to regenerate the lost tissue as the stromal and vascular (endothelial and smooth muscle cells) supports are critical to functioning. Due to the multipotential differentiation of adult stem cells, one can readily imagine organotypic replacement proceeding from essentially a singular population of stem cells.

The three organs most extensively studied are fat, bone, and heart. For fat, there exists the approach of direct transplantation of small fat organoids extracted from other, autologous locations (Kantanen et al. 2010; Stallworth and Wang 2010). However, the clinical outcome of using autologous fat is highly unpredictable as lack of revascularization leads to graft resorption (reviewed in [Rubin and Marra 2011]). Adipocyte stem cells on the other hand are a propitious alternative as they promote angiogenesis and vasculogenesis (Cao et al. 2005) in addition to being pericyte-like with respect to their interactions with endothelial cells (Traktuev et al. 2008).

Bone regeneration constitutes a greater challenge. After trauma, tumor resection, or surgical procedures, critical-size bone defects do not heal on their own and require indwelling prostheses. The use of autologous bone grafts that are adopted as treatment for such wounds is limited due to availability, donor-site morbidity, geometry, and cost. Research in rats (Song et al. 2011), rabbits (Zhao et al. 2011), dogs (Liao et al. 2011), and sheep (Boos et al. 2010) suggest that bone graft substitutes like tricalcium phosphate or degradable polymeric scaffolds combined with cells of regenerative potential like MSC increase both the osteoconductivity and the efficacy of bone formation.

Likewise, heart transplants are limited by donor supply, immunosuppression, and high cost. The use of stem cells as an alternative heart regeneration therapy is feasible in terms of differentiation into cardiomyocytes and safe also in the lung where these cells might migrate (Oguz et al. 2011; Wang et al. 2011). However, repeated studies with stem cells to regenerate heart tissue have failed due to death of the transplanted cells in the excessively harsh ischemic microenvironment; this death situation will subsequently be addressed in greater detail.

18.1.3 Adult Stem Cells: Current Use and Gaps in Therapy

One of the first issues with the use of stem cell therapy is obtaining a sufficient number of cells for tissue differentiation since stem cells are present in scarce numbers in the body. Obtaining high cell numbers involves expansion of the stem cells ex vivo with defined media like growth factors that need to remain neutral and not lead to differentiation of the stem cells (Tamama et al. 2006, 2010). However, even if the hurdle of cell numbers is overcome, despite the innate differentiation potential of stem cells, most of the therapeutic effects currently attributed to the cells are based on the ability to home to sites of injury, release trophic factors, and suppress inflammation (Prockop et al. 2010) instead of direct tissue generation. Current literature shows that MSCs (Park et al. 2010) and ASCs (Sadat et al. 2007; Ikegame et al. 2011) release growth factors including vascular endothelial growth factor (VEGF), IGF-1, HGF, and TGF-β, which protect injured tissue from hypoxia and serum deprivation apoptosis by activating and upregulating anti-apoptotic molecules and supporting new vessel formation (Park et al. 2010). MSCs also inhibit MMP signaling, prevent breakage of cell–matrix interactions and impede anoikis and apoptosis in surrounding cells (Scuteri et al. 2011). In addition, these cells express inflammatory cytokines like INFγ, TNFα, ILF1α, and ILF1β that suppress T-cell and other immune cell activity (Ren et al. 2008; Singer and Caplan 2011) aiding in immunosuppression.

All the trophic effects just described are thought to occur immediately following homing of the stem cells in vivo, probably in the first days to week of implantation since these cells release higher amounts of growth factors under stresses like hypoxia. However, these stem cells show markedly reduced cell numbers from the time of implantation until 7 days after implantation with no surviving cells after 2 weeks likely due to the same stressors at the graft site (Hoffmann et al. 2010; Rodrigues et al. 2010).

18.2 Adult Stem Cell Survival at Graft Site

18.2.1 Death of Stem Cells on Implantation

Although stem cells display several beneficial effects with regard to release of trophic factors and immunosuppression when tested on small animals, preclinical and clinical trials, few of these cells can be tracked beyond a few days of implantation. In one study involving MSC for treatment of inflammatory bowel disease, there were clinical improvements in all mice injected with MSC; however, only 40% of mice showed persistence (possibly engraftment) of MSC, with as little as 0.13% of MSC traceable in the intestine after 3 days (Semont et al. 2010). In another study where high numbers of ASC were injected for treatment of colitis and intestinal sepsis, there was a transient attachment of cells seen in the intestine at day 2, with very few detectable MSCs at day 6 (Gonzalez-Rey et al. 2009). Similarly, decreased numbers of MSCs have been observed after delivery to the ischemic kidney (Mias et al. 2008).

Both ASCs and MSCs rapidly die after injection/implantation into infarcted hearts, with less than 5% of the cells engrafting and fewer cells being identifiable over the first 14 days (van der Bogt et al. 2009; Noort et al. 2010). Also, MSCs transplanted as bone graft substitutes show decreased cell numbers after implantation. One study showed cell numbers decrease from 31.3% ± 2.3% on day 1, to 9.2% ± 1.1% on day 3, 0.3% ± 0.1% on day 7, and no identifiable MSC after 14 days. The decreased cell numbers correlated with increased cellular disintegration (Zimmermann et al. 2011).

The reduced stem cell numbers on implantation in the body has been consistent in studies across tissues, suggesting that there is reduced attachment and increased loss of these cells that could be occurring due to a combination of effects such as nutrient deprivation and hypoxia, increased cytokine production in the wound milieu, ischemia, enhanced reactive oxygen species (ROS), and oxidative stress in the tissue microenvironment (Menasche 2008).

18.2.2 Effects of Hypoxia and Serum Deprivation on Stem Cell Survival

MSCs reside in the bone milieu under conditions of 4%–7% oxygen (Kofoed et al. 1985). Studies suggest that when oxygen is dropped below 4% indicative of hypoxia in addition to subjecting MSCs to serum deprivation, cells display caspase-3-mediated programmed cell death with loss in membrane potential, cytochrome-c release, and accumulation of Bax in a p53-dependent manner, all mediators of the mitochondrial death pathway involving caspase-9, with very little death occurring via the death receptor pathway mediated by caspase-8 (Zhu et al. 2006). Additionally, MSCs in the presence of hypoxia generate increased levels of ROS due to increased expression of NAD(P)H oxidase and reduced catalase, display lower levels of the pro-survival protein survivin and increased Bcl-2/Bax ratios which culminates in cell death (Peterson et al. 2011). However, serum deprivation proves to be the stronger stimuli compared to hypoxia (Zhu et al. 2006) since glucose depletion in the presence of hypoxia leads to massive cell death in MSCs which is not seen in the presence of hypoxia alone (Deschepper et al. 2010). Additionally, recent studies show that MSCs are able to tolerate hypoxia for up to 72 h, following which they undergo caspase-mediated death, and this death is exacerbated on addition of the glycolysis inhibitor 2-deoxyglucose, suggesting that glycolysis may be a major pathway supporting survival in ischemic conditions (McGinley et al. 2011).

ASCs are also susceptible to hypoxic death (Follmar et al. 2006) and survive to a much lesser degree than bone-marrow-derived MSCs or cord-blood-derived stem cells, either in vitro in the presence of hypoxia or in vivo in infarcted mouse hearts (Gaebel et al. 2011) (Figure 18.1).

18.2.3 Effects of Cytokines

The wound environments in which adult stem cells are generally implanted are sites of high inflammatory cytokine activity. Similarly, implantation of any space-occupying substance into the body, such as polymeric scaffolds containing stem cells, generates a nonspecific immune response that the stem cells need to overcome and survive. MSCs are susceptible to FasL- and TRAIL-induced cell death, with TRAIL preferably killing fetal MSCs and FasL preferably killing adult MSCs (Fan et al. 2007; Gotherstrom et al. 2011). There have also been contrary reports wherein activation of neither of these pathways stimulates death in MSCs (Mazar et al. 2009; Szegezdi et al. 2009). To clarify these reports, studies in our lab show that cytokines such as FasL promote cell death in MSCs via caspase-3-mediated apoptosis and also

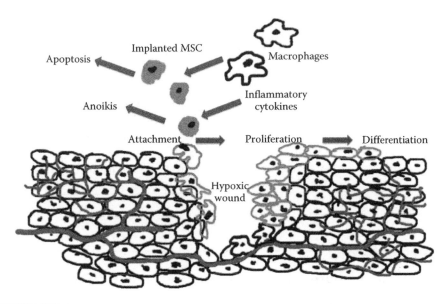

FIGURE 18.1
Stem cell expansion and survival at graft site. Stem cells, in therapeutic application, are generally implanted in ischemic wound regions or inflammatory tissue for tissue regeneration. However, the implanted cells need to attach to the microenvironment, proliferate to generate substantial cell numbers for differentiation into tissue, and differentiate. This progression is hampered by the harsh microenvironment, devoid of oxygen or nutrients, which increases oxidative stress and prevents attachment of the stem cells, leading to death by anoikis. Nonspecific immune responses brought about by invading immune cells such as macrophages increase pro-inflammatory cytokines, causing death of the stem cells by apoptosis. Most studies done with stem cells observe a massive reduction of the implanted cell numbers within 14 days, most likely occurring due to these mechanisms.

induce the production of ROS in MSC that contributes to apoptotic death in these cells (Rodrigues et al. in press).

18.2.4 Effects of Reactive Oxygen Species and Oxidative Stress

Stem cells are thought to possess high amounts of antioxidants glutathione and superoxide dismutase and concurrently low inherent levels of ROS (Urish et al. 2009). MDSCs, for example, display lower levels of oxidative stress-related death compared to their more differentiated phenotype, myoblasts (Oshima et al. 2005). However, this reduced baseline ROS does not prevent these cells from undergoing cell death in the presence of oxidative stress. Pretreatment of MDSCs with an ROS scavenger *N*-acetylcysteine (NAC) prior to implantation of cells in the heart increases cell survival which is dependent on signaling by the MAP kinase Erk, while on the contrary addition of a pro-oxidant di-ethyl maleate decreases survival (Drowley et al. 2010). Similarly, application of hydrogen peroxide, one of the major contributors of oxidative damage to MSC in vitro causes apoptosis by activation of p38MAPK in the early stages and JNK in the late stages; this activates caspase-3 via the mitochondrial pathway by causing Bax translocation to the mitochondrial membrane, releases cytochrome-c, and simultaneously causes activation of caspase-12 by ER-stress induction (Wei et al. 2010). Another effect seen due to paracrine ROS signaling on adult stem cells is anoikis. Treatment of MSC with hydrogen peroxide causes reduced attachment and cell spread by reducing focal adhesion-related molecules like p-FAK and p-Src as well as integrin-related molecules (Song et al. 2010). Contrarily, supplementing MSC with NAC increases attachment of MSC to the heart infarct region, provides a survival advantage, and revives infarcted hearts better. The various death-signaling pathways have been illustrated in Figure 18.2.

18.3 Ways of Improving Stem Cell Survival at Graft Site

18.3.1 Gene Therapy for Increased Survival

Introducing genes to enhance survival signaling in stem cells prior to implantation has been tested for prolonging the lifespan of stem cells in vivo. This mode of inducing survival is specific to the stem cells carrying the gene alone and does not affect cells of the surrounding tissue directly. In one study, adult rat MSCs were genetically modified ex vivo to overexpress the anti-apoptotic protein Bcl-2. These cells were injected into the rat myocardium bordering the infarct region. While there still was a reduction in cell numbers over 6 weeks, MSC overexpressing Bcl-2 showed a 2.2-fold increase in cell survival on day 4, 1.9-fold increase in survival after 6 weeks, and 1.2-fold increase over vector control cells after 6 weeks (Li et al. 2007). While these are miniscule improvements, the experiment did show some protection from death.

There have been other studies where MSC overexpressing GSK-3β when injected into infracted mice hearts showed increased survival due to increased production of VEGF-A (Cho et al. 2011). Another showed introduction of the human insulin growth factor 1 (IGF-1) gene in MSC increased IGF-1 production to nearly six times that of native cells, enhanced Erk signaling, and greater MSC survival. However, these modified MSC could differentiate only into the endodermal and mesodermal lineages and not into neuronal cells (Hu et al. 2008).

Lentiviral vector-mediated overexpression of pro-survival genes heat shock protein 27 (HSP27), HSP70, SOD1, and SOD3 in MSC has shown increased survival in the presence of hypoxia, glucose, and nutrient deprivation, with HSP70 overexpression most significantly increasing survival and decreasing caspase-3-mediated death (McGinley et al. 2011). Similarly in another study where rat MSCs overexpressing HSP20 were introduced in a rat left anterior

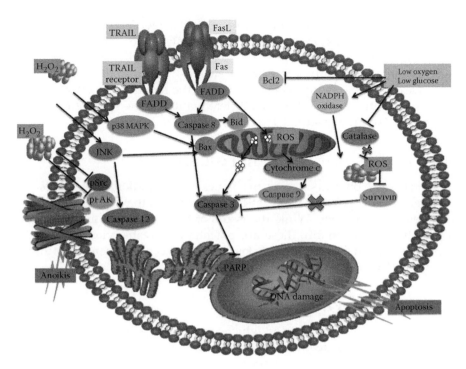

FIGURE 18.2

Adult stem cell death signaling. Hydrogen peroxide (H_2O_2) inhibits integrin signaling by inhibiting phosphorylation of FAK and Src, causing cells to detach and die by anoikis. H_2O_2 also activates the JNK and p38MAPK stress pathways, causing Bax to translocate to the mitochondria and initiate the mitochondrial death pathway involving loss of mitochondrial membrane potential, release of cytochrome-c, activation of caspase-9, and activation of the effector caspase-3. p38MAPK is involved in the early stages, while JNK is activated in the later stages of apoptosis. In addition, JNK activates the ER-stress-related caspase-12 death pathway. Pro-death cytokines such as TRAIL and FasL also activate cell death via caspase-8 and mitochondrial pathways leading to caspase-3 activation, in addition to generating ROS in the mitochondria, which also cause activation of caspase-3. Although both low glucose and hypoxia are used to mimic ischemic conditions in vitro, low glucose is the stronger death stimulus. In combination, these two conditions activate NADPH oxidase and inhibit catalase enzymes, which enhance ROS production in the cells, blocking the pro-survival protein survivin and Bcl-2, causing caspase-3 activation and cell death. DNA repair enzyme PARP is just one among several substrates of caspase-3, which the apoptotic effector cleaves and inactivates.

descending ligation model of infarct via intracardial injection, there was a twofold increase in survival in vivo on day 4, compared to MSC alone, owing to enhanced signaling of the pro-survival AKT pathway and increased production of the growth factors fibroblast growth factor-2 (FGF-2), VEGF, and IGF-1 (Wang et al. 2009).

Stable overexpression of the pro-survival protein AKT1 (protein kinase B) in swine bone marrow MSCs shows stable AKT activation for 2 weeks and increased attachment of the MSCs to swine myocardial walls after autograft into the infracted myocardium (Mangi et al. 2003; Rodrigues et al. 2010; Yu et al. 2010). All these studies, while showing only minor improvements, did demonstrate proof of principle to indicate that pro-survival gene therapy can be an advantageous option to increase stem cell viability in vivo. However, the issues of gene therapy remain and lie in the fact that the effects last for several generations, including postdifferentiation of the stem cells, and the integration itself may cause deleterious mutations.

18.3.2 Protein Factors for Increased MSC Survival

Presentation of proteins avoids the issues of integration with second site mutagenesis and possible neoplastic transformation that is inherent in gene therapeutic approaches. While the issues of timed released and concentration distributions remain, these are potentially solvable via cell engineering manipulations. A number of possible proteins have been forwarded to accomplish this task.

18.3.2.1 Heat Shock Proteins

HSPs are molecular chaperones, which act as checkpoints and either direct cells into survival or death under conditions of molecular stress. Treatment with recombinant human heat shock protein 90α (rhHSP90α) increases rat MSC survival in the presence of hypoxia and serum deprivation in a dose-dependent manner by increasing pro-survival proteins Bcl-2 and BclxL and decreasing the pro-apoptotic protein Bax. The change in expression of these proteins is thought to occur via interactions with toll-like receptor-4 (TLR-4), and the receptor tyrosine kinase ErBb-2. rhHSP90α binds ErBb-2 on its extracellular domain activating PI3K and Erk; similarly, it binds TLR-4 causing NFκB activation and production of nitric oxide. There is also PI3K–Akt activation via MyD88, reducing apoptosis in all cases (Gao et al. 2010).

18.3.2.2 Growth Factors/Angiopoetins/G-Protein-Coupled Receptor Signaling

Various growth factors have been queried for the increase in survival of stem cells based on their ability to activate pro-survival pathways including Erk and Akt. There is a long-appreciated role for these factors based on their trophic effects such as inducing blood vessel growth in the immediate surroundings, reducing hypoxia, and nutrient starvation. A number of studies with VEGF show ASC and MSC numbers increasing on pretreatment with VEGF or introduction with a VEGF peptide at ischemic sites due to increased Akt and induction of angiogenesis (Pons et al. 2008; Behr et al. 2010).

Treatment with neurotrophic factor and nerve growth factor shows increased MSC homing and survival after introduction in a traumatic brain injury model (Mahmood et al. 2002). Liver growth factor improves viability of NSCs grafted in rat brains in a study involving treatment of Parkinson's by raising levels of the anti-apoptotic protein Bcl-2 four times, without changing expression of Bax (Reimers et al. 2010). Epidermal growth factor receptor (EGFR) signaling is also known to play a major role in survival of NSCs by activation of downstream JAK–STAT and PI3K–AKT pathways (Grimm et al. 2009; Ayuso-Sacido et al. 2010; Tham et al. 2010).

In bone marrow MSC, restriction of the EGFR ligand EGF to the cell culture substratum causes restriction of EGFR receptors to the cell membrane, causing sustained EGFR and downstream Erk signaling, protecting MSC from FasL-mediated cell death. The use of EGF ligand in the soluble, free state, however, does not lead to the same survival effect due to internalization of the receptor and transient EGFR and Erk signaling (Fan et al. 2007). The advantage of using EGFR signaling for increasing MSC survival lies in the fact that EGFR signaling does not affect MSC differentiation while increasing proliferation and expansion, making it the ideal growth factor option for a lot of model systems (Tamama et al. 2006).

Another highly studied growth factor with respect to stem cells is FGF-2/b-FGF. Overexpression of FGF-2 in neural progenitor cells shows increased engraftment to the ischemic cortex and higher rates of survival (Jenny et al. 2009). Cord-blood-derived MSCs loaded onto a fibrin gel with FGF-2 and transplanted intramuscularly into the ischemic hindlimbs of mice display significantly lower apoptosis of MSC postimplantation. The same study showed MSC with FGF-2 enhanced expression of host-derived PDGFβ and NG2, which induces recruitment and homing of endothelial cells and pericytes, protecting muscles from ischemic degeneration (Bhang et al. 2011).

Angiopoetins are protein growth factors that stimulate angiogenesis and vasculogenesis by binding to their protein kinase Tie receptors and are known to have pro-survival effects on several cell types including neurons, endothelial cells, and cardiomyocytes. Angiopoetin 1 (Ang-1) was found to induce Tie phosphorylation in MSC, activate PI3K–AKT, increase Bcl-2/Bax ratios, and prevent caspase-9 and caspase-3 activation by hypoxia and serum deprivation (Liu et al. 2008).

Lysophosphatidic acid is a bioactive lipid that not only induces apoptosis in some cell types but also protects a number of cell types from hypoxia, serum deprivation, and drug-induced apoptosis, including MSC (Chen et al. 2008a,b). LPA binds the G-protein-coupled receptor LPA-1, which is the only LPA receptor expressed on MSC; activates Gi proteins, which activates downstream PI3K–AKT and Ras–Erk pathways; shuts Bax translocation to the mitochondria; and prevents death via hypoxia and serum-induced mitochondrial death pathway. In vivo studies in ischemic rat hearts with rat MSC pretreated with LPA survive much better than MSC without any prior conditioning, although there are reduced cell numbers in both cases. The number of LPA-treated MSC found in the ischemic myocardium after 1 h of implantation was $50.93\% \pm 11.43\%$ compared to $23.63\% \pm 2.24\%$ of untreated MSC; the numbers after 1 day were $34.96\% \pm 12.1\%$ of LPA-treated MSC compared to $14.75\% \pm 1\%$

of untreated MSC and only 1.38%±0.35% of LPA treated MSC compared to 0.69%±0.14% after 1 week. So also, the number of apoptotic cells in the peri-implant region after 1 day of implantation was 6%+2% with MSC treated with LPA and 16%+3.61% with untreated MSC, indicating that LPA has a major role to play in retaining stem cell numbers at the ischemic site (Liu et al. 2009a,b).

Growth factors like EGF therefore have the advantage of increasing stem cell numbers ex vivo and possibly in vivo, which makes for a defined media for the culture of the cells while maintaining their undifferentiated state and increases survival in vivo. The challenge with the use of growth factors is delivery and current bioengineering studies in regenerative medicine probe deeply into various biocompatible scaffolds and polymeric materials that can be used which will carry the growth factor in its bioactive form, restrict its activity to the stem cell compartment of regenerating tissue, and degrade as soon as its function is complete, minimizing any aberrations in the tissue.

18.3.2.3 Chemokines

Chemokines are a group of small-sized proteins that were initially discovered as chemoattractants but have since then been increasingly studied for their role in pro- and anti-inflammatory signaling and cellular migration. Pretreatment of MSC with stromal-derived factor-1 (SDF-1) which binds the chemokine receptor CXCR4 decreases cell death due to H_2O_2 in vitro and increases homing and survival of MSC in infarct and peri-infarct areas. The increase in survival is due to activation of PI3K–AKT survival pathway by CXCR4 signaling (Pasha et al. 2008). Also, MSC overexpressing the SDF-1 gene shows better engraftment and angiogenesis in a rat myocardial infarction model due to enhanced production of the pro-angiogenic growth factor VEGF and subsequent signaling by AKT and eNOS (Tang et al. 2009).

18.3.3 Hormones for Increased Survival

The pineal hormone melatonin has been used to increase survival of rat bone-marrow-derived MSC delivered to the ischemic kidney. Melatonin binds the melatonin receptor (both melatonin receptor 1 and melatonin receptor 2 are expressed in MSC) and activates the enzymes catalase and superoxide dismutase responsible for reducing ROS-mediated cell death. In addition, melatonin increases production of growth factors like FBF-2 and HGF and increases surrounding vascularity, improving conditions of survival for the MSC in the ischemic environment (Mias et al. 2008).

18.3.4 Anti-Ischemic Drugs for Increased Survival

Since the largest threat to stem cells once implanted at the site of injury is ischemia and loss of nutrients, one of the best ways to study survival would be to use methods to directly counteract ischemia. Two such studies with anti-ischemic drugs have been conducted. Trimetazidine (1-[2,3,4-trime-thoxybenzyl]piperazine or TMZ, also known as Vastarel) is a drug used to reduce metabolic damage induced due to ischemia by reducing the rate of fatty acid oxidation and increasing rates of anaerobic glycolysis (Lopaschuk et al. 2003), a means of metabolism found by several studies to be favorable to stem cells. MSCs preconditioned with TMZ show increases in HIF-1α, pAKT, and the pro-survival proteins Bcl-2 and survivin and show reduced cell death in the presence of oxidative damage caused by H_2O_2 (Wisel et al. 2009).

The prolyl hydroxylase inhibitor dimethyloxalylglycine (DMOG) is the other example of an anti-ischemic drug tested to increase MSC. The enzyme prolyl hydroxylase under conditions of normoxia allows HIF1α interaction with VHL, targeting the complex toward degradation. DMOG, by inhibiting prolyl hydroxylase, prevents the formation of HIF1α–VHL complex and allows HIF1α to shuttle to the nucleus, bind to p300–CBP, and transcribe genes involved in angiogenesis, energy metabolism, and survival. In addition, DMOG prevents translocation of apoptosis-inducing factor (AIF) from the mitochondria to the nucleus, preventing a caspase-3-independent apoptosis pathway; blocks release of cytochrome-c from the mitochondria; and activates PI3K-mediated survival. DMOG however does not affect the Erk pathway (Liu et al. 2009a,b).

18.3.5 Increasing Engraftment of MSC: Coating of MSC with Antibodies

Targeting stem cells to sites of inflammation, attachment to sites of inflammation, and survival at these sights pose major challenges. There have been studies where MSCs have been coated with antibodies to target the cells to sites of inflammation. MSC membranes coated with addressin antibodies target MSC to TNF-α secreting endothelial cells which are rich at sites of inflammation (Ko et al. 2009). This model has been studied in vivo in mice with inflammatory bowel disease. MSC coated with addressin antibodies displayed increased delivery and engraftment to the inflamed intestinal mucosa and exerted immunosuppressive effects to bring about survival of the mice (Ko et al. 2010). This study is a one of its kind focusing on increasing delivery and engraftment of MSC to inflammatory sites and can be used effectively along with methods to increase survival of MSC to bring about better effects.

18.3.6 Antioxidants/Reduced Temperatures/Mitochondrial Death Blockers

As indicated earlier, ROS rich at sites of injury pose a major threat to stem cells survival by initiating stress pathways and promoting cell deattachment. Also as mentioned earlier, pro-inflammatory cytokines are able to induce ROS increases in stem cells, making the cells responsible in promoting their own death. Studies with addition of the antioxidant NAC to MDSCs have indicated a surge in survival, which would be the case with a majority of other stem cell types (Drowley et al. 2010).

Immunosuppressants like cyclosporine A (CsA) have also been tested for enhancement of stem cell survival. CsA regulates membrane permeability in MSC like in several other cell types, prevents drop of membrane potential and release of cytochrome-c, and prevents activation of the mitochondrial apoptotic pathway. In addition, CsA deactivates the pro-apoptotic moiety BAD via calcineurin and protects MSC from hypoxia-/reoxygenation-induced death (Chen et al. 2008a,b).

The various survival pathways that have been studied in stem cells are described in Figure 18.3.

18.4 Conclusions and Future Directions

Reduction of stem cell numbers within the first 2 weeks of implantation remains a major hurdle in the application of stem cell therapy for regeneration of tissue. The increasing number of preclinical and clinical trials using stem cells that have been successful are due primarily to the ability of the stem cells to release trophic factors and cause immunosuppression, increasing vasculogenesis, increasing survival of surrounding cells, and preventing graft rejection. However, the prime potential of these cells, of differentiating into specific cell lineages, is lost as the cells die in the ischemic, oxidative, and inflammatory wound environment. The cells are lost in the presence of low oxygen, low nutrient concentrations, high levels of ROS, and pro-death cytokines. Thus, the field needs to develop ways to improve survival during the initial challenge period.

This chapter has brought together current literature on increasing attachment and survival of stem cells in the harsh graft environment. There have been successful gene transfer studies both in vitro and in vivo, under conditions of ischemia and oxidative stress where introduction of pro-survival genes have shown substantial increases in survival. However, gene therapy poses a threat of causing aberrations during gene integration, especially in a cell type whose main feature involves self-renewal. The use of protein factors on the other hand, overcomes the problem of neoplastic transformation via gene integrations. Growth factors, chemokines, and HSPs have been tested and have shown significant improvements in retaining cell numbers at sites of inflammation. However, these proteins have to be chosen in a manner that will improve both stem cell proliferation, to obtain sufficient numbers of these scarcely occurring cells, and survival, without affecting differentiation of the stem cells into an undesirable tissue lineage. Also, a limiting factor with the use of proteins currently is their mode of presentation. Bioengineering studies currently underway will be a major factor that

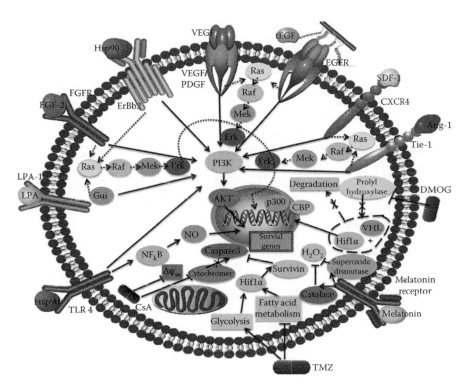

FIGURE 18.3

Adult stem cell survival signaling. HSP90 binds the orphan growth factor receptor ErBb-2 on its extracellular domain activating PI3K–AKT and Erk signaling pathways. It also binds TLR-4 causing both NFκB activation and nitric oxide (NO) synthesis and activation of PI3K–AKT. Growth factors, such as FGF-2 binding its receptor FGFR, VEGF binding its receptor VEGFR, PDGF bindings its receptor PDGFR, and Ang-1 binding its receptor Tie-1, activate the PI3K–AKT and MAPK–Erk pathways, which stimulate transcription of survival genes. Bioengineered models of growth factors tethered to a substratum in contact with the cell membrane such as a tethered epidermal growth factor (tEGF) restrict the activated growth factor receptor (in this case, EGFR) to the cell surface and cause sustained signaling of the downstream Erk pathway prolonging survival signaling. Similarly, bioactive lipids, including LPA binding its G-protein-coupled receptor LPA-1 and chemokines such as SDF-1 binding its receptor CXCR4, signal survival via the MAPK–Erk and AKT pathways. Hormones such as melatonin binding its receptor increase ROS-reducing enzymes catalase and superoxide dismutase, thus lowering hydrogen peroxide in cells and protecting cells from oxidative damage. Antiischemic drugs such as TMZ increase glycolysis in the cells and decrease fatty acid metabolism, causing increases in Hif1α and pro-survival proteins survivin and Bcl-2, which inhibit caspase-3 in addition to activating Akt. The prolyl hydroxylase inhibitor DMOG by inhibiting prolyl hydroxylase inhibits the complex formed by Hif1α which targets it to degradation, allowing Hif1α to shuttle to the nucleus and bind CBP and p300, allowing the transcription of pro-survival genes. Also shown in another drug, cyclosporine A that acts in the mitochondria, preventing the drop in membrane potential and release of cytochrome-c, thereby inhibiting the mitochondrial death pathway.

can bridge this gap. Development of biocompatible polymeric scaffolds that can function as a carrier of both stem cells and proteins, maintaining proteins in their active conformation and restricting the signaling of proteins to the stem and progenitor cell compartment alone for a desired duration of time until tissue repair has been established, will take cell therapy to newer heights. The use of growth factors with antioxidants to alleviate oxidative stress or drugs that counter ischemia might also be a useful way of improving stem cell survival in vivo. While these studies require further improvements prior to even preclinical testing of relevant defects, the recent findings with supportive and tethered factors hold promise that these challenges will be overcome.

Acknowledgments

The figures were prepared partly by Protein Lounge software and Doodle for iPad. This review was supported in part by grants from the U.S. NIGMS and NIDCR. We thank the members of Wells and Griffith laboratories for helpful discussions and suggestions.

References

Alvarez-Buylla, A. and Lim, D. A. 2004. For the long run: Maintaining germinal niches in the adult brain. *Neuron* 41(5): 683–686.

Ayuso-Sacido, A., Moliterno, J. A., Kratovac, S. et al. 2010. Activated EGFR signaling increases proliferation, survival, and migration and blocks neuronal differentiation in post-natal neural stem cells. *J Neurooncol* 97(3): 323–337.

Barker, N., Bartfeld, S., and Clevers, H. 2010. Tissue-resident adult stem cell populations of rapidly self-renewing organs. *Cell Stem Cell* 7(6): 656–670.

Behr, B., Tang, C., Germann, G. et al. 2010. Locally applied VEGFA increases the osteogenic healing capacity of human adipose derived stem cells by promoting osteogenic and endothelial differentiation. *Stem Cells* 29(2): 286–296.

Bhang, S. H., Lee, T. J., La, W. G. et al. 2011. Delivery of fibroblast growth factor 2 enhances the viability of cord blood-derived mesenchymal stem cells transplanted to ischemic limbs. *J Biosci Bioeng* 111(5): 584–589.

Bjerknes, M. and Cheng, H. 1999. Clonal analysis of mouse intestinal epithelial progenitors. *Gastroenterology* 116(1): 7–14.

Bjerknes, M. and Cheng, H. 2002. Multipotential stem cells in adult mouse gastric epithelium. *Am J Physiol Gastrointest Liver Physiol* 283(3): G767–G777.

Boos, A. M., Loew, J. S., Deschler, G. et al. 2010. Directly auto-transplanted mesenchymal stem cells induce bone formation in a ceramic bone substitute in an ectopic sheep model. *J Cell Mol Med* 15(6): 1364–1378.

Brayfield, C., Marra, K., and Rubin, J. P. 2010. Adipose stem cells for soft tissue regeneration. *Handchir Mikrochir Plast Chir* 42(2): 124–128.

Cao, Y., Sun, Z., Liao, L. et al. 2005. Human adipose tissue-derived stem cells differentiate into endothelial cells in vitro and improve postnatal neovascularization in vivo. *Biochem Biophys Res Commun* 332(2): 370–379.

Chen, J., Baydoun, A. R., Xu, R. et al. 2008a. Lysophosphatidic acid protects mesenchymal stem cells against hypoxia and serum deprivation-induced apoptosis. *Stem Cells* 26(1): 135–145.

Chen, T. L., Wang, J. A., Shi, H. et al. 2008b. Cyclosporine A pre-incubation attenuates hypoxia/reoxygenation-induced apoptosis in mesenchymal stem cells. *Scand J Clin Lab Invest* 68(7): 585–593.

Cho, J., Zhai, P., Maejima, Y. et al. 2011. Myocardial injection with GSK-3beta-overexpressing bone marrow-derived mesenchymal stem cells attenuates cardiac dysfunction after myocardial infarction. *Circ Res* 108(4): 478–489.

Choi, Y. S., Cha, S. M., Lee, Y. Y. et al. 2006. Adipogenic differentiation of adipose tissue derived adult stem cells in nude mouse. *Biochem Biophys Res Commun* 345(2): 631–637.

Deschepper, M., Oudina, K., David, B. et al. 2010. Survival and function of mesenchymal stem cells (MSCs) depend on glucose to overcome exposure to long-term, severe and continuous hypoxia. *J Cell Mol Med* 15(7): 1505–1514.

Drowley, L., Okada, M., Beckman, S. et al. 2010. Cellular antioxidant levels influence muscle stem cell therapy. *Mol Ther* 18(10): 1865–1873.

Duan, X., Kang, E., Liu, C. Y. et al. 2008. Development of neural stem cell in the adult brain. *Curr Opin Neurobiol* 18(1): 108–115.

Fan, V. H., Tamama, K., Au, A. et al. 2007. Tethered epidermal growth factor provides a survival advantage to mesenchymal stem cells. *Stem Cells* 25(5): 1241–1251.

Fausto, N. 1997. Hepatocytes break the rules of senescence in serial transplantation studies. Is there a limit to their replicative capacity? *Am J Pathol* 151(5): 1187–1189.

Follmar, K. E., Decroos, F. C., Prichard, H. L. et al. 2006. Effects of glutamine, glucose, and oxygen concentration on the metabolism and proliferation of rabbit adipose-derived stem cells. *Tissue Eng* 12(12): 3525–3533.

Fraser, J. K., Wulur, I., Alfonso, Z. et al. 2006. Fat tissue: An underappreciated source of stem cells for biotechnology. *Trends Biotechnol* 24(4): 150–154.

Gaebel, R., Furlani, D., Sorg, H. et al. 2011. Cell origin of human mesenchymal stem cells determines a different healing performance in cardiac regeneration. *PLoS One* 6(2): e15652.

Gao, F., Hu, X. Y., Xie, X. J. et al. 2010. Heat shock protein 90 protects rat mesenchymal stem cells against hypoxia and serum deprivation-induced apoptosis via the PI3K/Akt and ERK1/2 pathways. *J Zhejiang Univ Sci B* 11(8): 608–617.

Gonzalez-Rey, E., Anderson, P., Gonzalez, M. A. et al. 2009. Human adult stem cells derived from adipose tissue protect against experimental colitis and sepsis. *Gut* 58(7): 929–939.

Gotherstrom, C., Lundqvist, A., Duprez, I. R. et al. 2011. Fetal and adult multipotent mesenchymal stromal cells are killed by different pathways. *Cytotherapy* 13(3): 269–278.

Grimm, I., Messemer, N., Stanke, M. et al. 2009. Coordinate pathways for nucleotide and EGF signaling in cultured adult neural progenitor cells. *J Cell Sci* 122(Pt 14): 2524–2533.

Haimi, S., Suuriniemi, N., Haaparanta, A. M. et al. 2009. Growth and osteogenic differentiation of adipose stem cells on PLA/bioactive glass and PLA/beta-TCP scaffolds. *Tissue Eng Part A* 15(7): 1473–1480.

Hoffmann, J., Glassford, A. J., Doyle, T. C. et al. 2010. Angiogenic effects despite limited cell survival of bone marrow-derived mesenchymal stem cells under ischemia. *Thorac Cardiovasc Surg* 58(3): 136–142.

Hu, C., Wu, Y., Wan, Y. et al. 2008. Introduction of hIGF-1 gene into bone marrow stromal cells and its effects on the cell's biological behaviors. *Cell Transplant* 17(9): 1067–1081.

Huh, W. J., Pan, X. O., Mysorekar, I. U. et al. 2006. Location, allocation, relocation: Isolating adult tissue stem cells in three dimensions. *Curr Opin Biotechnol* 17(5): 511–517.

Ikegame, Y., Yamashita, K., Hayashi, S. I. et al. 2011. Comparison of mesenchymal stem cells from adipose tissue and bone marrow for ischemic stroke therapy. *Cytotherapy* 13(6): 675–685.

Jenny, B., Kanemitsu, M., Tsupykov, O. et al. 2009. Fibroblast growth factor-2 overexpression in transplanted neural progenitors promotes perivascular cluster formation with a neurogenic potential. *Stem Cells* 27(6): 1309–1317.

Jensen, K. B., Collins, C. A., Nascimento, E. et al. 2009. Lrig1 expression defines a distinct multipotent stem cell population in mammalian epidermis. *Cell Stem Cell* 4(5): 427–439.

Kantanen, D. J., Closmann, J. J., and Rowshan, H. H. 2010. Abdominal fat harvest technique and its uses in maxillofacial surgery. *Oral Surg Oral Med Oral Pathol Oral Radiol Endod* 109(3): 367–371.

Kenney, N. J., Smith, G. H., Lawrence, E. et al. 2001. Identification of stem cell units in the terminal end bud and duct of the mouse mammary gland. *J Biomed Biotechnol* 1(3): 133–143.

Kim, M., Choi, Y. S., Yang, S. H. et al. 2006. Muscle regeneration by adipose tissue-derived adult stem cells attached to injectable PLGA spheres. *Biochem Biophys Res Commun* 348(2): 386–392.

Ko, I. K., Kean, T. J., and Dennis, J. E. 2009. Targeting mesenchymal stem cells to activated endothelial cells. *Biomaterials* 30(22): 3702–3710.

Ko, I. K., Kim, B. G., Awadallah, A. et al. 2010. Targeting improves MSC treatment of inflammatory bowel disease. *Mol Ther* 18(7): 1365–1372.

Kofoed, H., Sjontoft, E., Siemssen, S. O. et al. 1985. Bone marrow circulation after osteotomy. Blood flow, pO_2, pCO_2, and pressure studied in dogs. *Acta Orthop Scand* 56(5): 400–403.

Li, W., Ma, N., Ong, L. L. et al. 2007. Bcl-2 engineered MSCs inhibited apoptosis and improved heart function. *Stem Cells* 25(8): 2118–2127.

Liao, H. T., Chen, C. T., Chen, C. H. et al. 2011. Combination of guided osteogenesis with autologous platelet-rich fibrin glue and mesenchymal stem cell for mandibular reconstruction. *J Trauma* 70(1): 228–237.

Liu, X., Hou, J., Shi, L. et al. 2009a. Lysophosphatidic acid protects mesenchymal stem cells against ischemia-induced apoptosis in vivo. *Stem Cells Dev* 18(7): 947–954.

Liu, X. B., Jiang, J., Gui, C. et al. 2008. Angiopoietin-1 protects mesenchymal stem cells against serum deprivation and hypoxia-induced apoptosis through the PI3K/Akt pathway. *Acta Pharmacol Sin* 29(7): 815–822.

Liu, X. B., Wang, J. A., Ogle, M. E. et al. 2009b. Prolyl hydroxylase inhibitor dimethyloxalylglycine enhances mesenchymal stem cell survival. *J Cell Biochem* 106(5): 903–911.

Lopaschuk, G. D., Barr, R., Thomas, P. D. et al. 2003. Beneficial effects of trimetazidine in ex vivo working ischemic hearts are due to a stimulation of glucose oxidation secondary to inhibition of long-chain 3-ketoacyl coenzyme a thiolase. *Circ Res* 93(3): e33–e37.

Mahmood, A., Lu, D., Wang, L. et al. 2002. Intracerebral transplantation of marrow stromal cells cultured with neurotrophic factors promotes functional recovery in adult rats subjected to traumatic brain injury. *J Neurotrauma* 19(12): 1609–1617.

Mangi, A. A., Noiseux, N., Kong, D. et al. 2003. Mesenchymal stem cells modified with Akt prevent remodeling and restore performance of infarcted hearts. *Nat Med* 9(9): 1195–1201.

Marra, K. G., Brayfield, C. A., and Rubin, J. P. 2011. Adipose stem cell differentiation into smooth muscle cells. *Methods Mol Biol* 702: 261–268.

Massirer, K. B., Carromeu, C., Griesi-Oliveira, K. et al. 2011. Maintenance and differentiation of neural stem cells. *Wiley Interdiscip Rev Syst Biol Med* 3(1): 107–114.

Mazar, J., Thomas, M., Bezrukov, L. et al. 2009. Cytotoxicity mediated by the Fas ligand (FasL)-activated apoptotic pathway in stem cells. *J Biol Chem* 284(33): 22022–22028.

McCullen, S. D., Zhu, Y., Bernacki, S. H. et al. 2009. Electrospun composite poly(L-lactic acid)/tricalcium phosphate scaffolds induce proliferation and osteogenic differentiation of human adipose-derived stem cells. *Biomed Mater* 4(3): 035002.

McDonald, S. A., Greaves, L. C., Gutierrez-Gonzalez, L. et al. 2008. Mechanisms of field cancerization in the human stomach: The expansion and spread of mutated gastric stem cells. *Gastroenterology* 134(2): 500–510.

McGinley, L., McMahon, J., Strappe, P. et al. 2011. Lentiviral vector mediated modification of mesenchymal stem cells and enhanced survival in an in vitro model of ischaemia. *Stem Cell Res Ther* 2(2): 12.

Menasche, P. 2008. Current status and future prospects for cell transplantation to prevent congestive heart failure. *Semin Thorac Cardiovasc Surg* 20(2): 131–137.

Mias, C., Trouche, E., Seguelas, M. H. et al. 2008. Ex vivo pretreatment with melatonin improves survival, proangiogenic/mitogenic activity, and efficiency of mesenchymal stem cells injected into ischemic kidney. *Stem Cells* 26(7): 1749–1757.

Miki, J. 2010. Investigations of prostate epithelial stem cells and prostate cancer stem cells. *Int J Urol* 17(2): 139–147.

Mills, J. C. and Shivdasani, R. A. 2011. Gastric epithelial stem cells. *Gastroenterology* 140(2): 412–424.

Nomura, S., Esumi, H., Job, C. et al. 1998. Lineage and clonal development of gastric glands. *Dev Biol* 204(1): 124–135.

Noort, W. A., Feye, D., Van Den Akker, F. et al. 2010. Mesenchymal stromal cells to treat cardiovascular disease: Strategies to improve survival and therapeutic results. *Panminerva Med* 52(1): 27–40.

Oguz, E., Ayik, F., Ozturk, P. et al. 2011. Long-term results of autologous stem cell transplantation in the treatment of patients with congestive heart failure. *Transplant Proc* 43(3): 931–934.

Oshima, H., Payne, T. R., Urish, K. L. et al. 2005. Differential myocardial infarct repair with muscle stem cells compared to myoblasts. *Mol Ther* 12(6): 1130–1141.

Park, K. S., Kim, Y. S., Kim, J. H. et al. 2010. Trophic molecules derived from human mesenchymal stem cells enhance survival, function, and angiogenesis of isolated islets after transplantation. *Transplantation* 89(5): 509–517.

Pasha, Z., Wang, Y., Sheikh, R. et al. 2008. Preconditioning enhances cell survival and differentiation of stem cells during transplantation in infarcted myocardium. *Cardiovasc Res* 77(1): 134–142.

Peterson, K. M., Aly, A., Lerman, A. et al. 2011. Improved survival of mesenchymal stromal cell after hypoxia preconditioning: role of oxidative stress. *Life Sci* 88(1–2): 65–73.

Pittenger, M. F., Mackay, A. M., Beck, S. C. et al. 1999. Multilineage potential of adult human mesenchymal stem cells. *Science* 284(5411): 143–147.

Planat-Benard, V., Menard, C., Andre, M. et al. 2004. Spontaneous cardiomyocyte differentiation from adipose tissue stroma cells. *Circ Res* 94(2): 223–229.

Pons, J., Huang, Y., Arakawa-Hoyt, J. et al. 2008. VEGF improves survival of mesenchymal stem cells in infarcted hearts. *Biochem Biophys Res Commun* 376(2): 419–422.

Prockop, D. J., Kota, D. J., Bazhanov, N. et al. 2010. Evolving paradigms for repair of tissues by adult stem/progenitor cells (MSCs). *J Cell Mol Med* 14(9): 2190–2199.

Reimers, D., Osuna, C., Gonzalo-Gobernado, R. et al. 2010. Liver growth factor promotes the survival of grafted neural stem cells in a rat model of Parkinson's disease. *Curr Stem Cell Res Ther* [Epub ahead of print].

Ren, G., Zhang, L., Zhao, X. et al. 2008. Mesenchymal stem cell-mediated immunosuppression occurs via concerted action of chemokines and nitric oxide. *Cell Stem Cell* 2(2): 141–150.

Rodrigues, M., Griffith, L. G., and Wells, A. 2010. Growth factor regulation of proliferation and survival of multipotential stromal cells. *Stem Cell Res Ther* 1(4): 32.

Rodrigues, M., Turner, O., and Stolz, D. 2012. Production of reactive oxygen species by multipotential stromal cells/mesenchymal stem cells upon exposure to FasL. *Cell Transplantation* (in press).

Rubin, J. P. and Marra, K. G. 2011. Soft tissue reconstruction. *Methods Mol Biol* 702: 395–400.

Sadat, S., Gehmert, S., Song, Y. H. et al. 2007. The cardioprotective effect of mesenchymal stem cells is mediated by IGF-I and VEGF. *Biochem Biophys Res Commun* 363(3): 674–679.

Scuteri, A., Ravasi, M., Pasini, S. et al. 2011. Mesenchymal stem cells support dorsal root ganglion neurons survival by inhibiting the metalloproteinase pathway. *Neuroscience* 172: 12–19.

Semont, A., Mouiseddine, M., Francois, A. et al. 2010. Mesenchymal stem cells improve small intestinal integrity through regulation of endogenous epithelial cell homeostasis. *Cell Death Differ* 17(6): 952–961.

Shackleton, M., Vaillant, F., Simpson, K. J. et al. 2006. Generation of a functional mammary gland from a single stem cell. *Nature* 439(7072): 84–88.

Shaker, A. and Rubin, D. C. 2010. Intestinal stem cells and epithelial-mesenchymal interactions in the crypt and stem cell niche. *Transl Res* 156(3): 180–187.

Singer, N. G. and Caplan, A. I. 2011. Mesenchymal stem cells: Mechanisms of inflammation. *Annu Rev Pathol* 6: 457–478.

Song, H., Cha, M. J., Song, B. W. et al. 2010. Reactive oxygen species inhibit adhesion of mesenchymal stem cells implanted into ischemic myocardium via interference of focal adhesion complex. *Stem Cells* 28(3): 555–563.

Song, K., Rao, N. J., Chen, M. L. et al. 2011. Enhanced bone regeneration with sequential delivery of basic fibroblast growth factor and sonic hedgehog. *Injury* 42(8): 796–802.

Stallworth, C. L. and Wang, T. D. 2010. Fat grafting of the midface. *Facial Plast Surg* 26(5): 369–375.

Stingl, J., Eirew, P., Ricketson, I. et al. 2006. Purification and unique properties of mammary epithelial stem cells. *Nature* 439(7079): 993–997.

Suh, H., Consiglio, A., Ray, J. et al. 2007. In vivo fate analysis reveals the multipotent and self-renewal capacities of Sox2+ neural stem cells in the adult hippocampus. *Cell Stem Cell* 1(5): 515–528.

Szegezdi, E., O'Reilly, A., Davy, Y. et al. 2009. Stem cells are resistant to TRAIL receptor-mediated apoptosis. *J Cell Mol Med* 13(11–12): 4409–4414.

Tamama, K., Fan, V. H., Griffith, L. G. et al. 2006. Epidermal growth factor as a candidate for ex vivo expansion of bone marrow derived mesenchymal stem cells. *Stem Cells* 24(3): 686–695.

Tamama, K., Kawasaki, H., and Wells, A. 2010. Epidermal growth factor (EGF) treatment on multipotential stromal cells (MSCs). Possible enhancement of therapeutic potential of MSC. *J Biomed Biotechnol* 2010: 795385.

Tang, J., Wang, J., Yang, J. et al. 2009. Mesenchymal stem cells over-expressing SDF-1 promote angiogenesis and improve heart function in experimental myocardial infarction in rats. *Eur J Cardiothorac Surg* 36(4): 644–650.

Tham, M., Ramasamy, S., Gan, H. T. et al. 2010. CSPG is a secreted factor that stimulates neural stem cell survival possibly by enhanced EGFR signaling. *PLoS One* 5(12): e15341.

Traktuev, D. O., Merfeld-Clauss, S., Li, J. et al. 2008. A population of multipotent CD34-positive adipose stromal cells share pericyte and mesenchymal surface markers, reside in a periendothelial location, and stabilize endothelial networks. *Circ Res* 102(1): 77–85.

Urish, K. L., Vella, J. B., Okada, M. et al. 2009. Antioxidant levels represent a major determinant in the regenerative capacity of muscle stem cells. *Mol Biol Cell* 20(1): 509–520.

Usas, A. and Huard, J. 2007. Muscle-derived stem cells for tissue engineering and regenerative therapy. *Biomaterials* 28(36): 5401–5406.

van der Bogt, K. E., Schrepfer, S., Yu, J. et al. 2009. Comparison of transplantation of adipose tissue- and bone marrow-derived mesenchymal stem cells in the infarcted heart. *Transplantation* 87(5): 642–652.

Wang, W., Jiang, Q., Zhang, H. et al. 2011. Intravenous administration of bone marrow mesenchymal stromal cells is safe for the lung in a chronic myocardial infarction model. *Regen Med* 6(2): 179–190.

Wang, X., Zhao, T., Huang, W. et al. 2009. Hsp20-engineered mesenchymal stem cells are resistant to oxidative stress via enhanced activation of Akt and increased secretion of growth factors. *Stem Cells* 27(12): 3021–3031.

Wei, H., Li, Z., Hu, S. et al. 2010. Apoptosis of mesenchymal stem cells induced by hydrogen peroxide concerns both endoplasmic reticulum stress and mitochondrial death pathway through regulation of caspases, p38 and JNK. *J Cell Biochem* 111(4): 967–978.

Werner, S., Krieg, T., and Smola, H. 2007. Keratinocyte-fibroblast interactions in wound healing. *J Invest Dermatol* 127(5): 998–1008.

Wisel, S., Khan, M., Kuppusamy, M. L. et al. 2009. Pharmacological preconditioning of mesenchymal stem cells with trimetazidine (1-[2,3,4-trimethoxybenzyl]piperazine) protects hypoxic cells against oxidative stress and enhances recovery of myocardial function in infarcted heart through Bcl-2 expression. *J Pharmacol Exp Ther* 329(2): 543–550.

Xu, Y., Balooch, G., Chiou, M. et al. 2007. Analysis of the material properties of early chondrogenic differentiated adipose-derived stromal cells (ASC) using an in vitro three-dimensional micromass culture system. *Biochem Biophys Res Commun* 359(2): 311–316.

Yang, L. and Peng, R. 2010. Unveiling hair follicle stem cells. *Stem Cell Rev* 6(4): 658–664.

Yu, Y. S., Shen, Z. Y., Ye, W. X. et al. 2010. AKT-modified autologous intracoronary mesenchymal stem cells prevent remodeling and repair in swine infarcted myocardium. *Chin Med J (Engl)* 123(13): 1702–1708.

Zhang, Y. V., Cheong, J., Ciapurin, N. et al. 2009. Distinct self-renewal and differentiation phases in the niche of infrequently dividing hair follicle stem cells. *Cell Stem Cell* 5(3): 267–278.

Zhao, M., Zhou, J., Li, X. et al. 2011. Repair of bone defect with vascularized tissue engineered bone graft seeded with mesenchymal stem cells in rabbits. *Microsurgery* 31(2): 130–137.

Zhu, W., Chen, J., Cong, X. et al. 2006. Hypoxia and serum deprivation-induced apoptosis in mesenchymal stem cells. *Stem Cells* 24(2): 416–425.

Zimmermann, C. E., Gierloff, M., Hedderich, J. et al. 2011. Survival of transplanted rat bone marrow-derived osteogenic stem cells in vivo. *Tissue Eng Part A* 17(7–8): 1147–1156.

Zuk, P. A., Zhu, M., Ashjian, P. et al. 2002. Human adipose tissue is a source of multipotent stem cells. *Mol Biol Cell* 13(12): 4279–4295.

19

Immunobiology of Biomaterial/ Mesenchymal Stem Cell Interactions

Peiman Hematti and Summer Hanson

CONTENTS

19.1 Wound Healing and Inflammation

Macrophages are key immune cells derived from monocyte precursors that have a vital role not only in innate and adaptive immunity but also tissue healing and remodeling as well. Normal healing is a complex sequence of events coordinating hemostasis, inflammation, and ultimately organized tissue regeneration (Martin 1997). Following injury, platelet aggregation initiates the clotting cascade, followed by infiltration of the wound bed with pro-inflammatory, cytokine-producing leukocytes, including neutrophils and macrophages. In later stages of healing, fibroblasts are recruited to the wound, depositing extracellular matrix (ECM) proteins such as collagen, fibronectin, and hyaluronic acid, leading to new tissue regeneration (Witte and Barbul 1997, Darby and Hewitson 2007). When considering tissue healing, the body has three primary objectives: to stop bleeding, to prevent infection, and to restore tissue function. These goals are illustrated, or addressed, in the characteristic phases of wound healing, including inflammation, proliferation, and remodeling. Even under optimal conditions, prudent wound care very often leads to contracture, loss of domain, and fibrosis or scarring. Therefore, there are many aspects of the healing process that may affect decision making, the use of an implant or biomaterial construct, and, ultimately, clinical outcome.

Inflammation is mediated by circulating monocyte-derived or tissue-resident macrophages recruited to the site of tissue injury. Such injuries could be from acute mechanical forces such as in trauma or surgery, chemical injuries, infectious agents, or chronic underlying pathology such as vascular insufficiency. During the inflammatory phase, hemostasis is achieved early, and the body works to clear the tissue of bacterial and other contaminants. A temporary fibrin matrix is established in the wound, and platelets attach to the fibrin meshwork to prevent further blood loss. Debridement occurs primarily via phagocytosis as inflammatory cells are drawn to the wound, first neutrophils, followed by macrophages. In addition, circulating serum proteins adsorb to nonself cell surfaces, a process known as opsonization, which targets foreign materials for phagocytosis. Macrophages recruited to wounded tissue engulf and degrade foreign material, send chemotactic signals to other circulating cells, and serve as antigen-presenting cells (APCs) in order to build the adaptive repertoire. This process proceeds over the first few days of injury or insult and is clinically characterized by swelling, redness, and tenderness at the site.

The proliferative phase, as the name implies, is characterized by an increase in cells recruited to the site in an effort to repair the wounded or missing tissue. This takes place over several weeks after injury. Tissue cells present at the periphery of the injury multiply and migrate in an attempt to repopulate the defect (i.e., epithelial cells in cutaneous wounds). During this phase of tissue repair, macrophages remain in tissue coordinating resolution of inflammation and debridement of the tissue bed from dead cells, apoptotic bodies, and foreign debris. Fibroblasts proliferate and deposit type I and type III collagen as the initial fibrin plug is replaced by a slightly more organized, intermediate granulation tissue, containing macrophages, fibroblasts, new vasculature, and ECM proteins.

In later stages, and after resolution of acute inflammation, stromal cells are being recruited to the site of injury which could lead to fibrosis, scar formation, and organ dysfunction. A similar process occurs when a biomaterial implant is introduced to the body. In these situations, the injured host tissue now encounters a new foreign body, and its interactions with this biomaterial could have significant impact on the natural history of the healing process.

Currently available treatment options address tissue injury; however, none of these methods are an ideal means to heal the injured tissue, and thus, new strategies are being developed to not only replace wounded tissues but also to regenerate the native tissue lost (Martin 1997). One can anticipate this task will require multimodality therapy to mediate both the inflammatory resolution and tissue reparative phases of wound healing. Macrophages, or mononuclear phagocytic cells, are ubiquitous in the body and essentially present in all tissues and ready to be provoked by injury. Similar to inflammatory macrophages found in a cutaneous wound or foreign body reaction to an implant are the Kupffer cells of the liver, leading to cirrhosis following insult or the osteoclasts responsible for bone degrading matrix at a fracture nonunion site. Indeed, these mechanisms of injury causing ischemia followed by inflammation and fibrosis (or scar) are common to many pathophysiological processes throughout body. While there are subtle differences in the various functional tissues (i.e., liver, kidney, or cardiac muscle), the general process of initial inflammation followed by body attempt to regenerate is similar to the connective tissues, vasculature, and skin, which are common sites of current clinically useful biomaterial implants, and the focus throughout this chapter.

19.2 Immunity and the Foreign Body Response

The immune system offers two primary pathways of protection: innate and adaptive immunity. Innate immunity can be considered the first line of defense against a wide variety of injuries. Through this pathway, foreign material is nonspecifically recognized primarily by phagocytic macrophages and thus cleared or destroyed from the host. In contrast, adaptive immunity is a dynamic protection, specific to pathogen exposure to T and B lymphocytes; in general, adaptive immunity plays a much smaller role in wound healing. Foreign body reactions, including reactions to implanted biomaterials, are much like early wound healing: circulating proteins and inflammatory cells are recruited to the foreign material, debris is cleared, and granulation tissue begins to form. However, instead of proceeding to tissue remodeling, the presence of a foreign material, such as a cell-based biomaterial implant, stimulates continued inflammation by tissue macrophages. Foreign body giant cell formation occurs when macrophages are challenged by an implant that is too large to engulf, so instead they fuse to form larger phagocytic cells. "Frustrated phagocytosis" and chronic inflammation then take place as these fused giant cells continue on the surface of the implanted biomaterial/device, which can lead to biomaterial resorption or device failure.

Over the last few decades, the heterogeneity of macrophage phenotype has become increasingly evident including their several important roles in tissue homeostasis (Mantovani et al. 2004). Classically activated macrophages respond to a variety of stimuli including tumor necrosis factor (TNF) α, interleukin (IL)-1, IL-12, and tissue hypoxia to initiate, and propagate, an effective innate immune response (Figure 19.1). Alternatively activated macrophages express a different set of cell surface markers and are thought to be more immunomodulatory and reparative. It is now apparent that there is a spectrum of macrophage profiles based on a variety of functional states, including classically activated macrophages for microbicidal activity, wound healing macrophages for tissue repair, and regulatory macrophages for anti-inflammatory activity (Mosser and Edwards 2008).

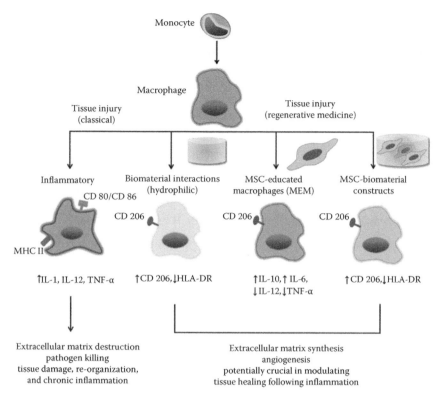

FIGURE 19.1
Illustration of the immunomodulatory effects of biomaterials and MSCs on macrophage phenotype and function. Upon stimulation such as tissue injury, macrophages can acquire a classically activated phenotype which is associated with pro-inflammatory cytokines (i.e., IL-1, IL-12, TNF-α), debris removal, and continued inflammation and tissue injury. Alternatively, there is evidence to support alternatively activated, anti-inflammatory phenotypes, can be acquired when monocytes are cultured on biomaterials with specific surface chemistry (i.e., hydrophilic) or cocultured with MSCs, with or without a biomaterial scaffold. These alternatively activated macrophages are associated with an increase in anti-inflammatory cytokines (i.e., IL-10, IL-6, TGF-β), ECM production, and tissue remodeling.

19.3 Biomaterial Constructs

Bioengineered scaffolds could provide a microenvironment that allows for nutrient diffusion as well as biochemical, physical, and cellular support that could stimulate proliferation, differentiation, and migration of implanted or tissue-resident cells. While many of these scaffolds are clinically used without cells for tissue augmentation or repair, such as hydroxyapatite, collagen, or hyaluronic acid, there is increasing interest in combining these materials with cells for in vivo tissue engineering strategies. Once implanted, biomaterial constructs quickly become covered with circulating proteins such as albumin, fibronectin, fibrinogen, and complement, to varying degrees based on the nature of the implanted material (Kou and Babensee 2011). These adsorbed proteins play a significant role in regulating host immune cells, including macrophages, response to biomaterials rather than the material itself. There are several factors associated with a material scaffold that can influence protein adsorption and cell differentiation or function, including chemical composition (hydrophobicity, chemical stimuli, or added pendant chains), scaffold geometry (2D vs. 3D), and spatial relationship (cell concentration, porosity) (Matheson et al. 2004, Shin 2007, McBane et al. 2011).

Different classes of materials are used in the manufacturing of medical devices or implants, including metals, ceramics, polymers (natural and synthetic), and composites or blends. Metals or metal alloys and ceramics are widely used in dental, cardiovascular, and orthopedic applications due to their stability and relatively low immunogenicity, while polymers, both natural and synthetic, are more suited toward soft tissue reconstruction or augmentation. Metallic-based constructs are considered inert in that they cause limited inflammation due to their composition; however, the combination of mechanical load, shear stress, and ischemia-reperfusion injury leads to chronic inflammation around the implant, promoting periprosthetic fibrosis as well as osteolytic breakdown and long-term device failure (Konttinen et al. 2005). There are new techniques under development to modify the surface chemistry and topography of metallic-based constructs, such as stainless steel and titanium, to better incorporate progenitor cells on the device surface for improved tissue–implant interface (Blanco et al. 2011). Ceramics, such as aluminum oxide and hydroxyapatite, are also useful in bone repair and regeneration due to the osteoconductive and inductive nature of these materials. They are a unique class of solid, inorganic compounds with relatively open, crystalline structures. The porous nature makes them less sturdy in load-bearing applications though more conducive to progenitor cell incorporation. Much of the interest in the development and modification of ceramic materials is based on improved osteointegration in bony implants; however, there is evidence to suggest that particulate matter in the microenvironment of the ceramic implants is recognized as debris and promotes an inflammatory reaction inducing phagocytic macrophages and potential bone breakdown (Fellah et al. 2010). There is ongoing work exploring the effect of combining these biomaterials with progenitor cells to modulate this inflammatory response.

Currently, the most clinically relevant class of biomaterials are the polymers, further divided as natural or synthetic, though composite materials will likely prove to be widely useful in soft tissue and bony repair as well as solid organ regeneration. Given the complexity and limited clinical use at this time of functional organ replacement and tissue engineering strategies, this chapter focuses on biomaterial polymers used in soft tissue regeneration. Like metallic and ceramic materials, synthetic polymers are constantly being developed and modified for better tissue integration, and the list of materials is seemingly endless. Examples range from the hydrophobic, nonabsorbing silicone, polyethylene, and polytetrafluoroethylene (PTFE) to the more polar poly-lactic and glycolic acid composites (PLA, PGA, and PLGA) to the water absorbing hydrogels such as poly(ethylene) glycol (PEG) and its various modified forms. Natural polymers, such as collagen, hyaluronic acid, and fibrin, are particularly useful in tissue engineering and regenerative medicine due to the similarity to ECM proteins and other naturally occurring in the body. These matrices offer a more supportive environment for progenitor cell-seeded constructs and guide attachment, proliferation, and differentiation in vitro. However, naturally occurring polymers are susceptible to normal physiology such as hydrolytic degradation, thereby limiting their stability in vivo. Thus, composite or blend materials are rapidly being developed with synthetic polymers that are nonimmunogenic and permit simple covalent incorporation of numerous chemical or biological moieties, including cell-adhesive peptide sequences and growth factors (Burdick and Anesth 2002) Hydrogel polymers, such as PEG diacrylate, are attractive for use in soft tissue engineering applications because they maintain form and structure for a longer length of time than collagen and hyaluronic acid, and also have high water content to provide an ECM-like environment (Stacey et al. 2009).

Extensive work is being conducted to characterize macrophage responses to commonly used biomaterials with the hope to further modulate the inflammatory response to biomaterials in a favorable manner (Figure 19.1) (Kou and Babensee 2011). For example, hydrophilic, anionic polymer surfaces induce apoptosis of adherent macrophages rather than fusion and foreign body giant cell formation (Brodbeck et al. 2001); furthermore, they have been associated with increased expression of the anti-inflammatory cytokine IL-10 and lower expression of the pro-inflammatory cytokine IL-8 (Brodbeck et al. 2003). While these in vitro examples illustrate the potential ability to influence immunity toward implantable materials, there has been little to suggest such trends will be observed in vivo. Chronic inflammation and fibrosis or capsule formation have shown variations in response to biomaterial size, surface chemistry, and topography, though these outcomes have yet to be adequately overcome (Kamath et al. 2008, Thevenot et al. 2008). Furthermore, the combination of functional cells encapsulated within the biomaterial offers another layer of complexity in the foreign body–inflammatory response based on viability, differentiation, or possible self-renewal.

19.4 Mesenchymal Stem Cells

Mesenchymal stem cells (MSCs) are adult, multipotent cells with many desirable properties for regenerative medicine (Hanson et al. 2010a,b). Although MSCs were originally derived from bone marrow (Battiwalla and Hematti

2009), it is now evident that MSCs reside within most adult connective tissues and organs (da Silva Meirelles et al. 2006, Hanson et al. 2010c, Lushaj et al. 2011). These cells are characterized by specific pattern of cell surface marker expression and multilineage differentiation capabilities (Dominici et al. 2006). Studies suggest that MSCs isolated from these diverse tissues possess similar biological characteristics, differentiation potential, and immunological properties (Puissant et al. 2005, Hoogduijn et al. 2007, Trivedi and Hematti 2008, Hanson et al. 2010c). Enthusiasm about MSCs for use in regenerative medicine has been fueled by evidence that these cells possess the ability to participate in the tissue repair process through a variety of paracrine mechanisms affecting tissue regeneration, vascularization, immune modulation, and inflammation (Figure 19.2) (Caplan 2007, Chamberlain et al. 2007, Dazzi and Horwood 2007, Phinney and Prockop 2007, Uccelli et al. 2007, Hematti 2008). In fact, one of the most intriguing properties of ex vivo expanded MSCs is their ability to affect the immune response through interaction with a broad range of immune cells. This has been the basis for rapid move of MSCs from in vitro and animal studies into human trials as a therapeutic modality for a diverse group of clinical applications over the last decade (Giordano et al. 2007, Horwitz 2008). MSCs have been investigated extensively in the context of hematopoietic stem cell

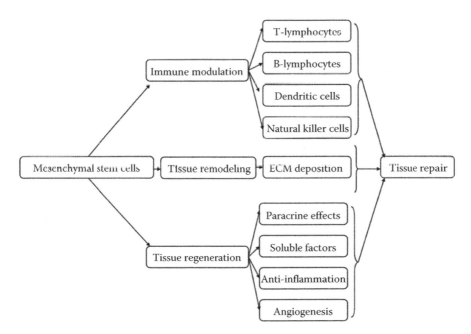

FIGURE 19.2
Illustration of the multiple mechanisms through which MSCs modulate tissue healing, inflammation, and regeneration. (Adapted and modified with permission from Battiwalla, M. and Hematti, P. 2009, Mesenchymal stem cells in hematopoietic stem cell transplantation, *Cytotherapy,* 11(5), 503–515, 2009.)

(HSC) transplantation, either for enhancement of HSC engraftment or for prevention and/or treatment of graft-versus-host disease (GVHD) (Le Blanc et al. 2008, Battiwalla and Hematti 2009, Tolar et al. 2010). GVHD is a major cause of morbidity and mortality after HSC transplantation, and despite major progress in prevention and/or treatment of GVHD over the last few decades, still it is considered a major challenge in the field. The use of MSCs in a percentage of patients has shown to be promising for prevention and/or treatment of GVHD. Nevertheless, much more research is needed before it can be determined what are the optimal doses, frequency, timing, and patient-specific clinical characteristics for MSC administration for this specific indication. Similarly, in the field of solid organ transplantation (SOT), rejection of the transplanted organ by recipient immune system leading to acute loss of the transplanted organ or long-term deterioration of its function is a major challenge (Orlando et al. 2010). Although long-term immunosuppression is the cornerstone of prevention of organ rejection after SOT, the use of these medications is associated with significant side effects, including a negative impact on the function of transplanted organ itself. MSCs, at least in theory, could provide a novel form of cellular therapeutics to favorably modulate immune responses toward transplanted organ (Hematti 2008). Indeed, the use of MSCs in SOT has just been initiated in the context of phase-I clinical trials (Perico et al. 2011, Popp et al. 2011).

Furthermore, culture-expanded MSCs derived from BM, and to a more limited extent, MSCs derived from other tissues such as fat, have also been used in several phase I-II trials for a variety of other indications including treatment of patients with osteogenesis imperfecta (Horwitz et al. 1999), metachromatic leukodystrophy and Hurler's disease (Koc et al. 2002), myocardial infarction (Chen et al. 2004), chronic obstructive pulmonary disease (Sueblinvong and Weiss 2009), amyotrophic lateral sclerosis (Deda et al. 2009), stroke (Lee et al. 2010), Crohn's disease (Taupin 2006), diabetes mellitus (Abdi et al. 2008), systemic sclerosis (Keysser et al. 2011), and systemic lupus erythematosus (Liang et al. 2010), among others. All of these studies have shown that systemic infusion of MSCs is very safe and potentially efficacious for treatment of, at least a percentage of patients, a wide range of disorders. The rapid progression from preclinical models to such a wide range of clinical scenarios can be attributed, to a large extent, to the unique immunomodulatory properties of MSCs that allows the use of MSCs from third party donors (in addition to autologous donors) without consideration of HLA matching. Such property makes these cells a potentially off-the-shelf cellular therapeutics and thus could avoid the cost and other logistical hurdles associated with production of autologous cells.

MSCs have been shown to migrate to the site of inflammation, stimulate resident progenitor cells in both proliferation and differentiation, promote recovery of injured cells through growth factor secretion and matrix remodeling, and exert immunomodulatory and anti-inflammatory effects, all of which make them a potential key player in wound healing and integration of tissue-engineered constructs (Caplan 2007, Chamberlain et al. 2007, Dazzi and

Horwood 2007, Phinney and Prockop 2007, Uccelli et al. 2007, Hematti 2008). In particular, given the ubiquitous role of macrophages in wound healing, initiation, and, subsequently, the resolution of inflammation, the recent findings on the effect of MSCs on immunophenotype of macrophages are of great significance (Kim and Hematti 2009). Indeed, there is an increasing interest in the potential use of MSCs for tissue engineering, particularly in combination with ECM-based biomaterial scaffolds. For example, bone marrow–derived MSCs have been used successfully for treatment of refractory wounds through local injection combined with biomaterials or skin grafts (Yoshikawa et al. 2008). MSC-derived chondrocytes, in combination with autologous epithelial cells, have also been used for reconstruction of decellularized cadaveric donor trachea (Macchiarini et al. 2008). Due to very attractive properties of MSCs, it is expected that there will be a huge expansion in utilization of MSCs for cellularizing biomaterials for reconstructive and regenerative purposes.

We have recently developed an in vitro model for investigating the immune response of human peripheral blood monocyte–derived macrophages when cocultured with human MSCs seeded in a hydrogel scaffold. Our group has investigated the effect of MSCs on immunophenotype of macrophages in the presence of an hyaluronic acid–hydrogel scaffold using a unique 3D coculture system (Hanson et al. 2011). Macrophages cultured on gels containing MSCs expressed an anti-inflammatory profile overall, compatible with the immunophenotype of alternatively activated macrophages (Figure 19.1). A unique aspect of this work is the allogeneic nature of the cells. In our study, the macrophages were derived from different donors than the six MSC lines used. This further verifies our previous observation that MSCs could play a significant role, in modulating the immunophenotype of macrophages (Kim and Hematti 2009). Interestingly, others have shown that while biologic scaffolds can induce an anti-inflammatory macrophage expression, adding a differentiated cellular component will induce a pro-inflammatory or classically activated phenotype in vivo (Brown et al. 2009, Valentin et al. 2009). Development of macrophages, with an anti-inflammatory immunophenotype, upon interaction with the MSC–hydrogel constructs could play a potentially significant role in tissue repair when using a cellular-biomaterial therapeutic approach. While there is huge effort in combining different type of cells with different biomaterials, incorporating MSCs into such cell/ biomaterial constructs could play a significant beneficial role through their unique immunomodulatory and anti-inflammatory properties.

19.5 Conclusions

Tissue healing is a complex process which involves an orchestrated interaction between many different types of cells including immune, vascular, and stromal cells in addition to the tissue-specific cells. The end result of the healing processes depends on a wide range of factors and is not always perfect. There is huge interest in using biomaterials in combination with stem cells, especially MSCs, for optimal healing of damaged tissues. MSCs, due to their multifaceted characteristics, including tissue trophic paracrine effects, immunomodulatory properties, and favorable anti-inflammatory effects, hold great promise in tissue bioengineering, such as MSC–hydrogel constructs. Further investigation of the role of MSCs in modulating interactions between biomaterials and host immune cells is highly warranted.

References

Abdi, R., Fiorina, P., Adra, C.N., Atkinson, M., and Sayegh, M.H. 2008, Immunomodulation by mesenchymal stem cells: A potential therapeutic strategy for type 1 diabetes, *Diabetes*, 57(7), 1759–1767.

Battiwalla, M. and Hematti, P. 2009, Mesenchymal stem cells in hematopoietic stem cell transplantation, *Cytotherapy*, 11(5), 503–515.

Brodbeck, W.G., Shive, M.S., Colton, E., Nakayama, Y., Matsuda, T., and Anderson, J.M. 2001, Influence of biomaterial surface chemistry on the apoptosis of adherent cells, *Journal of Biomedical Materials Research*, 55(4), 661–668.

Brodbeck, W.G., Voskerician, G., Ziats, N.P., Nakayama, Y., Matsuda, T., and Anderson, J.M. 2003, In vivo leukocyte cytokine mRNA responses to biomaterials are dependent on surface chemistry, *Journal of Biomedical Materials Research Part A*, 64(2), 320–329.

Brown, B.N., Valentin, J.E., Stewart-Akers, A.M., McCabe, G.P., and Badylak, S.F. 2009, Macrophage phenotype and remodeling outcomes in response to biologic scaffolds with and without a cellular component, *Biomaterials*, 30(8), 1482–1491.

Burdick, J.A. and Anesth, K.S. 2002, Photoencapsulation of osteoblasts in injectable RGD-modified PEG hydrogels for bone tissue engineering, *Biomaterials*, 23(22), 4315–4323.

Caplan, A.I. 2007, Adult mesenchymal stem cells for tissue engineering versus regenerative medicine, *Journal of Cellular Physiology*, 213(2), 341–347.

Chamberlain, G., Fox, J., Ashton, B., and Middleton, J. 2007, Concise review: Mesenchymal stem cells: Their phenotype, differentiation capacity, immunological features, and potential for homing, *Stem cells (Dayton, Ohio)*, 25(11), 2739–2749.

Chen, S.L., Fang, W.W., Qian, J., Ye, F., Liu, Y.H., Shan, S.J., Zhang, J.J., Lin, S., Liao, L.M., and Zhao, R.C. 2004, Improvement of cardiac function after transplantation of autologous bone marrow mesenchymal stem cells in patients with acute myocardial infarction, *Chinese Medical Journal*, 117(10), 1443–1448.

da Silva Meirelles, L., Chagastelles, P.C., and Nardi, N.B. 2006, Mesenchymal stem cells reside in virtually all post-natal organs and tissues, *Journal of Cell Science*, 119(Pt 11), 2204–2213.

Darby, I.A. and Hewitson, T.D. 2007, Fibroblast differentiation in wound healing and fibrosis, *International Review of Cytology*, 257, 143–179.

Dazzi, F. and Horwood, N.J. 2007, Potential of mesenchymal stem cell therapy, *Current Opinion in Oncology*, 19(6), 650–655.

Deda, H., Inci, M.C., Kurekci, A.E., Sav, A., Kayihan, K., Ozgun, E., Ustunsoy, G.E., and Kocabay, S. 2009, Treatment of amyotrophic lateral sclerosis patients by autologous bone marrow-derived hematopoietic stem cell transplantation: A 1-year follow-up, *Cytotherapy*, 11(1), 18–25.

Dominici, M., Le Blanc, K., Mueller, I., Slaper-Cortenbach, I., Marini, F., Krause, D., Deans, R., Keating, A., Prockop, D., and Horwitz, E. 2006, Minimal criteria for defining multipotent mesenchymal stromal cells. The International Society for Cellular Therapy position statement, *Cytotherapy*, 8(4), 315–317.

Fellah, B.H., Delorme, B., Sohier, J., Magne, D., Hardouin, P., and Layrolle, P. 2010, Macrophage and osteoblast responses to biphasic calcium phosphate microparticles. *Journal of Biomedical Material Research A*, 93(4), 1588–1595.

Giordano, A., Galderisi, U., and Marino, I.R. 2007, From the laboratory bench to the patient's bedside: An update on clinical trials with mesenchymal stem cells, *Journal of Cellular Physiology*, 211(1), 27–35.

Hanson, S.E., Bentz, M.L., and Hematti, P. 2010a, Mesenchymal stem cell therapy for nonhealing cutaneous wounds, *Plastic and Reconstructive Surgery*, 125(2), 510–516.

Hanson, S.E., Gutowski, K.A., and Hematti, P. 2010b, Clinical applications of mesenchymal stem cells in soft tissue augmentation, *Aesthetic Surgery Journal/The American Society for Aesthetic Plastic Surgery*, 30(6), 838–842.

Hanson, S.E., Kim, J., Johnson, B.H., Bradley, B., Breunig, M.J., Hematti, P., and Thibeault, S.L. 2010c, Characterization of mesenchymal stem cells from human vocal fold fibroblasts, *The Laryngoscope*, 120(3), 546–551.

Hanson, S.E., King, S.N., Kim, J., Chen, X., Thibeault, S.L., and Hematti, P. 2011, The effect of mesenchymal stromal cell-hyaluronic acid hydrogel constructs on immunophenotype of macrophages, *Tissue Engineering Part A*, 17(19–20), 2463–2471.

Hematti, P. 2008, Role of mesenchymal stromal cells in solid organ transplantation, *Transplantation Reviews (Orlando, Fla.)*, 22(4), 262–273.

Hoogduijn, M.J., Crop, M.J., Peeters, A.M., Van Osch, G.J., Balk, A.H., Ijzermans, J.N., Weimar, W., and Baan, C.C. 2007, Human heart, spleen, and perirenal fat-derived mesenchymal stem cells have immunomodulatory capacities, *Stem Cells and Development*, 16(4), 597–604.

Horwitz, E.M. 2008, Mesenchymal stromal cells moving forward, *Cytotherapy*, 10(1), 5–6.

Horwitz, E.M., Prockop, D.J., Fitzpatrick, L.A., Koo, W.W., Gordon, P.L., Neel, M., Sussman, M. et al. 1999, Transplantability and therapeutic effects of bone marrow-derived mesenchymal cells in children with osteogenesis imperfecta, *Nature Medicine*, 5(3), 309–313.

Kamath, S., Bhattacharyya, D., Padukudru, C., Timmons, R.B., and Tang, L. 2008, Surface chemistry influences implant-mediated host tissue responses, *Journal of Biomedical Materials Research Part A*, 86(3), 617–626.

Keysser, G., Christopeit, M., Fick, S., Schendel, M., Taute, B.M., Behre, G., Muller, L.P., and Schmoll, H.J. 2011, Treatment of severe progressive systemic sclerosis with transplantation of mesenchymal stromal cells from allogeneic related donors: Report of 5 cases, *Arthritis and Rheumatism*, 63(8), 2540–2542.

Kim, J. and Hematti, P. 2009, Mesenchymal stem cell-educated macrophages: A novel type of alternatively activated macrophages, *Experimental Hematology*, 37(12), 1445–1453.

Koc, O.N., Day, J., Nieder, M., Gerson, S.L., Lazarus, H.M., and Krivit, W. 2002, Allogeneic mesenchymal stem cell infusion for treatment of metachromatic leukodystrophy (MLD) and Hurler syndrome (MPS-IH), *Bone Marrow Transplantation*, 30(4), 215–222.

Kou, P.M. and Babensee, J.E. 2011, Macrophage and dendritic cell phenotypic diversity in the context of biomaterials, *Journal of Biomedical Materials Research Part A*, 96(1), 239–260.

Le Blanc, K., Frassoni, F., Ball, L., Locatelli, F., Roelofs, H., Lewis, I., Lanino, E. et al. & Developmental Committee of the European Group for Blood and Marrow Transplantation. 2008, Mesenchymal stem cells for treatment of steroid-resistant, severe, acute graft-versus-host disease: A phase II study, *Lancet*, 371(9624), 1579–1586.

Lee, J.S., Hong, J.M., Moon, G.J., Lee, P.H., Ahn, Y.H., Bang, O.Y., and STARTING collaborators 2010, A long-term follow-up study of intravenous autologous mesenchymal stem cell transplantation in patients with ischemic stroke, *Stem cells (Dayton, Ohio)*, 28(6), 1099–1106.

Liang, J., Zhang, H., Hua, B., Wang, H., Lu, L., Shi, S., Hou, Y., Zeng, X., Gilkeson, G.S., and Sun, L. 2010, Allogenic mesenchymal stem cells transplantation in refractory systemic lupus erythematosus: A pilot clinical study, *Annals of the Rheumatic Diseases*, 69(8), 1423–1429.

Lushaj, E.B., Anstadt, E., Haworth, R., Roenneburg, D., Kim, J., Hematti, P., and Kohmoto, T. 2011, Mesenchymal stromal cells are present in the heart and promote growth of adult stem cells in vitro, *Cytotherapy*, 13(4), 400–406.

Macchiarini, P., Jungebluth, P., Go, T., Asnaghi, M.A., Rees, L.E., Cogan, T.A., Dodson, A. et al. 2008, Clinical transplantation of a tissue-engineered airway, *Lancet*, 372(9655), 2023–2030.

Mantovani, A., Sica, A., Sozzani, S., Allavena, P., Vecchi, A., and Locati, M. 2004, The chemokine system in diverse forms of macrophage activation and polarization, *Trends in Immunology*, 25(12), 677–686.

Martin, P. 1997, Wound healing—Aiming for perfect skin regeneration, *Science (New York, N.Y.)*, 276(5309), 75–81.

Matheson, L.A., Santerre, J.P., and Labow, R.S. 2004, Changes in macrophage function and morphology due to biomedical polyurethane surfaces undergoing biodegradation, *Journal of Cellular Physiology*, 199(1), 8–19.

McBane, J.E., Ebadi, D., Sharifpoor, S., Labow, R.S., and Santerre, J.P. 2011, Differentiation of monocytes on a degradable, polar, hydrophobic, ionic polyurethane: Two-dimensional films vs. three-dimensional scaffolds, *Acta Biomaterialia*, 7(1), 115–122.

Mosser, D.M. and Edwards, J.P. 2008, Exploring the full spectrum of macrophage activation, *Nature Reviews Immunology*, 8(12), 958–969.

Orlando, G., Hematti, P., Stratta, R.J., Burke, G.W. 3rd, Di Cocco, P., Pisani, F., Soker, K., and Wood, K. 2010, Clinical operational tolerance after renal transplantation: Current status and future challenges, *Annals of Surgery*, 252(6), 915–928.

Perico, N., Casiraghi, F., Introna, M., Gotti, E., Todeschini, M., Cavinato, R.A., Capelli, C. et al. 2011, Autologous mesenchymal stromal cells and renal transplantation: A pilot study on safety and clinical feasibility, *Clinical Journal of the American Society of Nephrology*, 6(2), 412–422.

Phinney, D.G. and Prockop, D.J. 2007, Concise review: Mesenchymal stem/multipotent stromal cells: The state of transdifferentiation and modes of tissue repair—Current views, *Stem Cells (Dayton, Ohio)*, 25(11), 2896–2902.

Popp, F.C., Fillenberg, B., Eggenhofer, E., Renner, P., Dillman, J., Benseler, V., Schnitzbauer, A.A. et al. 2011, Safety and feasibility of third-party multipotent adult progenitor cells for immunomodulation therapy after liver transplantation—A phase I study (MISOT-I), *Journal of Transplantation Medicine*, 28(9), 124.

Puissant, B., Barreau, C., Bourin, P., Clavel, C., Corre, J., Bousquet, C., Taureau, C. et al. 2005, Immunomodulatory effect of human adipose tissue-derived adult stem cells: Comparison with bone marrow mesenchymal stem cells, *British Journal of Haematology*, 129(1), 118–129.

Shin, H. 2007, Fabrication methods of an engineered microenvironment for analysis of cell-biomaterial interactions, *Biomaterials*, 28(2), 126–133.

Stacey, D.H., Hanson, S.E., Lahvis, G., Gutowski, K.A., and Masters, K.A., 2009, *In vitro* adipogenic differentiation of preadipocytes varies with differentiation stimulus, culture dimensionality, and scaffold composition, *Tissue Engineering Part A*, 15(11) 3389–3399.

Sueblinvong, V. and Weiss, D.J. 2009, Cell therapy approaches for lung diseases: Current status, *Current Opinion in Pharmacology*, 9(3), 268–273.

Taupin, P. 2006, OTI-010 osiris therapeutics/JCR pharmaceuticals, *Current Opinion in Investigational Drugs (London, England: 2000)*, 7(5), 473–481.

Thevenot, P., Hu, W., and Tang, L. 2008, Surface chemistry influences implant biocompatibility, *Current Topics in Medicinal Chemistry*, 8(4), 270–280.

Tolar, J., Le Blanc, K., Keating, A., and Blazar, B.R. 2010, Concise review: Hitting the right spot with mesenchymal stromal cells, *Stem Cells (Dayton, Ohio)*, 28(8), 1446–1455.

Trivedi, P. and Hematti, P. 2008, Derivation and immunological characterization of mesenchymal stromal cells from human embryonic stem cells, *Experimental Hematology*, 36(3), 350–359.

Uccelli, A., Pistoia, V., and Moretta, L. 2007, Mesenchymal stem cells: A new strategy for immunosuppression?, *Trends in Immunology*, 28(5), 219–226.

Valentin, J.E., Stewart-Akers, A.M., Gilbert, T.W., and Badylak, S.F. 2009, Macrophage participation in the degradation and remodeling of extracellular matrix scaffolds, *Tissue Engineering Part A*, 15(7), 1687–1694.

Witte, M.B. and Barbul, A. 1997, General principles of wound healing, *The Surgical Clinics of North America*, 77(3), 509–528.

Yoshikawa, T., Mitsuno, H., Nonaka, I., Sen, Y., Kawanishi, K., Inada, Y., Takakura, Y., Okuchi, K., and Nonomura, A. 2008, Wound therapy by marrow mesenchymal cell transplantation, *Plastic and Reconstructive Surgery*, 121(3), 860–877.

20

Autologous Mesenchymal Stem Cells for Tissue Engineering in Urology

Guihua Liu, Chunhua Deng, and Yuanyuan Zhang

CONTENTS

20.1 Introduction

Various cell sources have been considered for tissue engineering and regenerative medicine. The most versatile category is pluripotent embryonic stem cells derived from human embryos (hES cells), which in principle can give rise to the more than 200 known adult human cell types derived from all three germ layers.[1] The sourcing of hES cells has raised ethical debates, however, and it would be difficult to obtain enough lines to permit histocompatibility matching for most potential recipients. Recent reports of methods to generate induced pluripotent stem (iPS) cells from specialized cells, such as skin fibroblasts, show that it may be possible to generate cell lines that would have the same broad potential for multilineage differentiation as hES cells from any individual.[2] However, undifferentiated hES and iPS cells have the potential to form teratomas. Therefore, ensuring the absence of residual stem cells in engineered tissue remains a hurdle to effective clinical implementation. Moreover, the cost of generating and characterizing individual iPS cell lines for each patient would be high, and the safety of iPS cells generated with oncogene-containing vectors remains to be established.

Cells from perinatal sources (e.g., amniotic fluid, umbilical cord blood, Wharton's jelly, and placenta)[3–6] and cells from fetal and adult tissues also are strong candidates for use in tissue engineering.[7,8] Among these, mesenchymal stem cells (MSCs) have generated great interest, and they have already been utilized in multiple clinical trials.[9–12] We provide evidence in our preliminary results that urine-derived stem cells (USCs) are closely related to MSCs. These MSCs were identified as plastic-adherent bone marrow stromal cell populations capable of forming colonies. Additionally, they differentiate into several specialized cell types of connective tissue, including osteoblasts (bone), chondrocytes (cartilage), adipocytes (fat cells), fibroblasts (periosteum), and adventitial reticular cells (bone marrow stroma). It is also claimed that bone marrow–derived MSCs can give rise to various additional mesodermal cell types, including smooth, skeletal, and cardiac muscle, endothelial cells, and perhaps even to cells of ectodermal (e.g., neurons) and endodermal (e.g., hepatocytes) origin. However, evidence that a single MSC can give rise to all of these cell types remains incomplete.[13] Subsequently, similar adherent multipotent cells have been identified in a wide range of prenatal and adult tissues (adipose, skin, bronchia, skeletal muscle, synovium, pancreas, kidney, amnion, placenta, umbilical cord, etc.), and it is now believed that MSCs are present throughout the body.[9,14,15] Some phenotypic differences have been described among MSCs from different sources, such as differing efficiencies or cofactor requirements in differentiation to particular specialized cell types.[16] MSCs also display heterogeneity in various biochemical parameters.[17] Furthermore, MSCs, as generally isolated, show a finite capacity for expansion in culture. These cells are typically limited to six or seven passages for the cells obtained from adult bone marrow or adipose tissues. Most MSC preparations probably contain at most only a small fraction of true stem cells, defined as self-renewing cells capable of differentiation into more than one specialized type, mixed with a range of stromal progenitor cells.[13,18]

In addition to their ability to give rise to useful specialized cell types, MSCs produce a plethora of growth factors, cytokines, and chemokines.[19] These appear to promote angiogenesis, inhibit fibrosis, stimulate survival and differentiation of other reparative cells, and regulate local immune responses. Importantly, MSCs can be transplanted across allogeneic barriers. Recognition of their immunomodulatory properties has led to clinical studies of allogeneic MSCs for the treatment of graft-versus-host disease and autoimmune conditions such as inflammatory bowel disease.[11,20]

MSC populations from multiple sources have been reported to consistently express a common set of cell surface markers, including CD13 (aminopeptidase N), CD29 (beta-1 integrin chain), CD44 (hyaluronan receptor), CD73 (SH3, ecto-5′-nucleotidase), CD90 (Thy-1), CD105 (SH2, endoglin), CD106 (VCAM-1), CD166 (ALCAM), and STRO-1 (a stromal cell marker). None of these markers by itself is indicative of the presence of MSCs, but the constellation seems characteristic.[14,21] In addition, human MSCs generally have been reported to be negative for CD45 (leukocyte common antigen, a hematopoietic marker), CD34 (gp105-120, a marker of early hematopoietic lineage cells and endothelial cells), CD31 (PECAM-1), lineage markers for various differentiated hematopoietic cells, CD117 (c-Kit, receptor for stem cell factor), HLA-DR, and CD133 (a marker of hematopoietic and many other adult stem cells).

A unifying view of MSCs that has emerged over the past 5 years identifies them as relatives of pericytes, cells associated with the walls of small blood vessels. Marker studies and transcriptional profiling initially suggested a relationship between MSCs and pericytes.[22,23] Remarkably, perivascular lineage cells (termed "mesoangioblasts" by some investigators) isolated from various tissues, including brain, skeletal muscle, and cardiac muscle, were then found to display MSC-like stem cell properties and also to possess significant regenerative capacity in disease models such as canine muscular dystrophy.[24–28] A recent paper from Péault and colleagues reports that cells isolated from many human tissues on the basis of expression of pericyte markers (CD146 = MCAM; NG2; PDGF receptor), and absence of mature hematopoietic, endothelial, and myogenic markers, also express the collection of surface markers characteristic of MSCs.[29,30] This group found that cells selected for CD146 expression, and absence of CD34 and CD56, are capable of extensive growth in culture, and clonal differentiation to myogenic, osteogenic, chondrogenic, and adipogenic lineages. They also show chemotactic responses to signals that can be associated with tissue injury. Perivascular cells with the same marker profile were identified in situ by immunofluorescent staining of blood vessels from multiple organs.[30] Thus, the conclusion that MSCs are a subpopulation of the pericyte cell family, associated with blood vessels throughout development and in the adult, has been strengthened.[31] Caplan, who originally introduced the term "MSCs," suggests that these cells play a crucial role in the repair of focal tissue injury by providing regenerative factors, regulating the inflammatory response, and also potentially differentiating into specialized cell types.[15] It is noteworthy that, despite overall similarities of markers and phenotype, MSCs are not homogenous throughout the body. Rather, the detailed behavior of MSCs/pericytes, particularly their relative capacity to give rise to specific differentiated cell types, can vary depending on the tissue of origin.[31]

20.2 Stem/Progenitor Cells Derived from the Genitourinary Tract

20.2.1 Urine

We recently demonstrated that it is possible to isolate and expand stem/progenitor cells from human voided urine[32–34] and urine from upper urinary tract.[35]

Approximately, 0.2% of cells collected from urine express markers characteristic of MSCs, can expand extensively in culture, and can differentiate toward
multiple bladder cell lineages as identified by the expression of urothelial,
smooth muscle, endothelial, and interstitial cell markers. We initially referred
to these cells as urine progenitor cells. However, our more recent experiments
indicated that urine-derived cells can give rise to additional specialized types,
including osteocytes, chondrocytes, and adipocytes. Furthermore, these cells
have self-renewal capability consistent with stem cells. There is now sufficient
evidence to provisionally designate them as USCs. It would be a simple, safe,
noninvasive, and low-cost approach to harvest cells.

Three types of cells exist in urine: differentiated, differentiating, and pro-
genitorstem cells. Most cells in urine are fully differentiated. They do not
attach to tissue culture plates. About 0.1% of cells in urine are differentiating cells, which do attach to plates and display the morphology and protein
markers of various bladder cell lineages. However, these cells do not expand
further after subculture. About 0.2% of the cells in urine have a phenotype
consistent with multipotent stem cells. USCs are easily cultured, appear
genetically stable after a number of passages, and maintain the ability to
give rise to more differentiated progeny.

USCs comprise an average of about 6 cells/100 mL urine (from 5 to 7
cells/100 mL urine). We have shown that, a few days after being placed in
a tissue culture well, a single cell forms a cluster of cells which appeared
small, compact, and uniform. A consistently high yield of cells was achieved
from each of these clonal lines. The cells reached confluence in about 2 weeks
when placed in a 3 cm diameter well at passage one. At passage 2, cells were
plated in 10 cm culture dishes, and a cell number of approximately 1 million
was reached in 3.5 weeks. Finally, in 6–7 weeks, the cultures expanded to
approximately 100 million cells at passage four. These cells displayed normal
exponential cell growth patterns, with a steady increase in cell numbers during a 10-day culture period. The average population doubling (PD) time was
31.3 h in mixed media. These urine-derived cells also showed the ability to
differentiate into various cell lineages, as described later, and were capable
of growing for 12–19 passages in vitro.

Cells from human urine specimens could be consistently cultured
long-term using a medium that we originally developed for culture of rat
urothelium.[36] However, the phenotype of the cultured human urine-derived
cells was not that of primary urothelial cells. The primary cultures from
urine did not show expression of the cytokeratins (CK7, CK13, and CK19/20),
which are characteristic of urothelium, nor did the cells express the urothe-
lial-specific protein uroplakin. After growth in medium containing higher
levels of epidermal growth factor (EGF), the cells were induced efficiently
to express the cytokeratin proteins and uroplakin.[32] However, after growth
in myogenic medium, the cultured cells expressed markers consistent with
smooth muscle, including alpha-smooth muscle actin (α-SM actin), desmin,
calponin, and myosin.[32] This led us to conclude that the urine-derived cells

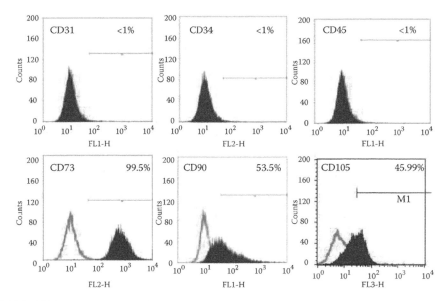

FIGURE 20.1
Analysis of surface marker expression in USC by FACS. A clone of USC (p4) was stained with CD markers expressed by mesenchymal as well as hematopoietic stem cells. All MSC markers (CD73, CD90, and CD105) stained positive whereas the hematopoietic markers (CD31, CD34, and CD45) were negative.

were progenitors (initially designated UPC) capable of giving rise to both urothelium and bladder smooth muscle. Furthermore, we found that the cells displayed a surface marker phenotype consistent with MSCs. Specifically, they expressed CD73, CD90, and CD105, and they were negative for both hematopoietic markers and endothelial markers including CD45, CD34, and CD31 (Figure 20.1). We concluded that the urine-derived progenitors were at least bipotential for the major bladder cell types. This result was surprising, because it was generally believed that muscle and epithelial cells in bladder represent separate cell lineages derived from mesoderm and endoderm, respectively. We have recently observed that the urine-derived cells express markers typical of MSCs and pericytes, and that they can differentiate to yield the characteristic cell lineages obtained from MSCs, namely, osteocytes, adipocytes, and chondrocytes.[37]

In our recent studies, USCs clones could be obtained from 85% of the urine samples tested. Fresh urine gave the highest rate of colony formation (67%) and urine stored at 4°C the lowest (30%). Urine from volunteers aged 13–40 gave the highest rate of clone recovery. Catheterization significantly enhanced the number of USCs in urine compared to spontaneously voided urine, possibly because catheterization resulted in cells being scraped off the inner bladder wall. Collecting triple urine samples increased the rate of clone formation.

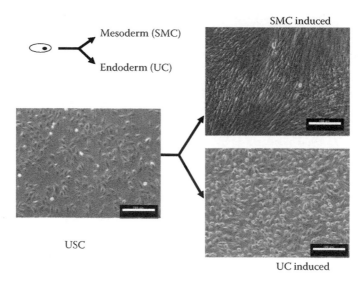

FIGURE 20.2
Morphology of differentiated uUSC. A single clone of uUSC(p3) differentiated to myogenic and uro-epithelial lineage by growth in lineage-specific differentiation medium for 14 days. A distinct change in cell shape from oval (non-treated control) to that of spindle shape (SMC induced) and cobblestone-like shape (uro-epithelial induced) was observed. Brightfield image scale bar = 100 µm.

More recently, we found that some urine-derived cells from the upper urinary tract possessed characteristics similar to voided USCs, that is, expansion capacity and bipotent differentiation to urothelium-like and SMC-like cells (Figure 20.2). These USCs obtained from upper urinary tract (uUSCs) can generate a large cell population from a single clone. We observed that the average expansion capacity of uUSCs is 44 ± 8.6 PDs (range 35–57 PD, $n = 5$). This implies that a single stem cell from the upper urinary tract, on average, can generate 1.8×10^{13} cells (2^{44}), within about 8 weeks. It is known that 1.4×10^9 cells are required for both SMCs and urothelial cells to create a tissue-engineered bladder.[2] To retain good bipotent differentiation capacity, we typically use USC below expansion passage 5 (p5). Under our optimized culture conditions, one cell clone of uUSCs can generate about 2×10^8 cells within 30 days, at p4. Our recent data showed about 150 mL of urine obtained from the upper urinary tract via nephrostomy tube contains 10 uUSCs clones. Expansion of the stem cells from this volume of urine potentially can yield about 2×10^9 cells. Thus, assuming efficient differentiation, uUSCs can provide an adequate number of cells to engineer a neobladder. Importantly, uUSCs are a reliable cell source, as cell clones can be obtained from almost every urine sample.[35]

In chronic bladder diseases, uUSCs might be a good cell source for bladder tissue regeneration because the cells from the upper urinary tract are normal. In addition, the risk of finding ureter, renal pelvic, or kidney cancer in bladder cancer patients could be eliminated with careful scanning by a series of examinations. These scanning examinations include urine cytology, imaging tests (such as intravenous pyelogram, bone scan, computed tomography scan, magnetic resonance imaging, and lung x-ray), cystoscopy/nephroureteroscopy, and tissue biopsy from upper urinary tract. In treatment of end-stage bladder diseases or muscle-invasive bladder cancer, using engineered bladder tissue with uUSCs as the cell source would be superior to current surgical procedures, that is, bladder reconstruction using intestinal segments. Risks of use of bowel segments include (1) tumorigenicity, as intestinal segments appear to be at an increased risk for malignancy, particularly adenocarcinoma, because of histological changes in the intestinal mucosa after long-term exposure to urine; and (2) complications such as stone formation and excess mucous secretion. Harvesting uUSCs from patients who already have a nephrostomy tube in place would be a simple and low-cost approach to obtaining cells for engineering bladder tissue. Therefore, cells derived from upper urinary tract urine might be a good source for bladder tissue engineering in patients with bladder cancer.[35]

There are many potential advantages to using USCs as a cell source for urological tissue engineering. First, cells can be easily harvested and grown in culture. USCs do not require enzyme digestion or culture on a layer of feeder cells to support cell growth. Second, since invasive surgical biopsy procedures are not necessary to harvest cells from urine, patient morbidity and potential complications, such as urethral or bladder trauma and urinary tract infections, are avoided. As USCs are autologous somatic cells, no ethical issues are involved in their use for tissue reconstruction, and no immune reaction to engineered implants should occur.

The quality of cells obtained from urine is similar to that of the biopsy-derived cells described earlier. When differentiated, USCs express all proteins characteristic of the various bladder cell lineages. Karyotype analysis has demonstrated that these cells are genetically stable. Importantly, there is a major cost advantage to using USCs—it costs about $50 to obtain cells from urine, versus about $5000 to isolate cells from a biopsy procedure. About 1.4×10^9 urothelial and smooth muscle cells (SMCs) are required for bladder tissue regeneration.[1] We estimate that three to four urine samples (about 40–45 USCs/600 mL urine) expanded for 4–5 weeks would yield a sufficient quantity of low-passage, healthy cells for clinical tissue engineering applications. This time frame is comparable to that required for expansion from a tissue biopsy (7–8 weeks).[38] USCs and the cells obtained through urological tissue biopsies come from the same urinary tract systems and have similar biological features. Therefore, collecting cells from urine could be an attractive alternative to the standard urological tissue biopsies currently used in cell therapy and tissue engineering.

20.2.2 Bladder

Human bladder stem/progenitor cells from urinary tract have been recently characterized and reported.[32,39] It has been shown experimentally that the bladder neck and trigone area have a higher density of urothelial progenitor cells,[40] and these cells are localized in the basal region.[41] In the past, it was possible to grow urothelial cells in the laboratory setting, but only with limited success. However, several protocols have been developed over the last two decades that have improved urothelial growth and expansion by enhancing culture conditions to support proliferation and differentiation of urothelial progenitor cells.[42–45] Now, normal human bladder epithelial and muscle cells can be efficiently harvested from surgical material, extensively expanded in culture, and their differentiation characteristics, growth requirements, and other biologic properties can be studied.[42,44–55] Additionally, human urothelial and muscle cells can attach and form sheets of cells onto nature collagen-based matrix and polymer scaffolds. The cell-matrix scaffold can then be implanted in vivo for repair bladder or urethra tissue defects. Histological analysis indicates that, within the cell-matrix construct, viable cells are able to self-assemble back into their respective tissue types, and they retain their native phenotype.[56]

Bladder tissue engineered using autologous cells has been used clinically. A small clinical study was conducted starting in 1998. Seven patients were treated using a collagen scaffold seeded with cells taken from biopsies of their own bladders, either with or without omental coverage, or a combined PGA-collagen scaffold seeded with cells and omental coverage. The patients reconstructed with the engineered bladder tissue created with the PGA-collagen cell-seeded scaffolds with omental coverage showed increased compliance, decreased end-filling pressures, increased capacities, and longer dry periods over time.[57] It is clear from this small study that the engineered bladders continued their improvement with time, suggesting continued development in vivo. Although the experience is promising, it is just a start, and the technology is not yet ready for wide dissemination, as further experimental and clinical studies are required.

20.2.3 Kidney

Kidney has long been considered an organ that is incapable of true regeneration. Furthermore, the question of whether or not the kidney contains adult stem cells remains controversial. However, increasing evidence of a regenerative response in the kidney has been observed following the injury resulting from both toxic and ischemic insults. These observations include evidence of renal progenitors of specific cell types involved in the formation of new renal tubular cells and the recovery of renal function recovery after ischemic injury. The presence of injured or dead cells following ischemia causes denudation of the tubular basement membrane, and sloughed cells and cellular debris fill tubular lumens. The kidney responds to the ischemic injury with a prompt regenerative response, resulting in regenerating tubules and improving kidney function. Although they remain elusive, the cells participating in renal regeneration are likely from pools of both exogenous and endogenous stem cells. The exogenous stem cells are probably largely derived from bone marrow and may be both hematopoietic and MSCs. In some studies, these cells appear to home to damaged sites in the injured kidney and form tubular epithelial cells following acute renal injury.[58–61] These MSCs might also produce growth factors such as IGF-1 to promote renal repair.[62] The endogenous stem cells are resident kidney stem cells found in the renal tubules and the papilla.[63,64] They are inactivated under physiological conditions. These stem cells posses the capacity to give rise to renal tubule cells following injury repair.[65,66] A recent study showed that repopulation of damaged renal tubules occurs primarily from proliferation of tubular epithelial cells and resident renal-specific stem cells, with some contribution of paracrine factors from bone marrow–derived MSCs.[67,68]

20.2.4 Testis

Primordial germ cells (PGC) are the embryonic progenitor cells of the gametes (spermatogonial stem cells [SSCs] and ova). In vivo, PGC colonize the gonadal ridge during early embryonic development and are then restricted to producing the gametes. However, if PGC are cultured in vitro in the presence of specific growth factors, they are able to form pluripotent embryonic germ cells (EGC) through a process that is relatively similar to that of nuclear reprogramming and generation of iPS cells in the laboratory (discussed later in this review).[69] These cells can contribute to all cellular lineages in chimeric embryos, including the germline. They also form teratomas when injected into immunocompromised animals.

For regenerative medicine purposes, PGC are not ideal, because they are derived from embryos and there are a number of controversial ethical issues surrounding the manipulation of human embryos. However, there has been much interest in isolating and describing SSCs in recent years. The presence of SSCs, which are derived from PGC in the testis, was originally inferred from the presence of ongoing spermatogenesis in the adult male. Recently, the cells believed to be the actual SSC were isolated from the adult testis of both mice and humans through selection of the markers STRA-8, GPR125, CD49fm, CD133, and others.[70–72] Interestingly, when SSCs from both mice and humans are cultured in specified media containing growth factors known to be required for maintenance of pluripotency of other types of stem cells, such as leukemia inhibitory factor (LIF) and glial cell line–derived neurotrophic factor (GDNF), they appear to convert to an ES cell-like state.[72] These converted cells have been termed adult germline stem cells (aGSCs), and they can differentiate into a number of somatic cell types encompassing all three embryonic germ layers when they are exposed to the same conditions used to differentiate ES cells. They also form teratomas when implanted in vivo. These results suggest that SSCs, which can be obtained through a small testicular biopsy, may be useful for the development of cell-based, autologous organ regeneration strategies. However, more research is required to overcome additional hurdles before this technology can be used clinically. In addition, since autologous regeneration strategies based on SSCs would only benefit males, researchers are working to identify and describe similar pluripotent cells that may reside in an ovarian niche for use in females.

20.3 Stem/Progenitor Cells Derived from Nonurological Tissues

Despite the convenience of using differentiated cells in tissue engineering applications, these cells have several shortcomings. These cells have a limited ability to grow in culture, and they tend to dedifferentiate in vitro, which may lead to insufficient numbers of cells. In addition, autologous bladder cells cannot be taken from patients with urinary tract malignancies. One solution to these problems is to prepare engineered tissues using stem cells from various sources. These types of stem/progenitor cells from nonurological tissue have been studied as cell sources for bladder regeneration and cell therapy for stress urinary incontinence.

20.3.1 Bone Marrow Mesenchymal Stem Cells

Adult MSCs, isolated from bone marrow, skeletal muscle, and adipose tissue, possess the capacity to differentiate into cells of connective tissue lineages, including muscle. Isolation and characterization of MSCs, and control of their myogenic differentiation derived from both preclinical and clinical studies have attracted attention to their potential use in urological regenerative medicine and tissue engineering.

Currently, the most effectively characterized types of multipotent stem cells are from bone marrow. Bone marrow mesenchymal stem cells (BMSCs) have been shown to differentiate into specialized cells, including hepatocytes,[73–75] neural cells,[76–79] and mainly mesodermal derivatives such as bone, cartilage, cardiac muscle, skeletal muscle, and fat. If BMSCs are placed on a proper biodegradable scaffold and implanted, they can act as anti-fibrotic, angiogenic, anti-apoptotic, and mitotic agents. Recently, BMSCs were evaluated as an alternative cell type for use in replacement of bladder SMCs when native bladder muscle tissue is unavailable. The potential of BMSCs to differentiate into cells with bladder SMC characteristics was assessed in vitro[80] and in different animal models.[81–83] Kanematsu et al.[83] showed that in vitro, both supernatants from cultured rat bladder cells (conditioned media) or media containing TGF-β and VEGF induced bone marrow cells to adopt an SMC phenotype. Recently, we have investigated the impacts of soluble growth factors, bladder extracellular matrix (ECM), and 3D dynamic culture on cell proliferation and differentiation of human BMSCs into SMCs.[80] Myogenic growth factors (PDGF-BB and TGF-1) alone, or combined either with bladder ECM or dynamic cultures, induced BMSCs to express smooth muscle–specific genes and proteins. Either ECM or the dynamic culture alone promoted cell proliferation but did not induce myogenic differentiation of BMSCs. A highly porous poly-L-lactic acid (PLLA) scaffold provided a 3D structure for maximizing the cell-matrix penetration, maintained myogenic differentiation of the induced BMSCs, and promoted tissue remolding with rich capillary formation in vivo. This study demonstrates that myogenic-differentiated BMSCs seeded on a nanofibrous PLLA scaffold can be used for cell-based tissue engineering for bladder cancer patients requiring cystoplasty.

In order to test this in vivo, bone marrow cells expressing green fluorescent protein were transplanted into lethally irradiated rats. Eight weeks following transplantation, bladder domes were replaced with acellular matrix grafts. Two weeks after the graft procedure, GFP expression in the matrices indicated that the transplanted marrow cells had repopulated the graft. By 12 weeks, these cells reconstituted the smooth muscle layer, with native SMCs also infiltrating the graft. In another rat study,[82] rapid regeneration of bladder SMCSs and urothelium occurred on BMSC-seeded collagen matrices, whereas fibrotic changes were observed in the nonseeded matrix group 3 months after bladder augmentation. In a large animal study,[81] BMSCs proliferated at the same rate as primary cultured bladder SMCs in vitro, and they had a similar histological appearance and contractile phenotype as primary cultured bladder SMCs. BMSCs had a significant contractile response to calcium-ionophore in vitro, and this response was similar to that seen in bladder SMCs but markedly different from fibroblasts. Immunohistochemical staining and Western blotting indicated that BMSCs expressed α-SM actin, but did not express desmin or myosin. In vivo, small intestinal submucosa (SIS) grafts seeded with BMSCs developed solid smooth-muscle bundle formations throughout the grafts, as did bladder cell-seeded SIS grafts. However, bladder tissue regeneration did not occur in animals that received cell-free scaffolding. These results indicate that BMSCs may provide an alternative cell source for bladder tissue engineering. This is relevant for patients with bladder malignancies who require bladder augmentation or replacement but do not have enough normal, nonmalignant bladder cells to use in tissue engineering applications.

Other MSCs such as skeletal muscle–derived progenitor cells[84–97] and adipose stem cells[85,95,96,98–102] have been investigated as potential candidates for cell-based tissue engineering and injection therapy in urology, and these studies are further described in Section 20.4.

20.3.2 Induced Pluripotent Stem Cells

iPS cells are a type of pluripotent stem cell that is artificially derived from a patient's own somatic cells (a non-pluripotent cell) by inducing a "forced" expression of certain genes. iPS cells were first produced in 2006 from mouse cells and then in 2007 from human cells. iPS cells are typically derived by transfecting stem cell–associated genes into non-pluripotent cells, such as adult fibroblasts. Transfection is typically achieved through viral vectors, such as retroviruses. Transfected genes include the master transcriptional regulators Oct-3/4 (Pouf51) and Sox2, although it is suggested that other genes may enhance the efficiency of induction. After 3–4 weeks, small numbers of transfected cells begin to become morphologically and biochemically similar to pluripotent stem cells, and these cells are typically isolated through morphological selection or through a reporter gene and/or antibiotic selection. This has been cited as an important advancement in stem cell research, as it may allow researchers to obtain pluripotent stem cells, which are important in research and potentially have therapeutic uses in urology, without the controversial use of embryos.

iPS cells are believed to be similar to natural pluripotent stem cells, such as ES cells in many respects, including expression of certain stem cell genes and proteins, chromatin methylation patterns, doubling time, embryoid body formation, teratoma formation, viable chimera formation, and potency and differentiability, but the full extent of their relation to natural pluripotent stem cells is still being assessed.

However, depending on the methods used, reprogramming of adult cells to obtain iPS cells may pose significant risks that currently limit the use of this technique in human therapy. For example, if viruses are used to genetically alter the cells, expression of oncogenes may potentially be triggered. In February 2008, a report published in the journal *Cell* announced the discovery of a technique that removed the need for oncogenes such as c-myc in induction of pluripotency, thereby increasing the potential use of iPS cells in human diseases. Even more recently, in April 2009, Sheng Ding in La Jolla, California, showed that the generation of iPS cells was possible without any genetic alteration of the adult cell.[103] Repeated treatment of the cells with certain proteins channeled into the cells via poly-arginine anchors was sufficient to induce pluripotency. The cells generated by this process are known as protein-iPS cells (piPS cells).

20.3.3 Human Amniotic Fluid Stem Cells

Human amniotic fluid cells are commonly used clinically as a diagnostic tool for the prenatal diagnosis of fetal genetic anomalies. Recently, increasing evidence demonstrated that fetal-derived stem cells can be isolated from amniotic fluid. These cells represent a novel class of pluripotent stem cells with intermediate characteristics between embryonic and adult stem cells, as they are capable of giving rise to lineages representative of all three germ layers but do not form teratomas when implanted in vivo.[3] These features, in addition to the absence of ethical concerns about their use, indicate that amniotic fluid stem (AFS) cells might be a promising cell source for tissue engineering and stem cell therapy. Perin et al.[104,105] have recently reported that AFS cells may be useful for kidney regeneration. In a series of studies, this group demonstrated that these pluripotent cells are able to differentiate into de novo kidney structures during organogenesis in vitro. Human male amniotic fluid cells were isolated between 12 and 18 weeks of gestation. AFS cells were isolated from these cultures and labeled with green fluorescent protein and Lac-Z-. Labeled human AFS cells were then microinjected into murine embryonic kidneys (12.5–18 days gestation), and these were maintained in a coculture system for 10 days. Histological analysis revealed that human AFS cells were able to contribute to the development of elemental kidney structures including renal vesicles, and C- and S-shaped bodies. Expression of the early kidney markers zona occludens-1, glial-derived neurotrophic factor, and claudin was confirmed by RT-PCR. Therefore, it is possible that AFS cells represent a potential cell source for future renal cell therapies.

20.4 Conclusions

Current advances in urological tissue engineering and stem cell-based therapy demonstrate that bladder and urethral tissues can be regenerated using autologous cells seeded onto biodegradable scaffolds. Vesicoureteral reflux and stress urinary incontinence can be corrected with injections of autologous stem cells contained in a hydrogel. However, many issues must be elucidated before these techniques can become widely used in the clinic. For example, the role of donor cells in tissue regeneration remains unclear, and it is not known whether the seeded stem cells proliferate and populate scaffold materials themselves, or if they stimulate to the activation, migration, proliferation, and differentiation of the local progenitor cells to complete the tissue regeneration. Additionally, an approach to promote angiogenesis and to facilitate innervation with a functional network of regenerated nerves will greatly improve tissue regeneration strategies to create a de novo urological organ.

References

1. Murry CE, Keller G. Differentiation of embryonic stem cells to clinically relevant populations: Lessons from embryonic development. *Cell.* 2008;**132**: 661–680.

2. Zhao R, Daley GQ. From fibroblasts to iPS cells: Induced pluripotency by defined factors. *J Cell Biochem.* 2008;**105**(4): 949–955.
3. De Coppi P, Bartsch G, Jr., Siddiqui MM et al. Isolation of amniotic stem cell lines with potential for therapy. *Nat Biotechnol.* 2007;**25**: 100–106.
4. Marcus AJ, Woodbury D. Fetal stem cells from extra-embryonic tissues: Do not discard. *J Cell Mol Med.* 2008;**12**: 730–742.
5. Moise KJ, Jr. Umbilical cord stem cells. *Obstet Gynecol.* 2005;**106**: 1393–1407.
6. Troyer DL, Weiss ML. Wharton's jelly-derived cells are a primitive stromal cell population. *Stem Cells.* 2008;**26**: 591–599.
7. Serakinci N, Keith WN. Therapeutic potential of adult stem cells. *Eur J Cancer.* 2006;**42**: 1243–1246.
8. Furth ME, Atala A. Current and future perspectives of regenerative medicine. In: Atala A, Lanza R, Thomson JA, Nerem RM, eds. *Principles of Regenerative Medicine.* Amsterdam, the Netherlands: Elsevier; 2008, pp. 2–15.
9. Vaananen HK. Mesenchymal stem cells. *Ann Med.* 2005;**37**: 469–479.
10. Caplan AI. Review: Mesenchymal stem cells: Cell-based reconstructive therapy in orthopedics. *Tissue Eng.* 2005;**11**: 1198–1211.
11. Le Blanc K, Ringden O. Immunomodulation by mesenchymal stem cells and clinical experience. *J Intern Med.* 2007;**262**: 509–525.
12. Giordano A, Galderisi U, Marino IR. From the laboratory bench to the patient's bedside: An update on clinical trials with mesenchymal stem cells. *J Cell Physiol.* 2007;**211**: 27–35.
13. Bianco P, Robey PG, Simmons PJ. Mesenchymal stem cells: Revisiting history, concepts, and assays. *Cell Stem Cell.* 2008;**2**: 313–319.
14. Musina RA, Bekchanova ES, Sukhikh GT. Comparison of mesenchymal stem cells obtained from different human tissues. *Bull Exp Biol Med.* 2005;**139**: 504–509.
15. da Silva Meirelles L, Caplan AI, Nardi NB. In search of the in vivo identity of mesenchymal stem cells. *Stem Cells.* 2008;**26**: 2287–2299.
16. Musina RA, Bekchanova ES, Belyavskii AV, Sukhikh GT. Differentiation potential of mesenchymal stem cells of different origin. *Bull Exp Biol Med.* 2006;**141**: 147–151.
17. Phinney DG. Biochemical heterogeneity of mesenchymal stem cell populations: Clues to their therapeutic efficacy. *Cell Cycle.* 2007;**6**: 2884–2889.
18. Phinney DG, Prockop DJ. Concise review: Mesenchymal stem/multipotent stromal cells: The state of transdifferentiation and modes of tissue repair—Current views. *Stem Cells.* 2007;**25**: 2896–2902.
19. Caplan AI, Dennis JE. Mesenchymal stem cells as trophic mediators. *J Cell Biochem.* 2006;**98**: 1076–1084.
20. Caplan AI. Adult mesenchymal stem cells for tissue engineering versus regenerative medicine. *J Cell Physiol.* 2007;**213**: 341–347.
21. Delorme B, Charbord P. Culture and characterization of human bone marrow mesenchymal stem cells. *Methods Mol Med.* 2007;**140**: 67–81.
22. Shi S, Gronthos S. Perivascular niche of postnatal mesenchymal stem cells in human bone marrow and dental pulp. *J Bone Miner Res.* 2003;**18**: 696–704.
23. Wieczorek G, Steinhoff C, Schulz R et al. Gene expression profile of mouse bone marrow stromal cells determined by cDNA microarray analysis. *Cell Tissue Res.* 2003;**311**: 227–237.

24. Brachvogel B, Moch H, Pausch F et al. Perivascular cells expressing annexin A5 define a novel mesenchymal stem cell-like population with the capacity to differentiate into multiple mesenchymal lineages. *Development.* 2005;**132**: 2657–2668.

25. Dore-Duffy P, Katychev A, Wang X, Van Buren E. CNS microvascular pericytes exhibit multipotential stem cell activity. *J Cereb Blood Flow Metab.* 2006;**26**: 613–624.

26. Sampaolesi M, Blot S, D'Antona G et al. Mesoangioblast stem cells ameliorate muscle function in dystrophic dogs. *Nature.* 2006;**444**: 574–579.

27. Dellavalle A, Sampaolesi M, Tonlorenzi R et al. Pericytes of human skeletal muscle are myogenic precursors distinct from satellite cells. *Nat Cell Biol.* 2007;**9**: 255–267.

28. Galvez BG, Sampaolesi M, Barbuti A et al. Cardiac mesoangioblasts are committed, self-renewable progenitors, associated with small vessels of juvenile mouse ventricle. *Cell Death Differ.* 2008;**15**: 1417–1428.

29. Crisan M, Huard J, Zheng B et al. Purification and culture of human blood vessel-associated progenitor cells. *Curr Protoc Stem Cell Biol.* 2008;Chapter 2: Unit 2B 2 1–2B 2 13.

30. Crisan M, Yap S, Casteilla L et al. A perivascular origin for mesenchymal stem cells in multiple human organs. *Cell Stem Cell.* 2008;**3**: 301–313.

31. Caplan AI. All MSCs are pericytes? *Cell Stem Cell.* 2008;**3**: 229–230.

32. Zhang Y, McNeill E, Tian H et al. Urine derived cells are a potential source for urological tissue reconstruction. *J Urol.* 2008;**180**: 2226–2233.

33. Bodin A, Bharadwaj S, Wu S, Gatenholm P, Atala A, Zhang Y. Tissue-engineered conduit using urine-derived stem cells seeded bacterial cellulose polymer in urinary reconstruction and diversion. *Biomaterials.* 2010;**31**: 8889–8901.

34. Wu S, Liu Y, Bharadwaj S, Atala A, Zhang Y. Human urine-derived stem cells seeded in a modified 3D porous small intestinal submucosa scaffold for urethral tissue engineering. *Biomaterials.* 2011;**32**: 1317–1326.

35. Bharadwaj S, Liu G, Shi Y et al. Characterization of urine-derived stem cells obtained from upper urinary tract for use in cell-based urological tissue engineering. *Tissue Eng Part A.* 2011;**17**(15–16): 2123–2132.

36. Zhang YY, Ludwikowski B, Hurst R, Frey P. Expansion and long-term culture of differentiated normal rat urothelial cells in vitro. *In Vitro Cell Dev Biol Anim.* 2001;**37**: 419–429.

37. Bharadwaj S, Wu S, Rohozinski J, Furth M, Atala A, Zhang Y. Multipotential Differentiation of Human Urine-Derived Stem Cells. *2nd World Congress on Tissue Engineering and Regenerative Medicine.* 2009; Daejeon, Korea, Vol. 6, p. S293.

38. Oberpenning F, Meng J, Yoo JJ, Atala A. De novo reconstitution of a functional mammalian urinary bladder by tissue engineering. *Nat Biotechnol.* 1999;**17**: 149–155.

39. Oottamasathien S, Williams K, Franco OE et al. Urothelial inhibition of transforming growth factor-beta in a bladder tissue recombination model. *J Urol.* 2007;**178**: 1643–1649.

40. Nguyen MM, Lieu DK, deGraffenried LA, Isseroff RR, Kurzrock EA. Urothelial progenitor cells: Regional differences in the rat bladder. *Cell Prolif.* 2007;**40**: 157–165.

41. Kurzrock EA, Lieu DK, Degraffenried LA, Chan CW, Isseroff RR. Label-retaining cells of the bladder: Candidate urothelial stem cells. *Am J Physiol—Renal Physiol.* 2008;**294**: F1415–F1421.
42. Cilento BG, Freeman MR, Schneck FX, Retik AB, Atala A. Phenotypic and cytogenetic characterization of human bladder urothelia expanded in vitro. *J Urol.* 1994;**152**: 665–670.
43. Scriven SD, Booth C, Thomas DF, Trejdosiewicz LK, Southgate J. Reconstitution of human urothelium from monolayer cultures. *J Urol.* 1997;**158**: 1147–1152.
44. Liebert M, Hubbel A, Chung M et al. Expression of mal is associated with urothelial differentiation in vitro: Identification by differential display reverse-transcriptase polymerase chain reaction. *Differentiation.* 1997;**61**: 177–185.
45. Puthenveettil JA, Burger MS, Reznikoff CA. Replicative senescence in human uroepithelial cells. *Adv Exp Med Biol.* 1999;**462**: 83–91.
46. Liebert M, Wedemeyer G, Abruzzo LV et al. Stimulated urothelial cells produce cytokines and express an activated cell surface antigenic phenotype. *Semin Urol.* 1991;**9**: 124–130.
47. Tobin MS, Freeman MR, Atala A. Maturational response of normal human urothelial cells in culture is dependent on extracellular matrix and serum additives. *Surg Forum.* 1994;**45**: 786.
48. Harriss DR. Smooth muscle cell culture: A new approach to the study of human detrusor physiology and pathophysiology. *Br J Urol.* 1995;**75**(Suppl 1): 18–26.
49. Freeman MR, Yoo JJ, Raab G et al. Heparin-binding EGF-like growth factor is an autocrine growth factor for human urothelial cells and is synthesized by epithelial and smooth muscle cells in the human bladder. *J Clin Investig.* 1997;**99**: 1028–1036.
50. Fauza DO, Fishman SJ, Mehegan K, Atala A. Videofetoscopically assisted fetal tissue engineering: Skin replacement. *J Pediatr Surg.* 1998;**33**: 357–361.
51. Fauza DO, Fishman SJ, Mehegan K, Atala A. Videofetoscopically assisted fetal tissue engineering: Bladder augmentation. *J Pediatr Surg.* 1998;**33**: 7–12.
52. Solomon LZ, Jennings AM, Sharpe P, Cooper AJ, Malone PS. Effects of short-chain fatty acids on primary urothelial cells in culture: Implications for intravesical use in enterocystoplasties.[see comment]. *J Lab Clin Med.* 1998;**132**: 279–283.
53. Lobban ED, Smith BA, Hall GD et al. Uroplakin gene expression by normal and neoplastic human urothelium. *Am J Pathol.* 1998;**153**: 1957–1967.
54. Nguyen HT, Park JM, Peters CA et al. Cell-specific activation of the HB-EGF and ErbB1 genes by stretch in primary human bladder cells. *In Vitro Cell Dev Biol Anim.* 1999;**35**: 371–375.
55. Rackley RR, Bandyopadhyay SK, Fazeli-Matin S, Shin MS, Appell R. Immunoregulatory potential of urothelium: Characterization of NF-kappaB signal transduction. *J Urol.* 1999;**162**: 1812–1816.
56. Atala A, Freeman MR, Vacanti JP, Shepard J, Retik AB. Implantation in vivo and retrieval of artificial structures consisting of rabbit and human urothelium and human bladder muscle. *J Urol.* 1993;**150**: 608–612.
57. Atala A, Bauer SB, Soker S, Yoo JJ, Retik AB. Tissue-engineered autologous bladders for patients needing cystoplasty.[see comment]. *Lancet.* 2006;**367**: 1241–1246.
58. Lin F. Renal repair: Role of bone marrow stem cells. *Pediatr Nephrol.* 2008;**23**: 851–861.

59. Lin F. Stem cells in kidney regeneration following acute renal injury. *Pediatr Res.* 2006;**59**: 74R–78R.
60. Krause D, Cantley LG. Bone marrow plasticity revisited: Protection or differentiation in the kidney tubule? *J Clin Invest.* 2005;**115**: 1705–1708.
61. Kale S, Karihaloo A, Clark PR, Kashgarian M, Krause DS, Cantley LG. Bone marrow stem cells contribute to repair of the ischemically injured renal tubule. *J Clin Invest.* 2003;**112**: 42–49.
62. Imberti B, Morigi M, Tomasoni S et al. Insulin-like growth factor-1 sustains stem cell mediated renal repair. *J Am Soc Nephrol.* 2007;**18**: 2921–2928.
63. Oliver JA, Maarouf O, Cheema FH, Martens TP, Al-Awqati Q. The renal papilla is a niche for adult kidney stem cells. *J Clin Invest.* 2004;**114**: 795–804.
64. Maeshima A, Sakurai H, Nigam SK. Adult kidney tubular cell population showing phenotypic plasticity, tubulogenic capacity, and integration capability into developing kidney. *J Am Soc Nephrol.* 2006;**17**: 188 198.
65. Duffield JS, Park KM, Hsiao LL et al. Restoration of tubular epithelial cells during repair of the postischemic kidney occurs independently of bone marrow-derived stem cells. *J Clin Invest.* 2005;**115**: 1743–1755.
66. Duffield JS, Bonventre JV. Kidney tubular epithelium is restored without replacement with bone marrow-derived cells during repair after ischemic injury. *Kidney Int.* 2005;**68**: 1956–1961.
67. Liu KD, Brakeman PR. Renal repair and recovery. *Crit Care Med.* 2008;**36**: S187–S192.
68. Lin F, Moran A, Igarashi P. Intrarenal cells, not bone marrow-derived cells, are the major source for regeneration in postischemic kidney. *J Clin Invest.* 2005;**115**: 1756–1764.
69. Nayernia K. Stem cells derived from testis show promise for treating a wide variety of medical conditions. *Cell Res.* 2007;**17**: 895–897.
70. Kanatsu-Shinohara M, Inoue K, Lee J et al. Generation of pluripotent stem cells from neonatal mouse testis. *Cell.* 2004;**119**: 1001–1012.
71. Seandel M, James D, Shmelkov SV et al. Generation of functional multipotent adult stem cells from GPR125+ germline progenitors. *Nature.* 2007;**449**: 346–350.
72. Conrad S, Renninger M, Hennenlotter J et al. Generation of pluripotent stem cells from adult human testis. *Nature.* 2008;**456**: 344–349.
73. Yang Y, Qu B, Huo JH, Wu SL, Zhang MY, Wang ZR. Serum from radiofrequency-injured livers induces differentiation of bone marrow stem cells into hepatocyte-like cells. *J Surg Res.* 2009;**155**: 18–24.
74. Saulnier N, Lattanzi W, Puglisi MA et al. Mesenchymal stromal cells multipotency and plasticity: Induction toward the hepatic lineage. *Eur Rev Med Pharmacol Sci.* 2009;**13**(Suppl 1): 71–78.
75. Mizuguchi T, Hui T, Palm K et al. Enhanced proliferation and differentiation of rat hepatocytes cultured with bone marrow stromal cells. *J Cell Physiol.* 2001;**189**: 106–119.
76. Yaghoobi MM, Mowla SJ. Differential gene expression pattern of neurotrophins and their receptors during neuronal differentiation of rat bone marrow stromal cells. *Neurosci Lett.* 2006;**397**: 149–154.
77. Xin H, Li Y, Chen X, Chopp M. Bone marrow stromal cells induce BMP2/4 production in oxygen-glucose-deprived astrocytes, which promotes an astrocytic phenotype in adult subventricular progenitor cells. *J Neurosci Res.* 2006;**83**: 1485–1493.

78. Shichinohe H, Kuroda S, Yano S et al. Improved expression of gamma-aminobutyric acid receptor in mice with cerebral infarct and transplanted bone marrow stromal cells: An autoradiographic and histologic analysis. *J Nucl Med.* 2006;**47**: 486–491.

79. Scintu F, Reali C, Pillai R et al. Differentiation of human bone marrow stem cells into cells with a neural phenotype: Diverse effects of two specific treatments. *BMC Neurosci.* 2006;**7**: 14.

80. Tian H, Bharadwaj S, Liu Y et al. Myogenic differentiation of human bone marrow mesenchymal stem cells on a 3D nano fibrous scaffold for bladder tissue engineering. *Biomaterials.* 2009;**31**: 870–877.

81. Zhang Y, Lin HK, Frimberger D, Epstein RB, Kropp BP. Growth of bone marrow stromal cells on small intestinal submucosa: An alternative cell source for tissue engineered bladder. *BJU Int.* 2005;**96**: 1120–1125.

82. Chung SY, Krivorov NP, Rausei V et al. Bladder reconstitution with bone marrow derived stem cells seeded on small intestinal submucosa improves morphological and molecular composition. *J Urol.* 2005;**174**: 353–359.

83. Kanematsu A, Yamamoto S, Iwai-Kanai E et al. Induction of smooth muscle cell-like phenotype in marrow-derived cells among regenerating urinary bladder smooth muscle cells. *Am J Pathol.* 2005;**166**: 565–573.

84. Strasser H, Marksteiner R, Margreiter E et al. Transurethral ultrasonography-guided injection of adult autologous stem cells versus transurethral endoscopic injection of collagen in treatment of urinary incontinence. *World J Urol.* 2007;**25**: 385–392.

85. Smaldone MC, Chen ML, Chancellor MB. Stem cell therapy for urethral sphincter regeneration. *Minerva Urol Nefrol.* 2009;**61**: 27–40.

86. Peyromaure M, Sebe P, Praud C et al. Fate of implanted syngenic muscle precursor cells in striated urethral sphincter of female rats: Perspectives for treatment of urinary incontinence. *Urology.* 2004;**64**: 1037–1041.

87. Nolazco G, Kovanecz I, Vernet D et al. Effect of muscle-derived stem cells on the restoration of corpora cavernosa smooth muscle and erectile function in the aged rat. *BJU Int.* 2008;**101**: 1156–1164.

88. Lu SH, Yang AH, Wei CF, Chiang HS, Chancellor MB. Multi-potent differentiation of human purified muscle-derived cells: Potential for tissue regeneration. *BJU Int.* 2010;**105**: 1174–1180.

89. Lu SH, Yang AH, Chen KK, Chiang HS, Chang LS. Purification of human muscle-derived cells using an immunoselective method for potential use in urological regeneration. *BJU Int.* 2009;**105**(11): 1598–1603.

90. Lu SH, Wei CF, Yang AH, Chancellor MB, Wang LS, Chen KK. Isolation and characterization of human muscle-derived cells. *Urology.* 2009;**74**: 440–445.

91. Lu SH, Cannon TW, Chermanski C et al. Muscle-derived stem cells seeded into acellular scaffolds develop calcium-dependent contractile activity that is modulated by nicotinic receptors. *Urology.* 2003;**61**: 1285–1291.

92. Hoshi A, Tamaki T, Tono K et al. Reconstruction of radical prostatectomy-induced urethral damage using skeletal muscle-derived multipotent stem cells. *Transplantation.* 2008;**85**: 1617–1624.

93. Cannon TW, Lee JY, Somogyi G et al. Improved sphincter contractility after allogenic muscle-derived progenitor cell injection into the denervated rat urethra. *Urology.* 2003;**62**: 958–963.

94. Rodriguez LV, Alfonso Z, Zhang R, Leung J, Wu B, Ignarro LJ. Clonogenic multipotent stem cells in human adipose tissue differentiate into functional smooth muscle cells. *Proc Natl Acad Sci USA.* 2006;**103**: 12167–12172.

95. Montzka K, Heidenreich A. Application of mesenchymal stromal cells in urological diseases. *BJU Int.* 2009;**150**(3): 309–312.

96. Jack GS, Zhang R, Lee M, Xu Y, Wu BM, Rodriguez LV. Urinary bladder smooth muscle engineered from adipose stem cells and a three dimensional synthetic composite. *Biomaterials.* 2009;**30**: 3259–3270.

97. Jack GS, Almeida FG, Zhang R, Alfonso ZC, Zuk PA, Rodriguez LV. Processed lipoaspirate cells for tissue engineering of the lower urinary tract: Implications for the treatment of stress urinary incontinence and bladder reconstruction. *J Urol.* 2005;**174**: 2041–2045.

98. Sakuma T, Matsumoto T, Kano K et al. Mature, adipocyte derived, dedifferentiated fat cells can differentiate into smooth muscle like cells and contribute to bladder tissue regeneration. *J Urol.* 2009;**182**: 355–365.

99. Liu J, Huang J, Lin T, Zhang C, Yin X. Cell-to-cell contact induces human adipose tissue-derived stromal cells to differentiate into urothelium-like cells in vitro. *Biochem Biophys Res Commun.* 2009;**309**(3): 931–936.

100. Lin G, Wang G, Banie L et al. Treatment of stress urinary incontinence with adipose tissue-derived stem cells. *Cytotherapy.* 2009;**12**(1): 88–95.

101. Lin G, Banie L, Ning H, Bella AJ, Lin CS, Lue TF. Potential of adipose-derived stem cells for treatment of erectile dysfunction. *J Sex Med.* 2009;**6**(Suppl 3): 320–327.

102. Cavarretta IT, Altanerova V, Matuskova M, Kucerova L, Culig Z, Altaner C. Adipose tissue-derived mesenchymal stem cells expressing prodrug-converting enzyme inhibit human prostate tumor growth. *Mol Ther.* 2009;**18**(1): 223–231.

103. Lin T, Ambasudhan R, Yuan X et al. A chemical platform for improved induction of human iPSCs. *Nat Methods.* 2009;**6**: 805–808.

104. Perin L, Sedrakyan S, Da Sacco S, De Filippo R. Characterization of human amniotic fluid stem cells and their pluripotential capability. *Methods Cell Biol.* 2008;**86**: 85–99.

105. Perin L, Giuliani S, Sedrakyan S, DA Sacco S, De Filippo RE. Stem cell and regenerative science applications in the development of bioengineering of renal tissue. *Pediatr Res.* 2008;**63**: 467–471.

21

Umbilical Cord Matrix Mesenchymal Stem Cells: A Potential Allogenic Cell Source for Tissue Engineering and Regenerative Medicine

Nirmal S. Remya and Prabha D. Nair

CONTENTS

21.1 Introduction

Mesenchymal stem cells or multipotent mesenchymal stromal cells (MSCs) have a great demand in tissue engineering and regenerative medicine because of their high in vitro expansion potential, self-renewal capacity, multipotentiality, and immunomodulatory properties (Deans and Moseley 2000). However, considering the limited availability, invasive collection procedure and donor age–related differences that influence both proliferative and differentiation capacity of autologous adult stem cells particularly bone marrow stromal cells which is "the gold standard," there is an increasing interest in the investigation of using fetal-derived MSCs as an allogenic stem cell source for the previously said applications (Romanov et al. 2003).

Fetal stem cells derived from the extra embryonic sources which are normally discarded at parturition are particularly appealing for clinical applications as they are not subject to the ethical concerns that surround the isolation of embryonic stem cells. The extra embryonic stem cell sources include amniotic membrane, amniotic fluid, and umbilical cord matrix, as well as the placenta. Considering the ease of isolation technique, yield of stem cells, and expansion potential, cells derived from the umbilical cord matrix seem to be promising in exhaustible source of MSCs for regenerative applications. These resident stem cells of extra embryonic tissues are generated early in the developmental process and hence may additionally possess enhanced potency.

The umbilical cord represents the link between mother and fetus during pregnancy. It is formed around the fifth week of development from extra embryonic mesoderm, and it functions throughout pregnancy to protect the umbilical vessels, which provide a bidirectional blood flow between the fetus and the placenta. Since they are formed after the development of blastula and contain remnants of the yolk sac and allantois, they are considered to be derived from the same zygote as the fetus. At term, the human umbilical cord weighs approximately 40 g, its length reaches to approximately 60–65 cm and has a mean diameter of 1.5 cm (Barbieri et al. 2008). It is composed of two arteries and one vein, surrounded by a unique connective tissue stroma rich in proteoglycans and mucopolysaccharides (Sobolewski et al. 1997). Umbilical cord itself contains fetal MSCs within their connective tissue, the Wharton's jelly (Wang et al. 2004), umbilical vein, sub-endothelial layer of cord veins (Romanov et al. 2003), and perivascular area (Sarugaser et al. 2005).

The connective tissue of the umbilical cord, the Wharton jelly, is a jelly-like material lying between the covering amniotic epithelium and the umbilical vessels. This extraembryonic mucous tissue helps in preventing compression, torsion, and bending of the cord blood vessels that provide circulatory support for the developing fetus. The composition of the extracellular matrix comprises collagens and proteoglycans where collagen fibers (mainly types 1, 3, and 7) form an interlacing network organized as a continuous skeleton that encases the umbilical vessels. Glycosaminoglycans (GAGs), predominantly hyaluronic acid, form a hydrated gel around the collagen fibrillar network maintaining the tissue architecture of the cord (Can and Karahuseyinoglu 2007). Within the abundant extracellular matrix of Wharton's jelly reside a set of fetal stem cells called umbilical cord matrix stem cells (UCMSCs) in relatively high numbers. In this chapter, we have tried to review the isolation and culture of umbilical cord MSC, structural and phenotypic characteristics, immunological properties, multi-differentiation potential, and their application in regenerative medicine.

21.2 Isolation of Umbilical Cord Matrix Stem Cells

The human umbilical cord, obtained preferably from caesarean sections, should be immediately transported to the laboratory in a sterile and cooled transfer medium such as balanced salt solution with 1X antibiotics. Several approaches to isolate cells have been reported (Mitchell et al. 2003; Sarugaser et al. 2005; Wang et al. 2004). Thorough cleaning to remove blood clots should be done to prevent RBC contamination. Before processing, it is advisable to remove vessels under sterile conditions which could be done by simply stripping them off from the surrounding tissue. Tissues are then mechanically chopped into smaller pieces to facilitate enzymatic digestion. Commonly used digestion protocols consist of using collagenase solutions with enzymes having strong collagenase activity as well as other proteinase including caseinase, clostripain, and trypsin. Type 1 collagenase has been widely used (Sarugaser et al. 2005). However, type 2 collagenase is reported to be more efficient in solubilizing the microfibrils than other types of collagenase as they have a greater clostripain activity (Karahuseyinoglu et al. 2007). Collagenase enzymes hydrolytically cleave the native collagen in their triple helix form. Many researchers use a combination of collagenase and hyaluronidase which facilitates the degradation of matrix ground substance and thus shortening the time required for isolation process (Jomura et al. 2007; Weiss et al. 2006). Hyaluronidases are enzymes that degrade hyaluronic acid, and this treatment seems to be beneficial as the matrix jelly is rich in hyaluronic acid. Duration of enzymatic treatment is critically important, as there is always the risk of excessive digestion which can damage the external lamina of the cells and other cell adhesion receptors preventing the cells from adhering to the culture substrate after isolation. An alternative digestion enzyme, dispase, can also be used. It is a neutral protease having rapid and effective but gentle fibronectinase and type 4 collagenase activity. In the conventional protocol (Wang et al. 2004) for isolation, a 16 h collagenase digestion is used; however, the long digestion duration time can lead to excessive cell damage and lower cell yield. Alternatively the digestion time can be minimized by giving a dynamic shaking in an orbital shaker or using a magnetic stirrer at 37°C. A cocktail of digestion enzymes can be optimized for getting higher yield of cells. The digested material is then filtered through a cheese cloth or any cell strainer having a pore size ranging from 70 to 100 µm to remove any unwanted tissue debris. Alternatively explant cultures can also be tried out by placing small sized ($<1\,cm^2$) tissue pieces undisturbed in a culture plate and regularly feeding with culture medium for about 2 weeks (Mitchell et al. 2003). A schematic representation of UCMSC isolation procedure is depicted in Figure 21.1.

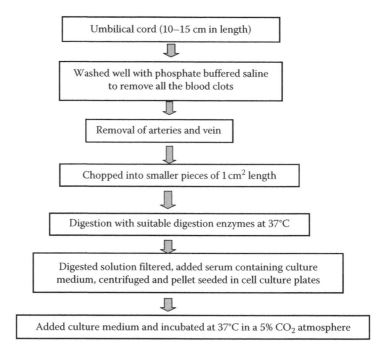

FIGURE 21.1
A schematic representation of UCMSC isolation procedure.

21.3 Growth Characteristics of Umbilical Cord Matrix Stem Cells

Mesenchymal stem cells in umbilical cord matrix (UCMSCs) are present in relatively high numbers, with an average of 400,000 cells isolated per cord (Karahuseyinoglu et al. 2007). This is significantly higher than the harvest obtained from adult bone marrow showing the advantage of using an alternative stem cell source for clinical applications. The freshly isolated cells (P0) display a fibroblast-like appearance over the first 3–4 days of culture. Once they gain 100% confluence in about 10–14 days, they typically appear as slender cells, and the morphology does not vary significantly up to P9 (Figure 21.2). The cells have a lag phase of 6–7 days with a log phase of another 7 days. Average population doubling time is in the range of $42.12 \pm 1.65\,h$ for P2 cells and is not significantly altered by passage number. UCMSC cells exhibit colony-forming ability, and the mean CFU-F is in the range 500–1000 mononuclear cells plated. They exhibit a stable karyotype between P1 and P9. Telomerase activity is usually present in the initial passages (P0 and P1) which may decline in the second and third passages and is significantly downregulated in later passages from P4. However, transplantation

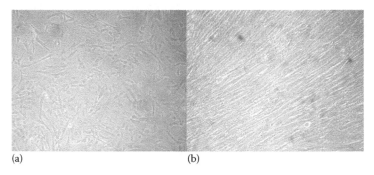

(a)　　　　　　　　　　　　　　(b)

FIGURE 21.2
(a) Human UCMSC after the 5th day of culture. (b) Human UCMSC at confluence.

experimental data have proven that the cells never get to a level of tumori-genic state (Fong et al. 2010). Reports have also shown that UCMSCs could be successfully cryopreserved and revived in culture (Petsa et al. 2009), which enhance the potential of UCMSC usage for therapeutic purposes.

UCMSCs resemble fibroblasts in structure and function. They possess moderate quantities of intra-cytoplasmic glycogen, lipid droplets, pro-colla-gen secretion granules, well-developed endoplasmic reticulum with dilated cisternae, golgi complexes, and numerous mitochondria. They express mus-cle-specific cytoskeletal filaments and hence are also sometimes termed as "myofibroblasts" or cells that exhibit some of the ultrastructural features of both smooth muscle cells and fibroblasts. Cells isolated are characterized based on their specific cell surface markers by fluorescence activated cell sorting (FACS) or by magnetic immunobead separation techniques (MACS). Table 21.1 summarizes the cell surface markers of human UCMSC detected by protein, mRNA or gene expression studies. UCMSCs express pluripotent stem cell markers Nanog, Oct-4, SSEA-3, SSEA-4, and Tra-1–60 at the gene and protein levels. The mRNA expression of Nanog, Oct-4, Sox-2, and Rex-1 was also maintained at least until P9 passage (Can and Karahuseyinoglu 2007).

21.4 Proteomic Profile of Umbilical Cord Matrix Stem Cells

Overall 60% of total proteome have been identified as proteins belonging to the cytoskeleton compartment and involved in protein biosynthesis, fold-ing and degradation. These proteins are responsible for the proliferative capacity and their quick adaptability of phenotype in response to adequate external stimuli. Antioxidants such as thioredoxin, peroredoxin, and gluta-thione transferases are also present, which help in preserving UCMSCs from oxidative injury during in vitro expansion. Proteins involved in nucleotide biosynthesis and signal transduction are poorly expressed in UCMSC.

TABLE 21.1

Cell Surface Markers of Human UCMSC

Marker	Positive/ Negative	Reference(s)
CD10	+	Weiss et al. (2006)
CD13	+	Weiss et al. (2006), Wu et al. (2007)
CD14	−	Karahuseyinoglu et al. (2007)
CD 29 (integrin beta1)	+	Wang et al. (2004), Weiss et al. (2006)
CD31 (PECAM)	−	Weiss et al. (2006)
CD33	−	Weiss et al. (2006)
CD34	−	Karahuseyinoglu et al. (2007)
CD38	−	Can and Karahuseyinoglu (2007)
CD44	+	Karahuseyinoglu et al. (2007), Sarugaser et al. (2005)
CD45	−	Karahuseyinoglu et al. (2007), Sarugaser et al. (2005)
CD49b (integrin alpha2)	+	Weiss et al. (2006)
CD49c (integrin alpha4)	+	Weiss et al. (2006)
CD49d (integrin alpha3)	+	Weiss et al. (2006)
CD51 (integrin alpha5)	+	Can and Karahuseyinoglu (2007), Weiss et al. (2006)
CD56	−	Weiss et al. (2006)
CD73 (SH3)	+	Karahuseyinoglu et al. (2007)
CD90 (Thy-1)	+	Can and Karahuseyinoglu (2007)
CD105 (endoglin, SH2)	+	Can and Karahuseyinoglu (2007), Karahuseyinoglu et al. (2007), Sarugaser et al. (2005)
CD123 (IL-3 receptor)	−	Sarugaser et al. (2005)
CD133	−	Weiss et al. (2006)
CD146	+	Can and Karahuseyinoglu (2007)
CD166 (*activated leukocyte adhesion molecule*)	+	Can and Karahuseyinoglu (2007)
CD235a (glycophorin A)	−	Sarugaser et al. (2005)
HLA-1	+	Sarugaser et al. (2005), Weiss et al. (2006)
HLA-DR	−	Lund et al. (2007)
HLA-DP	−	Lund et al. (2007)
HLA-DQ	−	Lund et al. (2007)
HLA A, B, C	+	Sarugaser et al. (2005)
HLA-G	−	Weiss et al. (2008)
Stro 1	−	Sarugaser et al. (2005)

21.5 Immune Properties of Umbilical Cord Matrix Stem Cells

Human UCMSCs from passages 4–9 have low immunogenicity and suppress the proliferation of activated immune cells. They are MHC class 2 negative, but exposure to interferon gamma stimulates increased MHC class 1 expression and induces MHC class 2 expression (Can and Karahuseyinoglu 2007).

UCMSCs additionally express IL-6 and VEGF genes, and the release of these hematopoietic and immunoregulatory cytokines in the medium has been recently shown to be important in imparting immunosuppressive properties to UCMSC (Weiss et al. 2008). UCMSCs like most other adult MSCs do not express co-stimulatory molecules like CD80, CD86, or CD 40 that contribute to immunogenicity. Additionally UCMSCs share some of the gene expression profile expressed by placental cells, as they are derived from postnatal tissue which is anatomically continuous with placenta. HLA-G, a human MHC class 1 molecule, and its soluble isoform HLA G6, selectively synthesized by the cytotrophoblastic cells at the maternal fetal interface, were seen expressed in UCMSC. In the normal fetus, this molecule is responsible for inhibiting the maternal dendritic NK cells, thereby preventing the fetal tissue from the attack of mother's immune system.

21.6 In Vitro Differentiation Potential of UCMSC: Application in Tissue Engineering

UCMSCs have multipotential to differentiate into cells of specific lineage under controlled conditions. As originated from extra embryonic mesoderm, differentiation to cells of mesenchymal lineages has been extensively studied. Table 21.2 summarizes the in vitro attempts done to demonstrate the multi-differentiation potential of UCMSC.

Pellet culture systems or 3D culture system is used to demonstrate chondrogenic potential of UCMSC. Chondrogenic differentiation of UCMSC has been described by many groups using chondrogenic growth factors such as transforming growth factor β1 (Wang et al. 2004) or transforming growth factor β3 (Wang et al. 2009b) or IGF, ascorbate-2-phosphate, and dexamethasone. Successful chondrogenic differentiation is indicated by the detection of the extra cellular matrix component GAG, by immunohistological staining, for example, of collagen II and aggrecan, or by verification of the expression of typical chondrocyte specific genes via PCR. When cultured in chondrogenic induction medium, shiny surfaced cell spheres resembling articular chondrocytes are formed within 3 weeks containing chondrocyte-like cells embedded in a mucopolysaccharide-rich stroma. A study by Wang et al. (2009a) reported that high-density cell seeding of UCMSC in 3D scaffolds promoted better biosynthesis of chondrocyte-specific matrix molecules, thereby enhancing mechanical integrity of the constructs (Wang et al. 2009a). Furthermore, in a comparative study of chondrogenic potential of human UCMSC and bone marrow mesenchymal stem cells (BMSC) by the same authors, even though efficient seeding and superior biosynthesis is observed in HUCMSC, a differentiation profile of large amount of type 1 collagen, small amount of type 2 collagen, and moderate aggrecan content

TABLE 21.2

Multi-Differentiation Potential of UCMSC

Differentiation Experiments	Specific Cell Types	Reference(s)
In vitro	Adipocyte	Wang et al. (2004)
	Chondrocyte	Wang et al. (2004)
	Osteocytes	Wang et al. (2004)
	Skeletal myocyte	Deans and Moseley (2000)
	Cardiomyocyte	Wang et al. (2004)
	Neuronal precursor	Mitchell et al. (2003)
	Endothelial cells	Wu et al. (2007)
	Islet-like clusters	Wu et al. (2009)
	Hepatocyte-like cells	Campard et al. (2008)
In vivo	Dopaminergic neuron	Weiss et al. (2006)
	Retinal photoreceptors	Lund et al. (2007)
	Islet-like clusters	Wu et al. (2009)
	Hepatocyte-like cells	Campard et al. (2008)
	Endothelial cells	Wu et al. (2007)
	Cardiomyocytes	Fan et al. (2010)
	Skeletal myocytes	Fan et al. (2010)

of UCMSC was observed when compared to the differentiation profile of BMSC under same culture conditions. So it is concluded that a modified set of signaling cues is required for optimal chondrogenesis of UCMSC (Wang et al. 2009b).

Successful adipogenic differentiation, defined by the appearance of cells containing intracellular lipid droplets, has been demonstrated by medium supplementation using dexamethasone, iso methyl butyl xanthine, indomethacin, and insulin (Wang et al. 2004). Compared to bone marrow, UCMSCs are capable of forming premature adipocytes with smaller multi-locular lipid droplets (Karahuseyinoglu et al. 2007). Adipocyte-specific genes, lipoprotein lipase, and plasminogen activator inhibitor-1 are seen expressed in UCMSC cultured in adipogenic medium. Oil red O staining is commonly applied to verify adipogenic differentiation.

Osteogenic potential is evaluated by the enhanced alkaline phosphatase expression and Ca mineralization, as demonstrated by von Kossa, or alizarin red staining. Expression of bone-specific markers such as osteopontin, osteonectin, osteocalcin, and bone sialoprotein 2 further confirms the differentiation of UCMSC to osteoblasts. Bone nodule formation is yet another phenomenon seen in osteogenically induced cells, as reported by Sarugaser et al. (2005). Osteogenic differentiation protocols for UCMSC use dexamethasone, β-glycerophosphate, and ascorbic acid as media supplements. Medium supplementation by BMPs has also been reported to enhance osteogenic differentiation. Wang et al. reported BMP2-mediated osteogenic differentiation of UCMSCs as a promising strategy for bone tissue engineering as expression level of osteogenic phenotypes in UCMSC under the effect of BMP2 is comparable to that of bone marrow MSC (Wang et al. 2010).

UCMSC differentiation to cardiomyocyte has also been reported by treating with 5 azacytidine or maintaining in cardiomyocyte-conditioned medium. 5 azacytidine, a chemical analogue of the cytosine nucleoside in the DNA and RNA hand helix, is the key chemical initiator of myogenic differentiation. Successful cardiomyocyte differentiation is demonstrated by the expression of cardiomyocyte markers N-cadherin and cardiac troponin I (Wang et al. 2004). Connexin 43, alpha actin and desmin, and other cardiomyocyte-related markers are also seen expressed by 5-azacytidine treatment. Troponin I, the inhibitory subunit of the troponin complex, is involved in cardiac muscle contraction, and F-actin is a cytoskeletal protein that maintains the structure of cardiac cell. Recently, differentiation of UCMSCs into skeletal myocytes expressing Myf5 and MyoD has been demonstrated in vitro (Fan et al. 2010)

Neuronal induction of UCMSC is first demonstrated by Mitchell et al. (2003) using a multistep neuronal induction procedure consisting of treatment with basic fibroblast growth factor, low serum media, DMSO, and butylated hydroxyanisole. Expression of neuron-specific enolase (NSE), a neural stem cell marker, is used to evaluate the neuronal potential of induced UCMSC cells. Several neuronal proteins including neuron-specific class 3 tubulin, neurofilament M, an axonal growth-cone-associated protein, and tyrosine hydroxylase are also seen expressed in differentiated cells. Markers for oligodendrocytes and astrocytes are also detected in neuronally induced UCMSC. A three-step neural induction protocol consisting of bFGF, beta-mercaptoethanol, neurotrophic factor-3 (NT-3), nerve growth factor (NGF), and brain-derived neurotrophic factor (BDNF) is also described for neuronal induction of UCMSC.

UCMSCs can be successfully induced into islet-like clusters in vitro through a four-stage differentiation protocol, as demonstrated by Chao et al. (2008b). These induced islet-like clusters are shown to contain human C-peptide and release human insulin in response to physiological glucose levels as well as express insulin and other pancreatic beta-cell-related genes. The capacity of UCMSCs to differentiate into hepatocyte-like clusters is also demonstrated by a three-step hepatogenic induction procedure wherein several of hepatic specific functions like glycogen storage, urea production, and expression of inducible cytochrome (CYP 3A4) enzyme are seen upregulated. However, the absence of markers of mature hepatocytes such as Hep Par 1 or HNF-4 shows that the differentiation is not complete (Campard et al. 2008).

UCMSCs have also got the potential to differentiate into endothelial cells when cultured in endothelial differentiation medium containing VEGF and bFGF. The successful endothelial differentiation is demonstrated by uptake of acetylated low-density lipoprotein and expression of endothelial-specific proteins such as platelet/endothelial cell adhesion molecule (PECAM) and

CD 34. Capillary tube formation also depicts the angiogenic property of differentiated endothelial cells (Wu et al. 2007).

21.7 In Vivo Transplantation Experiments with UCMSC: Applications in Cell-Based Therapy

There are only limited in vivo transplantation studies which show promising results while using UCMSC for regenerative treatment of cell-based disease states. A xenotransplantation experiment in which porcine UCMSCs transplanted into rat brain showed neither immune rejection nor teratoma formation due to transplanted material and also demonstrated the differentiation of undifferentiated UCMSC to neuronal precursor cells (Weiss et al. 2003).

A study by Ding et al. in which transplantation of human UCMSCs into the cortex of middle cerebral artery occlusion rat models showed improved neurological function, increased cortical neuronal activity, and promoted angiogenesis of the ischemic area. Other investigators also observed that transplanted UCMSCs survived for at least 5 weeks in the ischemic brain considerably accelerating neurological functional recovery. Intra-cerebral transplantation of human UCMSC for treating intracerebral hemorrhage in rat models also improved neurological function deficits, increased vascularity, and decreased injury volume, accelerating the neurological function recovery by the diseased rat models (Ding et al. 2007).

Experiments on effective transplantation of hUCMSC for treating spinal cord injuries were also attempted, and the results showed that human UCMSC promotes the regeneration of corticospinal fibers and locomotor recovery after spinal cord resection in rat models. The transplanted cells could be tracked live till 16 weeks and found to be secreting large amounts of human neutrophil-activating protein-2, NT-3, bFGF, and expressed glucocorticoid-induced tumor necrosis factor receptor as well as VEGF receptor in the host spinal cord that might have aided in spinal cord repair (Yang et al. 2008).

Cell-based therapies for Parkinson's disease explored a variety of stem cells for finding out a promising therapeutic strategy. Fu et al. could partially correct the lesion-induced amphetamine-evoked rotation by transplanting human UCMSCs into the striatum of Parkinsonian rat models (Fu et al. 2006). Weiss et al. could also demonstrate considerable change in apomorphine-induced rotations without evidence of formation of brain tumors as well as host immune rejection response on transplantation of undifferentiated UCMSCs into stratum of hemiparkinsonian rats without immune suppression (Weiss et al. 2006).

A recent study reported the successful induction of human UCMSC into islet-like clusters in vitro, and their in vivo transplantation into liver of streptozotocin-induced diabetic rats could significantly alter the hyperglycemic and glucose intolerant condition of diabetic rat models. A normalized stable blood glucose levels were maintained for over 9 weeks, without immunosuppressants (Chao et al. 2008b).There are also reports showing that UCMSC has the potential of maintaining the survival and function of islet-like cell clusters (Chao et al. 2008a). All these results suggest that human UCMSCs are a promising cell source for cell-based therapies for type 1 diabetes.

The differentiation capacity of UCMSC into hepatocyte-like clusters has also been investigated in vivo by injecting UCMSCs into the spleen of SCID mice with partial hepatectomy. It was observed that the transplanted UCMSCs expressed human albumin and AFP in recipient liver parenchyma and perivascular space (Campard et al. 2008). UCMSC transplantation into liver lesions of fibrosis-induced rat models also showed reduction in liver fibrosis with significantly lower levels of SGOT and SGPT after several weeks. Besides, expression of hepatic mesenchymal epithelial transition factor–phosphorylated type (MET-P) and hepatocyte growth factor were also seen upregulated. In another study, instead of differentiation to hepatocyte-like cells, the engrafted UCMSCs were found to secrete a variety of cytokines which might have helped in restoration of liver function (Tsai et al. 2009).

The therapeutic potential of UCMSC in myocardial infarction was also investigated in vivo by transplanting the cells in a rat myocardial infarction model. It was found that the transplantation had significantly improved cardiac function, increased vascularity, and decreased apoptosis as compared to control. Differentiation of UCMSCs into endothelial, smooth muscle as well as cardiomyocyte was also demonstrated by the expression of von Willebrand factor, smooth muscle actin and cardiac troponin-T in the transplanted cells (Fan et al. 2010).

In vivo studies on skeletal regeneration potential of UCMSCs are also reported where a subset of Wharton's jelly cells are transformed into multinucleated cells expressing skeletal muscle markers Myf5 and MyoD in vitro and following their injection into the previously damaged tibialis anterior muscle of rats. Successful muscle regeneration was noted on the defect site, as demonstrated by the co-localization of HLA-1 and sacromeric tropomyosin antigens (Troyer and Weiss 2008).

Endothelial differential potential of UCMSCs were also evaluated in vivo by transplanting UCMS cells into ischemic mouse models. It was observed that the cells proliferated and migrated from the local injection site and differentiated into a layer of endothelial cells in the lesion hind limb (Wu et al. 2007). A comparative study on endothelial differentiation potential by UCMSCs and bone marrow MSCs showed that UCMSCs have higher proliferative potential and higher expression of endothelial-specific markers after induction with specific endothelial induction factors. There are also reports that UCMSC transplantation into wire-injured femoral arteries in mice inhibited neointimal hyperplasia, thereby restoring endothelial integrity in injured vessels.

21.8 Banking of umbilical Cord Matrix Stem Cells: Challenges

The success of cord blood transplantation for treating hematopoietic disorders has contributed to the emergence of cord blood banks where cord blood units (CBUs) are collected, processed, and stored for later use. This has led to the establishment of public and private facilities for the same. It has been reported that at present there are >400,000 CBUs stored in banks for public use and >900,000 CBUs are stored in private banks (Ballen 2010). MSCs derived from Wharton's jelly of the discarded umbilical cord offer a low-cost, pain-free collection method of MSCs that may be cryogenically stored (banked) along with the umbilical CBU and then thawed to provide an in exhaustible stem cell source for the future therapeutic use in regenerative medicine. The challenges of cord stem cell banking include defining of standard good laboratory procedures for isolation, expansion, cryopreservation, and revival of functionally active UCMSCs in a large scale. The usage of stem cells in clinical application is also subject to the ethics and guidelines of the country of applicability. Unlike cord blood, which possesses only relatively low requirements for processing for banking, the procedures for harvesting, expansion, and cryopreservation of UCMSCs have to be standardized in order to meet the aforementioned. Thus, the huge potential of UCMSCs could be exploited for various therapeutic uses as well as in tissue engineering and regenerative medicine in future.

21.9 Conclusion and Future Perspectives

UCMSCs are derived from a noncontroversial tissue source that can be harvested with noninvasive procedures at low cost. Since UCMSCs are derived from a clinically discarded source, the problems and ethical concerns surrounding ESCs do not arise. These cells like bone marrow mesenchymal cells are plastic adherent, express markers of the mesenchymal cells such as CD10, CD13, CD29, CD44, CD90, and CD105 and lack markers of hematopoietic origin. Compared with other adult sources, which possess age-dependent differentiation and expansion potential, UCMSCs have high expansion and proliferation ability and multi-differentiation potential that makes it an attractive stem cell source for tissue engineering and other cell-based therapies. Despite the cells possessing high expansion potential due to high telomerase activity, they do not induce teratoma formation when transplanted intravenously or subcutaneously into SCID mice which enhance the application potential. Moreover, the cells do not require feeder layers or high-serum-containing medium which is a common cause of contamination. However, the lack of a definitive cell marker for identification and heterogeneous population at the isolation stage is the main concern arising when restricting the large-scale expansion and usage of UCMSCs for cell-based therapeutic purposes. Even though there are reports showing the promising immunosuppressive and immunomodulatory properties of UCMSCs, much more research is required before clinical translation. Long-term clinical phase trials are required to ensure that the phenotype of UCMSCs is not reverted to that of an embryonic stage. With their advantages of painless collection, abundant supply, lack of donor site morbidity and expandability, and better specific isolation, purification, and characterization procedures, UCMSC may soon become a promising alternative cell source for tissue engineering and regenerative medicine.

References

Ballen, K. 2010. Challenges in umbilical cord blood stem cell banking for stem cell reviews and reports. *Stem Cell Reviews and Reports* 6 (1):8

Barbieri, C., J. Cecatti, C. Souza, E. Marussi, and J. Costa. 2008. Inter- and intra-observer variability in sonographic measurements of the cross-sectional diameters and area of the umbilical cord and its vessels during pregnancy. *Reproductive Health* 5 (1):5.

Campard, D., P.A. Lysy, M. Najimi, and E. Marc Sokal. 2008. Native umbilical cord matrix stem cells express hepatic markers and differentiate into hepatocyte-like cells. *Gastroenterology* 134 (3):833.

Can A. and S. Karahuseyinoglu. 2007. Concise review: Human umbilical cord stroma with regard to the source of fetus-derived stem cells. *Stem Cells* 25 (11):2886.

Chao, K.C., K.F. Chao, C.-F. Chen, and S.H. Liu. 2008a. A novel human stem cell coculture system that maintains the survival and function of culture islet-like cell clusters. *Cell Transplantation* 17:657.

Chao, K.C., K.F. Chao, Y.S. Fu, and S.H. Liu. 2008b. Islet-like clusters derived from mesenchymal stem cells in Wharton's jelly of the human umbilical cord for transplantation to control type 1 diabetes. *PLoS ONE* 3 (1):e1451.

Deans, R.J. and A.B. Moseley. 2000. Mesenchymal stem cells: Biology and potential clinical uses. *Experimental hematology* 28 (8):875.

Ding, D.-C., W.-C. Shyu, M.-F. Chiang, S.-Z. Lin, Y.-C. Chang, H.-J. Wang, C.-Y. Su, and H. Li. 2007. Enhancement of neuroplasticity through upregulation of [beta]1-integrin in human umbilical cord-derived stromal cell implanted stroke model. *Neurobiology of Disease* 27 (3):339.

Fan, C.-G., Q.-J. Zhang, and J.-R. Zhou. 2010. Therapeutic potentials of mesenchymal stem cells derived from human umbilical cord. *Stem Cell Reviews and Reports* 7 (1):195.

Fong, C.-Y., A. Subramanian, A. Biswas, K. Gauthaman, P. Srikanth, M. Prakash Hande, and A. Bongso. 2010. Derivation efficiency, cell proliferation, freeze-thaw survival, stem-cell properties and differentiation of human Wharton's jelly stem cells. *Reproductive BioMedicine Online* 21 (3):391.

Fu, Y.-S., Y.-C. Cheng, M.-Y. Anya Lin, H. Cheng, P.-Mi. Chu, S.C. Chou, Y.-H. Shih, M.-H. Ko, and M.-S. Sung. 2006. Conversion of human umbilical cord mesenchymal stem cells in Wharton's jelly to dopaminergic neurons in vitro: Potential therapeutic application for parkinsonism. *Stem Cells* 24 (1):115.

Jomura, S., M. Uy, K. Mitchell, R. Dallasen, C.J. Bode, and Y. Xu. 2007. Potential treatment of cerebral global ischemia with Oct-4+ umbilical cord matrix cells. *Stem Cells* 25 (1):98.

Karahuseyinoglu, S., O. Cinar, E. Kilic, F. Kara, G.G. Akay, D.Ö. Demiralp, A. Tukun, D. Uckan, and A. Can. 2007. Biology of stem cells in human umbilical cord stroma: In situ and in vitro surveys. *Stem Cells* 25 (2):319.

Lund, R.D., S. Wang, B. Lu, S. Girman, T. Holmes, Y. Sauvé, D.J. Messina, I.R. Harris, A.J. Kihm, A.M. Harmon, F.-Y. Chin, A. Gosiewska, and S.K. Mistry. 2007. Cells isolated from umbilical cord tissue rescue photoreceptors and visual functions in a rodent model of retinal disease. *Stem Cells* 25 (4):1089.

Mitchell, K.E., M.L. Weiss, B.M. Mitchell, P. Martin, D. Davis, L. Morales, B. Helwig, M. Beerenstrauch, K. Abou-Easa, T. Hildreth, and D. Troyer. 2003. Matrix cells from Wharton's jelly form neurons and glia. *Stem Cells* 21 (1):50.

Petsa, A., S. Gargani, A. Felesakis, N. Grigoriadis, and I. Grigoriadis. 2009. Effectiveness of protocol for the isolation of Wharton's Jelly stem cells in large-scale applications. *In Vitro Cellular & Developmental Biology—Animal* 45 (10):573.

Romanov, Y.A., V.A. Svintsitskaya, and V.N. Smirnov. 2003. Searching for alternative sources of postnatal human mesenchymal stem cells: Candidate MSC-like cells from umbilical cord. *Stem Cells* 21 (1):105.

Sarugaser, R., D. Lickorish, D. Baksh, M.M. Hosseini, and J.E. Davies. 2005. Human umbilical cord perivascular (HUCPV) cells: A source of mesenchymal progenitors. *Stem Cells* 23 (2):220.

Sobolewski, K., E. Bańkowski, L. Chyczewski, and S. Jaworski. 1997. Collagen and glycosaminoglycans of Wharton's jelly. *Neonatology* 71 (1):11.

Troyer, D.L. and M.L. Weiss. 2008. Concise review: Wharton's jelly-derived cells are a primitive stromal cell population. *Stem Cells* 26 (3):591.

Tsai, P.-C., T.-W. Fu, Y.-M. Arthur Chen, T.-L. Ko, T.-H. Chen, Y.-H. Shih, S.-C. Hung, and Y.-S. Fu. 2009. The therapeutic potential of human umbilical mesenchymal stem cells from Wharton's jelly in the treatment of rat liver fibrosis. *Liver Transplantation* 15 (5):484.

Wang, L., N.H. Dormer, L.F. Bonewald, and M.S. Detamore. 2010. Osteogenic differentiation of human umbilical cord mesenchymal stromal cells in polyglycolic acid scaffolds. *Tissue Engineering Part A* 16 (6):1937–1948.

Wang, H.-S., S.-C. Hung, S.-T. Peng, C.-C. Huang, H.-M. Wei, Y.-J. Guo, Y.-S. Fu, M.-C. Lai, and C.-C. Chen. 2004. Mesenchymal stem cells in the Wharton's jelly of the human umbilical cord. *Stem Cells* 22 (7):1330.

Wang, L., K. Seshareddy, M.L. Weiss, and M.S. Detamore. 2009a. Effect of initial seeding density on human umbilical cord mesenchymal stromal cells for fibrocartilage tissue engineering. *Tissue Engineering Part A* 15 (5):1009–1017.

Wang, L., I. Tran, K. Seshareddy, M.L. Weiss, and M.S. Detamore. 2009b. A comparison of human bone marrow-derived mesenchymal stem cells and human umbilical cord-derived mesenchymal stromal cells for cartilage tissue engineering. *Tissue Engineering Part A* 15 (8):2259–2266.

Weiss, M.L., C. Anderson, S. Medicetty, K.B. Seshareddy, R.J. Weiss, I. VanderWerff, D. Troyer, and K.R. McIntosh. 2008. Immune properties of human umbilical cord Wharton's jelly-derived cells. *Stem Cells* 26 (11):2865.

Weiss, M.L., K.E. Mitchell, J.E. Hix, S. Medicetty, S.Z. El-Zarkouny, D. Grieger, and D.L. Troyer. 2003. Transplantation of porcine umbilical cord matrix cells into the rat brain. *Experimental Neurology* 182 (2):288.

Weiss, M.L., S. Medicetty, A.R. Bledsoe, R.S. Rachakatla, M. Choi, S. Merchav, Y. Luo, M.S. Rao, G. Velagaleti, and D. Troyer. 2006. Human umbilical cord matrix stem cells: Preliminary characterization and effect of transplantation in a rodent model of Parkinson's disease. *Stem Cells* 24 (3):781.

Wu, L.-F., N.-N. Wang, Y.-S. Liu, and X. Wei. 2009. Differentiation of Wharton's jelly primitive stromal cells into insulin-producing cells in comparison with bone marrow mesenchymal stem cells. *Tissue Engineering Part A* 15 (10):2865–2873.

Wu, K.H., B. Zhou, S.H. Lu, B. Feng, S.G. Yang, W.T. Du, D.S. Gu, Z.C. Han, and Y.L. Liu. 2007. In vitro and in vivo differentiation of human umbilical cord derived stem cells into endothelial cells. *Journal of Cellular Biochemistry* 100 (3):608.

Yang, C.-C., Y.-H. Shih, M.-H. Ko, S.-Y. Hsu, H. Cheng, and Y.-S. Fu. 2008. Transplantation of human umbilical mesenchymal stem cells from Wharton's jelly after complete transection of the rat spinal cord. *PLoS ONE* 3 (10):e3336.

22

Human Embryonic Stem Cells and Tissue Regeneration

Odessa Yabut, Carissa Ritner, and Harold S. Bernstein

CONTENTS

22.1 Introduction

Stem cells are defined by their ability to maintain long-term proliferation and self-renewal. Under specific conditions, stem cells also can differentiate into a diverse population of functionally specialized cell types. There are two main types of human stem cells classified according to their source and developmental potential: embryonic and adult, or tissue-specific, stem cells. Human embryonic stem cells (hESCs) are pluripotent cells that can differentiate into all types of somatic, and in some cases extraembryonic, tissues. Human adult stem cells are derived from nonembryonic tissues and are capable of generating specific cells from its organ or tissue of origin. Because of the unrestricted potential of hESCs, these cells provide a unique system for understanding human development and developing therapy for degenerative disease.

22.2 Derivation of Human Embryonic Stem Cells

hESCs were first derived from the inner cell mass of blastocyst-stage pre-implantation embryos (Figure 22.1). The inner cell mass is composed of pluripotent cells capable of differentiating into extraembryonic endoderm and the three germ layers that eventually generate all tissues of the embryo: ectoderm, mesoderm, and endoderm. To generate an hESC line, cells encompassing the inner cell mass are microsurgically removed and cultured in vitro under specific conditions designed to select cells with the capacity to expand in the undifferentiated state. Thomson et al. reported the first derivation of

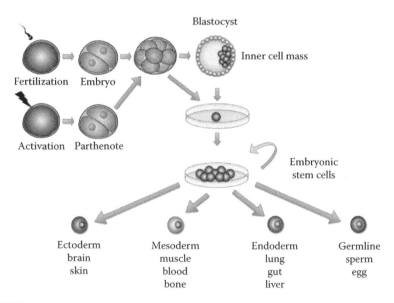

FIGURE 22.1
Generation of pluripotent hESC lines. Generation of hESC lines involves several steps. Donor embryos are first obtained after in vitro fertilization or by egg activation (parthenogenetic embryos) and allowed to develop in vitro. Pluripotent cells are then isolated either from the inner cell mass of preimplantation blastocysts or from 4-, 8-, or 16-cell stage morulae. Finally, isolated cells are plated in defined hESC medium with or without feeder cell layers to propagate and select for pluripotent cell populations. These processes have resulted in hESC lines able to generate tissues from all three embryonic germ layers and the germline.

pluripotent hESCs using this method (Thomson et al. 1998) and were quickly followed by a number of other groups (Cowan et al. 2004; Park et al. 2004; Reubinoff et al. 2000). To date, there are over 80 hESC lines that adhere to U.S. federal guidelines, many of which are widely used in basic and clinical research (http://grants.nih.gov/stem_cells/registry/current.htm).

hESC lines also have been derived from earlier stage embryos, including single blastomeres of four- or eight-cell stage embryos (Chung et al. 2008; Geens et al. 2009; Klimanskaya et al. 2006, 2007) and 16-cell morulae (Strelchenko et al. 2004; Strelchenko and Verlinsky 2006) (Figure 22.1). A single blastomere is considered totipotent and can produce an entire embryo. Thus, blastomere-derived hESCs resulting from the removal of a single blastomere from an early-stage embryo will, theoretically, not impede the ability of the remaining blastomeres to develop into a normal embryo. This could circumvent some of the ethical issues surrounding the use of embryonic tissue in biomedical research.

hESC lines also can be obtained from parthenogenetic embryos, which are generated when a single egg is fertilized in the absence of male sperm (Figure 22.1). Again, this could produce a source of hESCs without destruction of embryos. Parthenote-derived hESC lines have been generated through artificial fertilization of donor oocytes (Kim et al. 2007; Lin et al. 2007; Mai et al. 2007; Revazova et al. 2007). The ability to derive hESC lines from parthenote blastocysts is especially attractive not only because of their normal karyotype and their pluripotent properties but also because these lines contain homozygous major human lymphocyte antigen (HLA) alleles, which would circumvent immunological rejection involved in transplantation therapies (discussed later).

22.3 Defining Properties of Pluripotent Human Embryonic Stem Cells

It has been recognized that the derivation of hESC lines from different sources using different methods can introduce variability between lines. Thus, defining the essential properties and identifying features of hESCs is critical to their use. In this section, we will discuss the guidelines currently followed for characterizing new hESC lines.

22.3.1 Cell Morphology

Pluripotent hESCs maintain a specific cell morphology and density. hESCs have a high cell nucleus-to-cytoplasm ratio due to an enlarged nucleus and distinct nucleoli. Proliferating pluripotent hESCs form compact, spherical cell colonies when grown on mouse embryonic fibroblast feeder layers

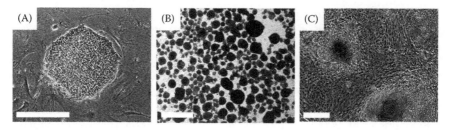

FIGURE 22.2
Typical undifferentiated and differentiating hESCs in culture. (A) A compact colony of proliferating pluripotent hESCs can be seen when cultured in defined medium on mouse embryonic fibroblasts. (B) Floating hEBs observed at 3 days after induction of differentiation. (C) Differentiating tissues, including cardiomyocytes, appear within adherent cultures at 48 h after plating hEBs onto a gelatin-coated culture dish. Bar, 25 μm.

(Figure 22.2A). Differentiating hESCs are easily distinguished by the loss of compact morphology and appearance of flattened cells that form at the edge of the colony. This can be avoided with regular supplementation of fresh medium and growth factors (Bodnar et al. 2004).

22.3.2 mRNA and microRNA Expression

A systematic study conducted by the International Stem Cell Initiative, a consortium of stem cell researchers from more than 15 countries, of 59 independently derived hESC lines was performed to identify a panel of molecular markers that are consistently and strongly expressed in pluripotent hESCs (Adewumi et al. 2007). These included developmentally regulated genes such as Nanog, POU domain class 5 homeobox 1 protein (POU5F/OCT4), teratocarcinoma-derived growth factor 1, DNA (cytosine-5-)-methyltransferase 3β, γ-aminobutyric acid A receptor β3, and growth differentiation factor 4. The study also established that the collective expression of stage-specific embryonic antigens 3 and 4, along with keratin sulfate (TRA-1-60, TRA-1-81, GDTM2, and GCT343) and protein (CD9 and Thy1) antigens, are reliable surface markers of pluripotent hESCs. Other characteristics of hESCs include the expression of the enzyme alkaline phosphatase, stem cell factor (or c-Kit ligand), and class 1 HLA.

Several groups have evaluated the expression profile of small, noncoding RNAs known as microRNAs (miRs) in hESCs (Bar et al. 2008; Laurent et al. 2008; Morin et al. 2008; Ren et al. 2009; Suh et al. 2004). miRs regulate translational efficiency of their target mRNAs (Guo et al. 2010). These studies identified a number of miR family clusters specifically expressed in pluripotent hESCs. Among these are miR-92b, the miR-302 cluster, miR-200c, the miR-368 and miR-154* clusters, miR-371, miR-372, miR-373*, miR-373, and the miR-515 cluster (Bar et al. 2008; Suh et al. 2004). Functional studies of some of these miRs, such as miR-302 and miR-92b, have established roles in self-renewal and pluripotency (Card et al. 2008; Sengupta et al. 2009). To date, a comprehensive comparison of miR expression profiles of all available hESC lines is still lacking. Such analysis is warranted, as this will identify miRs that are expressed across hESC lines, and could be used to select for pluripotent populations, evaluate newly derived hESC lines, and understand mechanisms that specifically regulate basic hESC biology.

22.3.3 Epigenetic Characteristics

Epigenetic mechanisms influence gene expression through heritable modifications in chromosomal or DNA structure, such as DNA methylation, histone modification, and X-chromosome inactivation. Similar to expression patterns of coding genes discussed earlier, the epigenetic properties of pluripotent hESCs have been used as molecular signatures to distinguish them from other cell types.

Chromatin structure of undifferentiated hESCs exists in an open conformation, making it readily accessible to the transcriptional machinery necessary for the maintenance of pluripotency (Gan et al. 2007). hESC lines also display DNA methylation profiles distinct from most other cell types (Bibikova et al. 2006). One study of 14 different lines revealed markedly reduced methylation patterns of CpG dinucleotides when compared to somatic cells. Further analysis revealed that the differential methylation of hESCs was specific to promoters of genes such as OCT4 and Nanog (Lagarkova et al. 2006) that are important for maintaining pluripotency. Thus, the unique epigenetic properties of hESCs likely promote maintenance of the pluripotent state and can be used as a signature of undifferentiated hESCs.

To date, almost all established female hESC lines analyzed exhibit partial or complete X-chromosome inactivation, a process that occurs as early as the blastocyst stage and leads to methylation of promoter regions. The states and levels of X-inactivation appear to differ between hESC lines and also between subcultures of hESC lines propagated by different labs (Dvash et al. 2010; Hall et al. 2008; Shen et al. 2008; Silva et al. 2008). This suggests that in addition to genetic heterogeneity, environment and culture conditions can lead to heterogeneous epigenetic states. The variability in X-chromosome inactivation could result in inconsistencies as hESCs are developed for therapeutic applications. Thus, generating an epigenetically naïve hESC line, in which X-chromosome inactivation or other epigenetic modifications have not yet occurred, is an important goal.

22.3.4 Pluripotency In Vitro and In Vivo

hESCs are defined in part by their capacity to differentiate, which can be tested in vivo and in vitro. A test of pluripotency in vitro involves determining the ability of hESCs to form human embryoid bodies (hEBs) when cultured in a non-adherent cell suspension in the absence of feeder cell layers

(Figure 22.2B). hEBs are spheroid colonies of differentiating hESCs that contain cells representative of all three embryonic germ layers (Itskovitz-Eldor et al. 2000). These can be differentiated in turn into specific tissues under specific culture conditions (Figure 22.2C), as discussed later.

The most commonly used in vivo method to test pluripotency involves the transplantation of undifferentiated hESCs into immunodeficient mice to induce the formation of teratomas (Gaur et al. 2010; King et al. 2009, 2011; Ritner and Bernstein 2010). Teratomas are benign tumors comprised of disorganized tissues characteristic of the three embryonic germ layers. Analysis of embryonic tissues found in teratomas from engrafted hESCs can be used to test their differentiation potential (Figure 22.3).

The ability of hESCs and hEBs to mimic in vitro and in vivo the events occurring during human embryonic development provides an unprecedented opportunity to understand the mechanisms involved in developmental processes, and important reagents for generating desired cell types suitable for cell therapy.

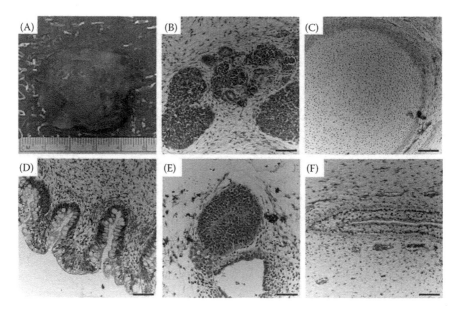

FIGURE 22.3
Teratoma formation provides an in vivo assay of hESC differentiation capacity. Proliferating cultures of hESCs were used to form teratomas by renal capsule grafting using established methods (Ritner and Bernstein 2010). (A) An explanted teratoma is shown. (B–F) Teratomas were sectioned and stained with hematoxylin and eosin to identify embryonic tissues. Representative tissues from all three embryonic germ layers can be seen, including mesoderm (B,C), endoderm (D), and ectoderm (E,F). (B) Nascent renal tubules and glomeruli within bed of primitive renal epithelium. (C) Cartilage surrounded by capsule of condensed mesenchyme. (D) Glandular intestinal structure. (E) Nascent neural tube. (F) Primitive squamous epithelium. Bar, 100 μm.

22.4 hESC-Derived Progenitors and Their Use in Cell-Based Therapies

Cellular deficiency or insufficiency is the root of diseases such as heart failure, diabetes, neurodegeneration, bone marrow failure, and spinal cord injury. For centuries, therapeutic approaches have been limited to the surgical removal of damaged tissues or treatment with pharmacological therapies to improve symptoms. Thus, the prospect of replacing damaged or missing cells with new functional cells has shifted the therapeutic paradigm toward restoring tissue function.

hESC-derived cell populations that would either replace damaged cells or induce neighboring cells to function normally provide a promising strategy for cell-based therapy. With hESCs, it is possible to generate lineage-restricted progenitors that are capable of differentiating into specialized post-mitotic cell types such as cardiomyocytes, pancreatic islet cells, chondrocytes, hematopoietic cells, endothelial cells, or neurons. The ability of hESCs to divide indefinitely makes these cells an inexhaustible large-scale source of specific progenitors. Current research studies are focused on identifying and refining ways for directing the differentiation of hESCs that will enrich for pure, homogenous populations of specific cell types. In the following sections, we will provide some examples of how differentiation of hESCs can be directed toward specific cell/tissue types, and the potential use of these cell types for clinical applications.

22.4.1 Ectodermal Derivatives

The dominant differentiation pathway in hESC cultures leads to the formation of ectoderm, which makes up cells of the nervous system and the epidermis (Table 22.1). hESC-derived neural progenitor cells are characterized by rosette-like neural structures that form in the presence of growth factors, fibroblast growth factor (FGF) 2 or epidermal growth factor (EGF), through either spontaneous differentiation of hESCs or after hEBs are plated onto adherent substrates (Reubinoff et al. 2001; Zhang et al. 2001). Neural rosettes have become the hallmark of hESC-derived neural progenitors, capable of differentiation into a broad range of neural cells in response to appropriate developmental signals. Many current studies are exploring ways to enhance the formation of neural rosettes in order to generate an enriched population of neural precursors. For example, stromal cells provide ectodermal signaling factors required for neural induction, as determined in animal model studies, and therefore promote the formation of neural rosettes (Kawasaki et al. 2000, 2002; Perrier et al. 2004).

Withdrawal of FGF2 and EGF and the addition of specific compounds can lead to the differentiation of neural rosettes into specific neural

TABLE 22.1

Derivation of Ectodermal Cells from hESCs

Cell Type	Method	Specific Factors and/or Conditions	Reference
Dopaminergic neurons	Coculture with stromal cells	FGF8 Shh	Perrier et al. (2004)
	Formation of neural rosettes	FGF8 Shh	Yan et al. (2005)
Schwann cells	Formation of neural rosettes	Ciliary neurotrophic factor Neuregulin 1β dbcAMP	Lee et al. (2007)
Motor neurons	Formation of neural rosettes	Retinoic acid Shh	Li et al. (2005)
Peripheral sympathetic and sensory neurons	Formation of neural rosettes	Withdrawal of FGF2/BDNF GDNF NGF Dibutyryl cyclic AMP	Lee et al. (2007)
RPE	Directed differentiation	Serum-free conditions Activin A Nicotinamide	Idelson et al. (2009)
Oligodendrocytes	Directed differentiation	B27 Thyroid hormone Retinoic acid FGF2 EGF Insulin	Nistor et al. (2005)
Basal keratinocytes	Direct differentiation in 3D culture	BMP4 Ascorbic acid	Guenou et al. (2009)

subtypes. Specifically, hESC-derived neural progenitors treated with FGF8 and sonic hedgehog (Shh) give rise to dopaminergic neurons (Yan et al. 2005), while treatment with Shh and retinoic acid induces motor neuron differentiation (Li et al. 2005). Neural crest stem cells derived from neural rosettes can differentiate into peripheral sympathetic and sensory neurons by withdrawing FGF2/EGF and adding BDNF, GDNF, NGF, and dbcAMP, or into Schwann cells in the presence of CNTF, neuregulin 1β, and dbcAMP (Lee et al. 2007). Neuroglial cells, such as oligodendrocytes, are generated with B27, thyroid hormone, retinoic acid, FGF2, EGF, and insulin (Nistor et al. 2005).

In 2010, Geron Corporation (Menlo Park, CA, United States) initiated the first clinical trial with hESCs in the United States using hESC-derived oligodendrocytes to treat acute spinal cord injuries (http://www.clinical-trials.gov/ct2/archive/NCT01217008). Oligodendrocytes are rapidly lost following acute spinal cord injury leading to demyelination and neuronal loss. In these trials, purified oligodendrocyte progenitor cells derived from hESCs will be injected into the spinal cord of paralyzed patients within 2 weeks of the acute injury. While this first trial is a safety study, the expectation is that these progenitor cells will terminally differentiate into oligodendrocytes and produce myelin, which insulates neuronal cell membranes and is critical for efficient conduction of nerve impulses. Thus, the transplantation of newly differentiated oligodendrocytes is expected to restore myelination of damaged neurons preventing further neuronal death and restoring function.

Retinal pigment epithelium (RPE) cells are another specific cell type derived from neuroectoderm. These support the neural retina by phagocytosing and renewing the photoreceptor outer segments of rhodopsin. Recent reports have shown that RPE can be induced from hESCs in the presence of nicotinamide and activin A under serum-free conditions (Idelson et al. 2009). hESC-derived pigmented cells exhibit the morphological and functional properties of RPE cells after transplantation in an animal model of macular degeneration, a disease caused by dysfunction and loss of RPE. These data have led to clinical trials using hESC-derived pigmented cells by Advanced Cell Technology (Santa Monica, CA, United States). For these trials, hESC-derived RPEs will be transplanted directly into the degenerating retinae of patients with Stargardt's macular dystrophy, a juvenile form of macular degeneration, or dry age-related macular degeneration, to rescue visual acuity.

22.4.2 Endodermal Derivatives

The lung, liver, and pancreas are comprised of cells derived from endoderm (Table 22.2). Directing the differentiation of hESCs toward definitive endoderm has allowed specific cell types, such as islet cells or hepatocytes, to be produced that ultimately may be used for treatment of diseases such as diabetes and liver disease. D'Amour et al. (2005) showed that selective induction of endoderm could be achieved through the addition of high concentrations of activin A, under low serum conditions, and in a stage-specific manner. Activin A mimics the action of Nodal, a ligand that activates transforming growth factor-β (TGFβ) signaling, which in turn leads to the induction of endoderm. The effect of activin A in inducing definitive endoderm is enhanced when additional factors such as Wnt3a (D'Amour et al. 2006) and Noggin (Sumi et al. 2008) are present, or when coupled with the suppression of the phosphoinositide 3-kinase pathway (McLean et al. 2007).

Induction of definitive endoderm can lead to the generation of specific progenitor populations following the addition of other factors. Among the most successful examples to date is the generation of pancreatic islet progenitors devised by Kroon et al. (2008). This was accomplished through the sequential exposure of hESCs to activin A and Wnt3A, followed by the addition of

TABLE 22.2

Derivation of Endodermal Cells from hESCs

Cell Type	Method	Specific Factors and/or Conditions	Reference
Hepatocytes	Directed differentiation	FGF BMP4 Hepatocyte growth factor Oncostatin M Dexamethasone	Agarwal et al. (2008), Cai et al. (2007)
Pancreatic islet progenitors	Directed differentiation	Activin A Wnt3A Keratinocyte growth factor/FGF7 Retinoic acid Cyclopamine Noggin	Kroon et al. (2008)
Lung alveolar cells	Genetic modification	Recombinant keratinocyte growth factor	Wang et al. (2007, 2010)

keratinocyte growth factor or FGF 7 to induce the formation of the primitive gut tube. Subsequently, retinoic acid, cyclopamine, and Noggin were added to inhibit hedgehog and TGFβ signaling and thus induce the differentiation of posterior foregut cells, the source of endocrine pancreas progenitors. These were cultured further to generate pancreatic endoderm cells. When these cells were engrafted in immunodeficient mice, they displayed the histological and structural characteristics of pancreatic islet cells and were able to sustain insulin production for at least 100 days (Kroon et al. 2008).

In a similar manner, hepatocytes can be obtained after differentiation of hESCs into definitive endoderm (Agarwal et al. 2008; Cai et al. 2007). A robust population of functional hepatocytes was generated with the sequential addition of low serum medium, collagen I matrix, and hepatic differentiation factors that include FGF, BMP4, hepatocyte growth factor, oncostatin M, and dexamethasone (Agarwal et al. 2008). These cells expressed known markers of mature hepatic cells, exhibited appropriate function, and were able to integrate and differentiate into mature liver cells when injected into mice with liver injury (Agarwal et al. 2008).

22.4.3 Mesodermal Derivatives

The differentiation of hESCs into mesoderm (Table 22.3) requires the activation of the TGFβ signaling pathway and has been accomplished through the stepwise and dosage-dependent addition of activin A, BMP4, and the growth factors, vascular endothelial growth factor (VEGF) and basic FGF (bFGF) (Evseenko et al. 2010). Mesodermal derivatives also have been successfully obtained by spontaneous differentiation of hESCs through hEB formation without first directing them toward mesoderm. Robust differentiation of

TABLE 22.3

Derivation of Mesodermal Cells from hESCs

Cell Type	Method	Specific Factors and/or Conditions	Reference
Dendritic cells	hEB formation	Serum-free conditions BMP4	Su et al. (2008)
Chondrocytes	hEB formation	Micromass of dissociated embryoid bodies BMP2	Toh et al. (2007)
	hEB formation	High-density culture of dissociated embryoid bodies Ascorbic acid Dexamethasone	Cohen et al. (2010)
	Directed differentiation on 3D scaffolds	Coculture with primary chondrocytes Poly-D, L-lactide scaffold	Vats et al. (2006)
Blood cells	Spin embryoid body formation	Serum-free conditions	Ng et al. (2005)
T and NK cells	Coculture with stromal cells	Coculture with stromal M210-B4 cells to enhance expansion of CD34+/CD45+ progenitors	(Woll et al. 2009)
Cardiomyocytes	hEB formation	Serum-free conditions bFGF	Kehat et al. (2001)
	Directed differentiation	Activin A BMP4	Laflamme et al. (2007)
	Directed differentiation	BMP4 BMP4/bFGF/activin A VEGF/DKK1 VEGF/DKK1/bFGF	Yang et al. (2008)
	Genetic modification	Cardiac-specific reporter	Ritner et al. 2011

hESCs into hematopoietic precursors that give rise to all blood cell types and components of the immune system has been achieved under serum-free conditions through spin hEB formation (Ng et al. 2005). Specific hematopoietic cells, such as functional dendritic cells, have been successfully differentiated from hESCs through spontaneous hEB formation under serum-free conditions with the addition of BMP4 at specific time points (Su et al. 2008). Hematopoietic progenitor cells that give rise to functional T and natural killer cells capable of targeting human tumor cells both in vitro and in vivo also have been derived from hESCs cocultured with stromal cells (Woll et al. 2009). Thus, the ability to differentiate hESCs into hematopoietic lineage cells promises to be useful in improving existing therapies that require blood cell transplantation, and in immune therapies that require induction of the immune response in an antigen-specific manner (Senju et al. 2010).

Cardiomyocytes represent another therapeutically important derivative of mesoderm that have been successfully generated from hESCs using several methods (Wong and Bernstein 2010). hESCs can spontaneously differentiate into cardiomyocytes through hEB formation under appropriate culture conditions. hEB-derived cardiomyocytes exhibit morphological, molecular, and electrophysiological properties similar to embryonic and adult cardiomyocytes (Kehat et al. 2001), and display quantifiable responses to physiological stimuli reminiscent of atrial, ventricular, and pacemaker/conduction tissue (He et al. 2003; Mummery et al. 2003; Ritner et al. 2011; Satin et al. 2004). Cardiomyocytes also have been generated by directed differentiation with activin A and BMP4; these cells successfully form specific cardiac lineages when transplanted in vivo (Laflamme et al. 2007). Another study used additional medium supplements that included VEGF, and the Wnt inhibitor, dickkopf homolog-1 (DKK1), followed by the addition of bFGF to promote efficient cardiomyocyte differentiation from hEBs (Yang et al. 2008). Success of these studies was measured by the expression of proteins specific for terminally differentiated cardiac muscle cells such as cardiac troponin T, atrial myosin light chain 2, and cardiac T-box transcription factors 5 and 20.

22.5 Tissue Engineering with hESCs

Tissue engineering utilizes biological substitutes to restore or maintain tissue function. As with cell transplantation, a successfully engineered tissue depends on the generation of the appropriate cell type that is able to provide normal cellular function. Thus, cells suitable for tissue engineering should have the ability to enter a desired differentiation program to produce a specific cell type, and be expandable in vitro to meet the needs of cell transplantation. hESCs provide much promise in tissue engineering and regeneration since hESCs can act as an inexhaustible in vitro source of differentiated cell types. The potential use of hESCs in tissue engineering include, but are not limited to, organ substitutes, vascularization, and ex vivo cartilage/bone construction.

Basal keratinocytes, the cells that make up the pluristratified epidermal layer of the skin, have been successfully differentiated from hESCs. Guenou et al. (2009) have shown that long-term culture of hESCs in defined medium supplemented with BMP4 and ascorbic acid leads to the directed differentiation of hESCs into basal keratinocytes. These cells express keratins 14 and 5, α6- and β4-integrins, collagen VII, and laminin 5 at levels comparable to postnatal keratinocytes. More importantly, these hESC-derived keratinocytes form a cohesive pluristratified epidermis when placed in 3D culture or when engrafted into immunodeficient mice. These findings prove the feasibility of using hESC-derived keratinocytes as a source of allograft for patients requiring skin restoration.

The use of hESCs to treat lung injury also has been an area of active investigation. A significant step toward directed differentiation of lung-specific cells was reported by Wang et al. (2007, 2010), in which genetically modified hESCs carrying lung-specific reporters under the control of promoters from tissue-specific genes, such as surfactant protein C, aquaporin 5, and T1α, resulted in purification of type I and type II alveolar epithelial cells. When engrafted into mice suffering from acute lung injury, these cells terminally differentiated in vivo into type I and type II alveolar epithelial cells and exhibited functional properties that included the capacity for gas exchange and histological amelioration of lung injury.

hESCs readily form connective tissue, such as bone or cartilage, as can be appreciated from teratoma formation assays (Figure 22.3). Thus, hESCs are a valuable source of cells suitable for connective tissue replacement therapy for a number of bone and joint diseases, such as osteoarthritis, which is characterized by the breakdown of cartilage within joints. Most successful and efficient protocols for directing chondrocyte differentiation from hESCs utilize 3D culture systems created by seeding hESCs at high density leading to formation of a pellet, or by introducing the cells into a synthetic 3D scaffold. Such systems enable cell–cell signaling between the undifferentiated hESCs and mature chondrocytes to stimulate homogeneous and sustained chondrogenic differentiation. For example, single-cell suspension of dissociated hEBs cultured as high-density micromass with BMP2 leads to efficient chondrocyte formation (Toh et al. 2007). hESCs cocultured with primary chondrocytes, or in the presence of osteogenic supplements and polymeric scaffolds, yield cartilaginous- or osteogenic-like cells (Bielby et al. 2004; Vats et al. 2006). More recently, feeder-free 3D culture systems have successfully derived multipotent connective tissue progenitors from hESCs yielding tendon-like structures (Cohen et al. 2010). The engraftment of these in vitro differentiated tendon structures in injured immunosuppressed mice restored ankle joint movements that rely on an intact Achilles tendon (Cohen et al. 2010). Furthermore, there is evidence that transplanted chondrogenic cells may exert a stimulatory effect through paracrine mechanisms that promote growth and repair of endogenous cells (Toh et al. 2010).

22.6 Current Challenges to Clinical Use

As discussed earlier, clinical trials of cell therapy with hESCs have begun. The International Stem Cell Banking Initiative was created by the International Stem Cell Forum, a group of national and international stem cell research funding agencies, to develop a set of best practices and principles for banking, testing, and distributing hESCs for clinical application

TABLE 22.4

Requirements for Standardization of Clinical-Grade hESCs

Requirement	Methods of Testing
Cell line identity	Short tandem repeat (STR) testing HLA
Sterility and pathogens	Bacteria/fungi/mycoplasma culture qPCR analysis for murine viral short interspersed elements (SINE)
Genetic/chromosomal stability	Single nucleotide polymorphism (SNP) analysis G-band karyotype analysis spreads Fluorescent in situ hybridization
Epigenetic stability	MicroRNA profiling Methylation analysis X-inactivation
Pluripotency	Teratoma formation SSEA-3/4, TRA-1-60, TRA-1-81 detection
Quality and differentiation ability	Gene expression profiling qPCR analysis Embryoid body formation
Functional assays	Potency Efficacy Lot-to-lot variability

(Crook et al. 2010). In the United States, the Food and Drug Administration also monitors these guidelines and has issued recommendations for the development of clinical trials of stem cell therapy (http://www.fda.gov/ BiologicsBloodVaccines/GuidanceComplianceRegulatoryInformation/ Guidances/Xenotransplantation/ucm074131.htm). These recommendations do not ensure the quality or efficacy of hESC-derived cells used for clinical application. Rather, these guidelines warrant that the cells used for therapy are reproducible and meet specific criteria to ensure patient safety (Table 22.4). The major safety concerns for the use of hESCs are discussed in the following sections.

22.6.1 Genetic Abnormalities

The earliest hESC lines derived are among the best characterized. However, they may not be the best lines for therapeutic applications as many of these lines were derived using animal products. Chromosomal and genomic instability has been detected among several of these early hESC lines, including loss of heterozygosity or copy-number variation in cancer-related genes (Lefort et al. 2008; Narva et al. 2010). Many of these mutations appeared to be induced by prolonged culture, since these changes were not observed in low passage cells. It has been proposed that such karyotypic aberrations occurred with adaptation to the original culture conditions used when the first few lines were being derived and expanded (Baker et al. 2007). These observations emphasize the need for complete characterization of hESC lines, particularly the effects of long-term culture, and the design of guidelines for designating therapeutic-grade hESCs.

22.6.2 Xenobiotic-Free Conditions

Many hESC lines currently in use have been exposed to animal products during their derivation and propagation. As such, these hESCs could harbor animal viruses and other unknown substances capable of eliciting a deleterious immune response in transplant recipients. Current hESC lines under development for clinical use must undergo extensive microbiological testing as recommended by the International Stem Cell Banking Initiative. In the United States, the Food and Drug Administration legally requires documentation of the source, potential genetically modified components, and pathogenic agents in any hESC-derived cell intended for therapeutic use. In addition, replacement media recently have been developed that would allow maintenance of hESCs in xenobiotic-free conditions. These include xenobiotic-free serum replacements such as knockout serum replacer (KSR; Invitrogen) or xenobiotic-free culture media such as HEScGRO (Millipore) or TeSR (STEMCELL).

Feeder-free culture systems also are being developed to reduce the risk of contamination with foreign agents when hESCs are cultured on animal feeder cell layers. Feeder-free and xenobiotic-free, defined culture media that consist of a combination of recombinant growth factors known to inhibit differentiation and maintain hESCs in the pluripotent state are now commercially available. However, some reports have associated feeder-free culture conditions with greater chromosomal instability and an increased risk of propagating genetically altered hESCs (Catalina et al. 2008). For this reason, most hESC labs practice a surveillance program for genomic instability in cultured lines (Gaur et al. 2010; Ritner et al. 2011) regardless of conditions used.

hESC lines derived using human feeder cells have been reported. For example, hESC lines have been successfully derived on human fibroblasts generated from neonatal foreskin (Ilic et al. 2009; Strom et al. 2010) and adult skin fibroblasts (Tecirlioglu et al. 2010). Some laboratories deriving new lines have moved exclusively to xenobiotic-free conditions (Genbacev et al. 2005). The ability to derive and maintain new hESC lines using human fibroblast feeder cells represents a significant step toward generating clinical-grade hESCs.

22.6.3 Enrichment, Directed Differentiation, and Purification Protocols

Germ layer tumorigenesis remains a primary safety concern when using pluripotent hESCs. As discussed earlier, in vivo transplantation of undifferentiated hESCs in immunodeficient mice results in teratoma formation.

Evidence of tumor-like growths also has been observed in differentiated hESC derivatives transplanted in vivo (Roy et al. 2006; Wernig et al. 2004). Thus, it is essential that candidate hESC derivatives intended for use in cell transplantation are free of tumorigenic cells. Another concern is the differentiation of hESC-derived cells into unwanted cell types. For example, the engraftment of inappropriate muscle cell subtypes into the myocardium could alter the electrical activity of recipient tissue, provoking arrhythmias (Gepstein et al. 2010). Thus, developing and further optimizing differentiation and purification protocols is necessary to minimize the generation of unwanted cell types for preclinical transplantation experiments and clinical therapy.

As discussed earlier in this chapter, enrichment of specific cell types can be achieved using molecules introduced at specific time points during culture. However, many of these methods yield only moderate enrichment that is not yet scalable for clinical application. It may be desirable to enrich first for partially differentiated, proliferative hESC intermediates with specific cell fates. These could then be expanded before further differentiation into cells for therapy. For example, the expression of the cell surface antigen, CD133, on proliferating hESCs identifies cells predestined toward a neuroectodermal fate (King et al. 2009). CD133-positive cells have been selected from cultures of undifferentiated hESCs and have been observed to differentiate primarily into neuroectodermal cells in vitro and in vivo (King et al. 2009).

In the absence of specific cell surface antigens like CD133 to identify tissue-specific precursors, molecular beacons have been used to select for specific subpopulations of hESCs. King et al. (2011) first demonstrated the utility of this system in isolating live Oct4-expressing pluripotent hESCs in a specific and high-throughput manner. Molecular beacons are single-stranded oligonucleotides that generate fluorescent signals when bound to their target mRNAs, making these cells detectable and selectable by fluorescence-activated cell sorting. More importantly, molecular beacons have a short lifespan within cells and do not alter the function or genomic structure of hESCs. Thus, this method can be used to enrich for desired hESC-derived cell populations or used to select against unwanted cell types, such as undifferentiated hESCs that could form tumors.

22.6.4 Circumventing Immune Rejection

Proliferating and differentiated hESCs express class I and II HLA as well as minor histocompatibility antigens at levels sufficient to activate the immune system (Bradley et al. 2002; Drukker et al. 2002). In addition, mismatch between hESC donor and recipient ABO blood group antigens could pose a significant barrier to hESC engraftment. While studies to determine the effects of ABO incompatibility on hESC transplantation are still lacking, this has long been a criterion for successful organ transplantation, and thus, it is likely that ABO incompatibility between hESC-donor cells and the recipient also would trigger immune rejection.

Ideally, having genetically identical donor and patient cells is the best way to circumvent immune rejection. Thus, there is high interest in developing and using somatic cell nuclear transfer to generate patient-specific hESC lines. Using this technique, the DNA obtained from either a patient's skin or muscle cell would be transferred into an unfertilized egg that has had its DNA removed. Subsequently, the egg is artificially fertilized and allowed to develop until it reaches the blastocyst stage to derive hESCs. The resulting hESC line would have an immunologic profile matching the patient and could be used for cell therapy. This technique has been conducted successfully in animals using species-specific ESCs, but the bona fide derivation of hESCs through somatic cell nuclear transfer has not yet been reported.

Another strategy is to generate hESC lines with the closest match to potential transplant patients. Suggestions have included engineering "universal donor hESCs," a blood antigen O cell in which the expression of HLA is suppressed, or chimeric hematopoietic cells derived from hESCs capable of inhibiting the immune response when cotransplanted with the desired hESC-derived cells (Drukker 2004). Alternatively, creating hESC banks that store lines representing HLA/ABO combinations that match the majority of potential patients has been proposed. Studies have provided estimates on how many hESC lines would be needed in order to support the needs of a specific population. Taylor et al. (2005) estimated that approximately 150 hESC lines could provide an HLA match for most of the population in the United Kingdom. Alternatively, approximately 10 parthenote-derived hESC lines that are homozygous for HLA types could be sufficient for a majority of the population. Studies by Nakajima et al. (2007) estimated that approximately 170 hESC lines, or 55 hESC lines with homozygous HLA types, would be sufficient for 80% of patients in the Japanese population. These findings demonstrate the feasibility of creating and maintaining an hESC bank with sufficient representation to support a large number of patients. However, in countries such as the United States, many more hESC lines would need to be established to serve its ethnically and genetically diverse population. Given the ethical issues and restrictions on hESC research, and the small number of approved hESC lines currently available, the creation of an hESC bank with a highly diverse collection of cell lines will undoubtedly face enormous challenges.

22.7 Conclusions

The field of hESC research has progressed significantly since their first derivation in 1998. The international scientific community has discovered the enormous potential of hESCs as newly derived lines continue to be developed and differentiation methods into various types of cells are optimized for scientific investigation and clinical use. It is clear that there remain major scientific challenges as well as ethical and legislative issues that must be addressed, especially in the United States. Certainly more questions will emerge as more is understood in the coming years. However, it is encouraging to see that clinical trials involving the use of hESCs in spinal cord injury and macular degeneration have begun. These studies will pave the way toward determining the therapeutic benefit of hESCs in regenerative medicine.

Acknowledgments

Work described in this chapter has been supported by grants from the National Institutes of Health (HL085377, HL007544), the California Institute for Regenerative Medicine (RC1-00104, RB3-05041), and the Muscular Dystrophy Association (186483) to H.S.B., and an institutional fellowship from the National Institutes of Health (HL007544) to O.Y.

References

Adewumi, O., B. Aflatoonian, L. Ahrlund-Richter, M. Amit, P. W. Andrews, G. Beighton, P. A. Choo et al. 2007. Characterization of human embryonic stem cell lines by the International Stem Cell Initiative. *Nat Biotechnol* 25 (7):803–816.

Agarwal, S., K. L. Holton, and R. Lanza. 2008. Efficient differentiation of functional hepatocytes from human embryonic stem cells. *Stem Cells* 26 (5):1117–11127.

Baker, D. E., N. J. Harrison, E. Maltby, K. Smith, H. D. Moore, P. J. Shaw, P. R. Heath, H. Holden, and P. W. Andrews. 2007. Adaptation to culture of human embryonic stem cells and oncogenesis in vivo. *Nat Biotechnol* 25 (2):207–215.

Bar, M., S. K. Wyman, B. R. Fritz, J. Qi, K. S. Garg, R. K. Parkin, E. M. Kroh et al. 2008. MicroRNA discovery and profiling in human embryonic stem cells by deep sequencing of small RNA libraries. *Stem Cells* 26 (10):2496–2505.

Bibikova, M., E. Chudin, B. Wu, L. Zhou, E. W. Garcia, Y. Liu, S. Shin et al. 2006. Human embryonic stem cells have a unique epigenetic signature. *Genome Res* 16 (9):1075–1083.

Bielby, R. C., A. R. Boccaccini, J. M. Polak, and L. D. Buttery. 2004. In vitro differentiation and in vivo mineralization of osteogenic cells derived from human embryonic stem cells. *Tissue Eng* 10 (9–10):1518–1525.

Bodnar, M. S., J. J. Meneses, R. T. Rodriguez, and M. T. Firpo. 2004. Propagation and maintenance of undifferentiated human embryonic stem cells. *Stem Cells Dev* 13 (3):243–253.

Bradley, J. A., E. M. Bolton, and R. A. Pedersen. 2002. Stem cell medicine encounters the immune system. *Nat Rev Immunol* 2 (11):859–871.

Cai, J., Y. Zhao, Y. Liu, F. Ye, Z. Song, H. Qin, S. Meng et al. 2007. Directed differentiation of human embryonic stem cells into functional hepatic cells. *Hepatology* 45 (5):1229–1239.

Card, D. A., P. B. Hebbar, L. Li, K. W. Trotter, Y. Komatsu, Y. Mishina, and T. K. Archer. 2008. Oct4/Sox2-regulated miR-302 targets cyclin D1 in human embryonic stem cells. *Mol Cell Biol* 28 (20):6426–6438.

Catalina, P., R. Montes, G. Ligero, L. Sanchez, T. de la Cueva, C. Bueno, P. E. Leone, and P. Menendez. 2008. Human ESCs predisposition to karyotypic instability: Is a matter of culture adaptation or differential vulnerability among hESC lines due to inherent properties? *Mol Cancer* 7:76.

Chung, Y., I. Klimanskaya, S. Becker, T. Li, M. Maserati, S. J. Lu, T. Zdravkovic et al. 2008. Human embryonic stem cell lines generated without embryo destruction. *Cell Stem Cell* 2 (2):113–117.

Cohen, S., L. Leshansky, E. Zussman, M. Burman, S. Srouji, E. Livne, N. Abramov, and J. Itskovitz-Eldor. 2010. Repair of full-thickness tendon injury using connective tissue progenitors efficiently derived from human embryonic stem cells and fetal tissues. *Tissue Eng Part A* 16 (10):3119–3137.

Cowan, C. A., I. Klimanskaya, J. McMahon, J. Atienza, J. Witmyer, J. P. Zucker, S. Wang, C. C. Morton, A. P. McMahon, D. Powers, and D. A. Melton. 2004. Derivation of embryonic stem-cell lines from human blastocysts. *N Engl J Med* 350 (13):1353–1356.

Crook, J. M., D. Hei, and G. Stacey. 2010. The International Stem Cell Banking Initiative (ISCBI): Raising standards to bank on. *In Vitro Cell Dev Biol Anim* 46 (3–4):169–172.

D'Amour, K. A., A. D. Agulnick, S. Eliazer, O. G. Kelly, E. Kroon, and E. E. Baetge. 2005. Efficient differentiation of human embryonic stem cells to definitive endoderm. *Nat Biotechnol* 23 (12):1534–1541.

D'Amour, K. A., A. G. Bang, S. Eliazer, O. G. Kelly, A. D. Agulnick, N. G. Smart, M. A. Moorman, E. Kroon, M. K. Carpenter, and E. E. Baetge. 2006. Production of pancreatic hormone-expressing endocrine cells from human embryonic stem cells. *Nat Biotechnol* 24 (11):1392–1401.

Drukker, M. 2004. Immunogenicity of human embryonic stem cells: Can we achieve tolerance? *Springer Semin Immunopathol* 26 (1–2):201–213.

Drukker, M., G. Katz, A. Urbach, M. Schuldiner, G. Markel, J. Itskovitz-Eldor, B. Reubinoff, O. Mandelboim, and N. Benvenisty. 2002. Characterization of the expression of MHC proteins in human embryonic stem cells. *Proc Natl Acad Sci USA* 99 (15):9864–9869.

Dvash, T., N. Lavon, and G. Fan. 2010. Variations of X chromosome inactivation occur in early passages of female human embryonic stem cells. *PLoS One* 5 (6):e11330.

Evseenko, D., Y. Zhu, K. Schenke-Layland, J. Kuo, B. Latour, S. Ge, J. Scholes, G. Dravid, X. Li, W. R. MacLellan, and G. M. Crooks. 2010. Mapping the first stages of mesoderm commitment during differentiation of human embryonic stem cells. *Proc Natl Acad Sci USA* 107 (31):13742–13747.

Gan, Q., T. Yoshida, O. G. McDonald, and G. K. Owens. 2007. Concise review: Epigenetic mechanisms contribute to pluripotency and cell lineage determination of embryonic stem cells. *Stem Cells* 25 (1):2–9.

Gaur, M., C. Ritner, R. Sievers, A. Pedersen, M. Prasad, H. S. Bernstein, and Y. Yeghiazarians. 2010. Timed inhibition of p38MAPK directs accelerated differentiation of human embryonic stem cells into cardiomyocytes. *Cytotherapy* 12 (6):807–817.

Geens, M., I. Mateizel, K. Sermon, M. De Rycke, C. Spits, G. Cauffman, P. Devroey, H. Tournaye, I. Liebaers, and H. Van de Velde. 2009. Human embryonic stem cell lines derived from single blastomeres of two 4-cell stage embryos. *Hum Reprod* 24 (11):2709–2717.

Genbacev, O., A. Krtolica, T. Zdravkovic, E. Brunette, S. Powell, A. Nath, E. Caceres et al. 2005. Serum-free derivation of human embryonic stem cell lines on human placental fibroblast feeders. *Fertil Steril* 83 (5):1517–1529.

Gepstein, L., C. Ding, D. Rehemedula, E. E. Wilson, L. Yankelson, O. Caspi, A. Gepstein, I. Huber, and J. E. Olgin. 2010. In vivo assessment of the electrophysiological integration and arrhythmogenic risk of myocardial cell transplantation strategies. *Stem Cells* 28 (12):2151–2161.

Guenou, H., X. Nissan, F. Larcher, J. Feteira, G. Lemaitre, M. Saidani, M. Del Rio et al. 2009. Human embryonic stem-cell derivatives for full reconstruction of the pluristratified epidermis: A preclinical study. *Lancet* 374 (9703):1745–1753.

Guo, H., N. T. Ingola, J. S. Weissman, and D. P. Bartel. 2010. Mammalian microRNAs predominantly act to decrease target mRNA levels. *Nature* 466 (7308):835–840.

Hall, L. L., M. Byron, J. Butler, K. A. Becker, A. Nelson, M. Amit, J. Itskovitz-Eldor, J. Stein, G. Stein, C. Ware, and J. B. Lawrence. 2008. X-inactivation reveals epigenetic anomalies in most hESC but identifies sublines that initiate as expected. *J Cell Physiol* 216 (2):445–452.

He, J. Q., Y. Ma, Y. Lee, J. A. Thomson, and T. J. Kamp. 2003. Human embryonic stem cells develop into multiple types of cardiac myocytes: Action potential characterization. *Circ Res* 93 (1):32–39.

Idelson, M., R. Alper, A. Obolensky, E. Ben-Shushan, I. Hemo, N. Yachimovich-Cohen, H. Khaner et al. 2009. Directed differentiation of human embryonic stem cells into functional retinal pigment epithelium cells. *Cell Stem Cell* 5 (4):396–408.

Ilic, D., G. Giritharan, T. Zdravkovic, E. Caceres, O. Genbacev, S. J. Fisher, and A. Krtolica. 2009. Derivation of human embryonic stem cell lines from biopsied blastomeres on human feeders with minimal exposure to xenomaterials. *Stem Cells Dev* 18 (9):1343–1350.

Itskovitz-Eldor, J., M. Schuldiner, D. Karsenti, A. Eden, O. Yanuka, M. Amit, H. Soreq, and N. Benvenisty. 2000. Differentiation of human embryonic stem cells into embryoid bodies compromising the three embryonic germ layers. *Mol Med* 6 (2):88–95.

Kawasaki, H., K. Mizuseki, S. Nishikawa, S. Kaneko, Y. Kuwana, S. Nakanishi, S. I. Nishikawa, and Y. Sasai. 2000. Induction of midbrain dopaminergic neurons from ES cells by stromal cell-derived inducing activity. *Neuron* 28 (1):31–40.

Kawasaki, H., H. Suemori, K. Mizuseki, K. Watanabe, F. Urano, H. Ichinose, M. Haruta et al. 2002. Generation of dopaminergic neurons and pigmented epithelia from primate ES cells by stromal cell-derived inducing activity. *Proc Natl Acad Sci USA* 99 (3):1580–1585.

Kehat, I., D. Kenyagin-Karsenti, M. Snir, H. Segev, M. Amit, A. Gepstein, E. Livne, O. Binah, J. Itskovitz-Eldor, and L. Gepstein. 2001. Human embryonic stem cells can differentiate into myocytes with structural and functional properties of cardiomyocytes. *J Clin Invest* 108 (3):407–414.

Kim, K., K. Ng, P. J. Rugg-Gunn, J. H. Shieh, O. Kirak, R. Jaenisch, T. Wakayama, M. A. Moore, R. A. Pedersen, and G. Q. Daley. 2007. Recombination signatures distinguish embryonic stem cells derived by parthenogenesis and somatic cell nuclear transfer. *Cell Stem Cell* 1 (3):346–352.

King, F. W., W. Liszewski, C. Ritner, and H. S. Bernstein. 2011. High-throughput tracking of pluripotent human embryonic stem cells with dual fluorescence resonance energy transfer molecular beacons. *Stem Cells Dev* 20 (3):475–484.

King, F. W., C. Ritner, W. Liszewski, H. C. Kwan, A. Pedersen, A. D. Leavitt, and H. S. Bernstein. 2009. Subpopulations of human embryonic stem cells with distinct tissue-specific fates can be selected from pluripotent cultures. *Stem Cells Dev* 18 (10):1441–1450.

Klimanskaya, I., Y. Chung, S. Becker, S. J. Lu, and R. Lanza. 2006. Human embryonic stem cell lines derived from single blastomeres. *Nature* 444 (7118):481–485.

Klimanskaya, I., Y. Chung, S. Becker, S. J. Lu, and R. Lanza. 2007. Derivation of human embryonic stem cells from single blastomeres. *Nat Protoc* 2 (8):1963–1972.

Kroon, E., L. A. Martinson, K. Kadoya, A. G. Bang, O. G. Kelly, S. Eliazer, H. Young et al. 2008. Pancreatic endoderm derived from human embryonic stem cells generates glucose-responsive insulin-secreting cells in vivo. *Nat Biotechnol* 26 (4):443–452.

Laflamme, M. A., K. Y. Chen, A. V. Naumova, V. Muskheli, J. A. Fugate, S. K. Dupras, H. Reinecke, C. Xu, M. Hassanipour, S. Police, C. O'Sullivan, L. Collins, Y. Chen, E. Minami, E. A. Gill, S. Ueno, C. Yuan, J. Gold, and C. E. Murry. 2007. Cardiomyocytes derived from human embryonic stem cells in pro-survival factors enhance function of infarcted rat hearts. *Nat Biotechnol* 25 (9):1015–1024.

Lagarkova, M. A., P. Y. Volchkov, A. V. Lyakisheva, E. S. Philonenko, and S. L. Kiselev. 2006. Diverse epigenetic profile of novel human embryonic stem cell lines. *Cell Cycle* 5 (4):416–420.

Laurent, L. C., J. Chen, I. Ulitsky, F. J. Mueller, C. Lu, R. Shamir, J. B. Fan, and J. F. Loring. 2008. Comprehensive microRNA profiling reveals a unique human embryonic stem cell signature dominated by a single seed sequence. *Stem Cells* 26 (6):1506–1516.

Lee, G., H. Kim, Y. Elkabetz, G. Al Shamy, G. Panagiotakos, T. Barberi, V. Tabar, and L. Studer. 2007. Isolation and directed differentiation of neural crest stem cells derived from human embryonic stem cells. *Nat Biotechnol* 25 (12):1468–1475.

Lefort, N., M. Feyeux, C. Bas, O. Feraud, A. Bennaceur-Griscelli, G. Tachdjian, M. Peschanski, and A. L. Perrier. 2008. Human embryonic stem cells reveal recurrent genomic instability at 20q11.21. *Nat Biotechnol* 26 (12):1364–1366.

Li, X. J., Z. W. Du, E. D. Zarnowska, M. Pankratz, L. O. Hansen, R. A. Pearce, and S. C. Zhang. 2005. Specification of motoneurons from human embryonic stem cells. *Nat Biotechnol* 23 (2):215–221.

Lin, G., Q. OuYang, X. Zhou, Y. Gu, D. Yuan, W. Li, G. Liu, T. Liu, and G. Lu. 2007. A highly homozygous and parthenogenetic human embryonic stem cell line derived from a one-pronuclear oocyte following in vitro fertilization procedure. *Cell Res* 17 (12):999–1007.

Mai, Q., Y. Yu, T. Li, L. Wang, M. J. Chen, S. Z. Huang, C. Zhou, and Q. Zhou. 2007. Derivation of human embryonic stem cell lines from parthenogenetic blastocysts. *Cell Res* 17 (12):1008–1019.

McLean, A. B., K. A. D'Amour, K. L. Jones, M. Krishnamoorthy, M. J. Kulik, D. M. Reynolds, A. M. Sheppard, H. Liu, Y. Xu, E. E. Baetge, and S. Dalton. 2007. Activin a efficiently specifies definitive endoderm from human embryonic stem cells only when phosphatidylinositol 3-kinase signaling is suppressed. *Stem Cells* 25 (1):29–38.

Morin, R. D., G. Aksay, E. Dolgosheina, H. A. Ebhardt, V. Magrini, E. R. Mardis, S. C. Sahinalp, and P. J. Unrau. 2008. Comparative analysis of the small RNA transcriptomes of Pinus contorta and Oryza sativa. *Genome Res* 18 (4):571–584.

Mummery, C., D. Ward-van Oostwaard, P. Doevendans, R. Spijker, S. van den Brink, R. Hassink, M. van der Heyden, T. Opthof, M. Pera, A. B. de la Riviere, R. Passier, and L. Tertoolen. 2003. Differentiation of human embryonic stem cells to cardiomyocytes: Role of coculture with visceral endoderm-like cells. *Circulation* 107 (21):2733–2740.

Nakajima, F., K. Tokunaga, and N. Nakatsuji. 2007. Human leukocyte antigen matching estimations in a hypothetical bank of human embryonic stem cell lines in the Japanese population for use in cell transplantation therapy. *Stem Cells* 25 (4):983–985.

Narva, E., R. Autio, N. Rahkonen, L. Kong, N. Harrison, D. Kitsberg, L. Borghese, J. Itskovitz-Eldor, O. Rasool, P. Dvorak, O. Hovatta, T. Otonkoski, T. Tuuri, W. Cui, O. Brustle, D. Baker, E. Maltby, H. D. Moore, N. Benvenisty, P. W. Andrews, O. Yli-Harja, and R. Lahesmaa. 2010. High-resolution DNA analysis of human embryonic stem cell lines reveals culture-induced copy number changes and loss of heterozygosity. *Nat Biotechnol* 28 (4):371–377.

Ng, E. S., R. P. Davis, L. Azzola, E. G. Stanley, and A. G. Elefanty. 2005. Forced aggregation of defined numbers of human embryonic stem cells into embryoid bodies fosters robust, reproducible hematopoietic differentiation. *Blood* 106 (5):1601–1603.

Nistor, G. I., M. O. Totoiu, N. Haque, M. K. Carpenter, and H. S. Keirstead. 2005. Human embryonic stem cells differentiate into oligodendrocytes in high purity and myelinate after spinal cord transplantation. *Glia* 49 (3):385–396.

Park, S. P., Y. J. Lee, K. S. Lee, H. Ah Shin, H. Y. Cho, K. S. Chung, E. Y. Kim, and J. H. Lim. 2004. Establishment of human embryonic stem cell lines from frozen-thawed blastocysts using STO cell feeder layers. *Hum Reprod* 19 (3):676–684.

Perrier, A. L., V. Tabar, T. Barberi, M. E. Rubio, J. Bruses, N. Topf, N. L. Harrison, and L. Studer. 2004. Derivation of midbrain dopamine neurons from human embryonic stem cells. *Proc Natl Acad Sci USA* 101 (34):12543–12548.

Ren, J., P. Jin, E. Wang, F. M. Marincola, and D. F. Stroncek. 2009. MicroRNA and gene expression patterns in the differentiation of human embryonic stem cells. *J Transl Med* 7:20.

Reubinoff, B. E., P. Itsykson, T. Turetsky, M. F. Pera, E. Reinhartz, A. Itzik, and T. Ben-Hur. 2001. Neural progenitors from human embryonic stem cells. *Nat Biotechnol* 19 (12):1134–1140.

Reubinoff, B. E., M. F. Pera, C. Y. Fong, A. Trounson, and A. Bongso. 2000. Embryonic stem cell lines from human blastocysts: Somatic differentiation in vitro. *Nat Biotechnol* 18 (4):399–404.

Revazova, E. S., N. A. Turovets, O. D. Kochetkova, L. B. Kindarova, L. N. Kuzmichev, J. D. Janus, and M. V. Pryzhkova. 2007. Patient-specific stem cell lines derived from human parthenogenetic blastocysts. *Cloning Stem Cells* 9 (3):432–449.

Ritner, C., and H. S. Bernstein. 2010. Fate mapping of human embryonic stem cells by teratoma formation. *J Vis Exp* 42:2036.

Ritner, C., S. S. Wong, F. W. King, S. S. Mihardja, W. Liszewski, D. J. Erle, R. J. Lee, and H. S. Bernstein. 2011. An engineered cardiac reporter cell line identifies human embryonic stem cell-derived myocardial precursors. *PLoS One* 6 (1):e16004.

Roy, N. S., C. Cleren, S. K. Singh, L. Yang, M. F. Beal, and S. A. Goldman. 2006. Functional engraftment of human ES cell-derived dopaminergic neurons enriched by coculture with telomerase-immortalized midbrain astrocytes. *Nat Med* 12 (11):1259–1268.

Satin, J., I. Kehat, O. Caspi, I. Huber, G. Arbel, I. Itzhaki, J. Magyar, E. A. Schroder, I. Perlman, and L. Gepstein. 2004. Mechanism of spontaneous excitability in human embryonic stem cell derived cardiomyocytes. *J Physiol* 559 (Pt 2):479–496.

Sengupta, S., J. Nie, R. J. Wagner, C. Yang, R. Stewart, and J. A. Thomson. 2009. MicroRNA 92b controls the G1/S checkpoint gene p57 in human embryonic stem cells. *Stem Cells* 27 (7):1524–1528.

Senju, S., S. Hirata, Y. Motomura, D. Fukuma, Y. Matsunaga, S. Fukushima, H. Matsuyoshi, and Y. Nishimura. 2010. Pluripotent stem cells as source of dendritic cells for immune therapy. *Int J Hematol* 91 (3):392–400.

Shen, Y., Y. Matsuno, S. D. Fouse, N. Rao, S. Root, R. Xu, M. Pellegrini, A. D. Riggs, and G. Fan. 2008. X-inactivation in female human embryonic stem cells is in a nonrandom pattern and prone to epigenetic alterations. *Proc Natl Acad Sci USA* 105 (12):4709–4714.

Silva, S. S., R. K. Rowntree, S. Mekhoubad, and J. T. Lee. 2008. X-chromosome inactivation and epigenetic fluidity in human embryonic stem cells. *Proc Natl Acad Sci USA* 105 (12):4820–4825.

Strelchenko, N., and Y. Verlinsky. 2006. Embryonic stem cells from morula. *Methods Enzymol* 418:93–108.

Strelchenko, N., O. Verlinsky, V. Kukharenko, and Y. Verlinsky. 2004. Morula-derived human embryonic stem cells. *Reprod Biomed Online* 9 (6):623–629.

Strom, S., F. Holm, R. Bergstrom, A. M. Stromberg, and O. Hovatta. 2010. Derivation of 30 human embryonic stem cell lines—Improving the quality. *In Vitro Cell Dev Biol Anim* 46 (3–4):337–344.

Su, Z., C. Frye, K. M. Bae, V. Kelley, and J. Vieweg. 2008. Differentiation of human embryonic stem cells into immunostimulatory dendritic cells under feeder-free culture conditions. *Clin Cancer Res* 14 (19):6207–6217.

Suh, M. R., Y. Lee, J. Y. Kim, S. K. Kim, S. H. Moon, J. Y. Lee, K. Y. Cha et al. 2004. Human embryonic stem cells express a unique set of microRNAs. *Dev Biol* 270 (2):488–498.

Sumi, T., N. Tsuneyoshi, N. Nakatsuji, and H. Suemori. 2008. Defining early lineage specification of human embryonic stem cells by the orchestrated balance of canonical Wnt/beta-catenin, Activin/Nodal and BMP signaling. *Development* 135 (17):2969–2979.

Taylor, C. J., E. M. Bolton, S. Pocock, L. D. Sharples, R. A. Pedersen, and J. A. Bradley. 2005. Banking on human embryonic stem cells: Estimating the number of donor cell lines needed for HLA matching. *Lancet* 366 (9502):2019–2025.

Tecirlioglu, R. T., L. Nguyen, K. Koh, A. O. Trounson, and A. E. Michalska. 2010. Derivation and maintenance of human embryonic stem cell line on human adult skin fibroblast feeder cells in serum replacement medium. *In Vitro Cell Dev Biol Anim* 46 (3–4):231–235.

Thomson, J. A., J. Itskovitz-Eldor, S. S. Shapiro, M. A. Waknitz, J. J. Swiergiel, V. S. Marshall, and J. M. Jones. 1998. Embryonic stem cell lines derived from human blastocysts. *Science* 282 (5391):1145–1147.

Toh, W. S., E. H. Lee, X. M. Guo, J. K. Chan, C. H. Yeow, A. B. Choo, and T. Cao. 2010. Cartilage repair using hyaluronan hydrogel-encapsulated human embryonic stem cell-derived chondrogenic cells. *Biomaterials* 31 (27):6968–6980.

Toh, W. S., Z. Yang, H. Liu, B. C. Heng, E. H. Lee, and T. Cao. 2007. Effects of culture conditions and bone morphogenetic protein 2 on extent of chondrogenesis from human embryonic stem cells. *Stem Cells* 25 (4):950–960.

Vats, A., R. C. Bielby, N. Tolley, S. C. Dickinson, A. R. Boccaccini, A. P. Hollander, A. E. Bishop, and J. M. Polak. 2006. Chondrogenic differentiation of human embryonic stem cells: The effect of the micro-environment. *Tissue Eng* 12 (6):1687–1697.

Wang, D., D. L. Haviland, A. R. Burns, E. Zsigmond, and R. A. Wetsel. 2007. A pure population of lung alveolar epithelial type II cells derived from human embryonic stem cells. *Proc Natl Acad Sci USA* 104 (11):4449–4454.

Wang, D., J. E. Morales, D. G. Calame, J. L. Alcorn, and R. A. Wetsel. 2010. Transplantation of human embryonic stem cell-derived alveolar epithelial type II cells abrogates acute lung injury in mice. *Mol Ther* 18 (3):625–634.

Wernig, M., F. Benninger, T. Schmandt, M. Rade, K. L. Tucker, H. Bussow, H. Beck, and O. Brustle. 2004. Functional integration of embryonic stem cell-derived neurons in vivo. *J Neurosci* 24 (22):5258–5268.

Woll, P. S., B. Grzywacz, X. Tian, R. K. Marcus, D. A. Knorr, M. R. Verneris, and D. S. Kaufman. 2009. Human embryonic stem cells differentiate into a homogeneous population of natural killer cells with potent in vivo antitumor activity. *Blood* 113 (24):6094–6101.

Wong, S. S., and H. S. Bernstein. 2010. Cardiac regeneration using human embryonic stem cells: Producing cells for future therapy. *Regen Med* 5 (5):763–775.

Yan, Y., D. Yang, E. D. Zarnowska, Z. Du, B. Werbel, C. Valliere, R. A. Pearce, J. A. Thomson, and S. C. Zhang. 2005. Directed differentiation of dopaminergic neuronal subtypes from human embryonic stem cells. *Stem Cells* 23 (6):781–790.

Yang, L., M. H. Soonpaa, E. D. Adler, T. K. Roepke, S. J. Kattman, M. Kennedy et al. 2008. Human cardiovascular progenitor cells develop from a KDR+ embryonic-stem-cell-derived population. *Nature* 453 (7194):524–528.

Zhang, S. C., M. Wernig, I. D. Duncan, O. Brustle, and J. A. Thomson. 2001. In vitro differentiation of transplantable neural precursors from human embryonic stem cells. *Nat Biotechnol* 19 (12):1129–1133.

23

Clinical Applications of Mesenchymal Stem Cell–Biomaterial Constructs for Tissue Reconstruction

Summer Hanson and Peiman Hematti

CONTENTS

23.1 Introduction

Millions of reconstructive procedures are performed each year to address a variety of defects (American Society of Plastic Surgeons 2010). Within the reconstructive armamentarium are techniques spanning from simple closure to complex, composite tissue transplantation moving skin, fat, muscle, and bone from one part of the body to another. This involves revascularization of the blood supply of the transferred tissues and is associated with additional site of surgery (donor site) with its own complications and pain. While the current techniques of reconstruction are relatively reliable, they are not without limitation. Thus, there continues to be aggressive investigation into tissue-engineered constructs in many clinical applications.

In recent years, numerous biomaterials, and several cell-based products, have emerged on the market with Food and Drug Administration (FDA) approval. The goal of such treatment strategies as bioengineered scaffolds and cellular therapies is to act as "smart band-aids," to replace senescent resident cells and reestablish the anatomy and physiology (Panuncialman and Falanga 2007, Herdrich et al. 2008). Tissue-engineered dressings such as Dermagraft®, derived from human foreskin fibroblasts seeded on a polyglactin mesh, and Apligraf®, a bilayered product consisting of dermal fibroblasts and bovine collagen under human keratinocytes, are currently available with FDA approval for treatment of diabetic foot ulcers and venous leg ulcers (Ehrenreich and Ruszczak 2006). Integra®, another FDA-approved biomaterial, is an example of a sheet scaffold composed of shark-derived chondroitin sulfate and bovine collagen, used as a dermal replacement in full-thickness skin loss. Finally, acellular dermal matrix (ADM, Alloderm® or Flex HD®) is a human dermal tissue derivative that is processed to remove the cellular component, leaving a nonimmunogenic ECM-based scaffold for new tissue ingrowth. Despite these advances in wound care and reconstruction, there continues to be a need for more effective tissue replacement, or perhaps more importantly tissue restoration, that many believe may be addressed by cell-based treatment modalities (Hanson et al. 2010a,b).

Novel cell therapies involving the transplantation of progenitor or stem cells to the patient are being developed in combination with biomaterial scaffolds based on many physiologically relevant ECM proteins. When considering cells for use in such developmental applications, there are several interesting options. There is literature to support the use of terminally differentiated cells in certain applications such as autologous chondrocyte transplantation as a mean to replace articular cartilage in arthritis (Ochs et al. 2011); a review of such tissue-specific cell therapy is not included here as these techniques essentially replace "like with like." Rather, there is a huge interest in using multipotent or pluripotent stem cells to address tissue defects as sort of a reconstructive alchemy, generating completely new tissue instead of fibrosis or scar.

23.2 Mesenchymal Stromal/Stem Cells

Mesenchymal stem cells (MSCs) are adult, multipotent progenitor cells characterized by specific cell surface marker expression and differentiation potential along adipogenic, osteogenic, and chondrogenic lineages that were originally isolated from bone marrow (Dominici et al. 2006, Hanson et al. 2010a,b). While there is evidence suggesting MSCs have the ability to affect many regenerative processes through secretion of trophic and angiogenic paracrine factors, perhaps a more intriguing property is the immunomodulatory properties of MSCs (Figure 23.1). These properties have allowed systemic or local administration of MSCs derived from autologous, HLA-matched allogeneic or non-HLA-matched third-party donors in a wide variety of clinical studies (Battiwalla and Hematti 2009). The ability to use these

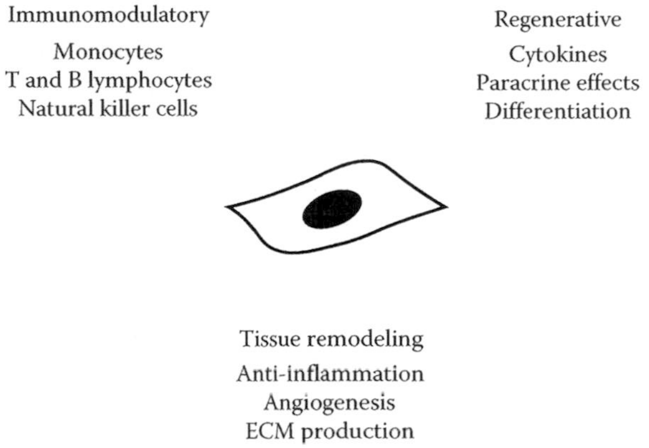

Immunomodulatory

Monocytes
T and B lymphocytes
Natural killer cells

Regenerative

Cytokines
Paracrine effects
Differentiation

Tissue remodeling
Anti-inflammation
Angiogenesis
ECM production

FIGURE 23.1
Functional mechanisms of MSCs involved in reconstruction. There are several functional mechanisms identified with culture-expanded MSCs that are favorable when developing cell-based therapies for wound healing and regenerative medicine. These affect the inflammatory, remodeling, and regenerative pathways of the healing process potentially offering a multi-modal cell-based therapy.

cells without respect to HLA typing allows these cells to be potentially used as "off the shelf" cellular therapeutics. Thus, during the last decade, adult tissue–derived MSCs have rapidly moved from in vitro and preclinical animal studies into human trials as a therapeutic modality for a diverse group of clinical applications (Giordano et al. 2007). It is now evident that MSCs, in addition to bone marrow, reside within most adult tissues and organs (da Silva Meirelles et al. 2006). Importantly, studies suggest that MSCs isolated from these diverse tissues possess similar biological characteristics, immunological properties, and differentiation potential, making many of them a suitable cell type for reconstructive applications (Puissant et al. 2005, Hoogduijn et al. 2007, Hanson et al. 2010c). Of particular interest in reconstructive surgery is the isolation of cells with characteristics similar to bone marrow–derived MSCs from the stromal vascular fraction (SVF) of adipose tissue (Zuk et al. 2001). The adipose tissue–derived MSCs are of particular interest to plastic surgeons due to its ease of isolation from tissues that are otherwise being discarded.

23.3 Clinical Applications

When considering tissue engineering strategies in particular, it is likely that a scaffold or structural support is necessary. Scaffolds provide a microenvironment that allows for nutrient diffusion as well as biochemical, physical, and cellular stimuli that guides proliferation, differentiation, and migration. Several reviews highlight the strategies which have been employed to create adipose, bone, or cartilage tissue including de novo tissue induction, injectable composite systems, and scaffold-based tissue regeneration (Gomillion and Burg 2006). Early clinical utility of combining MSCs with a "scaffold" for clinical reconstruction is demonstrated in the application of MSC-rich cell lipoaspirate added to autologous adipose grafts as soft tissue fillers (Matsumoto et al. 2006, Yoshimura et al. 2008a) or bone marrow aspirate added to cancellous bone graft for small bony defects or fracture healing (Connolly et al. 1989, 1991).

23.4 Soft Tissue and Wound Healing

To improve "graft take" and therefore predictability and efficacy of autologous fat grafting, Matsumoto et al. developed a novel method of concurrent transfer of lipoaspirated fat with adipose-derived progenitor cells termed "cell-assisted lipotransfer" (CAL) (Matsumoto et al. 2006). In this technique, a portion of the lipoaspirated fat is processed to isolate the heterogeneous mixture of cells of the SVF; the remaining lipoaspirate is processed for fat grafting, serving as a biological scaffold for the MSC-containing SVF cells. The foundation of this technology is that the additional cells will improve graft survival and reduce postoperative atrophy or resorption through enhanced angiogenesis and cell self-renewal (Yoshimura et al. 2008a,b).

Local delivery of MSCs into cutaneous wounds has been reported in several case series. Typically this involves direct injection of autologous cells followed by staged reconstruction (Hanson et al. 2010a). Falanga et al. investigated a unique delivery system using fibrin glue and MSCs in both acute (n = 4) and chronic (n = 6) wound settings (Falanga et al. 2007). Autologous culture-expanded bone marrow–derived MSCs were combined with fibrin spray for topical application. The acute wounds studied were full-thickness cutaneous defects created by surgical excision. In the setting of chronic wounds, the authors chose lower extremity wounds that were present for greater than 1 year and were refractory to standard treatments including topical growth factors and bioengineered skin substitutes. Acute wounds were healed within 8 weeks, while chronic wounds were significantly improved or healed completely by 16–20 weeks. Biopsies of the tissues treated with topical MSCs and fibrin glue demonstrated higher concentration of CD29+ cells, one of the surface markers found on MSCs. These results indicate fibrin glue potentially provides a delivery system to maintain MSCs in the acute wound bed, but allows for migration out of the fibrin matrix as healing progresses.

Yoshikawa et al. (2008) reported 20 patients with various non-healing wounds (i.e., burns, lower extremity ulcers, and decubiti) treated with ex vivo culture-expanded autologous bone marrow–derived MSCs, and a dermal replacement (Pelnac®), with or without autologous skin graft. The authors report that 18 of the 20 wounds appeared to be healed completely with the cell-composite graft transfer and the addition of MSCs facilitated regeneration of the native tissue by histologic examination. However, these authors only used a relatively low number of cells that were available at the end of passage 0 and did not report on the characterization of cultured cells. This is especially important since passage 0, when the culture flasks are confluent after initial plating, potentially contains many other types of cells, including macrophages, which would affect wound healing as well. While these reports demonstrate the heterogeneity of the type of wounds treated with MSCs, they also illustrate the variations in culture and application techniques that limit the current body of evidence in support of MSC therapy.

23.5 Ligament and Tendon

Similar to adipose transfer and cartilage repair, the relatively avascular and acellular nature of tendons and ligaments challenges healing of these tissues. This is further complicated by mechanical stress and motion. Current advances in tendon healing focus on the use of bioactive molecules such as hyaluronic acid (HA), transforming growth factor-β (TGF-β), and platelet-rich plasma (PRP). There are several in vivo studies of anterior cruciate ligament repair using MSCs directly applied to the wounded ligament or in combination with a cadaveric ligament (Joshi et al. 2009). Furthermore, MSC injection is used in veterinary applications to repair ligamentous injuries or inflammation. Such reports have shown an increase in vascularity and tissue formation as well as improved biomechanical functions (Joshi et al. 2009, Woo 2009). An additional application of MSCs in ligament and tendon reconstruction includes tissue-engineered constructs which could be mechanically stressed in a bioreactor prior to implantation (Butler et al. 2009a,b). To our knowledge there are no clinical reports of MSCs in such injury though one can anticipate this is not far off in the spectrum of cell-based therapeutics in reconstructive applications.

23.6 Cartilage

The relatively acellular and poorly vascularized nature of cartilage makes it not only susceptible to injury but also difficult to heal. The structural components of cartilage tissue include predominantly collagen II, HA, proteoglycans, and glycoproteins surrounded by a peripheral rim of perichondrium. Articular cartilage in particular is a unique functional tissue allowing for both shock absorption and lubrication for joint movement. For decades surgical procedures have been aimed at repairing articular cartilage to restore motion or reduce pain. For this reason there is interest in developing cell-based therapies for cartilage regeneration.

The first reported cases using culture-expanded, autologous MSCs to repair cartilage defects were reported by Wakitani et al. (2002, 2004). The group treated two patients with patellar defects with MSCs seeded on a hydrogel of type I collagen. These constructs were implanted into the cartilage defects and covered with a periosteal flap. Within 6 months, both patients noted an improvement in their pain and activity. Biopsies at 1 and 2 years show that the tissues were repaired with fibrocartilage. Additional cases of this MSC–collagen–periosteal flap methodology have been reported in a variety of osteochondral defects (Wakitani et al. 2007). In total, the authors have used this combination of ex vivo expanded bone marrow MSCs with a type I collagen sponge for implantation into cartilage defects in over 40 patients. The authors now have over 10 year follow-up from these original cases and show no evidence of symptom recurrence, clinical problem, infection, or abnormal pathology (Wakitani 2007).

In the first randomized, prospective controlled study, this technique was applied to osteoarthritis in patients undergoing high tibial osteotomies (HTO) (Wakitani et al. 2002). Twenty-four patients undergoing HTO for medial compartment osteoarthritis were randomly assigned to receive autologous MSC transplantation or cell-free controls. Bone marrow aspirate was harvested from the iliac crest, and cells were expanded for approximately 30 days. MSCs were then embedded in a type I collagen gel and maintained in culture an additional night. At the time of surgery, those patients randomized to receive cell transplantation had a segment of disease articular surface abraded and treated with the collagen gel–MSC construct followed by a periosteal cover (average 14 × 35 mm). The control group underwent a similar procedure with a cell-free collagen sponge implanted and covered with local periosteum. The mean follow-up was 16 months. The authors report a significant improvement in scores assessing pain, function, and muscle strength before and after the procedure in both groups. There was no difference in clinical outcomes between the cell transplantation and control group, indicating either treatment is appropriate. Second-look arthroscopic procedures were performed at an average of 42 weeks after surgery on nine of the cell-transplanted patients and showed firm, regular white cartilage in the area of sponge transplantation. Similar arthroscopy was performed in six of the control patients showing softer, yellowish, irregular cartilage in the area of collagen sponge implantation. Additionally, tissue biopsies were obtained and showed hyaline-like cartilage in the cell-treated group and fibrocartilage in the control group. The authors observed that the arthroscopic and histological grading scores were better in the cell transplantation group compared to controls.

A large defect resulting from methicillin-resistant *Staphylococcus aureus* septic arthritis following trauma was treated in a similar manner by Adachi et al. (2005). The patient presented with severe knee pain and limited range of motion. Imaging showed an osteochondral defect of the medial femoral condyle measuring $20 \times 25 \times 25$ mm causing pain and an audible click in the knee joint. First, bone marrow aspirate was taken from the tibia under local anesthetic, and MSCs were expanded to passage 3. A porous calcium hydroxyapatite sheet was suspended in the defect, and the autologous MSCs were infiltrated directly in to the matrix scaffold. Partial weight-bearing was initiated at 3 weeks, at which time the patient's range of motion improved, his pain improved, and the audible click had resolved. Arthroscopy and tissue biopsy taken 1 year after the procedure demonstrated fibrous tissue with a deep layer extracellular matrix expressing glycosaminoglycans with integration into the underlying bone. Despite the lack of conclusive evidence supporting the clinical efficacy of these cell therapy applications, these pioneering studies have shown that use of ex vivo culture-expanded MSCs is safe and practical. Furthermore, such studies could provide the platform for other novel cellular therapeutics such as embryonic stem cell derivatives (Trivedi and Hematti 2008).

23.7 Bone

There is evidence to suggest that biomaterial scaffolds clinically used for their osteoconductive properties, such as hydroxyapatite or other ceramics, can further be enhanced by adding MSCs. Clinically, bone marrow aspirate has been added to fracture sites for decades to promote healing or treat fracture nonunion (Connolly et al. 1989, 1991). Additionally, autologous bone marrow has been combined with a variety of scaffolds to address bony defects such as cysts or segmental defects (Tiedeman et al. 1995, Jager et al. 2009). Culture-expanded bone marrow MSCs seeded on hydroxyapatite scaffolds were used to treat large segment defects (4–7 cm) in the extremities stabilized with external fixation (Quarto et al. 2001). The three patients described in this report had complete bone fusion at approximately 6 months and stable durability at an average follow-up of 6 years. Morishita et al. reported that autologous, ex vivo expanded MSCs that were cultured within hydroxyapatite pellets were used to address defects resulting from benign tumors of the extremities (Morishita et al. 2006). The authors showed evidence of osseointegration maintained at more than 2 year follow-up. In another case, a novel population of progenitor cells derived from periosteum (with characteristics similar to MSCs) were implanted with a 3D piece of porous coral for replacement of the distal phalanx of the thumb following traumatic avulsion (Vacanti et al. 2001). Follow-up biopsy showed both lamellar bone and ossified tissue; perhaps more importantly the patient had good function of this thumb.

Engineered bone constructs are gaining support in select maxillofacial applications, as an alternative to autologous bone grafting as well for small defects or cysts, fracture support, or congenital anomalies. Perhaps most promising is to augment deficient bone stock for dental implants. The use of dental implants is limited by the volume of viable bone in the posterior maxillary floor, and often an autologous bone graft or synthetic bone mineral is used to supplement the maxillary ridge for future implantation. One study compared 12 consecutive patients undergoing sinus floor augmentation with each patient serving as their own control (Rickert et al. 2010). Both sides of the patients' mouth were augmented with a hydroxyapatite-based scaffold with one side receiving "standard therapy" of the scaffold combined with autologous bone graft, while the other received the scaffold combined with autologous bone marrow MSCs (P0). The authors demonstrated significantly more new bone formation in the area treated with MSCs compared to standard bone graft and no differences in overall healing, complication rate, or ability to place dental implants between either group. Shayesteh et al. demonstrated similar success in six consecutive patients treated with hydroxyapatite–β tricalcium phosphate ceramic loaded with culture-expanded autologous bone marrow MSCs (P2-3) (Shayesteh et al. 2008). Demineralized bone matrix combined with culture-expanded MSCs was used to provide graft material for alveolar defects in a report of two patients with unilateral complete cleft palates. There were no acute complications in the two patients reported. In each of these reports, MSCs were added to osteoinductive scaffolds and implanted into defects in their undifferentiated state, demonstrating new bone formation as the grafts are incorporated.

An alternative strategy is to culture MSCs on the scaffolds in osteogenic media to induce bone development de novo and then implant the constructs. Meijer et al. studied one such protocol (Meijer et al. 2008). Bone marrow–derived MSCs were harvested via bone marrow aspirate from the iliac crest, expanded to P3, seeded onto hydroxyapatite particles, and cultured for an additional 7 days in osteogenic culture media. The bone substitutes were placed directly into the defect site and covered with a local periosteal flap. Four months postimplantation, the patients had dental implants placed and biopsies taken. There was bone formation observed in three of the six patients. The other three patients failed to show new bone formation. Interestingly, constructs from each of the patients were implanted in a subcutaneous pocket on the backs of athymic mice at the time of their original surgery as well. At 6 weeks time, all of the constructs exhibited new bone formation on histological analysis, although this was not observed in each of the clinical correlates. This study illustrates the unpredictable nature of translational research when developing complex tissue-engineered constructs, particularly when using rodents as preclinical models.

23.8 Composite Reconstruction

Tissue engineering strategies using MSCs have been explored more recently in select complex clinical scenarios as well. Macchiarini et al. published their results with the transplantation of a tissue-engineered tracheal segment, the first of its kind (Macchiarini et al. 2008) in a young woman with bronchomalacia and expiratory collapse refractory to standard treatment and reconstruction. Autologous bronchial epithelial cells were isolated from biopsies and expanded in supplemented media, while autologous chondrocytes were derived from bone marrow MSCs treated with chondrogenic media. The scaffold was a decellularized allograft from a cadaveric trachea. It was seeded with epithelial cells on the inner surface, chondrocytes on the outer surface, and then maintained in a bioreactor to introduce biomechanical cues that would need to be tolerated by the engineered construct. This ex vivo generated graft was then implanted end-to-end to normal tracheal tissue to replace the diseased segment. According to the authors, the lung immediately ventilated well. Subsequent evaluations showed no indication of inflammation related to the donor trachea, normal pulmonary function tests, and an immeasurable improvement in quality of life.

23.9 Conclusions

The likelihood of tissue regeneration in situ is limited by the size of the defect, the nature of the injury, the mechanical or functional requirements of the tissue, and the surrounding soft tissue envelop. MSC therapy offers an alternative in which cells with multipotent differentiation capabilities are transferred to the site of injury. While MSCs offer an ease of harvest and expansion of a nonimmunogenic cell line, their use in complex tissue reconstruction can be guided by the characteristics of a biomaterial scaffold and the local tissue environment. The clinical cases and series reported here offer promising outcomes and applications of MSC-based therapy in a variety of tissue defects. Well-designed randomized clinical trials are the next phase in order to scrutinize the true potential of these highly complex therapeutic modalities in this area of regenerative medicine. Further work is necessary to determine the fate of the transplanted cells and constructs in terms of engraftment, inflammation, mechanical function, and durability given the unique tissues of interest.

While the outcomes are varied among the clinical applications, these reports are promising, as they all show the safety of these cells. MSCs, alone or in combination with biomaterials, offer novel potential solutions for a wide array of challenging clinical disorders, ushering in a new era in regenerative medicine.

References

Adachi, N., Ochi, M., Deie, M., and Ito, Y. 2005, Transplant of mesenchymal stem cells and hydroxyapatite ceramics to treat severe osteochondral damage after septic arthritis of the knee, *The Journal of Rheumatology*, 32(8): 1615–1618.

American Society of Plastic Surgeons. 2010, *2010 National Plastic Surgery Statistics*. Available at: www.plasticsurgery.org/News-and-Resources/2010-Statistics.html (Accessed March 5, 2012).

Battiwalla, M. and Hematti, P. 2009, Mesenchymal stem cells in hematopoietic stem cell transplantation, *Cytotherapy*, 11(5): 503–515.

Butler, D.L., Goldstein, S.A., Guldberg, R.E., Guo, X.E., Kamm, R., Laurencin, C.T., McIntire, L.V., Mow, V.C., Nerem, R.M., Sah, R.L., Soslowsky, L.J., Spilker, R.L. & Tranquillo, R.T. 2009a, The impact of biomechanics in tissue engineering and regenerative medicine, *Tissue Engineering, Part B: Reviews*, 15(4): 477–484.

Butler, D.L., Hunter, S.A., Chokalingam, K., Cordray, M.J., Shearn, J., Juncosa-Melvin, N., Nirmalanandhan, S., and Jain, A. 2009b, Using functional tissue engineering and bioreactors to mechanically stimulate tissue-engineered constructs, *Tissue Engineering, Part A*, 15(4): 741–749.

Connolly, J.F., Guse, R., Tiedeman, J., and Dehne, R. 1989, Autologous marrow injection for delayed unions of the tibia: A preliminary report, *Journal of Orthopaedic Trauma*, 3(4): 276–282.

Connolly, J.F., Guse, R., Tiedeman, J., and Dehne, R. 1991, Autologous marrow injection as a substitute for operative grafting of tibial nonunions, *Clinical Orthopaedics and Related Research*, (266): 259–270.

da Silva Meirelles, L., Chagastelles, P.C., and Nardi, N.B. 2006, Mesenchymal stem cells reside in virtually all post-natal organs and tissues, *Journal of Cell Science*, 119(Pt 11): 2204–2213.

Dominici, M., Le Blanc, K., Mueller, I., Slaper-Cortenbach, I., Marini, F., Krause, D., Deans, R., Keating, A., Prockop, D., and Horwitz, E. 2006, Minimal criteria for defining multipotent mesenchymal stromal cells. The International Society for Cellular Therapy position statement, *Cytotherapy*, 8(4): 315–317.

Ehrenreich, M. and Ruszczak, Z. 2006, Update on tissue-engineered biological dressings, *Tissue Engineering*, 12(9): 2407–2424.

Falanga, V., Iwamoto, S., Chartier, M., Yufit, T., Butmarc, J., Kouttab, N., Shrayer, D., and Carson, P. 2007, Autologous bone marrow-derived cultured mesenchymal stem cells delivered in a fibrin spray accelerate healing in murine and human cutaneous wounds, *Tissue Engineering*, 13(6): 1299–1312.

Giordano, A., Galderisi, U., and Marino, I.R. 2007, From the laboratory bench to the patient's bedside: An update on clinical trials with mesenchymal stem cells, *Journal of Cellular Physiology*, 211(1): 27–35.

Gomillion, C.T. and Burg, K.J. 2006, Stem cells and adipose tissue engineering, *Biomaterials*, 27(36): 6052–6063.

Hanson, S.E., Bentz, M.L., and Hematti, P. 2010a, Mesenchymal stem cell therapy for nonhealing cutaneous wounds, *Plastic and Reconstructive Surgery*, 125(2): 510–516.

Hanson, S.E., Gutowski, K.A., and Hematti, P. 2010b, Clinical applications of mesenchymal stem cells in soft tissue augmentation, *Aesthetic Surgery Journal/The American Society for Aesthetic Plastic Surgery*, 30(6): 838–842.

Hanson, S.E., Kim, J., Johnson, B.H., Bradley, B., Breunig, M.J., Hematti, P., and Thibeault, S.L. 2010c, Characterization of mesenchymal stem cells from human vocal fold fibroblasts, *The Laryngoscope*, 120(3): 546–551.

Herdrich, B.J., Lind, R.C., and Liechty, K.W. 2008, Multipotent adult progenitor cells: Their role in wound healing and the treatment of dermal wounds, *Cytotherapy*, 10(6): 543–550.

Hoogduijn, M.J., Crop, M.J., Peeters, A.M., Van Osch, G.J., Balk, A.H., Ijzermans, J.N., Weimar, W., and Baan, C.C. 2007, Human heart, spleen, and perirenal fat-derived mesenchymal stem cells have immunomodulatory capacities, *Stem Cells and Development*, 16(4): 597–604.

Jager, M., Jelinek, E.M., Wess, K.M., Scharfstadt, A., Jacobson, M., Kevy, S.V., and Krauspe, R. 2009, Bone marrow concentrate: A novel strategy for bone defect treatment, *Current Stem Cell Research & Therapy*, 4(1): 34–43.

Joshi, S.M., Mastrangelo, A.N., Magarian, E.M., Fleming, B.C., and Murray, M.M. 2009, Collagen-platelet composite enhances biomechanical and histologic healing of the porcine anterior cruciate ligament, *The American Journal of Sports Medicine*, 37(12): 2401–2410.

Macchiarini, P., Jungebluth, P., Go, T., Asnaghi, M.A., Rees, L.E., Cogan, T.A., Dodson, A., Martorell, J., Bellini, S., Parnigotto, P.P., Dickinson, S.C., Hollander, A.P., Mantero, S., Conconi, M.T. & Birchall, M.A. 2008, Clinical transplantation of a tissue-engineered airway, *Lancet*, 372(9655): 2023–2030.

Matsumoto, D., Sato, K., Gonda, K., Takaki, Y., Shigeura, T., Sato, T., Aiba-Kojima, E., Iizuka, F., Inoue, K., Suga, H. & Yoshimura, K. 2006, Cell-assisted lipotransfer: Supportive use of human adipose-derived cells for soft tissue augmentation with lipoinjection, *Tissue Engineering*, 12(12): 3375–3382.

Meijer, G.J., de Bruijn, J.D., Koole, R., and van Blitterswijk, C.A. 2008, Cell based bone tissue engineering in jaw defects, *Biomaterials*, 29(21): 3053–3061.

Morishita, T., Honoki, K., Ohgushi, H., Kotobuki, N., Matsushima, A., and Takakura, Y. 2006, Tissue engineering approach to the treatment of bone tumors: Three cases of cultured bone grafts derived from patients' mesenchymal stem cells, *Artificial Organs*, 30(2): 115–118.

Ochs, B.G., Muller-Horvat, C., Albrecht, D., Schewe, B., Weise, K., Aicher, W.K., and Rolauffs, B. 2011, Remodeling of articular cartilage and subchondral bone after bone grafting and matrix-associated autologous chondrocyte implantation for osteochondritis dissecans of the knee, *The American Journal of Sports Medicine*, 39(4): 764–773.

Panuncialman, J. and Falanga, V. 2007, The science of wound bed preparation, *Clinics in Plastic Surgery*, 34(4): 621–632.

Puissant, B., Barreau, C., Bourin, P., Clavel, C., Corre, J., Bousquet, C., Taureau, C., Cousin, B., Abbal, M., Laharrague, P., Penicaud, L., Casteilla, L. & Blancher, A. 2005, Immunomodulatory effect of human adipose tissue-derived adult stem cells: Comparison with bone marrow mesenchymal stem cells, *British Journal of Haematology*, 129(1): 118–129.

Quarto, R., Mastrogiacomo, M., Cancedda, R., Kutepov, S.M., Mukhachev, V., Lavroukov, A., Kon, E., and Marcacci, M. 2001, Repair of large bone defects with the use of autologous bone marrow stromal cells, *The New England Journal of Medicine*, 344(5): 385–386.

Rickert, D., Sauerbier, S., Nagursky, H., Menne, D., Vissink, A., and Raghoebar, G.M. 2010, Maxillary sinus floor elevation with bovine bone mineral combined with either autogenous bone or autogenous stem cells: A prospective randomized clinical trial, *Clinical Oral Implants Research*, 22(3): 251–258.

Shayesteh, Y.S., Khojasteh, A., Soleimani, M., Alikhasi, M., Khoshzaban, A., and Ahmadbeigi, N. 2008, Sinus augmentation using human mesenchymal stem cells loaded into a beta-tricalcium phosphate/hydroxyapatite scaffold, *Oral Surgery, Oral Medicine, Oral Pathology, Oral Radiology, and Endodontics*, 106(2): 203–209.

Tiedeman, J.J., Garvin, K.L., Kile, T.A., and Connolly, J.F. 1995, The role of a composite, demineralized bone matrix and bone marrow in the treatment of osseous defects, *Orthopedics*, 18(12): 1153–1158.

Trivedi, P. and Hematti, P. 2008, Derivation and immunological characterization of mesenchymal stromal cells from human embryonic stem cells, *Experimental Hematology*, 36(3): 350–359.

Vacanti, C.A., Bonassar, L.J., Vacanti, M.P., and Shufflebarger, J. 2001, Replacement of an avulsed phalanx with tissue-engineered bone, *The New England Journal of Medicine*, 344(20): 1511–1514.

Wakitani, S. 2007, Present status and perspective of articular cartilage regeneration, *Yakugaku Zasshi: Journal of the Pharmaceutical Society of Japan*, 127(5): 857–863.

Wakitani, S., Imoto, K., Yamamoto, T., Saito, M., Murata, N., and Yoneda, M. 2002, Human autologous culture expanded bone marrow mesenchymal cell transplantation for repair of cartilage defects in osteoarthritic knees, *Osteoarthritis and Cartilage/OARS, Osteoarthritis Research Society*, 10(3): 199–206.

Wakitani, S., Mitsuoka, T., Nakamura, N., Toritsuka, Y., Nakamura, Y., and Horibe, S. 2004, Autologous bone marrow stromal cell transplantation for repair of full-thickness articular cartilage defects in human patellae: Two case reports, *Cell Transplantation*, 13(5): 595–600.

Wakitani, S., Nawata, M., Tensho, K., Okabe, T., Machida, H., and Ohgushi, H. 2007, Repair of articular cartilage defects in the patello-femoral joint with autologous bone marrow mesenchymal cell transplantation: Three case reports involving nine defects in five knees, *Journal of Tissue Engineering and Regenerative Medicine*, 1(1): 74–79.

Woo, S.L. 2009, Tissue engineering: Use of scaffolds for ligament and tendon healing and regeneration, *Knee Surgery, Sports Traumatology, Arthroscopy: Official Journal of the ESSKA*, 17(6): 559–560.

Yoshikawa, T., Mitsuno, H., Nonaka, I., Sen, Y., Kawanishi, K., Inada, Y., Takakura, Y., Okuchi, K., and Nonomura, A. 2008, Wound therapy by marrow mesenchymal cell transplantation, *Plastic and Reconstructive Surgery*, 121(3): 860–877.

Yoshimura, K., Sato, K., Aoi, N., Kurita, M., Hirohi, T., and Harii, K. 2008a, Cell-assisted lipotransfer for cosmetic breast augmentation: Supportive use of adipose-derived stem/stromal cells, *Aesthetic Plastic Surgery*, 32(1): 48–55; discussion 56–57.

Yoshimura, K., Sato, K., Aoi, N., Kurita, M., Inoue, K., Suga, H., Eto, H., Kato, H., Hirohi, T., and Harii, K. 2008b, Cell-assisted lipotransfer for facial lipoatrophy: Efficacy of clinical use of adipose-derived stem cells, *Dermatological Surgery*, 34(9): 1178–1185.

Zuk, P.A., Zhu, M., Mizuno, H., Huang, J., Futrell, J.W., Katz, A.J., Benhaim, P., Lorenz, H.P., and Hedrick, M.H. 2001, Multilineage cells from human adipose tissue: Implications for cell-based therapies, *Tissue Engineering*, 7(2): 211–228.

24

Clinical Aspects of the Use of Stem Cells and Biomaterials for Bone Repair and Regeneration

Roger A. Brooks

CONTENTS

24.1 Introduction

In common with all other tissues in the body, bone is produced and maintained by stem cells and their progeny. These cells confer a limited capacity for self-repair on bone tissue as seen with fracture repair (McKibbin 1978) and the filling of small defects (Gosain et al. 2000) and can also be exploited clinically in limb-lengthening techniques using distraction osteogenesis (Aronson and Rock 1997). There are, however, many situations where help is required for successful bone repair, and this is most commonly provided by bone grafts. Bone grafting is a very common procedure; it was estimated that more than 1.5 million bone grafting procedures were carried out worldwide in 2002, and it is likely that this number has grown considerably (Hubble 2002). Bone graft is used to augment fracture healing, particularly in situations where self-repair may be compromised (Marino and Ziran 2010) and to fill a variety of contained and uncontained defects produced as a consequence of trauma or surgery, for example, to remove tumors, and these can involve substantial lengths of bone (Malizos et al. 2004). There are also a number of other very specific bone grafting techniques used to stimulate bone repair: impaction grafting, to support bone formation around replacement joints, and arthrodesis, used to relieve pain in damaged joints by producing a bone bridge across the joint. This is done particularly in the ankle and the spine with spinal fusion being a common surgical procedure (Deyo et al. 2005).

The principle of bone grafting is to provide an osteoconductive surface which will allow bone to grow from the graft recipient into and through the defect where bone repair is required. This is followed by a process of remodeling, leading to the appropriate distribution of new bone and the resorption of the graft. The major types of bone graft in use are autograft, fresh-frozen and freeze-dried allograft, and demineralized bone. There are, however, important differences between these graft materials that lead to differences in outcome, and these are, in part, a consequence of the cellular component of the graft (Schwarz et al. 1991, Graham et al. 2010). Autograft is generally accepted as the most effective bone repair material, but it has a limited supply and requires surgical harvest, producing comorbidity in the patient. Allograft bone is available in larger quantities, and the techniques of freezing and freeze drying allow storage prior to use; however, clinical outcome is more variable (McGarvey and Braly 1996). While freezing produces death and breakup of the cellular component of the graft, it does not fully remove the immunogenicity of the graft (Strong et al. 1996, VandeVord et al. 2005). In addition, the loss of cells removes the osteogenic component which is conferred on autograft by the presence of mesenchymal progenitor cells (MPCs) present on the bone surfaces. Demineralized bone matrix (DBM) contains no cells, and while this removes antigenicity and can provide osteoconductive and osteoinductive properties, there is no osteogenic component. Research and development of bone graft substitute materials has been extensive, and some of these materials have entered routine clinical practice (Bueno and Glowacki 2009). Typically, they provide a scaffold material with a surface that allows a good osteoconductive response with various additional benefits ranging from biodegradability, intrinsic osteoinductivity and the ability to be combined with osteoinductive growth factors. However, as with allograft and demineralized bone, there are no endogenous cells, and there is a lack of the osteogenic response they engender. Used alone, these materials rely on the cells present in the implant site and the surrounding bone and those brought in by invading blood vessels to enable a repair response. Some of these materials are able to elicit bone formation, although this is generally less than with autograft; however, recently, there has been evidence presented that certain ceramic materials are able to produce equivalent repair to that seen with autogeneic bone (Yuan et al. 2010).

In theory, it should be possible to confer the osteoinductive and osteogenic properties of autograft bone onto allograft, demineralized bone, and bone graft substitute materials by the addition of a cellular component. It is known that the principal cell involved in bone formation within the graft is the osteoblast, and these cells derive from cells of the mesenchymal lineage present on bone surfaces and within the bone marrow (BM) (Aubin 1998). There are a small number of long-lived multipotent cells present within the BM fitting the description of stem cells. Similar cells are also present in other tissues including the periosteum, umbilical cord, and adipose tissue (Liu et al. 2009). There are also many more cells within BM described as MPCs and that are capable of differentiating into osteoblasts. These cells can be harvested and have been investigated for their potential to augment bone grafts and bone graft substitutes.

This chapter describes what constitutes a stem cell in the context of bone repair and the various sources of both mesenchymal stem cells (MSCs) and more differentiated progenitor cells, and also presents the methods and controls necessary for the production of cells for clinical use. An overview of the biomaterials used as bone graft substitutes in combination with stem cells will be given. It will then concentrate on reviewing the clinical studies that have been published using BM and mesenchymal cells for bone repair, and finally, the potential for the future clinical use of stem cells will be assessed.

24.2 Stem Cells for Bone Repair

Any discussion of the use of stem cells and their progeny in bone repair necessitates a description of what is meant by a stem cell and clarification of terminology. Stem cells are, by definition, self-renewing cells that are able to give rise to other cells within the lineage while maintaining their stem cell characteristics (Alison and Islam 2009). Traditionally, it has been considered that this was due to asymmetric division of stem cells where the stem cell gave rise to a copy of itself and a daughter cell that was no longer a stem cell but produced more committed progenitor cells through rounds of proliferation which, following further expansion, resulted in the differentiated cells responsible for tissue function. However, recently, it has been shown, at least in the intestinal epithelial crypt, that there is a pool of stem cells that can divide to give two copies of themselves or two committed cells in a stochastic process (Lopez-Garcia et al. 2011) and this may be the general model.

24.2.1 Mesenchymal Cells

In bone, the stem cell that ultimately gives rise to the bone-forming osteoblast is the MSC. They are present in low numbers within the BM as part of the MPC population which is itself a subset of the wider group of marrow stromal cells which also contains support cells for the other cells within the BM. Marrow stromal cells can be separated from BM by plastic adherence and have often been called MSCs; they also have the same acronym, and this can lead to confusion. A good case has been made for the use of the term skeletal stem cell for the non-hematopoietic stem cell in BM (Tare et al. 2010); however, in this chapter, MSC will be used for the low-abundance stem cells, and MPC will be used for the multipotent mesenchymal cell population that is isolated and expanded in culture for tissue engineering applications, and the term bone marrow stromal cell (BMSC) will be used for the wider population. In this context, it is also important to clarify that clinical studies seeking to use cell augmentation for bone repair also use unfractionated BM which will contain both MSCs and BMSCs and concentrated BM where the relative proportion of stem and stromal cells will be increased (Hernigou et al. 2005).

24.2.2 Mesenchymal Stem Cell Characterization

There is no specific marker of the MSC in BM; however, the MPC population can be identified by the presence of certain cell surface markers and the absence of others (Tare et al. 2010). There has been controversy in the literature over the surface markers needed to delineate MPCs, and there is heterogeneity in MPC populations depending on the methods used for isolation and culture (Myers et al. 2010, Shenaq et al. 2010); however, a consensus has been drawn up by International Society for Cellular Therapy for minimum criteria to be used when characterizing these cells. These are that cells are adherent to plastic when in cell culture; that they do not express the hematopoietic markers CD11b, CD14, CD34, and CD45, the B-lymphocyte antigens CD19 and CD79a, and HLA-DR but do express the markers CD73, CD90, and CD105 (endoglin); and that they are capable of differentiating into cells of three major mesenchymal lineages: chondrocytes, adipocytes, and osteoblasts (Dominici et al. 2006). The melanoma adhesion molecule (CD146) is also considered to be an important marker (Sorrentino et al. 2008). Other reviews offer a more extensive list of markers that have been used to identify MPCs (Garcia-Gomez et al. 2010, Tare et al. 2010).

24.2.3 Multipotentiality

When plated on plastic at clonal density in culture, MPCs are able to give rise to characteristic colonies termed colony-forming units (CFU-F) (Owen and Friedenstein 1988). By altering the culture conditions, the MPCs in the colonies are able to differentiate into cells of the chondrogenic, adipogenic, and osteogenic lineages in vitro (Pittenger et al. 1999), and muscle cell differentiation has also been shown (Wakitani et al. 1995). This multipotentiality clearly arises from a common precursor; however, there is also evidence of some plasticity with the ability of trans-differentiation between osteogenic and adipogenic phenotype at least in vitro (Nuttall et al. 1998, Park et al. 1999). The value of MPCs for tissue engineering and regenerative medicine is enhanced by their ability to differentiate into a number of tissues where repair is required; this is true in the heart and in BM but is particularly so in the musculoskeletal system.

24.2.4 Stem Cell Niche

The identification of the MSC and specification of its location in BM have proved to be very difficult, in part, because of the lack of a specific MSC marker. However, there are a number of pointers as to its location and from these have developed the idea of the stem cell niche. The MSC niche was first proposed by Shi and Gronthos who used a panel of antibodies to screen BM MSCs and showed the presence of alpha-smooth muscle actin, CD146, and in some cells, expression of the pericyte-associated antigen 3G5 but not the endothelial marker von Willebrand factor (Shi and Gronthos 2003). This strongly suggested a perivascular location for MSCs in BM. Subcutaneous transplantation of CD146-positive CFU-F cells into immunodeficient nude mice showed that these cells were able to recapitulate the formation of an MSC niche taking part in the production of osteoblasts and a bone compartment within which sinusoids formed. The cells also gave rise to sinusoidal adventitial reticular cells. The production of these stromal cells preceded the establishment of hematopoiesis, and they also elaborated the protein angiopoietin-1, important for both angiogenesis and hematopoiesis.

24.2.5 MSCs in Other Tissues

A sub-endothelial pericyte location for the MSC may help to explain another interesting aspect of these adult stem cells. MSCs were originally identified in BM (Friedenstein et al. 1976); however, it is now clear that cells with similar characteristics reside in many tissues in the body. Adipose tissue, skeletal muscle, periosteal membrane, and umbilical cord have been particularly well studied because, alongside BM, they offer the potential to provide cells for cell therapy and tissue engineering applications. Evidence has been provided for the presence of MSC-like cells in most tissues in the body (Meirelles et al. 2006). This can be explained by these cells residing in the pericyte compartment alongside vascular capillaries (Crisan et al. 2008). Adipose tissue is potentially a very valuable source of MPCs for bone repair and regeneration. There is a relatively accessible source of this tissue in most individuals that could provide a supply of autologous cells. Since the first description of multilineage progenitor cells deriving from stem cells in adipose tissue (Zuk et al. 2001), there has been a large amount of research detailing the properties of these cells. They were shown to be present as a higher fraction of nucleated cells present than is seen in BM and share many of the properties of the BM cells (Zuk et al. 2002). They are considered to be derived from the perivascular population and have a particularly good angiogenic potential (Song et al. 2010). These factors indicate that they are promising candidates for musculoskeletal applications and specifically in the repair of bone. The periosteum is a site of new bone formation throughout life, producing increases in long bone diameter in addition to being the source of cells for the fracture callus, and long before the presence of multipotential MPCs was known, cells isolated from this source had been shown to have both chondrogenic and osteogenic potential (Nakahara et al. 1990). The demonstration that they could be grown at clonal density, expanded through many passages, and could be differentiated down the chondrogenic, osteogenic, adipogenic, and myogenic lineages both in vitro and in vivo indicated that they, too, have the potential to provide cells for augmentation of bone repair strategies (De Bari et al. 2006). A particularly interesting source of MPCs is the umbilical cord. The presence of hematopoietic stem cells in cord blood and their ability to engraft BM (Vormoor et al. 1994) led to the setting up of public and private cord blood banks where samples would be available for unrelated HLA-matched recipients and for autologous or family use, respectively (Navarrete and Contreras 2009). The demonstration of MPCs in cord blood (Erices et al. 2000) indicates that cells could also be banked for tissue regeneration requirements including bone repair (Seshareddy et al. 2008). MPC numbers in cord blood are low, but these cells are also present in cord perivascular regions and in Wharton's jelly at a higher cell fraction (Troyer and Weiss 2008). The phenotypes of these cells vary depending on the source (Secco et al. 2009), but they fulfill the criteria of MPCs. How cells from the different cord compartments perform clinically remains to be seen.

24.2.6 Allogeneic Mesenchymal Cells and Immune Privilege

A standard approach to the use of MPCs to augment bone grafts and bone graft substitute materials is to use autologous cells. This requires harvesting the tissue containing the MPCs, and although it may require a surgical procedure prior to the grafting operation, it has the benefits of no immunological response and no risks of viral transfer. A potential benefit of MPCs would be the ability to use allogeneic cells as these could be available off the shelf to surgeons for use in orthopedic procedures. This possibility is made more attractive by the immune status of MPCs; they have several characteristics supporting an immunoprivileged position and an immunosuppressive action. They have been shown to have a non-immunogenic phenotype with no expression of CD40, CD80, and CD86, and thus, there is a lack of the co-stimulatory signals required for T-lymphocyte activation via major histocompatibility complex Class I (Tse et al. 2003, Chamberlain et al. 2007). Lymphocytes exposed to MPCs become anergic and are not responsive to further challenges (Beyth et al. 2005). MPCs are also able to suppress T-lymphocyte proliferation responses to lymphocyte cognate antigens and dendritic cell function (Di Nicola et al. 2002, Krampera et al. 2003, Ramasamy et al. 2007). While these actions have been seen experimentally, evidence of a lack of immune response following the clinical use of allogeneic cells for bone repair is not available. There is also evidence that cells differentiated from MPCs in vivo are immunogenic (Huang et al. 2010). These results indicate caution in the clinical use of allogeneic MPCs for augmentation of bone repair.

24.3 Biomaterials for Tissue Engineering of Bone

The development of biomaterials for the repair of bone has had a long and successful history with a number of different materials being produced. By 2008, in the United States, there were 54 companies producing 103 products licensed for use by the FDA (Engelhardt 2008). These materials fall into several well-known groups, biologically derived materials including DBM (Zhang et al. 1997, Iwata et al. 2002), collagen (Geiger et al. 2003), and materials inspired by the mineral phase of bone including the bioactive ceramics, for example, hydroxyapatite (HA), β-tricalcium phosphate (βTCP) (Dorozhkin 2010), bioglass (Baino and Vitale-Brovarone 2011), calcium sulfate (Thomas and Puleo 2009), and various combinations of these (Du et al. 1999, Urban et al. 2007, Scotchford et al. 2011). The major factors distinguishing the different products available are the formulation and configuration of the product. For example, this may be in the form of blocks, granules, putty, or injectable cement (LeGeros et al. 2008, Low et al. 2010). These materials have variously been shown to have osteoconductive and osteoinductive properties and have, until recently, been used without the addition of cells. In many cases, the use of these materials leads to successful bone repair. The ability to augment and enhance the repair and regenerative properties of materials by the addition of an osteogenic component is obviously attractive, and these materials form the starting point for clinical evaluation of MPC augmentation. A much larger range of materials has shown promise for bone repair as potential carriers for stem cells and, in particular, MPCs. These include synthetic polymers such as the polyesters, for example, polycaprolactone and polylactide (Nair and Laurencin 2007); and natural molecules such as hyaluronic-based hydrogels, fibrin, and alginate; and composites based on these materials and in combination with more traditional ceramic materials to provide improved mechanical properties (Hutchens et al. 2006). The complex natural environment of stem cells has provided inspiration for the development of more complex materials mimicking extracellular matrix including self-assembled peptides and nanocomposites and modification of the surface chemistry of materials to include cell adhesion motifs (George and Ravindran 2010, Harvey et al. 2010, Mata et al. 2010, Paletta et al. 2010).

The surface of the material is clearly important in affecting the behavior of cells. Surface topography has been shown to be important for bone formation (Buser et al. 1991), and the ability to modify the surface at the nanoscale is now also known to be important (Sjostrom et al. 2009). The addition of cell adhesion peptides to the surface has been used to improve cell attachment, with the fibronectin sequence arginine–glycine–aspartic acid (RGD) being particularly well investigated (Comisar et al. 2007, Martino et al. 2009). The fabrication of complex structures has been made possible by the development of three-dimensional printing and stereolithography techniques (Tamimi et al. 2009, Kim et al. 2010) with surface structural modification using electrostatic atomization deposition (Li et al. 2010b). A further interesting fabrication technique to provide materials with micro- and nanoscale topography while altering the mechanical properties is electrospinning; potentially, these electrospun materials can mimic the fibrillar structure of the extracellular matrix (Yoshimoto et al. 2003, Nam et al. 2011).

An important factor in both the addition of stem cells to a biomaterial for implantation in bone and in the subsequent response of the bone to the implant is the porosity; this can be introduced and seen at the macro-, micro-, and nanoscales. Pore sizes of 300–400 μm diameter and interconnected pore size greater than 50 μm are considered suitable (Lu et al. 1999, Kuboki et al. 2001, Murphy et al. 2010). This degree of porosity is necessary for neovascularization of the material during bone formation, although the use of granules can also provide suitable spaces for new blood vessels. Microporosity has also been shown to be important in stimulating bone formation and this was thought to be primarily by altering surface topography (Hing et al. 2005). Biodegradability is another key material property. Many traditional materials are nondegradable or poorly degradable and become integrated into the surrounding bone; however, bone regeneration requires replacement of the material with bone tissue that will then be remodeled. The difficulty is producing a material that degrades at a rate to match or co-ordinate with the rate of new bone formation. Some ceramic materials are inherently soluble, with the degree of solubility being dependent on the chemistry; in general, these materials are not rapidly replaced by bone. Collagen-based materials will be broken down by matrix metalloproteinases at a rate dependent upon the degree of cross-linking (Yahyouche et al. 2011). Polymer materials can also be designed to include MMP-sensitive sites for controlled degradation (Lutolf et al. 2003). For several polymer materials, degradation occurs by hydrolysis, and for the polyesters, this will result in the generation of acidic degradation products (Yang et al. 2009). The effects of degradation on MPCs and on the surrounding bone should therefore be considered.

24.4 Clinical Use of Stem Cell Augmentation for Bone Repair

The goal of research and development work into MSCs and progenitors and the biomaterial scaffolds with which they can be used is to prepare for the use of stem-cell-augmented bone grafts and bone graft substitutes in the clinic. Clinical use is predicated on the need for new treatments and alternatives to existing treatments for a range of orthopedic challenges including fracture nonunion, repair of large segmental defects, and the ability to replace the use of autograft bone with its concomitant morbidity and improve on the outcomes seen when using allograft. The clinical use of stem-cell-augmented therapy is still in its early stages, and there is much to learn but the studies carried out to date, mainly involving single-case studies, and small clinical series provide information which should inform future work. With the increasing production of MPCs for clinical use and, in particular, their expansion in culture prior to reimplantation, there has developed a regulatory framework for their clinical use and their safe production.

24.4.1 Clinical Regulation

Regulation on the clinical use of MPCs for augmentation of bone repair clearly depends upon the national regulations applicable in the countries concerned. This chapter will briefly describe regulations in Europe and the United States. In Europe, MPCs are classified as advanced therapy medicinal products (ATMPs) under Directive 2001/83/EC and European regulation No. 1394/2007 which include gene therapy products, somatic cell therapy medicinal products and tissue-engineered products including where they are used to augment approved implant materials. Cells or tissues are considered engineered if they have undergone substantial manipulation relevant for regeneration, repair or replacement, and the cells are not intended to be used for the same functions in the recipient as in the donor. Manipulation does not, however, include cell separation, concentration or purification, and therefore, BM and concentrated BM prepared perioperatively are not covered. There is an exception made for custom-made (hospital) ATMPs under medical prescription for an individual patient. These regulations also comply with rules on human cell and tissue donation, procurement and testing under directive 2004/23/EC and on human blood and blood components (2002/98/EC) and emphasize the traceability of cells which potentially could be expanded and used in allogeneic procedures.

In the United States, the regulation of human cells, tissues, and cell and tissue-based products covers MPCs and is designed particularly to prevent the introduction, transmission and spread of communicable diseases and must comply with good tissue practice under the Code of Federal Regulations Title 21, Part 271 (21 CFR Part 1271). A biological product can only be used clinically under a biologics license, which requires evidence of safety and efficacy. During development, the sponsor can work under the investigational new drug regulations (21 CFR Part 312).

In all cases, new clinical procedures using MPCs will require authorization through Local Research Ethics Committees. Further details on the regulations affecting the clinical use of stem cells and their products can be found elsewhere (Sensebe et al. 2010).

24.4.2 Cell Harvesting and Production

The production of a suitable cell population for MPC augmentation is dependent on the source of the cells and degree of processing that the cells are going to undergo prior to implantation. Unseparated autologous BM simply requires the selection of a suitable source, usually the iliac crest, and obtaining a suitable volume; large volumes are over-diluted with blood, and low volumes produce variable yields of MPCs (Fennema et al. 2009). BM may also be concentrated to increase the relative proportion of MPCs to total cells. Density centrifugation is the most common method using the Smart Prep2, Harvest Technologies or the Cobe 2991 cell processor, Cobe Laboratories Inc. The separation of MPCs from BM, adipose tissue, or cord blood and their subsequent expansion in culture require the application of good manufacturing process (GMP) rules. Production follows a written specification detailing procurement, cell isolation and expansion methods, harvesting, and delivery or preservation. At each stage in the process, there is testing against defined standards. This is to ensure the safety and efficacy of the MPC preparation. Details of these processes are available (Bieback et al. 2010); however, two important issues are highlighted here. There is a perceived need for improvement to the culture medium used for the expansion of MPCs. Traditionally, fetal calf serum (FCS) has been used as an essential supplement, and it is available as a GMP grade suitable for clinical use; however, FCS potentially carries the risk of transfer of infectious or immunogenic proteins, and the presence of antibodies to FCS have been found following stem cell transplantation although these were considered clinically insignificant (Sundin et al. 2007). Pooled human platelet lysate and a combination of human serum with a cocktail of cytokines and other supplements both proved effective in achieving good expansion of MPCs and maintaining multipotentiality without FCS (Bieback et al. 2009, Pytlik et al. 2009). Concern has also been raised following reports of malignant transformation of both adipose-derived and BM MPCs in long-term culture (Rubio et al. 2005, Rosland et al. 2009). It is now clear that, in both cases, this was due to contamination with tumor cell lines (De la Fuente et al. 2010, Garcia et al. 2010, Torsvik et al. 2010) and it emphasizes the requirement for great stringency in the culture of MPCs to prevent contamination.

24.4.3 Cell Therapy of Fracture Nonunion

The first use of stem cells to augment bone grafting was reported in 1978 when autologous BM, isolated from the iliac crest, was used with deproteinized bovine bone to stimulate repair in a variety of cases of fracture non union, delayed union, trauma, and arthrodesis (Salama and Weissman 1978). In total, 26 patients were treated, and the results were reported as good on the basis of radiographs and clinical examination; however, the follow-up time was only 6 months. The injection of autologous BM and DBM into the fracture callus in cases of nonunion and, in a further study, into bone defect sites was reported (Connolly et al. 1991, Tiedeman et al. 1995). These studies reported union in 18 of 20 patients in the first study and, in the second study, healing in 30 of 39 when followed up at 19 months. The authors concluded that the use of BM and DBM was as effective as autograft without the disadvantages. A particularly important study for understanding the use of MPCs in orthopedic procedures investigated the use of concentrated BM injected directly into the fracture site in 60 cases of nonunion (Hernigou et al. 2005). BM was harvested from the iliac crest and concentrated using a Cobe 2991 cell processor. In each patient, the number of CFU-F was determined in the harvested marrow and in the concentrate, which showed a 4.2-fold increase. Bone formation was determined from CT images at 4 months and compared to preoperative scans. Fifty-three patients achieved union, and importantly, the CFU-F number was significantly lower in the seven patients who did not heal. There was also a correlation between the volume of mineralized callus formed and the number of progenitors injected. This study indicates that a minimum number of progenitors are required to stimulate bone repair. Numbers are variable between patients and there are reports of a decline in CFU-F with age (Muschler et al. 2001) although other studies have not replicated this (Oreffo et al. 1998, McCann et al. 2010) and have also reported a significant positive correlation between CFU-F number and body mass index in males.

24.4.4 First Augmentation of Bone Graft Substitute Materials

Reports describing the use of culture-expanded MPCs to augment bone repair were published in the *New England Journal of Medicine* in 2001. The first study used autologous culture-expanded MPCs seeded into porous HA blocks to treat three patients with 4–7 cm segmental defects of the tibia, ulnar, and humerus in which other treatments had failed (Quarto et al. 2001). The defects were stabilized with external fixators. The outcome was assessed using radiographs, CT, and angiography, and union of the graft with host bone was seen in all cases between 5 and 7 months. Follow-up of these cases between 6 and 7 years following surgery showed good implant integration and no cases of fracture (Marcacci et al. 2007). The second pioneering study replaced the phalanx of the thumb in a trauma patient with a construct consisting of porous coral HA seeded with culture-expanded periosteal cells in a calcium alginate gel (Vacanti et al. 2001). The thumb healed well after surgery, and radiographs and densitometry showed good bone formation, integration, and a gradual reduction in density to equal that of the right thumb. Good function was obtained as measured by pinch strength at 28 months.

24.4.5 Augmentation of Maxillary Sinus Elevation

Maxillary sinus floor elevation using autograft bone is a common procedure to create sufficient bone volume for dental implantation. In attempts to overcome donor site morbidity, bone graft substitutes have been used but suffer from prolonged healing times. This procedure is thus ideal for the use of cell augmentation, and a number of clinical studies have been conducted. A preliminary study in two patients using cells released by digestion of periosteum from the mandible combined with degradable polymer fleece composed of poly(lactic-glycolic acid)–poly(ethylene glycol) copolymer and polydioxanone (Ethisorb) showed good graft stability on x-rays, and biopsies showed good trabecular bone formation (Schmelzeisen et al. 2003). This was followed by a more extensive study comparing 35 patients given the periosteal cell-polymer construct and 41 controls grafted with autologous bone (Voss et al. 2010). Although x-rays at 2 months showed a tight bone-implant interface in all cases, there were 11 instances of complications and subsequent dental implant losses in the cell group compared to 1 in the controls, and the authors concluded that autologous bone was more reliable. A number of other papers have reported cases and case series for maxillary sinus floor elevation using culture-expanded MPCs and porous HA or biphasic HA/βTCP. Radiographs in these studies showed good sinus elevation and osseointegration, but no control groups were included (Yamada et al. 2004, 2008, Shayesteh et al. 2008). Recently, two studies compared cell-augmented grafts for sinus augmentation with controls. One study compared allograft supplemented with culture-expanded MPCs to allograft alone (Gonshor et al. 2011) in 21 patients. Histology and histomorphometry on biopsies taken at an average follow-up time of 3.7 months showed significantly more bone and less residual graft material in the grafts supplemented with progenitors. In the second study, cells were prepared by BM concentration using a SmartPrep centrifuge (Rickert et al. 2010). Twelve patients had BioOss HA combined with cells on one side and autograft on the contralateral side. Biopsies were taken at an average of 14.8 weeks. Significantly more bone was seen in the cell-treated group; however, while resorption of autograft was evident, no resorption was seen in the HA group. These studies clearly indicate the potential benefits of cell augmentation in this procedure.

24.4.6 Stem Cell Augmentation in a Range of Orthopedic Procedures

There have been a number of clinical studies described in the past 8 years in which concentrated BM and both adipose-derived and BM-derived culture-expanded MPCs have been used in a variety of procedures to stimulate bone formation. These usually involved a small number of cases and were essentially proof of principle studies in which the response to the cell augmentation is described. A preliminary study of the value of BM-expanded MPCs in distraction osteogenesis was conducted in three patients, two with achondroplasia and one with congenital pseudarthrosis, and showed normal time to achieve the target length with no complications (Kitoh et al. 2004). The authors then continued to a larger study in 46 patients undergoing distraction osteogenesis for limb lengthening for short stature or because limbs were of an unequal length (Kitoh et al. 2007). Seventeen patients were treated with one or two injections of MPCs in a platelet gel into the osteotomy site, one at the start of distraction and the second at the start of consolidation. Twenty nine patients had no injections as a control group. Delays in consolidation were only seen in the control, and the number of complications was significantly less in the treated group. Bone defects as a result of tumor removal or trauma have been targeted. Three patients had defects filled with MPCs culture-expanded for 4 weeks, combined with porous HA blocks and granules (Morishita et al. 2006). At 3 months, radiographs and CT showed good graft incorporation, and although there were no controls, it was suggested that the results were better than those reported in the literature using HA alone. Adipose-derived MPCs were used to treat a large calvarial defect in a 7-year-old 2 years after the injury; they were placed in the defect in fibrin glue, and CT showed near-complete repair at 3 months (Lendeckel et al. 2004). One of the least positive of all the reports of clinical studies of MPC augmentation used iliac-crest-expanded MPCs in porous HA (ProOsteon) to fill various defects in the mandible in six patients (Meijer et al. 2008). Only three of six showed good bone formation in biopsies taken at 4 months, and in two cases, bone was only present close to the implant interface. A comparative study looked at the ability to repair bone defects with HA Orthoss granules or a porcine collagen sponge, both augmented with BM concentrate prepared using a SmartPrep 2 (Jager et al. 2011). The number of CFU-F was increased from $2.8/cm^2$ to $4.1/cm^2$ by concentration. Twelve patients were assigned to collagen and 27 to HA. Thirty-six of the 39 patients showed good bone healing on x-rays, but the healing was more rapid with HA (17.3 weeks) compared to collagen (22.4 weeks). An interesting study was carried out by Ohgushi et al., who isolated and cultured MPCs from iliac crest BM on the porous surfaces of ankle prostheses for 2 weeks (Ohgushi et al. 2005). Joint replacement was carried out in three patients using this cell-augmented prosthesis, and at 2 year follow-up, there had been no complications, and x-rays showed radio-dense regions at the interface where the cells had been seeded.

Spinal arthrodesis is a procedure where autograft bone is often used. As an alternative, cell augmentation was used in a study of 41 patients undergoing lumbar or thoracic spinal fusion for degenerative disc disease or thoracolumbar fracture (Gan et al. 2008). BM concentrate produced using a Cobe 2991 cell processor showed an average 4.3-fold increase in CFU-F and was combined with macroporous βTCP granules. At a mean follow-up of 34.5 months, CT images showed good fusion in 95.1% of cases; there were, however, no controls.

In impaction grafting, allograft bone granules are used. In a study of two patients, morselized allograft was combined with autologous BM and used for impaction grafting around a hip prosthesis (Tilley et al. 2006). After the impaction, a sample of the graft bed was taken and processed for histology. Staining showed that there was good adherence of cells to the allograft and that they had survived the impaction process. The patients showed good recovery with initial graft incorporation and no complications.

In the clinical studies described in the preceding sections, all the procedures have used autologous cells. This is unsurprising as the regulations surrounding the production and testing of allogeneic MPCs for human use are more stringent. They are, however, being developed in a number of centers, and there is one clinical study reporting their use (Rush et al. 2009). The MPCs were seeded in a demineralized bone allograft (Trinity Evolution) and used in 23 patients for nonunion, mal-union and segmental bone loss in the foot and ankle. There were no reported complications to the use of this material, and there was new bone formation and union observed in 91.3% of cases.

24.4.7 Large Grafts and the Importance of Angiogenesis

Bone is a highly vascular tissue, and one of the requirements for bone repair and successful incorporation of bone grafts and bone graft substitutes is stimulation of neovascularization (Fang et al. 2005, Kanczler and Oreffo 2008). The perivascular location of MSCs suggests their close links to the vasculature. MPCs may contribute to vascularization in two ways. First, there is some evidence that they can differentiate into an endothelial cell phenotype and could contribute to blood vessel formation (Zhang et al. 2010). It is more likely, however, that these cells are able to stimulate angiogenesis by the release of paracrine factors. MPCs from BM, when placed in a relatively hypoxic environment, as would be found in a grafted bone defect, increase expression of both mRNA and protein for the potent angiogenic cytokine vascular endothelial growth factor (VEGF) (Potier et al. 2007). Adipose-derived MPCs also upregulate VEGF and the angiogenic cytokine basic fibroblast growth factor in response to hypoxia (Lee et al. 2009a, He et al. 2010). The production of these locally acting cytokines may help to stimulate neovascularization of the graft. There are situations where the extent of bone damage is so large that there would be insufficient vascularization to support bone formation throughout a graft. Repair of these large defects can be achieved with vascularized autografts such as the vascularized fibula (Innocenti et al. 2009). MPCs have been used in a novel strategy to prepare vascularized tissue-engineered constructs for large defect repair. This procedure was first used to prepare a vascularized graft for the repair of a defect of the mandible greater than 7 cm long created following removal of a tumor (Warnke et al. 2004). In a two-stage procedure, autologous BM was harvested from the iliac crest and added to HA BioOss blocks and collagen containing bone morphogenic protein-7 in a titanium mesh cage designed to fit the defect. The construct was placed in the latissimus dorsi muscle for 7 weeks to develop a vascular supply. The vascularized graft was then carefully removed and grafted into the jaw. After 11 days, skeletal scintigraphy showed evidence of new bone formation throughout the graft and integration. Long-term follow-up was not possible as the patient died of unrelated factors. A second example was recently reported using a very similar procedure for bone repair after hemimaxillectomy following removal of a large keratocyst (Mesimaki et al. 2009). In this case, autologous culture-expanded MPCs and βTCP granules were used. Vascularization was carried out in the rectus abdominis muscle for 8 months, after which it was implanted into the maxilla. CT and x-rays showed good bony integration at follow-up times to 12 months.

24.4.8 Clinical Trials

The studies reported, to date, on the clinical use of BM and MPCs in augmentation of bone repair cover a range of study design. There are a number that could be considered as phase 1 trials where evidence was obtained for the safety of the procedure in a small number of patients. Two groups went on from the initial pilot study to recruit further patients in an expanded series. Only three studies have compared stem-cell-augmented treatment with a control group using a standard, currently used treatment as would be seen in a phase 3 trial (Kitoh et al. 2007, Rickert et al. 2010, Voss et al. 2010), and even here, the numbers were small. A further study compared two graft substitute materials augmented with concentrated BM (Jager et al. 2011). There is clearly a need for larger phase 3 trials following on from safety and efficacy studies to adequately assess the benefits of stem cell augmentation using a randomized control design. A review of trials currently recorded at www.clinicaltrials.gov shows only one phase 3 randomized control trial currently planned. This is to compare Trinity allogeneic MPCs in allograft with DBM for the repair of bone defects following the removal of benign tumors. Clinical trials are, of course, expensive, and there are ethical issues when randomizing patients to treatments that may have less than ideal outcomes. However, suitable standard treatments are available as controls for most orthopedic conditions requiring bone repair using autograft, allograft, bone graft substitutes or the bone morphogenic protein products InFuse and OP-1. In part, this could be responsible for the slow progress toward full clinical trials; when acceptable treatments are available, there is difficulty in convincing sponsors to fund large trials which may, at best, show only small incremental benefits. The potential benefits of stem cell augmentation are, however, important and may include an increased rate of bone repair, the promotion of a suitable time course of implant degradation and bone remodeling and a decrease in the number of patients with a poor clinical outcome, for example, failure to achieve union in cases of poor fracture healing. Hopefully, progress will be made toward conducting trials, particularly in cases of fracture nonunion, large segmental defects, and spinal fusion where outcomes can be poor.

24.5 Summary and Future

The use of stem and progenitor cells to augment bone graft and bone graft substitute materials is still a relatively new therapeutic option, and much remains to be learned to determine in what situations they can be best applied, how treatments can be improved and what mechanisms underlie the bone repair response. There is a huge and rapidly growing literature detailing the cell biology of MSCs and their progeny from in vitro experiments, and this provides important insights into the potential of stem cells and how their behavior is modified by interaction with new and existing biomaterials, controlled by cytokines and affected by changes to the physical environment, for example, mechanical factors or changes in oxygen tension. Their interactions with other cells can also be studied. This work will continue to inform our understanding of how the MPCs behave when used clinically for bone repair. The most commonly used MPCs are obtained from BM and harvested from the iliac crest. Adipose-derived and periosteal MPCs also show potential. In the future, MPCs from cord blood may be used either as a tissue-matched allogeneic population or as autologous cells from individual banked samples. There are a variety of different protocols used for the isolation and expansion of MPC populations. The production of standardized protocols is an option although this is premature as it is not yet clear which procedures work optimally in terms of producing the best yield of osteoprogenitors from different sources. There is, however, a requirement for improved methods for the characterization of MSCs and their differentiated progeny which will allow a better understanding of the specific characteristics of populations that produce the best outcome, both experimentally and clinically.

A population of circulating MPCs have been identified in adult blood (Kuznetsov et al. 2001); however, their low number (Kuznetsov et al. 2007) has not made them suitable candidates for use in bone repair strategies. The identification of the perivascular location of MSCs in BM and other tissue has, however, led to the suggestion that these cells are ideally placed to respond to circulating cytokines and chemokines and relocate to sites of injury where they could participate in the repair process (Fox et al. 2007). The homing of MPCs has since been demonstrated in animal models where injected cells preferentially locate within damaged tissue (Kumagai et al. 2008, Lee et al. 2009b). However, whether this facility can be used clinically in bone repair therapies remains to be seen.

Along with improved understanding of mesenchymal stem and progenitor cells, there is a need for improvements to the biomaterials used. Traditional, allograft, DBM and bioceramics have shown great utility in many applications from bone defect filling to impaction grafting. They have also been shown to be acceptable substrates for the addition of BM and MPCs. Improvements to allow control of biodegradability during bone formation is important as many of the materials in current use resorb poorly or not at all and other materials are too soluble or can produce inflammatory responses. In addition, currently used materials do not provide mechanical strength necessary for load bearing applications. The challenge of producing an osteoconductive or osteoinductive, mechanically strong material which shows predictable degradation in clinical situations is not going to be easy.

A large number of preclinical animal studies have been conducted to evaluate the benefits of MPC augmentation on the repair of bone defects. Unlike the reported clinical studies, these invariably follow a randomized control design. A review of these models in a variety of species and skeletal site concluded that, in all cases, scaffolds seeded with MPCs showed improvements compared to unseeded controls (Cancedda et al. 2007). This strongly suggests a sound basis for their clinical use. Care must, however, be taken in extrapolating these results; there may be publication bias and there are a number of other potentially confounding factors to consider. Patients are often elderly, and aging is known to impair bone repair, at least, in respect to fracture repair (Mehta et al. 2010), and the animals used are usually young and do not have concomitant disease, for example, osteoporosis. In addition, animal studies are generally designed to control variability and are therefore likely to overestimate clinical outcome (Muschler et al. 2010), leading to difficulties in translation. In spite of these caveats, animal implantation will continue to provide important evidence of the potential for new strategies of stem cell augmentation.

Experimental studies have been conducted to investigate genetically modified MPCs, in particular, with members of the bone morphogenetic protein family and with vascular endothelial growth factor. These have demonstrated enhanced bone formation and angiogenesis in a number of small animal models. In the future, this approach may be further developed so that it can be used clinically to enhance bone formation; however, this is dependent on the development of safe nonviral vectors and prevention of insertional mutagenesis and uncontrolled gene expression.

Clinical studies have so far been encouraging. While most do not meet the standards needed for demonstration of improvement over existing therapies, they have shown that the use of MPC augmentation is safe and produces acceptable outcomes. This has been shown across a range of orthopedic conditions and using BM, BM concentrate, and culture-expanded MPCs isolated from periosteum, BM, and adipose tissue. Preparations from these different sources have not been compared clinically; however, there have been laboratory investigations showing differences in MPC yield, gene expression, and differentiation potential of MPCs isolated from different sources (Wagner et al. 2005, Rebelatto et al. 2008), indicating that there may be differences in their utility for different clinical applications. An important question is to what extent the MPCs contribute directly to bone formation within the implant site and how much of the repair process is due to paracrine effects. MPCs are known to produce a large range of factors capable of modifying the behavior of other cells (Caplan and Dennis 2006). The potential benefits of autologous MPCs to promote bone repair are undoubted, and clinical augmentation strategies using these cells are going to be used increasingly in the future. There is also continuing hope that allogeneic MPCs will prove useful. The immunological characteristics of these cells are in accordance with this. However, evidence of the absence of immunological responses when these cells are used clinically to augment bone grafts and bone graft substitutes in addition to their efficacy in stimulating bone repair is required before their promise can be realized.

Finally, a review on stem cells would not be complete without mentioning the potential for pluripotent cells for bone repair. The seminal work of Evans and Kaufman, leading to production of embryonic stem cells (Evans and Kaufman 1981) and the isolation of equivalent cells from human tissue (Thomson et al. 1998), raised the hope of unlimited cells for tissue repair. These cells are pluripotent and capable of elaborating all extraembryonic tissues including bone; however, they suffer from ethical and immunological problems. The more recent development of induced pluripotent stem cells (iPS) has renewed the optimism that eventually these cells will be available for tissue repair (Takahashi et al. 2007). These cells can be prepared from adult differentiated cells by the introduction of a small number of genetic modifications resulting in pluripotent stem cells which are immunologically compatible with the donor. Mouse iPS have been shown to differentiate into cells displaying the characteristics of osteoblasts, both in vitro and in vivo (Li et al. 2010a, Bilousova et al. 2011). However, these cells currently suffer from a number of drawbacks that would prevent their use clinically (Teo and Vallier 2010). Gene insertion by viral gene transfer is required for efficient reprogramming, increasing the risk of insertional mutagenesis, and c-myc overexpression may increase the risk of tumor formation. Differentiated cells are required for tissue engineering applications, and there is currently difficulty in ensuring the production of a fully differentiated population of cells from iPS cultures. This is critically important as undifferentiated cells are associated with the development of teratomas in vivo. These drawbacks and the variability of each iPS cell line currently make the delivery of patient-specific therapy using these cells impossible and ensure that, for the foreseeable future, mesenchymal stem and progenitor cells will continue to be the most suitable therapeutic option.

Acknowledgment

The author wishes to acknowledge funding from the National Institute for Health Research.

References

Alison, M. R. and Islam, S. 2009. Attributes of adult stem cells. *Journal of Pathology* 217: 144–160.

Aronson, J. and Rock, L. 1997. Limb-lengthening, skeletal reconstruction, and bone transport with the Ilizarov method. *Journal of Bone and Joint Surgery—American Volume* 79A: 1243–1258.

Aubin, J. E. 1998. Advances in the osteoblast lineage. *Biochemistry and Cell Biology—Biochimie Et Biologie Cellulaire* 76: 899–910.

Baino, F. and Vitale-Brovarone, C. 2011. Three-dimensional glass-derived scaffolds for bone tissue engineering: Current trends and forecasts for the future. *Journal of Biomedical Materials Research Part A* 97A: 514–535.

Beyth, S., Borovsky, Z., Mevorach, D. et al. 2005. Human mesenchymal stem cells alter antigen-presenting cell maturation and induce T-cell unresponsiveness. *Blood* 105: 2214–2219.

Bieback, K., Hecker, A., Kocaomer, A. et al. 2009. Human alternatives to fetal bovine serum for the expansion of mesenchymal stromal cells from bone marrow. *Stem Cells* 27: 2331–2341.

Bieback, K., Kinzebach, S., and Karagianni, M. 2010. Translating research into clinical scale manufacturing of mesenchymal stromal cells. *Stem Cells International.* DOI:10.4061/2010/193519.

Bilousova, G., Jun, D. H., King, K. B. et al. 2011. Osteoblasts derived from induced pluripotent stem cells form calcified structures in scaffolds both in vitro and in vivo. *Stem Cells* 29: 206–216.

Bueno, E. M. and Glowacki, J. 2009. Cell-free and cell-based approaches for bone regeneration. *Nature Reviews Rheumatology* 5: 685–697.

Buser, D., Schenk, R. K., Steinemann, S. et al. 1991. Influence of surface characteristics on bone integration of titanium implants—A histomorphometric study in miniature pigs. *Journal of Biomedical Materials Research* 25: 889–902.

Cancedda, R., Giannoni, P., and Mastrogiacomo, M. 2007. A tissue engineering approach to bone repair in large animal models and in clinical practice. *Biomaterials* 28: 4240–4250.

Caplan, A. I. and Dennis, J. E. 2006. Mesenchymal stem cells as trophic mediators. *Journal of Cellular Biochemistry* 98: 1076–1084.

Chamberlain, G., Fox, J., Ashton, B., and Middleton, J. 2007. Concise review: Mesenchymal stem cells: Their phenotype, differentiation capacity, immunological features, and potential for homing. *Stem Cells* 25: 2739–2749.

Comisar, W. A., Kazmers, N. H., Mooney, D. J., and Linderman, J. J. 2007. Engineering RGD nanopatterned hydrogels to control preosteoblast behavior: A combined computational and experimental approach. *Biomaterials* 28: 4409–4417.

Connolly, J. F., Guse, R., Tiedeman, J., and Dehne, R. 1991. Autologous marrow injection as a substitute for operative grafting of tibial nonunions. *Clinical Orthopaedics and Related Research* 266: 259–270.

Crisan, M., Yap, S., Casteilla, L. et al. 2008. A perivascular origin for mesenchymal stem cells in multiple human organs. *Cell Stem Cell* 3: 301–313.

De Bari, C., Dell'accio, F., Vanlauwe, J. et al. 2006. Mesenchymal multipotency of adult human periosteal cells demonstrated by single-cell lineage analysis. *Arthritis and Rheumatism* 54: 1209–1221.

De La Fuente, R., Bernad, A., Garcia-Castro, J., Martin, M. C., and Cigudosa, J. C. 2010. Spontaneous human adult stem cell transformation (retraction of vol. 65, p. 3035, 2005). *Cancer Research* 70: 6682.

Deyo, R. A., Gray, D. T., Kreuter, W., Mirza, S., and Martin, B. I. 2005. United States trends in lumbar fusion surgery for degenerative conditions. *Spine* 30: 1441–1445.

Di Nicola, M., Carlo-Stella, C., Magni, M. et al. 2002. Human bone marrow stromal cells suppress T-lymphocyte proliferation induced by cellular or nonspecific mitogenic stimuli. *Blood* 99: 3838–3843.

Dominici, M., Le Blanc, K., Mueller, I. et al. 2006. Minimal criteria for defining multipotent mesenchymal stromal cells. The International Society for Cellular Therapy position statement. *Cytotherapy* 8: 315–317.

Dorozhkin, S. V. 2010. Bioceramics of calcium orthophosphates. *Biomaterials* 31: 1465–1485.

Du, C., Cui, F. Z., Zhu, X. D., and De Groot, K. 1999. Three-dimensional nano-HAP/collagen matrix loading with osteogenic cells in organ culture. *Journal of Biomedical Materials Research* 44: 407–415.

Engelhardt, S. A. 2008. Bone graft materials in orthopaedics: 50+ companies vie for a piece of the $1.9bb pie. *Orthopaedic Product News*, May/June: Available online at http://www.orthoworld.com/site/docs/opn/us_opn-2008-06.pdf (Accessed 2011).

Erices, A., Conget, P., and Minguell, J. J. 2000. Mesenchymal progenitor cells in human umbilical cord blood. *British Journal of Haematology* 109: 235–242.

Evans, M. J. and Kaufman, M. H. 1981. Establishment in culture of pluripotential cells from mouse embryos. *Nature* 292: 154–156.

Fang, T. D., Salim, A., Xia, W. et al. 2005. Angiogenesis is required for successful bone induction during distraction osteogenesis. *Journal of Bone and Mineral Research* 20: 1114–1124.

Fennema, E. M., Renard, A. J. S., Leusink, A., Van Blitterswijk, C. A., and De Boer, J. 2009. The effect of bone marrow aspiration strategy on the yield and quality of human mesenchymal stem cells. *Acta Orthopaedica* 80: 618–621.

Fox, J. M., Chamberlain, G., Ashton, B. A., and Middleton, J. 2007. Recent advances into the understanding of mesenchymal stem cell trafficking. *British Journal of Haematology* 137: 491–502.

Friedenstein, A. J., Gorskaja, U. F., and Kulagina, N. N. 1976. Fibroblast precursors in normal and irradiated mouse hematopoietic organs. *Experimental Hematology* 4: 267–274.

Gan, Y. K., Dai, K. R., Zhang, P. et al. 2008. The clinical use of enriched bone marrow stem cells combined with porous beta-tricalcium phosphate in posterior spinal fusion. *Biomaterials* 29: 3973–3982.

Garcia, S., Bernad, A., Martin, M. C. et al. 2010. Pitfalls in spontaneous in vitro transformation of human mesenchymal stem cells. *Experimental Cell Research* 316: 1648–1650.

Garcia-Gomez, I., Elvira, G., Zapata, A. G. et al. 2010. Mesenchymal stem cells: Biological properties and clinical applications. *Expert Opinion on Biological Therapy* 10: 1453–1468.

Geiger, M., Li, R. H., and Friess, W. 2003. Collagen sponges for bone regeneration with rhBMP-2. *Advanced Drug Delivery Reviews* 55: 1613–1629.

George, A. and Ravindran, S. 2010. Protein templates in hard tissue engineering. *Nano Today* 5: 254–266.

Gonshor, A., Mcallister, B. S., Wallace, S. S., and Prasad, H. 2011. Histologic and histomorphometric evaluation of an allograft stem cell-based matrix sinus augmentation procedure. *International Journal of Oral & Maxillofacial Implants* 26: 123131.

Gosain, A. K., Song, L. S., Yu, P. R. et al. 2000. Osteogenesis in cranial defects: Reassessment of the concept of critical size and the expression of TGF-beta isoforms. *Plastic and Reconstructive Surgery* 106: 360–371.

Graham, S. M., Leonidou, A., Aslam-Pervez, N. et al. 2010. Biological therapy of bone defects: The immunology of bone allo-transplantation. *Expert Opinion on Biological Therapy* 10: 885–901.

Harvey, E. J., Henderson, J. E., and Vengallatore, S. T. 2010. Nanotechnology and bone healing. *Journal of Orthopaedic Trauma* 24: S25–S30.

He, J. W., Genetos, D. C., Yellowley, C. E., and Leach, J. K. 2010. Oxygen tension differentially influences osteogenic differentiation of human adipose stem cells in 2D and 3D cultures. *Journal of Cellular Biochemistry* 110: 87–96.

Hernigou, P., Poignard, A., Beaujean, F., and Rouard, H. 2005. Percutaneous autologous bone-marrow grafting for nonunions—Influence of the number and concentration of progenitor cells. *Journal of Bone and Joint Surgery—American Volume* 87A: 1430–1437.

Hing, K. A., Annaz, B., Saeed, S., Revell, P. A., and Buckland, T. 2005. Microporosity enhances bioactivity of synthetic bone graft substitutes. *Journal of Materials Science—Materials in Medicine* 16: 467–475.

Huang, X. P., Sun, Z., Miyagi, Y. et al. 2010. Differentiation of allogeneic mesenchymal stem cells induces immunogenicity and limits their long-term benefits for myocardial repair. *Circulation* 122: 2419–2429.

Hubble, M. J. W. 2002. Bone grafts. *Surgical Technology International* 10: 261–265.

Hutchens, S. A., Benson, R. S., Evans, B. R., O'Neill, H. M., and Rawn, C. J. 2006. Biomimetic synthesis of calcium-deficient hydroxyapatite in a natural hydrogel. *Biomaterials* 27: 4661–4670.

Innocenti, M., Abed, Y. Y., Beltrami, G. et al. 2009. Biological reconstruction after resection of bone tumors of the proximal tibia using allograft shell and intramedullary free vascularized fibular graft: Long-term results. *Microsurgery* 29: 361–372.

Iwata, H., Sakano, S., Itoh, T., and Bauer, T. W. 2002. Demineralized bone matrix and native bone morphogenetic protein in orthopaedic surgery. *Clinical Orthopaedics and Related Research* 395: 99–109.

Jager, M., Herten, M., Fochtmann, U. et al. 2011. Bridging the gap: Bone marrow aspiration concentrate reduces autologous bone grafting in osseous defects. *Journal of Orthopaedic Research* 29: 173–180.

Kanczler, J. M. and Oreffo, R. O. C. 2008. Osteogenesis and angiogenesis: The potential for engineering bone. *European Cells & Materials* 15: 100–114.

Kim, K., Yeatts, A., Dean, D., and Fisher, J. P. 2010. Stereolithographic bone scaffold design parameters: Osteogenic differentiation and signal expression. *Tissue Engineering Part B—Reviews* 16: 523–539.

Kitoh, H., Kitakoji, T., Tsuchiya, H. et al. 2004. Transplantation of marrow-derived mesenchymal stem cells and platelet-rich plasma during distraction osteogenesis—A preliminary result of three cases. *Bone* 35: 892–898.

Kitoh, H., Kitakoji, T., Tsuchlya, H., Katoh, M., and Ishiguro, N. 2007. Transplantation of culture expanded bone marrow cells and platelet rich plasma in distraction osteogenesis of the long bones. *Bone* 40: 522–528.

Krampera, M., Glennie, S., Dyson, J. et al. 2003. Bone marrow mesenchymal stem cells inhibit the response of naive and memory antigen-specific T cells to their cognate peptide. *Blood* 101: 3722–3729.

Kuboki, Y., Jin, Q. M., and Takita, H. 2001. Geometry of carriers controlling phenotypic expression in BMP-induced osteogenesis and chondrogenesis. *Journal of Bone and Joint Surgery—American Volume* 83A: S105–S115.

Kumagai, K., Vasanji, A., Drazba, J. A., Butler, R. S., and Muschler, G. F. 2008. Circulating cells with osteogenic potential are physiologically mobilized into the fracture healing site in the parabiotic mice model. *Journal of Orthopaedic Research* 26: 165–175.

Kuznetsov, S. A., Mankani, M. H., Gronthos, S. et al. 2001. Circulating skeletal stem cells. *Journal of Cell Biology* 153: 1133–1139.

Kuznetsov, S. A., Mankani, M. H., Leet, A. I. et al. 2007. Circulating connective tissue precursors: Extreme rarity in humans and chondrogenic potential in guinea pigs. *Stem Cells* 25: 1830–1839.

Lee, S. W., Padmanabhan, P., Ray, P. et al. 2009b. Stem cell-mediated accelerated bone healing observed with in vivo molecular and small animal imaging technologies in a model of skeletal injury. *Journal of Orthopaedic Research* 27: 295–302.

Lee, E. Y., Xia, Y., Kim, W. S. et al. 2009a. Hypoxia-enhanced wound-healing function of adipose-derived stem cells: Increase in stem cell proliferation and up-regulation of VEGF and bFGF. *Wound Repair and Regeneration* 17: 540–547.

Legeros, R. Z., Daculsi, G., and Legeros, J. P. 2008. Bioactive bioceramics. In *Orthopedic Biology and Medicine: Musculoskeletal Tissue Regeneration: Biological Materials and Method*, W. S. Pitrzak, ed., pp. 153–181. Totowa, NJ: Humana Press Inc.

Lendeckel, S., Jodicke, A., Christophis, P. et al. 2004. Autologous stem cells (adipose) and fibrin glue used to treat widespread traumatic calvarial defects: Case report. *Journal of Cranio-Maxillofacial Surgery* 32: 370–373.

Li, F., Bronson, S., and Niyibizi, C. 2010a. Derivation of murine induced pluripotent stem cells (IPS) and assessment of their differentiation toward osteogenic lineage. *Journal of Cellular Biochemistry* 109: 643–652.

Li, X., Koller, G., Huang, J. et al. 2010b. A novel jet-based nano-hydroxyapatite patterning technique for osteoblast guidance. *Journal of the Royal Society Interface* 7: 189–197.

Liu, Z. J., Zhuge, Y., and Velazquez, O. C. 2009. Trafficking and differentiation of mesenchymal stem cells. *Journal of Cellular Biochemistry* 106: 984–991.

Lopez-Garcia, C., Klein, A. M., Simons, B. D., and Winton, D. J. 2011. Intestinal stem cell replacement follows a pattern of neutral drift. *Science* 330: 822–825.

Low, K. L., Tan, S. H., Zein, S. H. S. et al. 2010. Calcium phosphate-based composites as injectable bone substitute materials. *Journal of Biomedical Materials Research Part B—Applied Biomaterials* 94B: 273–286.

Lu, J. X., Flautre, B., Anselme, K. et al. 1999. Role of interconnections in porous bioceramics on bone recolonization in vitro and in vivo. *Journal of Materials Science—Materials in Medicine* 10: 111–120.

Lutolf, M. P., Lauer-Fields, J. L., Schmoekel, H. G. et al. 2003. Synthetic matrix metalloproteinase-sensitive hydrogels for the conduction of tissue regeneration: Engineering cell-invasion characteristics. *Proceedings of the National Academy of Sciences of the United States of America* 100: 5413–5418.

Malizos, K. N., Zalavras, C. G., Soucacos, P. N., Beris, A. E., and Urbaniak, J. R. 2004. Free vascularized fibular grafts for reconstruction of skeletal defects. *Journal of the American Academy of Orthopaedic Surgeons* 12: 360–369.

Marcacci, M., Kon, E., Moukhachev, V. et al. 2007. Stem cells associated with macroporous bioceramics for long bone repair: 6-to 7-year outcome of a pilot clinical study. *Tissue Engineering* 13: 947–955.

Marino, J. T. and Ziran, B. H. 2010. Use of solid and cancellous autologous bone graft for fractures and nonunions. *Orthopedic Clinics of North America* 41: 15–26.

Martino, M. M., Mochizuki, M., Rothenfluh, D. A. et al. 2009. Controlling integrin specificity and stem cell differentiation in 2D and 3D environments through regulation of fibronectin domain stability. *Biomaterials* 30: 1089–1097.

Mata, A., Geng, Y., Henrikson, K. J. et al. 2010. Bone regeneration mediated by biomimetic mineralization of a nanofiber matrix. *Biomaterials* 31: 6004–6012.

McCann, R. M., Marsh, D. R., Horner, A., and Clarke, S. A. 2010. Body mass index is more predictive of progenitor number in bone marrow stromal cell population than age in men: Expanding the predictors of the progenitor compartment. *Tissue Engineering Part A* 16: 889–896.

Mcgarvey, W. C. and Braly, W. G. 1996. Bone graft in hindfoot arthrodesis: Allograft vs autograft. *Orthopedics* 19: 389–394.

Mckibbin, B. 1978. Biology of fracture healing in long bones. *Journal of Bone and Joint Surgery—British Volume* 60: 150–162.

Mehta, M., Strube, P., Peters, A. et al. 2010. Influences of age and mechanical stability on volume, microstructure, and mineralization of the fracture callus during bone healing: Is osteoclast activity the key to age-related impaired healing? *Bone* 47: 219–228.

Meijer, G. J., De Bruijn, J. D., Koole, R., and Van Blitterswijk, C. A. 2008. Cell based bone tissue engineering in jaw defects. *Biomaterials* 29: 3053–3061.

Meirelles, L. D. S., Chagastelles, P. C., and Nardi, N. B. 2006. Mesenchymal stem cells reside in virtually all post-natal organs and tissues. *Journal of Cell Science* 119: 2204–2213.

Mesimaki, K., Lindroos, B., Tornwall, J. et al. 2009. Novel maxillary reconstruction with ectopic bone formation by GMP adipose stem cells. *International Journal of Oral and Maxillofacial Surgery* 38: 201–209.

Morishita, T., Honoki, K., Ohgushi, H. et al. 2006. Tissue engineering approach to the treatment of bone tumors: Three cases of cultured bone grafts derived from patients' mesenchymal stem cells. *Artificial Organs* 30: 115–118.

Murphy, C. M., Haugh, M. G., and O'Brien, F. J. 2010. The effect of mean pore size on cell attachment, proliferation and migration in collagen-glycosaminoglycan scaffolds for bone tissue engineering. *Biomaterials* 31: 461–466.

Muschler, G. F., Nitto, H., Boehm, C. A., and Easley, K. A. 2001. Age- and gender-related changes in the cellularity of human bone marrow and the prevalence of osteoblastic progenitors. *Journal of Orthopaedic Research* 19: 117–125.

Muschler, G. F., Raut, V. P., Patterson, T. E., Wenke, J. C., and Hollinger, J. O. 2010. The design and use of animal models for translational research in bone tissue engineering and regenerative medicine. *Tissue Engineering Part B—Reviews* 16: 123–145.

Myers, T. J., Granero-Molto, F., Longobardi, L. et al. 2010. Mesenchymal stem cells at the intersection of cell and gene therapy. *Expert Opinion on Biological Therapy* 10: 1663–1679.

Nair, L. S. and Laurencin, C. T. 2007. Biodegradable polymers as biomaterials. *Progress in Polymer Science* 32: 762–798.

Nakahara, H., Bruder, S. P., Goldberg, V. M., and Caplan, A. I. 1990. In vivo osteo-chondrogenic potential of cultured-cells derived from the periosteum. *Clinical Orthopaedics and Related Research* 259: 223–232.

Nam, J., Johnson, J., Lannutti, J. J., and Agarwal, S. 2011. Modulation of embryonic mesenchymal progenitor cell differentiation via control over pure mechanical modulus in electrospun nanofibers. *Acta Biomaterialia* 7: 1516–1524.

Navarrete, C. and Contreras, M. 2009. Cord blood banking: A historical perspective. *British Journal of Haematology* 147: 236–245.

Nuttall, M. E., Patton, A. J., Olivera, D. L., Nadeau, D. P., and Gowen, M. 1998. Human trabecular bone cells are able to express both osteoblastic and adipocytic pheno-type: Implications for osteopenic disorders. *Journal of Bone and Mineral Research* 13: 371–382.

Ohgushi, H., Kotobuki, N., Funaoka, H. et al. 2005. Tissue engineered ceramic arti-ficial joint—Ex vivo osteogenic differentiation of patient mesenchymal cells on total ankle joints for treatment of osteoarthritis. *Biomaterials* 26: 4654–4661.

Oreffo, R. O. C., Bennett, A., Carr, A. J., and Triffitt, J. T. 1998. Patients with primary osteoarthritis show no change with ageing in the number of osteogenic precur-sors. *Scandinavian Journal of Rheumatology* 27: 415–424.

Owen, M. and Friedenstein, A. J. 1988. Stromal stem-cells—Marrow-derived osteo-genic precursors. *Ciba Foundation Symposia* 136: 42–60.

Paletta, J. R. J., Bockelmann, S., Walz, A. et al. 2010. RGD-functionalisation of PLLA nanofibers by surface coupling using plasma treatment: Influence on stem cell differentiation. *Journal of Materials Science—Materials in Medicine* 21: 1363–1369.

Park, S. R., Oreffo, R. O. C., and Triffitt, J. T. 1999. Interconversion potential of cloned human marrow adipocytes in vitro. *Bone* 24: 549–554.

Pittenger, M. F., Mackay, A. M., Beck, S. C. et al. 1999. Multilineage potential of adult human mesenchymal stem cells. *Science* 284: 143–147.

Potier, E., Ferreira, E., Andriamanalijaona, R. et al. 2007. Hypoxia affects mesenchymal stromal cell osteogenic differentiation and angiogenic factor expression. *Bone* 40: 1078–1087.

Pytlik, R., Stehlik, D., Soukup, T. et al. 2009. The cultivation of human multipotent mesenchymal stromal cells in clinical grade medium for bone tissue engineering. *Biomaterials* 30: 3415–3427.

Quarto, R., Mastrogiacomo, M., Cancedda, R. et al. 2001. Repair of large bone defects with the use of autologous bone marrow stromal cells. *New England Journal of Medicine* 344: 385–386.

Ramasamy, R., Fazekasova, H., Lam, E. W. F. et al. 2007. Mesenchymal stem cells inhibit dendritic cell differentiation and function by preventing entry into the cell cycle. *Transplantation* 83: 71–76.

Rebelatto, C. K., Aguiar, A. M., Moretao, M. P. et al. 2008. Dissimilar differentiation of mesenchymal stem cells from bone marrow, umbilical cord blood, and adipose tissue. *Experimental Biology and Medicine* 233: 901–913.

Rickert, D., Sauerbier, S., Nagursky, H. et al. 2010. Maxillary sinus floor elevation with bovine bone mineral combined with either autogenous bone or autogenous stem cells: A prospective randomized clinical trial. *Clinical Oral Implants Research* 22: 251–258.

Rosland, G. V., Svendsen, A., Torsvik, A. et al. 2009. Long-term cultures of bone marrow-derived human mesenchymal stem cells frequently undergo spontaneous malignant transformation (this article contains errors due to a cross contamination of the cell lines we used). *Cancer Research* 69: 5331–5339.

Rubio, D., Garcia-Castro, J., Martin, M. C. et al. 2005. Spontaneous human adult stem cell transformation (retracted article. See vol. 70, p. 6682, 2010). *Cancer Research* 65: 3035–3039.

Rush, S. M., Hamilton, G. A., and Ackerson, L. M. 2009. Mesenchymal stem cell allograft in revision foot and ankle surgery: A clinical and radiographic analysis. *Journal of Foot & Ankle Surgery* 48: 163–169.

Salama, R. and Weissman, S. L. 1978. Clinical use of combined xenografts of bone and autologous red marrow—Preliminary-report. *Journal of Bone and Joint Surgery—British Volume* 60: 111–115.

Schmelzeisen, R., Schimming, R., and Sittinger, M. 2003. Making bone: Implant insertion into tissue-engineered bone for maxillary sinus floor augmentation—A preliminary report. *Journal of Cranio-Maxillofacial Surgery* 31: 34–39.

Schwarz, N., Schlag, G., Thurnher, M. et al. 1991. Fresh autogeneic, frozen allogeneic, and decalcified allogeneic bone-grafts in dogs. *Journal of Bone and Joint Surgery—British Volume* 73: 787–790.

Scotchford, C. A., Shataheri, M., Chen, P. S. et al. 2011. Repair of calvarial defects in rats by prefabricated, degradable, long fibre composite implants. *Journal of Biomedical Materials Research Part A* 96A: 230–238.

Secco, M., Moreira, Y. B., Zucconi, E. et al. 2009. Gene expression profile of mesenchymal stem cells from paired umbilical cord units: Cord is different from blood. *Stem Cell Reviews* 5: 387–401.

Sensebe, L., Krampera, M., Schrezenmeier, H., Bourin, P., and Giordano, R. 2010. Mesenchymal stem cells for clinical application. *Vox Sanguinis* 98: 93–107.

Seshareddy, K., Troyer, D., and Weiss, M. L. 2008. Method to isolate mesenchymal-like cells from Wharton's jelly of umbilical cord. *In Methods in cell Biology: Stem Cell Culture,* J. P. Mather (Ed.), Vol. 86, pp 101–119, San Diego, CA: Elsevier Academic Press Inc.

Shayesteh, Y. S., Khojasteh, A., Soleimani, M. et al. 2008. Sinus augmentation using human mesenchymal stem cells loaded into a beta-tricalcium phosphate/hydroxyapatite scaffold. *Oral Surgery, Oral Medicine, Oral Pathology, Oral Radiology, and Endodontology* 106: 203–209.

Shenaq, D. S., Rastegar, F., Petkovic, D. et al. 2010. Mesenchymal progenitor cells and their orthopedic applications: Forging a path towards clinical trials. *Stem Cells International Online.* DOI:10.4061/2010/519028.

Shi, S. and Gronthos, S. 2003. Perivascular niche of postnatal mesenchymal stem cells in human bone marrow and dental pulp. *Journal of Bone and Mineral Research* 18: 696–704.

Sjostrom, T., Dalby, M. J., Hart, A. et al. 2009. Fabrication of pillar-like titania nanostructures on titanium and their interactions with human skeletal stem cells. *Acta Biomaterialia* 5: 1433–1441.

Song, S. Y., Chung, H. M., and Sung, J. H. 2010. The pivotal role of VEGF in adipose-derived-stem-cell-mediated regeneration. *Expert Opinion on Biological Therapy* 10: 1529–1537.

Sorrentino, A., Ferracin, M., Castelli, G. et al. 2008. Isolation and characterization of CD146(+) multipotent mesenchymal stromal cells. *Experimental Hematology* 36: 1035–1046.

Strong, D. M., Friedlaender, G. E., Tomford, W. W. et al. 1996. Immunologic responses in human recipients of osseous and osteochondral allografts. *Clinical Orthopaedics and Related Research* 326: 107–114.

Sundin, M., Ringden, O., Sundberg, B. et al. 2007. No alloantibodies against mesenchymal stromal cells, but presence of anti-fetal calf serum antibodies, after transplantation in allogeneic hematopoietic stem cell recipients. *Haematologica—The Hematology Journal* 92: 1208–1215.

Takahashi, K., Tanabe, K., Ohnuki, M. et al. 2007. Induction of pluripotent stem cells from adult human fibroblasts by defined factors. *Cell* 131: 861–872.

Tamimi, F., Torres, J., Gbureck, U. et al. 2009. Craniofacial vertical bone augmentation: A comparison between 3D printed monolithic monetite blocks and autologous onlay grafts in the rabbit. *Biomaterials* 30: 6318–6326.

Tare, R. S., Kanczler, J., Aarvold, A. et al. 2010. Skeletal stem cells and bone regeneration: Translational strategies from bench to clinic. *Proceedings of the Institution of Mechanical Engineers Part H—Journal of Engineering in Medicine* 224: 1455–1470.

Teo, A. K. K. and Vallier, L. 2010. Emerging use of stem cells in regenerative medicine. *Biochemical Journal* 428: 11–23.

Thomas, M. V. and Puleo, D. A. 2009. Calcium sulfate: Properties and clinical applications. *Journal of Biomedical Materials Research Part B—Applied Biomaterials* 88B: 597–610.

Thomson, J. A., Itskovitz-Eldor, J., Shapiro, S. S. et al. 1998. Embryonic stem cell lines derived from human blastocysts. *Science* 282: 1145–1147.

Tiedeman, J. J., Garvin, K. L., Kile, T. A., and Connolly, J. F. 1995. The role of a composite, demineralized bone matrix and bone marrow in the treatment of osseous defects. *Orthopedics* 18: 1153–1158.

Tilley, S., Bolland, B., Partridge, K. et al. 2006. Taking tissue-engineering principles into theater: Augmentation of impacted allograft with human bone marrow stromal cells. *Regenerative Medicine* 1: 685–692.

Torsvik, A., Rosland, G. V., Svendsen, A. et al. 2010. Spontaneous malignant transformation of human mesenchymal stem cells reflects cross-contamination: Putting the research field on track (correction letter for *Cancer Res.* 2009 Jul 1, 69(13), 5331–5339). *Cancer Research* 70: 6393–6396.

Troyer, D. L. and Weiss, M. L. 2008. Concise review: Wharton's jelly-derived cells are a primitive stromal cell population. *Stem Cells* 26: 591–599.

Tse, W. T., Pendleton, J. D., Beyer, W. M., Egalka, M. C., and Guinan, E. C. 2003. Suppression of allogeneic T-cell proliferation by human marrow stromal cells: Implications in transplantation. *Transplantation* 75: 389–397.

Urban, R. M., Turner, T. M., Hall, D. J., Inoue, N., and Gitelis, S. 2007. Increased bone formation using calcium sulfate-calcium phosphate composite graft. *Clinical Orthopaedics and Related Research* 459: 110–117.

Vacanti, C. A., Bonassar, L. J., Vacanti, M. P., and Shuttlebarger, J. 2001. Replacement of an avulsed phalanx with tissue-engineered bone. *New England Journal of Medicine* 344: 1511–1514.

Vandevord, P. J., Nasser, S., and Wooley, P. H. 2005. Immunological responses to bone soluble proteins in recipients of bone allografts. *Journal of Orthopaedic Research* 23: 1059–1064.

Vormoor, J., Lapidot, T., Pflumio, F. et al. 1994. Immature human cord-blood progenitors engraft and proliferate to high-levels in severe combined immunodeficient mice. *Blood* 83: 2489–2497.

Voss, P., Sauerbier, S., Wiedmann-Al-Ahmad, M. et al. 2010. Bone regeneration in sinus lifts: Comparing tissue-engineered bone and iliac bone. *British Journal of Oral & Maxillofacial Surgery* 48: 121–126.

Wagner, W., Wein, F., Seckinger, A. et al. 2005. Comparative characteristics of mesenchymal stem cells from human bone marrow, adipose tissue, and umbilical cord blood. *Experimental Hematology* 33: 1402–1416.

Wakitani, S., Saito, T., and Caplan, A. I. 1995. Myogenic cells derived from rat bone-marrow mesenchymal stem-cells exposed to 5-azacytidine. *Muscle & Nerve* 18: 1417–1426.

Warnke, P. H., Springer, I. N. G., Wiltfang, J. et al. 2004. Growth and transplantation of a custom vascularised bone graft in a man. *Lancet* 364: 766–770.

Yahyouche, A., Zhidao, X., Czernuszka, J. T., and Clover, A. J. P. 2011. Macrophage-mediated degradation of crosslinked collagen scaffolds. *Acta Biomaterialia* 7: 278–286.

Yamada, Y., Nakamura, S., Ito, K. et al. 2008. Injectable tissue-engineered bone using autogenous bone marrow-derived stromal cells for maxillary sinus augmentation: Clinical application report from a 2–6-year follow-up. *Tissue Engineering Part A* 14: 1699–1707.

Yamada, Y., Ueda, M., Hibi, H., and Nagasaka, T. 2004. Translational research for injectable tissue-engineered bone regeneration using mesenchymal stem cells and platelet-rich plasma: From basic research to clinical case study. *Cell Transplantation* 13: 343–355.

Yang, Z. J., Best, S. M., and Cameron, R. E. 2009. The influence of alpha-tricalcium phosphate nanoparticles and microparticles on the degradation of poly(d,l-lactide-co-glycolide). *Advanced Materials* 21: 3900–3904.

Yoshimoto, H., Shin, Y. M., Terai, H., and Vacanti, J. P. 2003. A biodegradable nanofiber scaffold by electrospinning and its potential for bone tissue engineering. *Biomaterials* 24: 2077–2082.

Yuan, H. P., Fernandes, H., Habibovic, P. et al. 2010. Osteoinductive ceramics as a synthetic alternative to autologous bone grafting. *Proceedings of the National Academy of Sciences of the United States of America* 107: 13614–13619.

Zhang, G., Drinnan, C. T., Geuss, L. R., and Suggs, L. J. 2010. Vascular differentiation of bone marrow stem cells is directed by a tunable three-dimensional matrix. *Acta Biomaterialia* 6: 3395–3403.

Zhang, M., Powers, R. M., and Wolfinbarger, L. 1997. A quantitative assessment of osteoinductivity of human demineralized bone matrix. *Journal of Periodontology* 68: 1076–1084.

Zuk, P. A., Zhu, M., Ashjian, P. et al. 2002. Human adipose tissue is a source of multipotent stem cells. *Molecular Biology of the Cell* 13: 4279–4295.

Zuk, P. A., Zhu, M., Mizuno, H. et al. 2001. Multilineage cells from human adipose tissue: Implications for cell-based therapies. *Tissue Engineering* 7: 211–228.

25

Clinical Translation of Tissue Engineering and Regenerative Medicine Technologies

Alejandro Nieponice

CONTENTS

25.1 Introduction

Regenerative medicine has made a substantial progress in the last decade with more technologies ready to help patients. Translation of these therapies from the benchtop to the bedside is not always a straight pathway, and many issues need to be addressed to achieve a successful outcome. The science, the industrial partner, the clinical partner, the regulations, and the implementation within a health-care center are the key aspects to be strictly analyzed and combined before attempting clinical translation. Scientific merit forms the basis of any new medical therapy as it gives rationale for its existence and it is usually the easiest part of the translational pathway. Identifying the right industrial partner can often be challenging since tissue engineering tools are frequently sophisticated and expensive to execute with little appeal for the immediate anxiety of the business-oriented population. The clinical partners, responsible for calling the indication of the medical therapy, are a critical step of this chain and need to be involved from the early stages of the process to guarantee a smooth transition into patients. Regulations are still emerging while the area of regenerative medicine evolves and, opposed to other medical devices or drugs, these technologies involve a complexity of steps that create many regulatory hurdles that can hamper the most powerful scientific discovery. Finally, implementation of regenerative medicine within a health-care center outside of a clinical trial is still a challenge, and the right space or model needs to be created. A comprehensive picture of this translational process is presented in this chapter with a detailed analysis of all the key aspects introduced earlier.

25.2 Science

25.2.1 Methodology and Translational Feasibility

Current technologies in tissue engineering and regenerative medicine (TERM) can range from simple medical devices or cell injections to a complex combination of cells and scaffolds that can make clinical translation a true challenge. In the past decade, we have seen several examples of each option that have reached patients with different degrees of success based on the complexity of the approach. Undoubtedly, those technologies that do not involve complex manipulation and can provide off-the-shelf availability have shown a much greater impact in the clinical setting. The best example for that category is extracellular matrix (ECM) scaffolds. Medical products derived from this technology have helped millions of patients worldwide in a wide range of clinical applications [1,2]. ECM scaffolds can be obtained from different organs and tissue sources that are subjected to a complete decellularization and posterior lyophilization and can be designed with different shapes and mechanical properties. Scaffold degradation is critical for the constructive remodeling of the ECM observed in a variety of body systems. In vivo studies have shown that ECM scaffolds degrade within 3 months after implantation to repair load bearing tissues [3,4]. In vitro studies have shown that ECM degradation products recruit a variety of established progenitor cell lines [5–7], and hybrid in vivo/in vitro study showed that UBM-ECM that was used for repair of the Achilles tendon in mice and subsequently harvested at early time points after implantation also recruited progenitor cells [8]. In addition to the chemotactic response, degradation

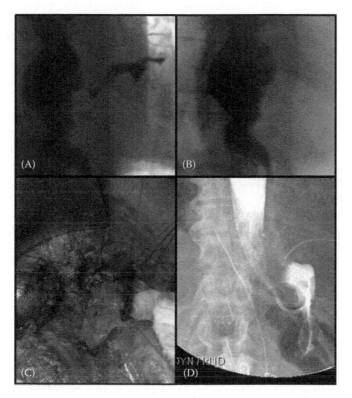

FIGURE 25.1
(A) A 55 year old patient developed a bronchoesophageal fistula as a postoperative complication. (B) After three sessions of ECM deployment, the tract was closed and the patient resumed a normal life. (C) A 58 year old patient with a large esophageal tear after removal of a polypropylene mesh inclusion. The left anterior wall of the esophagus is reconstructed with an ECM patch (Matristem, Acell Inc., Indiana, the United States). (D) A postoperative barium swallow at day 5 shows intact esophageal transit.

products of ECM scaffolds have been shown to promote angiogenesis [9] and provide bacteriostasis [10], which supports the angiogenesis and bacterial resistance that ECM scaffolds have shown in vivo [3,11–13] In our group, we have successfully used ECM scaffolds in patients for esophageal repair and reconstruction [14,15]. We have successfully treated esophageal leaks, reconstructed the esophageal wall (Figure 25.1), and prevented stricture after extensive endoscopic resection to avoid esophagectomy [16].

On a different approach, autologous stem-cell injections are being widely studied to treat infinity of clinical problems, and its uncontrolled use and commercialization have been object of many debates and ethical dilemmas. Clinical efficacy in large patient series and with solid control populations is difficult to obtain, making it a big hurdle for regulatory purposes. Two of the main clinical applications reported for this field include myocardial repair for patients with myocardial infarction or congestive heart failure and osteoarticular applications like osteoarthritis or osteochondral defects [17–20].

Another relevant approach that has proved translation to the clinical setting is the use of cell sheets without the need of scaffolds [21]. By using thermoresponsive dishes, cell sheets were created for a number of medical conditions, corneal repair being the one with greatest impact in the clinical community. Nishida et al. have reported the use of autologous oral mucosa cell sheets to repair the epithelial layer of the cornea in patients with bilateral corneal stem-cell deficiency [22,23]. Clinical trials are still ongoing, and massive application of this technology is still limited by the complexity of the fabrication process.

Combining biological scaffolds and stem cells in one product is definitely the most complex approach in regenerative medicine and as such is very rare to find it widely accepted in the clinical community outside a few clinical trials or case series reports. However, some technologies have accomplished all the necessary steps undergoing a very difficult pathway and have reached the patients with encouraging results. A good example of that is the work of Shin'oka et al., who have reported the use of a biological scaffold seeded with bone marrow stem cells as a vascular conduit replacement in pediatric cardiovascular surgery [24]. His group has treated more than 50 patients with long-term follow-up showing safety and efficacy of this approach. In the same area, L'Heureux et al. have developed a cell-based arterial conduit where the cells produce their own ECM in vitro that will later serve as the scaffold to be remodeled in vivo [25–27]. They have carefully evolved from a successful preclinical testing to a controlled clinical application for arteriovenous fistulae grafts. Finally, the use of scaffolds and cells for hollow organ reconstruction has reached clinical translation through the group of Atala and colleagues who have reported one of the landmarks in TERM with the reconstruction of a urinary bladder to treat patients with neurological dysfunction of that organ [28–30].

In summary, all the approaches mentioned earlier have transitioned into the clinical setting to some extent. However, the number of patients treated in each case seems to correlate more with the complexity of the methodology involved than with the relevance of the problem they are trying to address.

25.2.2 Understanding the Needs and the "Gold Standard"

Another important consideration when thinking of clinical translation is understanding the needs to treat a particular disease. At this level, two scenarios can be identified. In the first one, the new technology is trying to address an unsolved problem where there is no "gold standard" to treat a particular condition. This is the less common situation and usually the toughest to address from the scientific point of view but where biotechnology is expected to yield a significant contribution in the years to come.

One clear example of this would be the treatment of an incurable disease like Parkinson. All therapies available aim to control the symptoms rather than to cure the disease. As a consequence, any therapy proving to restore the normal function would have a tremendous impact and would be rapidly adopted by the clinical community.

On the other hand, the second and most common scenario usually involves a problem that has a suboptimal but well-established standard therapy. Jumping from suboptimal to optimal can often be challenging. A thorough understanding of the "gold standard" therapy for a particular disease should be mandatory before thinking of a new alternative. One example that illustrates this scenario is the coronary artery bypass graft surgery that has a considerable success rate in the long-term follow-up. Improving graft survival would be the most intuitive endpoint to demonstrate, but it may be impossible to do as many study years would be needed to reach a serious statistical evidence. In contrast, there are several aspects of this surgery that can be addressed in the short term such as acute thrombosis or conduit availability in reoperations that would help a technology move forward. These alternatives only become evident when we explore the clinical problem in depth.

25.3 Industrial Partner

Industrial partners are a key aspect of the translational pathway. The most successful therapy can be deemed unwisely if handled inappropriately by the industry. At the same time, the technology needs to have a profitable side to gain the interest of the industrial partner. Without the industry involvement, the chances for a TERM technology of reaching the patients are very scarce. Accordingly, it is very important that scientists work together with industry from the early stages to understand mutual needs and facilitate the process. This is a concept that is not always adopted by basic scientists that tend to work on their side first, thinking that industry comes later. Although this is true in the natural occurrence of events, it is very helpful to have an early feedback when designing a particular experiment. This can save a lot of time and money in the long run.

25.3.1 Large Companies vs. Start-Up in Biotechnology

Both large and start-up companies have advantages and disadvantages in biotechnology. As opposed to pharmaceutical or medical device companies, biotech industry outcomes are less predictable. They often face situations that can be related to manufacturing, regulatory, or reimbursement that are not established elsewhere and can be a critical challenge to sustainability.

In the case of start-up companies, funding is usually the main limitation. Although they have the advantage of having a dedicated team and with a particular technology being their only priority, the length of the development process until critical reimbursement can prevent the company from proving the concept before going bankrupt. This has been the case in the early 2000s for many of the tissue engineering endeavors that had raised a lot of promise some years before and ended up failing due to financial issues.

On the opposite side, large pharmaceutical or medical device companies that explore the TERM market have the advantage of less limited funding that can sustain a technology much longer until it proves effective. But the limitation in these cases is usually the lack of focus or effort that big companies put onto these projects as they do not represent a substantial portion of their income.

25.3.2 Market Size Matters

A very good idea with a lot of potential for a particular patient can easily be truncated by the market size. Unfortunately, if the problem we are trying to address is not life threatening where a company can charge unlimited amounts for a single therapy and still be profitable, the size of the patient population will limit the adoption of the idea by the industrial partner. Esophageal surgery is a good example. With only 15,000 esophagectomies performed each year in the United States, any therapy aiming to treat or prevent complications from this procedure would have a great impact on the patient (complications can be devastating) but would never be eagerly adopted by the industry as there is no business model. The only chance of those technologies to reach clinical translation is to be the side of a more profitable core project. In the same way, a technology that has a large market size and can be highly profitable very rapidly will probably be adopted and pushed to the market by the industry regardless of its scientific merit.

25.3.3 Challenging the Reimbursement Pathway

As described earlier, the manufacturing or processing of several TERM technologies can be very complex, lengthy, and expensive. In the business world, return on investment (ROI) is a measure of the efficiency from each investment and is calculated as profit divided by investment. An expensive manufacturing cost increases the required investment, necessitating higher prices to deliver an ROI that will be acceptable to investors. For a TERM technology, high profitability will only be reached by high reimbursement per unit since widespread adoption is unlikely in the early stages. The health-care system will only provide high reimbursement in those cases where other options are not available or clearly suboptimal or where a high benefit–cost ratio can be proven. The latter is not always easy to demonstrate during the early adoption of a medical therapy, and a contingency plan should be considered until reimbursement can be established as expected. A common pathway we are starting to observe in biotech companies is to find an alternative product somehow related to the core technology that has less clinical impact but higher reimbursement chances. This usually helps the company to survive through the development of the core technology but has the potential risk of making it lose focus on the primary mission.

25.4 Clinical Partners

25.4.1 Early Involvement in the Preclinical Process

Historically, there has been a large gap between the basic research and the clinical science. A technology can spend years in the laboratory before a clinician hears about it. There is a clear need for that to change if we want to succeed in bringing new therapies to patients. The input or feedback that a clinician who is actively practicing can provide to the experimental design is remarkable. While the basic scientist knows all the details about the cutting-edge technology, the clinician will provide tools that will make the translation smoother. The concept of making it work first and then adjusting it to make it clinically relevant can often be disappointing as the complexity usually increases to obtain the results and then it is hard to go back and change it. The early involvement of the clinical partner is particularly important in the preclinical testing. The animal model to be utilized, the surgical technique, and the endpoint analysis are key steps where there is a mutual benefit for the clinical and the basic scientist. The latter can gain accurate feedback of the performance and emphasize clinical relevance while the former can get early training and familiarization with the technology that will help him translate it in the next phase.

25.4.2 Identifying the Right Partner

Identifying the right partner is also a critical step. For grant purposes, there is a tendency to choose the most renown practitioner with the longest CV possible. This is a valid option to obtain the initial funding, but it is often followed by a disconnection between the partners where the clinician is too busy to spend time in the lab and the basic scientist continues to move forward without interaction. Sometimes, the clinical partner will send the most junior trainee to the lab to justify his commitment with the project. This is also counterproductive because decision making is usually necessary during the experiments and the trainee does not necessarily have the expertise to make the right call. This may end up in unnecessary money use and repeated experiments.

Understanding the technology from its early infancy is also very relevant to the physician. When it comes to the point of clinical translation, the practitioner will be challenged not only to know how to provide the new therapy but also to understand its performance. Failure to do so can lead to inappropriate use and misjudgments that can hinder the overall process.

25.5 Regulations

The multiple combinations of cells and materials involved in TERM technologies have created a big impact in the standard regulations available for medical products that were initially established to fulfill either a pharmacologic agent or a medical device approval. Evolution of the science led in the past years to a rapid adjustment of regulatory agencies by creating new offices dedicated exclusively to biological products [31–33]. Reviewing existing regulatory guidelines for TERM exceeds the scope of this chapter and is clearly published in scientific journals or books.

The main purpose of this section is to emphasize the need of understanding these regulations when designing the experiments to facilitate the translational process later on.

25.6 Ethics

The main debate within TERM revolves around the use of embryonic stem cells and its view by some religious groups that relate this to the abortion. This debate will likely be present as long as we have different religious thoughts, so it will be forever. A separate book chapter could be written just on this, but it is not the aim of this book. There are two options when facing this challenge. One of them is refraining from using embryonic stem cells. Adult stem cells are free from any ethical concern and so are the newly described iPS (induced pluripotent stem cells) with similar therapeutic potentials. The second option is pursuing its use, participating in a debate, and working under the guidelines that rule a particular society, respecting all opinions and understanding disagreement as a positive social tool that should never be discriminated.

25.7 Health-Care Institutions

Adoption of TERM technologies by health-care institutions remains a challenge. While off-the-shelf technologies similar to medical devices like ECM scaffolds can be easily acquired and used, combination products or cell-based products are more difficult to introduce. There is little doubt that clinically successful cell-based therapies will be autologous in origin. This involves not only the cell manufacturing process but also the cell extraction from the patient. Transportation and manipulation involve several regulatory hurdles that add increasing costs and may challenge reimbursement as described earlier [34,35]. Moreover, cell-based therapies tend to be tailored to the patient needs, making it difficult to standardize a treatment for all patients. It is not clear yet whether TERM companies involving cell-based products will be able to ship and distribute from one single factory, maintain product viability, and still be profitable. The need of having a space within the health-care institution appears as a valid alternative.

25.7.1 Clinical Translation Unit Concept

In order to facilitate the translation of cell-based therapies from the benchtop to the bedside, we have created the Clinical Translation Unit (CTU). The first of its kind, the unit consists of a state-of-the art cell culture facility integrated completely in the operative room (OR) (Figure 25.2). The main concept of the unit is that the cells never leave the OR. This allows the physicians to bring the patient, extract and isolate the required cells, expand them in culture if necessary, and then bring back the patient for final implantation. It also has the advantage of being conceived inside an already sterile environment, facilitating compliance with GLP guidelines and decreasing the associated building cost. This concept may allow biotech companies who own a particular license to interact with the health-care institution and provide together the desired therapy with mutual benefits. The institution would be able to bring cutting-edge therapies to its patients, and the companies would not need to support a GMP facility and subsequent distribution.

25.7.2 Tissue Engineering and Regenerative Medicine Division

A concept like the CTU would think the TERM paradigm similarly to the transplant programs in the hospital. The latter involve a multidisciplinary team with specialized technicians and highly skilled professionals who can handle a particular therapy. Following a similar pathway, a division within the hospital can be created that will specialize in TERM and will have the necessary personnel to interact with the patients, the companies, and the health-care institution. Although this remains as wishful thinking,

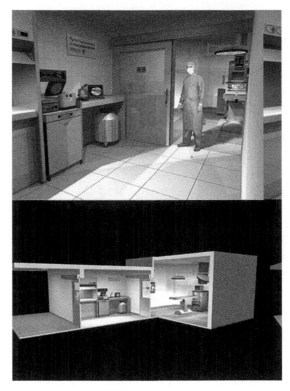

FIGURE 25.2
The CTU concept. A state-of-the-art facility built inside the OR to avoid cell transportation and facilitate implementation of cell-based therapies.

rapid evolution of biotechnology may transform it in a real scenario in the near future.

25.8 Conclusions

Clinical translation of TERM technologies is a complex and challenging pathway that requires multidisciplinary interaction from the early stages and through all its phases. There has been much expectation from the biotechnology field by the clinical world in the last decade. While several new therapies hold a great deal of promise, failure to delivery has raised some skepticism that will need to be overcome by accurate moves and clear endpoints. This will help the basic scientist, the clinician, the industrial partner, and the health-care provider to understand the need of interaction to bring the patient the final product.

References

1. Badylak, S.F., Xenogeneic extracellular matrix as a scaffold for tissue reconstruction. *Transpl Immunol*, 2004. **12**(3–4): 367–377.
2. Badylak, S.F., The extracellular matrix as a biologic scaffold material. *Biomaterials*, 2007. **28**(25): 3587–3593.
3. Gilbert, T.W. et al., Degradation and remodeling of small intestinal submucosa in canine Achilles tendon repair. *J Bone Joint Surg Am*, 2007. **89**(3): 621–630.
4. Record, R.D. et al., In vivo degradation of ^{14}C-labeled small intestinal submucosa (SIS) when used for urinary bladder repair. *Biomaterials*, 2001. **22**(19): 2653–2659.
5. Brennan, E.P. et al., Chemoattractant activity of degradation products of fetal and adult skin extracellular matrix for keratinocyte progenitor cells. *J Tissue Eng Regen Med*, 2008. **2**(8): 491–498.
6. Crisan, M. et al., A perivascular origin for mesenchymal stem cells in multiple human organs. *Cell Stem Cell*, 2008. **3**(3): 301–313.
7. Reing, J.E. et al., Degradation products of extracellular matrix affect cell migration and proliferation. *Tissue Eng Part A*, 2009. **15**(3): 605–614.
8. Beattie, A.J. et al., Chemoattraction of progenitor cells by remodeling extracellular matrix scaffolds. *Tissue Eng*, 2008. doi: 10.1089/ten.tea.2008.0162.
9. Li, F. et al., Low-molecular-weight peptides derived from extracellular matrix as chemoattractants for primary endothelial cells. *Endothelium*, 2004. **11**(3–4): 199–206.
10. Brennan, E.P. et al., Antibacterial activity within degradation products of biological scaffolds composed of extracellular matrix. *Tissue Eng*, 2006. **12**(10): 2949–2955.
11. Badylak, S.F. et al., Morphologic study of small intestinal submucosa as a body wall repair device. *J Surg Res*, 2002. **103**(2): 190–202.
12. Jernigan, T.W. et al., Small intestinal submucosa for vascular reconstruction in the presence of gastrointestinal contamination. *Ann Surg*, 2004. **239**(5): 733–738; discussion 738–740.
13. Shell, D.H., 4th et al., Comparison of small intestinal submucosa and expanded polytetrafluoroethylene as a vascular conduit in the presence of gram-positive contamination. *Ann Surg*, 2005. **241**(6): 995–1001; discussion 1001–1004.
14. Nieponice, A., T.W. Gilbert, and S.F. Badylak, Reinforcement of esophageal anastomoses with an extracellular matrix scaffold in a canine model. *Ann Thorac Surg*, 2006. **82**(6): 2050–2058.
15. Nieponice, A. et al., An extracellular matrix scaffold for esophageal stricture prevention after circumferential EMR. *Gastrointest Endosc*, 2009. **69**(2): 289–296.
16. Badylak, S.F. et al., Esophageal preservation in five male patients after endoscopic inner-layer circumferential resection in the setting of superficial cancer: a regenerative medicine approach with a biologic scaffold. *Tissue Eng Part A*, **17**(11–12): 1643–1650.
17. Smits, P.C., Myocardial repair with autologous skeletal myoblasts: A review of the clinical studies and problems. *Minerva Cardioangiol*, 2004. **52**(6): 525–535.
18. Wollert, K.C. and H. Drexler, Clinical applications of stem cells for the heart. *Circ Res*, 2005. **96**(2): 151–163.

19. Centeno, C.J. et al., Regeneration of meniscus cartilage in a knee treated with percutaneously implanted autologous mesenchymal stem cells. *Med Hypotheses*, 2008. **71**(6): 900–908.
20. Centeno, C.J. et al., Increased knee cartilage volume in degenerative joint disease using percutaneously implanted, autologous mesenchymal stem cells. *Pain Physician*, 2008. **11**(3): 343–353.
21. Ohki, T. et al., Treatment of oesophageal ulcerations using endoscopic transplantation of tissue engineered autologous oral mucosal epithelial cell sheets in a canine model. *Gut*, 2006. **55**(12): 1704–1710.
22. Nishida, K. et al., Functional bioengineered corneal epithelial sheet grafts from corneal stem cells expanded ex vivo on a temperature-responsive cell culture surface. *Transplantation*, 2004. **77**(3): 379–385.
23. Nishida, K. et al., Corneal reconstruction with tissue-engineered cell sheets composed of autologous oral mucosal epithelium. *N Engl J Med*, 2004. **351**(12): 1187–1196.
24. Shin'oka, T. et al., Midterm clinical result of tissue-engineered vascular autografts seeded with autologous bone marrow cells. *J Thorac Cardiovasc Surg*, 2005. **129**(6): 1330–1338.
25. L'Heureux, N., T.N. McAllister, and L.M. de la Fuente, Tissue-engineered blood vessel for adult arterial revascularization. *N Engl J Med*, 2007. **357**(14): 1451–1453.
26. L'Heureux, N. et al., Technology insight: The evolution of tissue-engineered vascular grafts–from research to clinical practice. *Nat Clin Pract Cardiovasc Med*, 2007. **4**(7): 389–395.
27. L'Heureux, N. et al., Human tissue-engineered blood vessels for adult arterial revascularization. *Nat Med*, 2006. **12**(3): 361–365.
28. Atala, A., Creation of bladder tissue in vitro and in vivo. A system for organ replacement. *Adv Exp Med Biol*, 1999. **462**: 31–42.
29. Atala, A., Engineering tissues and organs. *Curr Opin Urol*, 1999. **9**(6): 517–526.
30. Atala, A., Tissue engineering for the replacement of organ function in the genitourinary system. *Am J Transplant*, 2004. **4**(Suppl 6): 58–73.
31. Carpenter, M.K. and L.A. Couture, Regulatory considerations for the development of autologous induced pluripotent stem cell therapies. *Regen Med*, **5**(4): 569–579.
32. Sethe, S.C., The implications of "advanced therapies" regulation. *Rejuvenation Res*, **13**(2–3): 327–328.
33. van Zanten, J. and P. de Vos, Regulatory considerations in application of encapsulated cell therapies. *Adv Exp Med Biol*, **670**: 31–37.
34. Nozaki, T. et al., Transportation of transplantable cell sheets fabricated with temperature-responsive culture surfaces for regenerative medicine. *J Tissue Eng Regen Med*, 2008. **2**(4): 190–195.
35. Heng, B.C. et al., The cryopreservation of human embryonic stem cells. *Biotechnol Appl Biochem*, 2005. **41**(Pt 2): 97–104.

Index